Autonomous Driving Handbook

Yu Huang · Zijiang Yang

Autonomous Driving Handbook

清華大学出版社
Tsinghua University Press

Springer

Yu Huang
Roboraction.AI
Sunnyvale, CA, USA

Zijiang Yang
School of Computer Science
and Technology
University of Science and Technology
of China
Hefei, Anhui, China

ISBN 978-981-95-3724-2 ISBN 978-981-95-3725-9 (eBook)
https://doi.org/10.1007/978-981-95-3725-9

Jointly published with Tsinghua University Press
The print edition is not for sale in Mainland China. Customers from Mainland China please order the print book from: Tsinghua University Press.

The original submitted manuscript has been translated into English. The translation was done using artificial intelligence. A subsequent revision was performed by the author(s) to further refine the work and to ensure that the translation is appropriate concerning content and scientific correctness. It may, however, read stylistically different from a conventional translation.

This Springer imprint is published by the registered company Springer Nature Singapore Pte Ltd.
The registered company address is: 152 Beach Road, #21-01/04 Gateway East, Singapore 189721, Singapore

Foreword

This book systematically introduces the cutting-edge technical theories in the field of autonomous driving today. Starting from the background knowledge of autonomous driving, it deeply discusses the software and hardware platforms, perception, mapping, localization, planning decisions, control, simulation, safety, integration and validation and verification, ethic and legal aspects, and introduces the applications of automatic parking systems, vehicle networking technology, neural rendering, diffusion (GenAI) model and large-scale model-based application. The book consists of 18 chapters: Chap. 1 is an overview of the autonomous driving system (development structure, scene classification, data loop and large-scale model); Chap. 2 briefly introduces the basic theory of autonomous driving, namely computer vision and deep learning; Chaps. 3 and 4 analyze the software and hardware platforms of autonomous driving, including sensors, computing platforms, software architecture, and operating systems; Chaps. 5–9 introduce the perception, mapping, localization, planning, and control modules of autonomous driving, respectively; Chap. 10 focuses on the simulation module of autonomous driving; Chap. 11 discusses the safety model of autonomous driving; Chap. 12 introduces the technical integration and validation and verification (V&V); Chap. 13 discusses a special case of autonomous driving—the automatic parking system; Chap. 14 introduces vehicle networking technology (vehicle–road collaboration and vehicle platooning); Chaps. 15 and 16 introduce two recent hot topics related to autonomous driving, namely 3D scene neural rendering (mainly NeRF and 3D Gaussian Splatting) and content generation and policy learning with diffusion models; Chap. 17 discusses the breakthrough of AI, large language model and large multi-modality model, applied for autonomous driving; Finally, Chap. 18 introduces the ethics and legal issues in autonomous driving.

This book is suitable for readers with a certain background, such as those with theoretical base and practical experience in computer vision and machine learning (even deep learning). This book provides a large amount of first-hand information on cutting-edge autonomous driving technology, covering many aspects of developing autonomous driving. It is hoped that this book can inspire and touch the front-line developers of autonomous driving, broaden their thinking when encountering problems and difficulties, and find solutions.

Preface

The Curtain of Autonomous Driving

The dream of autonomous driving for human beings even predates the birth of the vehicle. In the sixteenth century, Da Vinci tried to use high-strength springs to empower vehicles to drive automatically on planned routes. Modern society began to pay serious attention to autonomous driving through the autonomous driving challenge—Grand Challenge, launched by the Defense Advanced Research Projects Agency (DARPA) in the United States in 2004. In 2005, the champion of the autonomous off-road challenge was Stanford University's Stanley; in 2007, the champion of the autonomous urban challenge was Carnegie Mellon University's Boss.

DARPA's competition not only promoted academic research in universities but also gave birth to several leading autonomous driving companies. In 2009, Google acquired Stanford University's autonomous driving team, the predecessor of Google subsidiary Waymo. In 2015, Uber gradually incorporated Carnegie Mellon University's autonomous driving team. In the same year, Tesla released Autopilot, which seems to be L4 autonomous driving in name, but it is actually L2 driving assistant. The DARPA competition also made a company, Velodyne, which produces 64-line mechanical LiDAR.

The Current Status of Technology Development

The development of autonomous driving basically follows two technological paths. One is the leapfrog development path, which directly develops L4 level autonomous driving systems. This is mainly pursued by high-tech companies like Google and Baidu, which do not have a deep understanding of the automotive industry and directly apply the model of robot development to autonomous vehicles. This technological path often disregards cost, using high-definition maps and high-precision inertial navigation localization (IMU). However, it is difficult to operate in places without maps (no need to interface with navigation maps, directly using high-definition maps for planning, lacking understanding of the computational complexity of downloading high-definition maps for large-scale navigation areas and global planning). They install the best and most expensive sensors, such as LiDAR (at that time, the depth learning technology of vision was relatively primitive), and use the strongest computing platform (since it is not mass-produced, some L4 autonomous driving companies directly use industrial computers as support, without considering the difficulty of platform migration). In the face of the landing of Robotaxi, although it has not targeted any scenario, in the test of highways and city streets, based on reliable perception performance, it indeed provides the strongest planning and decision-making ability under complex traffic conditions. The planning and decision-making algorithms developed based on the data that we see now mostly come from L4 companies, and several of them have even held competitions in this area (based on their own open-source datasets). In the construction of the simulation platform, L4 companies are far ahead (Tesla basically already has the ability to land and develop L4), including data replay and single-point testing visualization capabilities. In terms of high-definition maps, the technology of L4 companies is also relatively mature. The sensor data collected by L4 companies, including LiDAR and vehicle localization trajectories, is also higher than that of general L2 companies.

The other technological path is progressive development, that is, starting from the L2 assisted autonomous driving system, and then gradually adapting to more complex traffic environments, transitioning from highways and elevated roads to scenarios with ramp and toll stations, and finally entering urban streets and trails. The level of the autonomous driving system also gradually evolves to L2+, L3, L3+, and L4 levels (recently, a new trend has been seen, that is, L4 autonomous driving development companies, L2 autonomous driving OEMs, and Tier 1 companies cooperate to jointly develop L2+ mass-produced vehicles). The progressive development path is generally adopted by OEMs and Tier 1 companies. They will first consider cost, vehicle regulations, and the definition of the ODD (Operational Design Domain) of mass-produced users. Initially, Mobileye was the main supplier, and later the development models of Tesla and Nvidia became the mainstream. Due to cost constraints, this path often uses cameras as the main sensor, supplemented by millimeter-wave radar accepted by vehicle companies. Relatively speaking, the cutting-edge level of this path (such as Tesla) has obvious advantages in the vision perception technology it uses. Because a large number of false alarms appear in the radar during fusion, Tesla even abandoned it in the autonomous driving perception module. Previously, parking autonomous driving used ultrasonic sensors, which are gradually combined with fisheye cameras to provide parking assistance, memorable parking, and even valet parking applications. Tesla also uses ultrasound to provide perception in congested traffic scenarios. Perhaps due to the cost pressure, the progressive development path is generally an "emphasized perception, de-emphasized high-definition map" mode, and even end-to-end integration of perception, online mapping and localization have been has been deployed in the BEV/Occupancy network model (such as Tesla). As an industry-leading company, Tesla has gradually improved the data-driven development tool chain, achieving a data loop including data curation, data labeling, simulation, model update, scenario test evaluation, and model deployment. Based on the long-tail problem of autonomous driving and the uncertainty problem of AI models, Google also has a similar framework, but Tesla has pushed it to the extreme, achieving a virtuous cycle on mass-produced user vehicles, and has now also launched the L4 autonomous driving version FSD.

It is worth mentioning that driven by Tesla, not only has the perception system of open-set object recognition, namely the occupancy network, been realized, but the end-to-end autonomous driving system has become the mainstream, and the previous modular "perception-prediction-planning" pipeline is being replaced by a unified neural network model. In particular, with the emergence of large language models (LLMs), world models and the progress of the training plus reasoning scaling law, in the research of end-to-end autonomous driving, there will be a hierarchical dual system that satisfies fast control for maneuvers and deep slow thinking for decision-making, which can completely solve the residual problems of autonomous driving with the help of embodied AI. Autonomous driving basically exists in two development stages, 1.0 and 2.0. The 1.0 stage of autonomous driving uses a variety of sensors to form perception input, such as LiDAR, vision cameras, radar, inertial IMU, wheel speedometers, and GPS/differential GPS. Each sensor has differences in perception capabilities, and a multi-modal sensor fusion architecture is used specifically, currently mostly adopting a late-fusion strategy, filtering the results of each sensor in related tasks once to achieve complementary or redundant effects. There are two paths in this aspect, one is to rely on LiDAR and high-definition maps, which is expensive and mainly used by Robotaxi and other L4 companies; the other is to mainly use emphasized vision, de-emphasized high-definition map method, which is low in cost and is mostly the idea of L2/L2+ autonomous driving companies for mass production. Both paths will have very traditional post-processing steps (especially vision), and a lot of debugging work and problems also come from this. In addition, the planning and decision-making of this stage mostly use rule based methods, in fact, there is no data-driven model, such as open-source Autoware and Baidu Apollo. Since the operating scenarios of L4 companies are in some specific areas with high-definition maps, the sensor accuracy invested in perception is relatively high, and explorations of data for planning and decision-making models training have been made; Relatively speaking, L2/L2+ companies have not yet established a data-driven development mode for planning and decision making modules, and mostly adopt solutions based on the traditional optimization theories. They generally start from high-way scenarios and upgrade to the "ramp-to-ramp" model like Tesla, rarely supporting complex scenarios of urban autonomous driving (such as roundabout intersections and unprotected left turns). The stage of autonomous driving 2.0 should be marked by data-driven, and there is also a significant improvement in the perception framework of the 1.0 stage. The data-driven development model tends to design and train end-to-end models. For planning decisions, it requires a large amount of driving data to learn the driving behavior of "experienced drivers", including behavior cloning in imitation learning, and estimating the joint distribution of behavior-strategy through model-based reinforcement learning (M-RL), no longer relying on the optimal theory for problem solving under various constraints. Trajectory prediction is an important prelude, requiring good modeling of intelligent interactive behavior and analysis of existing uncertainty effect. For perception, the 2.0 era needs to consider which part of traditional vision or signal processing (filtering) is replaced by machine learning models, truly adopting a development model that solves problems by collecting the related data. For example, Tesla's recent BEV (Bird's Eye View) and Occupancy Network are directly implemented through deep learning models to output the required information, rather than using traditional vision and fusion theory to process the model output twice. Sensor fusion theory has also been upgraded from late-fusion to feature fusion in the model and even early data fusion (if there is certain prior knowledge of synchronization and calibration). Here we can see that the Transformer network plays an important role in this perception framework, and it also puts higher requirements on the computing platform. Based on the demand for this data-driven autonomous driving platform, the design idea of large-scale models is also introduced. Because of the acquisition of a large amount of data, including efficient data curation, automatic labeling, and simulation technology assistance, it is necessary to maintain a large teacher model on the server side to support the training and iterative update of the student small model deployed to the vehicle side. As mentioned earlier, large language models (LLMs) and world models will drive greater progress in the autonomous driving 2.0 stage, especially the dual system of fast and slow thinking, which will expand from the application of embodied intelligence to the field of autonomous driving.

Challenges and Future of Autonomous Driving

One of the challenges facing the transformation of the automotive industry is the transition to the new model of software-defined vehicles. The software code of autonomous driving will exceed 500 million lines, one of the most complex software systems. In comparison, the Boeing 787 has 6.5 million lines of code, Android has 15 million lines of code, Windows 11 has 50 million lines of code, Facebook has 62 million lines of code, and modern vehicles have 100 million lines of code. At the same time, the software of intelligent vehicles is the culmination of modern software, including real-time embedded software, cloud computing software, consumer software, simulation software, AI algorithms, and large-scale multi-agent collaborative system software. Moreover, such a large and complex software system has extremely high requirements for safety and reliability, because software failure can lead to life-threatening serious consequences. Therefore, intelligent vehicle software needs to use the most advanced theories and practices of software engineering, testing and verification. At the same time, the progress of autonomous driving depends on the development of artificial intelligence technology. Although artificial intelligence has defeated humans in Go (a black-white play game), Go is played in a closed environment, and the play board format and rules have not changed for more than 2000 years. Machine learning technology trained on data has difficulty solving the endless open scenes in the driving environment. For example, a nursery rhyme that children can easily understand—"I found a penny on the side of the road and handed it to the policeman," artificial intelligence technology needs to solve many problems to understand. For example, why it is reasonable to pick up a penny and unreasonable to pick up a billion yuan, why to give it to the police, and so on. The large amount of known unsafe accumulated by people in the grow-up process, plus the unknown unsafe of the open traffic scene, is a long-tail problem that is difficult to solve. At present, large language models that collect a lot of human knowledge are expected to solve this residual problem through deep thinking and reasoning capabilities (including chains of thoughts, trees of thoughts, and reflection with memory).

Although there are many challenges to achieving autonomous driving, the new generation of intelligent vehicles can bring a nearly zero-emission, zero-accident, zero-cost transportation system to society. Since 2020, the mass-market electric vehicles with a price of about 15,000 US dollars and a single charge driving range of more than 400 km can completely benchmark the performance and use cost of mainstream gasoline vehicles, marking that electric vehicles have reached the critical point of surpassing gasoline vehicles. In the next 10 years, the popularization of electric vehicles will reduce global oil consumption by about 30%, and the zero-emission goal of the automotive industry is just around the corner. In addition, with the improvement of safety brought by autonomous driving, the vision of zero-accident travel is finally reached. At present, 1.35 million people die from traffic accidents every year in the world, which is equivalent to one person dying from a traffic accident every 24 s, and the cost related to traffic accidents accounts for about 0.5% of the global GDP. According to the latest research by Waymo, its current autonomous driving technology can avoid 75% of collision accidents and reduce 93% of serious injuries, which is higher than the 62.5 and 84% of the ideal human driver model. Moreover, the maintenance and driving costs of electric vehicles can be significantly reduced compared to gasoline vehicles, even close to "zero." First of all, electric vehicles have less than 200 moving parts, while gasoline vehicles have more than 2000. Firstly, the maintenance cost of electric vehicles during their lifecycle is less than half of that of gasoline vehicles. The motor drive and body design of electric vehicles can support 1.6 million kilometers of operation. Under normal driving conditions, the battery can support a driving distance of 500,000–800,000 km. It can also continue to be used by replacing the battery, which is more than 10 times the lifespan of gasoline vehicles. The depreciation cost within its lifecycle can also be reduced to 1/10 of the original. Based on battery swapping technology, Battery as a Service (BaaS) provides another dimension to reduce travel costs, especially in the field of commercial vehicles. Battery swapping services can reduce service interruptions due to charging time. Battery warehouses can reduce battery procurement costs through large-scale standardization, and can also optimize battery maintenance and care. Battery swapping can also better utilize the low prices of the power grid. In summary, the depreciation cost, maintenance cost, and usage cost of electric vehicles are all equivalent to 1/10 of traditional gasoline vehicles. In the scenario of shared travel, the cost per kilometer is reduced from about 1 yuan today to 0.1 yuan. Another major expense of shared travel is the software usage fee. In the future, as the autonomous driving mode matures, according to the current Tesla autonomous driving subscription model, it is 1400 yuan per month. Assuming the monthly mileage of shared travel is 6000 km, the cost of the AI driver per kilometer is about 0.02 yuan. Combining the above data, through electrification and autonomous driving, the cost of shared travel can be reduced from 1.5 yuan today to 0.12 yuan, and the cost of 100 km of travel is 12 yuan. If multiple people travel together, the cost is further reduced, which is basically the same as the cost of using water, electricity, and internet services.

Purpose of Book Writing

As the development of autonomous driving is in full swing, more and more companies, engineers, scientists, and students are entering or planning to enter this field. The author hopes to introduce readers to the cutting-edge technology and theory of autonomous driving today.

This book mainly includes the following content.

- A brief introduction to the general theory of autonomous driving, including the levels of autonomous driving, two different development structures (modular and end-to-end), classification of application scenarios of autonomous driving, the concept of data closed loop and the AI breakthrough by large-scale models.
- Before introducing the cutting-edge technology, the basic theories involved are laid out, such as computer vision, image processing, optimization theory, machine learning, and deep learning.
- Before starting the focus of this book, readers need to have a preliminary understanding of the hardware and software platforms of autonomous driving. The hardware part includes sensors (cameras, LiDAR, millimeter-wave radar, ultrasonic radar, and vehicle networking), computing platforms, wire-controlled chassis, and electronic and electrical architecture. The software part includes software architecture (such as AUTOSAR and aSPICE), software development V-model, and operating system.
- The real highlight of this book is Chaps. 5–9, namely perception, HD maps, localization, planning, and control. The perception part involves various sensor calibration, monocular object ranging, depth map estimation, 3D obstacle detection and tracking, sensor fusion (data level and task level), lane line detection, traffic sign detection and recognition, traffic signal light detection and recognition, binocular vision, driving area segmentation, human body posture estimation, driver monitoring system, and the new generation BEV/Occupancy network perception framework. The HD map part introduces the structure of the HD map, the construction of the semantic map, the map based on lane lines, and the SLAM based on deep learning. The localization module and the HD map module are inseparable,

including localization based on lane line maps, map localization based on LiDAR, localization of hybrid sensor fusion, and localization methods based on deep learning. The planning part involves basic planning theory, modeling and prediction of driving behavior, modeling and prediction of pedestrian behavior, and imitation learning based on deep learning. The control method is relatively traditional, but in addition to reviewing the classic vehicle kinematics and dynamics models, control theory (such as PID and MPC), it also discusses the control theory based on deep learning.

- This book also discusses an important link in autonomous driving research: simulation, introducing sensor simulation, traffic simulation models, vehicle/pedestrian simulation models, visualization platforms, digital twins and safety-critical scenario data generation, etc.
- As an inseparable part of autonomous driving, safety theory is involved in this book. It covers the famous NHTSA safety elements, the functional safety standard ISO 26262 and the expected functional safety standard ISO 21448 (SOTIF) of the International Standardization Organization, network security, safety hazards and solutions of autonomous driving. It also includes Intel Mobileye's Responsibility-Sensitive Safety model.
- The seamless integration and collaborative work of various autonomous driving technologies are the basis for the normal operation of the system. At the same time, how to verify the safety and reliability of these systems in various complex environments has become a key challenge in the development of autonomous driving. This book discusses system architecture design, module integration methods, testing and V&V processes, functional and safety evaluation, and performance indicator systems.
- At present, one of the scenarios where autonomous driving technology is closest to commercial landing is automatic parking. This book specifically discusses the calibration of the parking system's vision system, parking space detection, motion planning, sensor fusion (ultrasonic and surround view camera system), and autonomous parking system. It particularly discusses the map building of the parking lot and the parking planning method based on deep learning.
- Vehicle networking is an important aid to autonomous driving and can be seen as an "extension" of the sensor. This book also discusses the application of vehicle networking technology, such as vehicle-road collaboration, vehicle collaborative perception, and planning and control of vehicle platooning.
- This book separately analyzes and discusses two currently popular technologies. The first is neural rendering, which includes the basics of the Neural Radiance Field (NeRF) model, algorithm acceleration, generalization, and expansion techniques, as well as 3D Gaussian splatting which combines the point rendering and explicit (not implicit as NeRF) representation; the second is the diffusion model, an optimal content generation model, introducing its basic theory, improvement methods, and applications in image synthesis, image-to-image translation, text-to-image generation, and policy/planning learning.
- This book also notices the recent rapid development of large language models in the field of AI, and specifically discusses the pre-training and fine-tuning

of large language models, the emerging capabilities like in-contextual learning (ICL), chains of thoughts (CoT) and instruction following, the extension to visual-language models and multi-modality models, and introduces the theories of world models, embodied intelligence, and AI agents, especially the recent reasoning/test-time scaling law, to discuss the application of large scale models in autonomous driving.
- Finally, this book discusses the ethical and legal issues of autonomous driving. When faced with moral dilemmas, how can autonomous vehicles make ethical choices? When an autonomous vehicle has an accident, who should be blamed? Is it the manufacturer, the software developer, or the user of the vehicle? In addition, the data collection and processing of autonomous driving systems also brings challenges in privacy protection and network security.

Targeted Readers

Readers of this book are expected to have a foundational background in computer vision and machine learning (including deep learning), along with some practical experience. This book is not intended as an entry-level introduction to autonomous driving; rather, it aims to engage deeply with the complex challenges facing the field and provide readers with a comprehensive perspective. By presenting a wealth of first-hand information on cutting-edge technologies, the book seeks to inspire practitioners and researchers at the forefront of autonomous driving, broaden their perspectives, and help them identify new solutions when confronted with problems and difficulties.

Sunnyvale, USA — Yu Huang
Hefei, China — Zijiang Yang

Praise for *Autonomous Driving Handbook*

"Dr. Huang and Prof. Yang are top scientists in the field of autonomous driving and related areas. This work comprehensively and profoundly expounds this field, from basic theory to the research frontier. At this critical juncture of the development of autonomous driving technology, the publication of the present book is particularly timely. For readers with a good scientific background, especially those who are interested in engaging in autonomous driving research and development, this book will be an indispensable treasure. I highly recommend it!"

—Andrew Chi-Chih Yao, *Dean of the Institute for Interdisciplinary Information Sciences at Tsinghua University, Academician of the Chinese Academy of Sciences, Turing Award Winner (2000)*

"Autonomous driving presents one of the greatest software engineering challenges of our generation: building complex, safety-critical, AI-enabled systems that we can trust with human lives. In this handbook, Yu Huang and Zijiang Yang have successfully mapped this vast terrain with remarkable clarity. They combine a rigorous treatment of the classical pipeline, perception, mapping, localisation, planning, and control, with an insightful discussion of emerging paradigms, from large-scale AI models to embodied intelligence. What I particularly value is the authors' recognition of safety, validation, and verification as central to progress in the field, not afterthoughts. This book will be an indispensable resource for engineers, researchers, and students seeking both a detailed technical foundation and a vision for the future of autonomous driving."

—James/Jim Woodcock, *Professor of Software Engineering, University of York, retired in 2024, Fellow of the Royal Academy of Engineering in 2011*

"From the Internet industry to the intelligent automobile industry, my entrepreneurial experience has made me believe that autonomous driving technology will give cars a new product definition, bring more innovations, and become a carrier and connection point for many emerging technologies. I hope that the masterpiece of Dr. Yu Huang and Prof. Zijiang Yang can add strength to more people on the difficult road of the development of the intelligent automobile industry toward autonomous driving. If you keep thinking about it, there will be echoes."

—Xiang Li, *Founder, Chairman and CEO of Li Auto*

"Relying on a new generation of chips and software to make autonomous driving a reality will change the way humans travel, but related technological innovations and implementations are still in their infancy. I hope that the joint work of Dr. Yu Huang and Prof. Zijiang Yang can attract more technology personnel to the vast sea of stars of autonomous driving."

—Kai Yu, *Founder and CEO of Horizon Robotics*

"When Google launched its autonomous driving project within Google X—many people, myself included, anticipated an epic success story akin to Boeing's 707 commercial jet project, which revolutionized human transportation. Unfortunately, the journey has proved far more winding than we imagined. Despite numerous impressive technological breakthroughs, it has taken much longer for autonomous driving companies to scale and achieve substantial profitability. One major reason lies in long-tail errors—the unpredictable, diverse, and infrequent edge cases that can still cause system failures or unsafe behavior. Today, the technological landscape is changing rapidly, driven by advances in end2end learning, new data simulation, world models, and multimodal large language models (LLMs). This is an ideal time to revisit the classical foundations of autonomous driving and explore emerging paradigms. In this context, Dr. Yu and Dr. Yang's book provides an excellent discussion along both dimensions, and I highly recommend it as a valuable reference for researchers and practitioners in the autonomous driving field."

—Dr. Liangliang Cao, *Director of Gemini and Principal Engineer, Google*

"*Autonomous Driving Handbook* offers one of the most comprehensive views I've seen on the evolution of intelligent mobility. Huang and Yang move beyond the technical layers—sensors, maps, planning—to frame autonomy as the convergence of data, AI, and real-world systems. What stands out is their perspective on the next phase of development: a transition from rule-based design to adaptive, learning-driven intelligence. The book captures how vehicles are becoming dynamic agents in a connected ecosystem rather than isolated machines. For leaders thinking about the intersection of AI, infrastructure, and automation, this is a valuable reference. It highlights not just where the technology is, but where strategic vision needs to be."

—Dr. James Wu, *VP and General Manager of Drive Sim and DeepMap, Nvidia*

"Autonomous driving, a thrilling and complex field within Robotics and Physical AI, profoundly impacts transportation and automotive industries, and human well-being. It is a rapidly evolving discipline that integrates Computer Vision, Machine Learning, Control Theory, Mechatronics, and more.

This comprehensive book explores high-level principles, key hardware and software components, and various detailed topics in autonomous driving. It serves as an indispensable resource for both beginners and experienced professionals, offering essential knowledge and a contextual framework. A key feature is its use of current and influential examples, such as Multi-Modal End-to-end Driving Models and breakthroughs in 3D Gaussian Splatting, to illustrate theoretical concepts with practical applications.

I highly recommend this book for anyone aiming to master the many facets of autonomous driving and contribute to realizing its full potential."

—Chang Yuan, Ph.D., *Distinguished Engineer, Nvidia*

"This handbook of autonomous driving offers a remarkably practical and up-to-date guide to autonomous driving, closely reflecting the architectures of the most advanced systems being developed in the industry today. It masterfully bridges the classic 'perception-planning-control' pipeline with the 'Autonomous Driving 2.0' paradigm driven by end-to-end learning and large-scale AI. I particularly value its deep dive into how cutting-edge models—including LLMs, Diffusion Models, and NeRF—are being integrated into real-world applications. This is more than a theoretical overview; it's an essential reference for any engineer or researcher actively building the next generation of autonomous systems."

—Dr. Xianming Liu, *Senior Director of Autonomous Driving, XPeng Motors*

"This book stands out as a rare and invaluable encyclopedia for both aspiring engineers and researchers entering the field, as well as seasoned professionals seeking to broaden their perspectives across subdomains of autonomous driving. Authored by two accomplished experts and scientists deeply involved in the field, the book strikes an effective balance between breadth and depth. Within a single volume, it systematically introduces the major domains of autonomous driving and provides detailed explanations of representative and influential technologies in each. One of the book's most remarkable features is its comprehensive and well-organized list of references—hundreds of foundational papers that collectively form the cornerstone of the autonomous driving knowledge base. The *Autonomous Driving Handbook* aspires to serve as both a definitive reference and a guiding compass for future learners, inspiring continued exploration and innovation in this rapidly advancing field."

—Dr. Shiyu Song, *Perception Tech Lead and Manager at Waymo*

"Whether you are a student entering the field, a software engineer building perception systems, or a seasoned researcher exploring the next generation of AI, this book 'handbook of autonomous driving' is your definitive guide to autonomous driving. It provides a robust theoretical foundation in computer vision and machine learning before diving deep into the practical application of every component in the autonomous vehicle stack. The chapters on Perception and Planning are exceptionally thorough, and the inclusion of modern architectures like Transformers and BEV makes the content immediately applicable to current industry challenges. Moreover, by dedicating significant space to emerging technologies like Gaussian Splatting, Diffusion Models, and World Models, the author ensures this book will remain a valuable resource for years to come. It's a must-read for anyone looking to build, understand, or innovate in the world of autonomous systems."

—Zhenhua Yu, Ph.D., *ex-Tesla FSD Staff AI Engineer*

"The *Autonomous Driving Handbook* provides a comprehensive and forward-looking overview of the field, uniting theoretical foundations with practical system design. Dr. Yu Huang and Prof. Zijiang Yang present a clear and integrated account of the entire autonomy pipeline, from perception and mapping to safety and validation. As a computer vision researcher, I particularly appreciate the book's synthesis of classical vision methods with emerging paradigms such as neural rendering, diffusion models, and large-scale AI systems. It is an indispensable reference for researchers and practitioners advancing intelligent autonomous systems."

—Junsong Yuan, *Professor of Computer Science and Engineering, State University of New York at Buffalo, USA, Fellow of Institute of Electrical and Electronics Engineers in 2021, Fellow of International Association for Pattern Recognition in 2018*

"Dr. Yu Huang and Prof. Zijian Yang are leading figures in the field of intelligent systems and autonomous driving. Their new book *Autonomous Driving Handbook* represents one of the most comprehensive and technically rigorous treatments of this rapidly evolving domain. The work bridges foundational theory, spanning computer vision, deep learning, mapping, planning, and control, with practical system engineering aspects such as hardware–software co-design, simulation, safety validation, and large-scale deployment. What I find particularly commendable is how the authors integrate classical robotics and modern generative AI paradigms—such as neural rendering (NeRF), 3D gaussian splatting, diffusion, and large language models—into a unified framework for embodied autonomy. The book provides both conceptual depth and implementation-level clarity, offering an invaluable reference for researchers, engineers, and graduate students committed to advancing safe, interpretable, and scalable autonomous driving systems."

—Zhangyang Atlas Wang, *Temple Foundation Endowed Associate Professor, The University of Texas at Austin, Research Director, XTX Markets*

"What distinguishes this book is its bridge from today's robust autonomy stack to where the field is heading. Clear and unapologetically practical, it treats autonomous driving as an end-to-end systems problem—connecting algorithms to software platforms, data pipelines, evaluation at scale, and the realities of integration, simulation, and road testing—making safety, validation, and verification first-class engineering constraints. Beyond mature modules, the authors lay out the implications of foundation models, agentic systems, generative simulation and synthetic data, and neural rendering for perception, prediction, and planning. The result is a rigorous, forward-leaning guide that helps engineers and researchers reason about capability, scalability, and verifiability—and ship deployable stacks today while preparing for the learning-centric, data-driven future of autonomy."

—Dr. Humphrey Shi, *VP of Nvidia, Associate Professor of School of Interactive Computing and Machine Learning Center, Georgia Tech*

"As someone who works in computer vision, deep learning, and edge computing, I strongly recommend *Autonomous Driving Handbook*. The book connects foundational CV/DL theory to realworld vehicle deployment, guiding the reader from sensing and perception through prediction, mapping, and localization to planning and control, and it concludes with the practical task of running these models efficiently on automotive hardware. I especially value its rigorous treatment of simulation and V&V (including ISO 26262/SOTIF), alongside forwardlooking chapters on neural rendering (NeRF/3DGS), diffusionbased planning, and the emerging role of foundation/world models and multimodal agents. The result is a rare endtoend perspective that serves both researchers pushing the state of the art and engineers shipping reliable edge systems."

—Dr. Ning Xu, *Sr. Fellow and VP of Advanced R&D | Leading Technology Innovations, Adeia Inc.*

"The last few years have witnessed revolutionary progress in autonomous driving. This timely book introduces the diverse range of topics that have contributed to this progress. It is suitable for both scientists interested in design principles and engineers keen on understanding the implementation challenges. Particularly impressive is its self-contained and comprehensive treatment of perception, planning, and control in the context of autonomous driving."

—Rajeev Alur, *Zisman Family Professor, Director of ASSET Center for Trustworthy AI, Department of Computer and Information Science, University of Pennsylvania*

Contents

1 Overview of Autonomous Driving Systems 1
1.1 Classification of Autonomous Driving 1
1.2 Modular Development Structure 3
1.3 End-to-End Development Structure 4
1.4 Autonomous Driving Scenarios 5
1.5 Data Closed Loop 7
1.6 Large Scale Model 9
1.7 Summary 10
References 11

2 Basic Theory of Autonomous Driving 13
2.1 Computer Vision 14
2.1.1 Computer Vision Bottom Layer 15
2.1.2 Middle Layer in Computer Vision 19
2.1.3 High Level Computer Vision 26
2.2 Image Processing 27
2.2.1 Image Signal Processor 27
2.2.2 Image Filtering 28
2.2.3 Image Enhancement 29
2.3 Optimization Theory 30
2.4 Machine Learning 34
2.4.1 Support Vector Machine 37
2.4.2 Random Forest 39
2.5 Deep Learning 41
2.5.1 Convolutional Neural Network 42
2.5.2 Generative Adversarial Networks 50
2.5.3 Recurrent Neural Network 53
2.5.4 Transformer Network 54
2.6 Neural Network Model Compression and Acceleration 57
2.6.1 Parameter Pruning and Sharing 57
2.6.2 Low-Rank Factorization 58

2.6.3 Transferred/Compact Convolutional Filter ... 58
2.6.4 Knowledge Distillation ... 59
2.6.5 MobileNets ... 60
2.7 Architecture of Efficient Transformer and its Application ... 62
2.8 Summary ... 64
References ... 65

3 Hardware Platform for Autonomous Driving ... 69
3.1 Sensors ... 69
3.1.1 Camera ... 70
3.1.2 Lidar ... 71
3.1.3 Millimeter Wave Radar ... 74
3.1.4 Ultrasonic Radar ... 75
3.1.5 Inertial Navigation ... 75
3.1.6 GPS ... 75
3.1.7 Vehicle Networking ... 76
3.2 Vehicle Control ... 77
3.2.1 VCU/ECU/MCU/HCU ... 77
3.2.2 CAN Bus ... 77
3.2.3 ESP ... 78
3.2.4 EPS ... 78
3.3 Computing Platform ... 78
3.3.1 NVIDIA Drive AGX Xavier and Orin ... 79
3.3.2 Mobileye's EQ-X ... 79
3.3.3 TI's TDA4VM ... 79
3.3.4 Qualcomm's Snapdragon Ride ... 80
3.4 Steer-By-Wire Chassis ... 80
3.5 Electronic and Electrical Architecture ... 82
3.6 Summary ... 82
References ... 83

4 The Software Platform for Autonomous Driving ... 85
4.1 AUTOSAR ... 86
4.2 Software Development V-Model ... 87
4.3 aSPICE Software Development Process ... 88
4.4 ISO 26262 Software Development Process ... 89
4.5 NVIDIA DriveWorks Software Platform ... 89
4.6 Vehicle Operating System ... 90
4.6.1 ROS ... 91
4.6.2 QNX ... 92
4.6.3 BOSCH Iceoryx ... 93
4.6.4 DORA ... 94

4.7 Autonomous Driving Cloud Platform ... 96
4.7.1 Open Source Distributed Deep Learning Framework ... 96
4.7.2 Example Architecture of Autonomous Driving Cloud ... 98
4.8 DevOps and MLOps ... 100
4.9 Summary ... 100
References ... 101

5 Perception Module of Autonomous Driving ... 103
5.1 Sensor Calibration ... 104
5.1.1 Camera Calibration ... 104
5.1.2 Lidar Calibration ... 105
5.1.3 Hand–Eye Calibration ... 106
5.1.4 Camera-GPS/IMU Calibration ... 107
5.1.5 LiDAR–Camera Calibration ... 107
5.1.6 Camera–Radar Calibration ... 116
5.1.7 Lidar–IMU Calibration ... 116
5.2 Obstacle Ranging of Monocular Vision ... 118
5.3 Depth Map Estimation of Monocular Vision ... 123
5.3.1 Traditional Methods ... 124
5.3.2 Deep Learning Methods ... 127
5.4 Monocular Vision 3D Obstacle Detection ... 136
5.5 Object Tracking ... 146
5.5.1 Single Object Tracking ... 146
5.5.2 Multi-object Tracking ... 149
5.5.3 Deep Learning-Based Object Tracking ... 150
5.6 Sensor Fusion ... 154
5.6.1 Data Level ... 156
5.6.2 Task Level ... 163
5.7 Lane Line Detection ... 170
5.8 Traffic Sign Detection and Recognition ... 179
5.9 Traffic Light Detection and Recognition ... 181
5.10 Drivable Area Segmentation ... 184
5.11 Binocular Vision Perception ... 186
5.11.1 Stereo Matching ... 187
5.11.2 Binocular Online Calibration ... 190
5.11.3 Binocular Vision Perception System ... 191
5.11.4 Deep Learning Binocular Disparity Estimation ... 192
5.12 Human Pose Estimation ... 198
5.13 Driver Monitoring System ... 205
5.14 BEV's Visual Perception System ... 211
5.14.1 Homography-Based BEV ... 212
5.14.2 Depth-Based BEV ... 214

5.14.3 BEV Based on MLP ... 218
5.14.4 Transformer-Based BEV ... 221
5.14.5 Extension of BEV Framework ... 228
5.14.6 Problems with BEV ... 232
5.15 Occupancy Network of Visual Perception ... 233
5.15.1 Definition of Occupancy Network ... 234
5.16 Summary ... 241
References ... 241

6 High-Definition Map of Autonomous Driving ... 253
6.1 HD Maps ... 253
6.1.1 Map Formats ... 257
6.2 Semantic Map ... 258
6.3 High-Definition Map Based on Lane Lines ... 261
6.4 Deep Learning-Based SLAM Methods ... 265
6.5 Summary ... 268
References ... 268

7 Localization Module for Autonomous Driving ... 271
7.1 Lane Map-Based Localization ... 273
7.2 Lidar-Based Localization ... 277
7.2.1 Normal Distribution Transform Localization ... 277
7.2.2 Particle Filter Localization ... 278
7.2.3 Histogram Filter Localization ... 278
7.3 Localization Based on Sensor Fusion ... 279
7.4 Localization Method Based on Deep Learning ... 282
7.5 Summary ... 285
References ... 286

8 Planning Module for Autonomous Driving ... 289
8.1 Basic Planning Theory ... 289
8.1.1 Frenet Coordinate System ... 294
8.1.2 EM Planner ... 295
8.2 Driving Behavior Model and Prediction ... 298
8.2.1 Driving Behavior Prediction ... 298
8.2.2 Learning Driving Behavior ... 309
8.3 Pedestrian Behavior Model and Prediction ... 312
8.4 Driving Behavior Cloning/Learning ... 316
8.5 Summary ... 330
References ... 330

9 Control Module for Autonomous Driving ... 333
9.1 Kinematic and Dynamic Model of the Vehicle ... 334
9.2 Traditional Control Algorithms ... 339
9.2.1 Classic PID Control ... 339
9.2.2 LQR Control ... 341
9.2.3 Model Predictive Control ... 342

9.3 Control Methods for Path and Trajectory Stability 344
9.3.1 Control of Path Stability 344
9.3.2 Control of Trajectory Stability 347
9.4 Vehicle Control Based on Deep Learning 348
9.4.1 End-to-End Autonomous Driving Control 350
9.4.2 Motion Control Based on Machine Learning 352
9.5 Summary 364
References 365

10 Simulation Module of Autonomous Driving 367
10.1 Sensors 368
10.1.1 Camera 368
10.1.2 LiDAR 375
10.1.3 Millimeter-Wave Radar 378
10.1.4 Other Sensors 380
10.2 Traffic Model 381
10.3 Vehicle and Pedestrian Models 382
10.4 Data Visualization Model 385
10.4.1 XVIZ 385
10.4.2 streetscape.gl 386
10.5 Road Network Simulation 386
10.5.1 Road and Environment Modeling 387
10.5.2 Customizable Props and Signs 387
10.5.3 Functional Road Network 387
10.6 Construction of Scenario Library and Autonomous Driving Test 388
10.6.1 Scenario Definition 388
10.6.2 Classification of Simulation Test Methods 388
10.6.3 Construction of Scenario Library 389
10.6.4 Scenario Description Language OpenSCENARIO 390
10.7 Safety-Critical Scenario Generalization Based on Digital Twin 391
10.8 Summary 400
References 400

11 Safety Model 403
11.1 Basic Concepts 405
11.2 NHTSA 406
11.3 ISO 26262 406
11.4 ISO/PAS 21448 SOTIF 408
11.5 Intel Mobileye RSS Safety Model 409
11.6 Network Security 410
11.7 Safety Solutions for Autonomous Driving Systems 411
11.8 Summary 415
References 415

12 System Integration, Validation and Verification (V&V) 417
12.1 System Integration Methods 417
12.2 System V&V Methods and Processes 420
12.3 Safety V&V 424
12.4 Performance Evaluation and Metrics 426
12.5 The Formal Method of V&V 429
12.6 Summary 434
References 434

13 Automatic and Autonomous Parking 437
13.1 Basic Introduction to the Automatic Parking System 438
13.2 Calibration of Surround-View Vision System 440
13.3 Perception System of Fisheye Camera 441
13.4 Parking Spot Detection 445
13.5 Motion Planning for Parking 446
13.6 Sensor Fusion in Parking 447
13.7 Automatic Valet Parking 448
13.8 Semantic Maps and Localization in Parking Lots 450
13.9 Deep Learning Planning Techniques for Autonomous Parking 452
13.10 Summary 457
References 457

14 Vehicle Networking 459
14.1 Intelligent Connected Vehicles 460
14.2 Vehicle Networking and Social Internet of Things 464
14.3 Edge Computing 467
14.4 Vehicle-Roadside Collaboration Technology 468
14.5 Collaborative Perception in Autonomous Driving 474
14.6 Cooperative Planning and Control of Vehicle Platooning 478
14.7 Summary 484
References 484

15 Neural Rendering Technology 487
15.1 Original NeRF 488
15.2 Acceleration Methods for NeRF 492
15.2.1 AutoInt 493
15.2.2 PlenOctree 496
15.2.3 Plenoxel 498
15.3 Rendering of Dynamic Scenes 499
15.4 NeRF Relighting Techniques 502
15.5 Generalization Techniques of NeRF 507
15.5.1 GRAF 507
15.5.2 GIRAFFE 509

15.6 Techniques for Quality Improvement 512
15.6.1 Mip-NeRF 512
15.6.2 NeRF-in-the-Dark 516
15.7 Gaussian Splatting Techniques 518
15.8 NeRF and 3DGS for Autonomous Driving 522
15.9 Summary 523
References 523

16 Diffusion Models 525
16.1 Score-Based Generative Networks 526
16.2 Denoising Diffusion Probabilistic Model 528
16.3 DDIM 531
16.4 SDE 534
16.5 Image/Video Synthesis 535
16.5.1 Image 535
16.5.2 Video 540
16.5.3 New View Synthesis 540
16.6 Image-to-Image Translation 543
16.7 Text-to-Image/Video Synthesis 545
16.8 Improvements to the Diffusion Model 549
16.9 Diffusion Model for Policy and Planning Learning 551
16.10 Summary 554
References 555

17 Large Scale Models in Autonomous Driving 559
17.1 Pre-training and Fine-Tuning of Large Language Model 560
17.1.1 Foundation Model and Large Language Model 560
17.1.2 Pre-training of Large Language Model 560
17.1.3 Fine-Tuning of Large Language Model 563
17.2 Prompting, ICL and CoT of Large Language Model 564
17.2.1 Prompting 564
17.2.2 In-Context Learning 565
17.2.3 Chain-of-Thoughts and Extension 565
17.3 Vision-Large Language Model 566
17.4 Multi-modality Large Language Model 568
17.5 World Model and Embodied Intelligence 568
17.5.1 World Model 568
17.5.2 Embodied Intelligence 569
17.6 Inference-Time Optimization for LLM Reasoning 570
17.7 Large Scale Model Applied for Autonomous Driving 572
17.7.1 A Language Agent Driver 572
17.7.2 DriveVLM-Dual: Combination of VLM and Traditional Pipeline 573

17.7.3 Senna: Combining VLM and E2E Model 575
17.7.4 Google's EMMA 576
17.7.5 World Model: DrivingWorld 578
17.8 Summary 579
References 579

18 Ethics and Legal Aspects 583
18.1 Driving Dilemmas 583
18.2 Ethical Issues 584
18.3 Legal Issues 586
18.4 Privacy and Data Security 588
18.5 Moral and Ethical Impacts on Society 591
18.6 Summary 594
References 594

List of Figures

Fig. 1.1 Simplified SAE autonomous driving classification 2
Fig. 1.2 Modular structure of the autonomous driving system 3
Fig. 1.3 Structure of NVIDIA's PilotNet end-to-end autonomous driving learning system 4
Fig. 1.4 Center camera as input data 4
Fig. 1.5 Structure of Comma AI's OpenPilot 5
Fig. 1.6 Speed-complexity relationship of autonomous driving scenarios 7
Fig. 1.7 Tesla's autonomous driving data closed loop framework 8
Fig. 1.8 Google Waymo's autonomous driving data closed loop framework 8
Fig. 2.1 Malik's 3R computer vision architecture 14
Fig. 2.2 Bayer pattern 27
Fig. 2.3 BLF Filtering Example 28
Fig. 2.4 NLM filtering example 29
Fig. 3.1 Internal structure of CCD 70
Fig. 3.2 Internal structure of CMOS 71
Fig. 3.3 Principle of Lidar 72
Fig. 4.1 DORA reference architecture 94
Fig. 4.2 DORA software architecture 95
Fig. 5.1 The 3D occupancy annotation pipeline 235
Fig. 5.2 FB-Occ framework 236
Fig. 5.3 TPVFormer 237
Fig. 5.4 MonoScene 238
Fig. 5.5 Fast-Occ pipeline 239
Fig. 5.6 The overall architecture of OpenOccupancy 239
Fig. 5.7 Vision-based 3D occupancy perception network methods 240
Fig. 6.1 Layered structure of HD maps from Lyft 254
Fig. 6.2 The process of high-definition mapping by the Japanese company ZENRIN 256
Fig. 8.1 The UniAD framework 327

Fig. 8.2 VAD architecture 328
Fig. 8.3 FusionAD architecture 328
Fig. 8.4 GenAD framework 328
Fig. 8.5 VADv2 architecture 329
Fig. 8.6 SparseDrive framework 330
Fig. 9.1 The control flow of PID 339
Fig. 9.2 The structure of MPC 343
Fig. 10.1 The overall framework of EPE 372
Fig. 10.2 Overview of the adversarial scenario generation pipeline 394
Fig. 12.1 Categories of system integration methods 418
Fig. 12.2 The system verification pipeline 422
Fig. 12.3 The system validation pipeline 423
Fig. 12.4 Categories of system performance evaluation metrics 428
Fig. 13.1 The modular architecture of an automatic parking system 437
Fig. 13.2 The whole system pipeline of AVP-SLAM 450
Fig. 14.1 Intersection of CAV 461
Fig. 14.2 CAV's converging road scene 461
Fig. 15.1 The illustration of NeRF 488
Fig. 15.2 The illustration of 3DGS 519
Fig. 16.1 Graph models of diffusion inference model (top) and non-Markov inference model (down) 532
Fig. 16.2 Schematic diagram of the SDE generating a score-based generative model 534
Fig. 16.3 Schematic diagram of CDM generating class conditional 256×256 images 539
Fig. 16.4 Diffusion model's 3D UNet architecture 540
Fig. 16.5 Imagen visualization 547
Fig. 16.6 The overview of multi-agent motion trajectory prediction based on the diffusion model 552
Fig. 16.7 The overall architecture of DiffusionDrive 553
Fig. 16.8 The overview of diffusion planner 554
Fig. 17.1 Foundation model 560
Fig. 17.2 Agent-driver 572
Fig. 17.3 The pipeline of DriveVLM and DriveVLM dual 574
Fig. 17.4 The data mining and annotation pipeline 575
Fig. 17.5 Senna architecture 576
Fig. 17.6 Illustration of EMMA 577
Fig. 17.7 DrivingWorld pipeline 578
Fig. 18.1 Ethical issues of decision-making in autonomous driving 585
Fig. 18.2 Legal issues: Liability of autonomous driving vehicles 587

List of Tables

Table 2.1 MobileNets architecture with layer configurations, filter shapes, and input sizes 61
Table 4.1 ISO 26262 software architecture design guidelines method 90
Table 4.2 ISO 26262 software architecture level error detection mechanism 90
Table 5.1 OverFeat model parameters 182
Table 5.2 Model structure parameters of the traffic light classifier 183
Table 5.3 Structural parameters of the spatial transformer network 184
Table 5.4 Parameters of the GC-Net model 195

Chapter 1
Overview of Autonomous Driving Systems

There are some ambiguities in the concept of autonomous driving systems, such as how to define an autonomous driving system, why there are so many sub-modules in the development of autonomous driving, and what constitutes the implementation of autonomous driving. This chapter provides a general introduction to autonomous driving systems, helping readers understand the classification of autonomous driving, clarify the differences between modular development and end-to-end learning development, and understand the application scenarios of autonomous driving.

Section 1.1 introduces the classification of autonomous driving, Sect. 1.2 introduces the modular development structure, Sect. 1.3 introduces the less commonly used end-to-end development structure, Sect. 1.4 introduces the classification of autonomous driving implementation scenarios, Sect. 1.5 introduces the data loop, and finally, Sect. 1.6 gives an overview of Large Scale Model's application in autonomous driving.

1.1 Classification of Autonomous Driving

Mobileye, Tesla, and Google Waymo are all researching autonomous driving, but how to define an autonomous driving system and the differences between them involve the classification of automotive autonomous driving.

There are currently two sets of standards in the industry [1]: The first set is established by the U.S. Department of Transportation's NHTSA (National Highway Traffic Safety Administration). The second set is established by SAE International. The difference between these two sets of standards mainly lies in the definitions of L4 and L5, generally based on SAE.

SAE divides the degree of automobile automation into six levels, as shown in Fig. 1.1.

Y. Huang and Z. Yang, *Autonomous Driving Handbook*,
https://doi.org/10.1007/978-981-95-3725-9_1

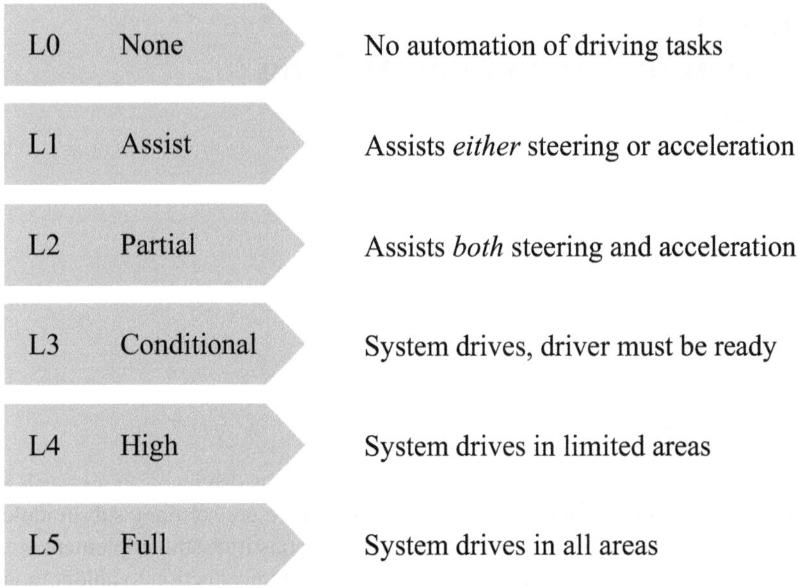

Fig. 1.1 Simplified SAE autonomous driving classification

- L0 (driver only, no automation): The car has no autonomy; all tasks are fully operated by the driver.
- L1 (assisted drive): The car is still controlled by the driver, but there are some driving assistance functions in the overall vehicle design.
- L2 (partial automation): Includes some automatic functions, such as acceleration and steering, but the driver cannot disengage and still needs to monitor the environment.
- L3 (conditional automation): Still requires a driver, but does not need to monitor the environment. The driver needs to be ready to take over when warned.
- L4 (high automation): The system performs all driving operations under certain conditions; the driver takes over when possible.
- L5 (full automation): The system performs all driving operations under any conditions; the driver does not need to take over.

The ADAS (Advanced Driver Assistance Systems) product provided by Mobileye belongs to L2, while the Autopilot currently implemented by Tesla belongs to L2∼L3, Audi A8's traffic jam pilot (TJP) system for congested traffic situations (speed less than 60km) belongs to L3, and the autonomous driving developed by Google WAYMO, Baidu, General Motors Cruise AI, and Ford Argo AI are classified as L4. Currently, L5 is not possible and is the ultimate ideal goal. To put it vividly, the levels of autonomous driving are, L1 is feet-off, L2 is hands-off, L3 is eyes-off, and L4 is mind-off.

1.2 Modular Development Structure

There are two types of development structures for autonomous driving systems: end-to-end "black-box" development structure and modular development structure [1, 2]. This section introduces the second type of development structure, while Sect. 1.3 discusses the first type of development structure. However, the modular development structure is still the mainstream at present. The modular structure of the autonomous driving system is shown in Fig. 1.2.

The hardware platform includes sensors, vehicle networking, and actuators.

1. Sensors: Devices that provide external environmental data, such as cameras, lidar, millimeter-wave radar, GPS, Inertial Measurement Units (IMU), etc.
2. Vehicle Networking (V2X): Used to receive and transmit information, mainly exchanging location and speed data with other vehicles around.
3. Actuators: The interface for vehicle control. For wire-controlled vehicles, the inputs to the actuators are three signals: the accelerator pedal, the brake pedal, and the steering wheel.

The software platform includes perception (HD maps, localization), planning, and control.

1. Perception: Understanding the surrounding environment through data collected by sensors, such as obstacles, depth, lane lines, traffic signs, traffic lights, segmentation situations, etc.
    - HD map: Provides lane-level maps.
    - Localization: Determines its own position based on the match between the HD map and the sensor data collected on-site.
2. Planning includes three levels: mission (path) planning, behavior planning (decision-making), and motion planning.
    - Mission planning: Find the shortest or fastest road from the starting point to the end point.
    - Behavior planning: Make decisions among various driving behaviors, such as lane changing, overtaking, turning, accelerating/decelerating, parking, etc.

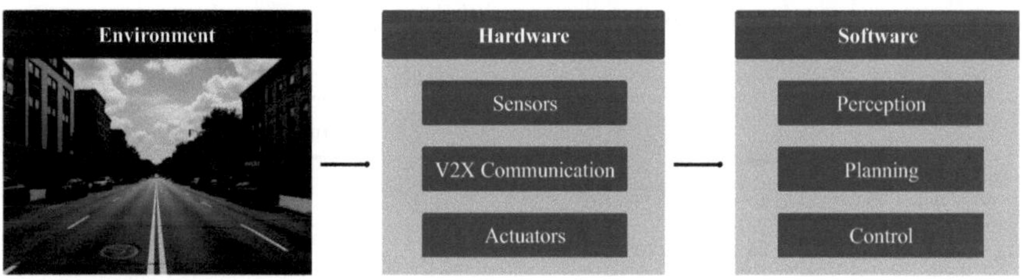

Fig. 1.2 Modular structure of the autonomous driving system

This decision depends on traffic rules, driving habits, and traffic conditions (obstacles).

- Motion planning: A type of local planning, used to determine specific lanes and actions, generate reference paths, and time-referenced trajectories.

3. Control is connected to motion planning, also known as motion control, tracks the error of executing the plan, feeds it back to the controller to stabilize to the reference path and reference trajectory. Correspondingly, control also includes two parts: path tracking and short-term trajectory tracking.

1.3 End-to-End Development Structure

The end-to-end autonomous driving system is like a "black box", such a simplified development structure hides the difficulty of learning. Its input is all sensor data, and the output is the vehicle control signal. Currently, typical examples following this development structure include NVIDIA's PilotNet, Comma AI's open-source OpenPilot, and the Software 2.0 proposed by Tesla's Autopilot director.

As shown in Fig. 1.3, NVIDIA's PilotNet uses three cameras (left, right, middle) to capture external environmental data and input it into the deep learning Convolutional Neural Network (CNN) model, and the output is the steering wheel turning angle for vehicle control [3].

However, during inference/testing, PilotNet only uses the center camera as input data to control the vehicle's steering wheel [4], as shown in Fig. 1.4.

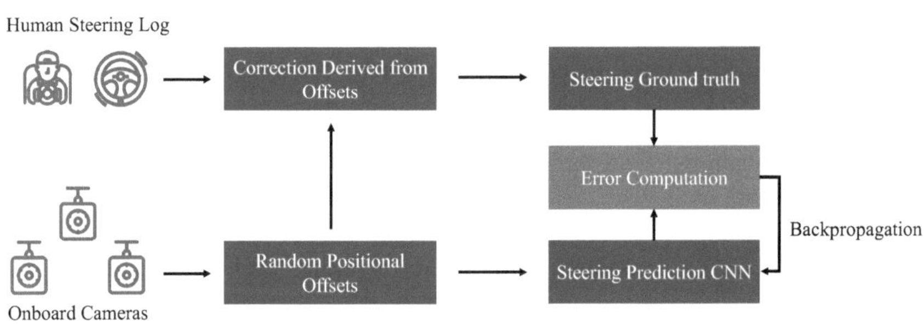

Fig. 1.3 Structure of NVIDIA's PilotNet end-to-end autonomous driving learning system

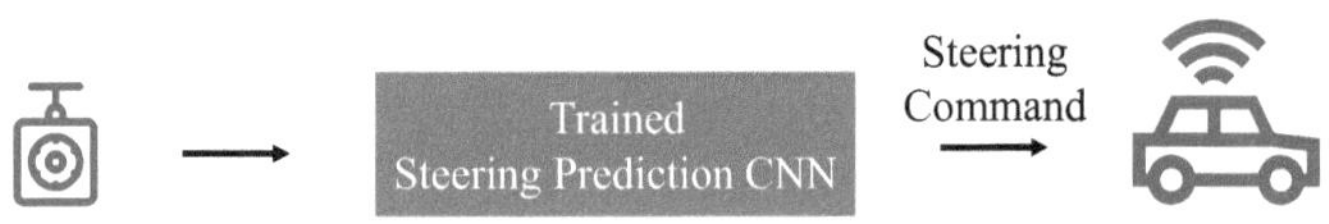

Fig. 1.4 Center camera as input data

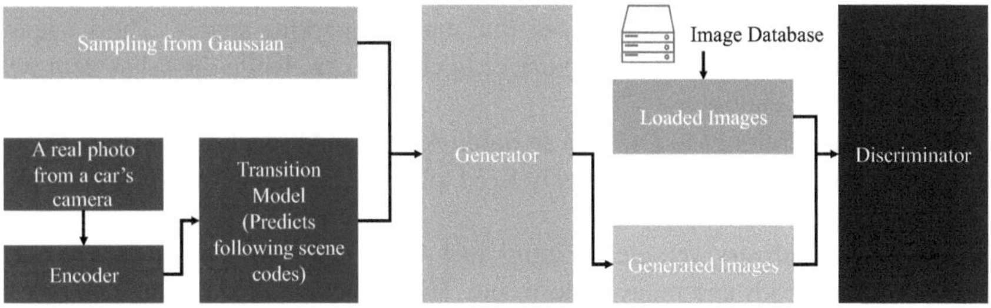

Fig. 1.5 Structure of Comma AI's OpenPilot

As shown in Fig. 1.5, the learning structure of Comma AI's OpenPilot includes two parts [5].

One is to use a combination of Generative Adversarial Network (GAN) and Variational Autoencoder (VAE) as the main training body. The second is to use an action-conditioned Recurrent Neural Network (RNN) model to learn a transition model through 15 frames of video data, which serves as the core of future inference driving actions.

However, in actual on-vehicle experiments, the vehicle must have two auxiliary driving functions, namely: ①ACC (Adaptive Cruise Control): It can be confirmed that it is controlled by the throttle and brake pedal. ②Assisted steering system: It can be confirmed that it is controlled by the steering wheel.

Basically, for the structure of Software 2.0 proposed by the director of Tesla Autopilot [6], the system inputs all sensor data, that is, 8 cameras, 12 ultrasonic radars, 1 mm-wave radar, and inertial data. The output is control signals, namely acceleration and deceleration, and steering wheel. So far, there is no evidence that Tesla has completed the development of such a practical system.

However, this approach is not accepted by most autonomous driving developers. They generally adopt an end-to-end development structure that relies heavily on machine learning or deep learning. The main reason for not being adopted is that there are too many edge cases, deep learning model training requires a large amount of data, and it is not easy to control the learning process.

1.4 Autonomous Driving Scenarios

The scenarios of autonomous driving can be divided into the following scenarios according to difficulty [1].

1. Autonomous/Automatic parking: Autonomous parking is L4, while automatic parking is generally L2∼L3. The lowest level of automatic parking has no vehicle control and only provides parking suggestions, such as virtual reference trajectories. L2 automatic parking provides parking space detection and determines planning and control functions. Simple ultrasonic radar only detects one parking

space, while the visual system can find multiple parking spaces. The L2 automatic parking system allows the driver to choose one, while the L3 system will automatically choose.
2. Park shuttle: Belongs to L4, such as Baidu Apollo and French startup EasyMiles' EZ-10, the driving path is relatively fixed, planning decisions are simple, mainly running in closed-loop roads.
3. Unmanned delivery vehicle: Belongs to L4, the difficulty lies in the crowded community roads and many pedestrians. The advantage is low speed, no one in the car. Meituan, JD.com, Alibaba's Cainiao, and Silicon Valley startup Nuro all have this scenario.
4. Highway commercial truck cruising: Belongs to L3∼L4, its advantages are mainly on highways, the business model is simple, it has advantages in reducing driver costs and reducing driver fatigue. The disadvantage is that the warning time for high-speed driving needs to be long, and the vehicle is large and long, and the surrounding monitoring is difficult. Google, startup TuSimple, and Plus.ai have layouts in the autonomous driving research and development of this scenario.
5. Highway Traffic Jam Assist (HTJAP): Belongs to L3, the advantage is slow speed, the disadvantage is traffic congestion, and extreme situations (corner case) need to be prevented; Audi A8 has achieved this function, for which 54-line lidar Scala were installed.
6. Highway Autopilot (HAP): Tesla Autopilot, belongs to L2+, provides automatic follow-up and lane keeping functions on highways, the disadvantage is insufficient warning time, the advantage is that it is easier to achieve autonomous driving in a relatively closed environment, currently expanding to street scenes, and adding the ability to detect and recognize traffic lights.
7. Urban Traffic Jam Assist (UTJAP): Belongs to L4, there are no related products on the market, only research and development testing, such as Cruise's research and development in San Francisco and Argo's research and development in Miami.
8. Unmanned taxi (equivalent to urban automatic cruising (urban driving)): Belongs to L4, only in trial operation, such as Google's Waymo One developed in Phoenix.

Figure 1.6 is a dimension analysis for speed and computational complexity, which vividly shows some typical scenarios. Among them, racing circuit driving is a scene that has not been widely concerned. It has the fastest speed, but it is still applied in a closed park, and the difficulty is still below urban driving.

There is currently no very universal model for the development of autonomous driving. The reason is that the technology is not mature, and commercial landing needs to consider different application scenarios. In different application scenarios, some problems are simplified, but they still face unique challenges. If these application scenarios are compared to the various root systems of this big tree of autonomous driving, then the technological development in various application scenarios is like deeply "rooting" the root system into the soil of commercial landing.

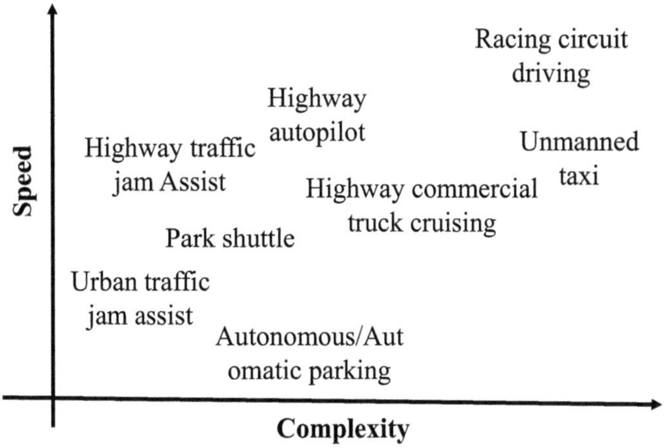

Fig. 1.6 Speed-complexity relationship of autonomous driving scenarios

1.5 Data Closed Loop

The reason for combining autonomous driving and data closed loop is that the autonomous driving project has been recognized as a task to solve the long tail effect of data distribution [7]. The occasional extreme cases are one of the sources of data for upgrading data-driven algorithm models.

What are the core technologies and modules that constitute the data closed loop of autonomous driving? First, the algorithms and modules of autonomous driving are data-driven, and secondly, there are reasonable and effective methods to use the continuous data.

Figure 1.7 shows the data closed-loop framework of Tesla's autonomous driving, including determining inaccurate examples, unit testing and augmentation, (data) annotation and cleaning, (model) training and (re)deployment.

Figure 1.8 shows Google Waymo's autonomous driving data closed loop framework. The specific process is to first collect data (sensor logs), then select through data mining and active learning, and generate annotated data through (human) annotators or (machine) automatic annotation, then develop models, perform automatic parameter optimization, then verify and deploy, and finally release on the vehicle side.

Data collection and cleaning/filtering is the primary task of the data loop, which is divided into two parts: the vehicle end and the cloud/server end.

Generally, the vehicle end will adopt different strategies during the research and development stage and the mass production stage. In the research and development stage, the data collection vehicle will generally store the collected data directly in the form of a hard disk, either manually connecting the hard disk to the server or through a special data transmission channel; during the mass production stage, data will be

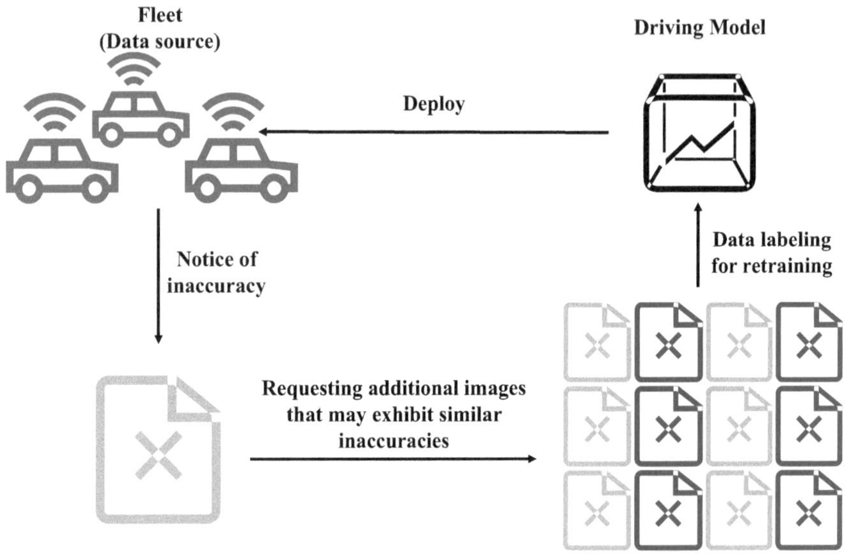

Fig. 1.7 Tesla's autonomous driving data closed loop framework

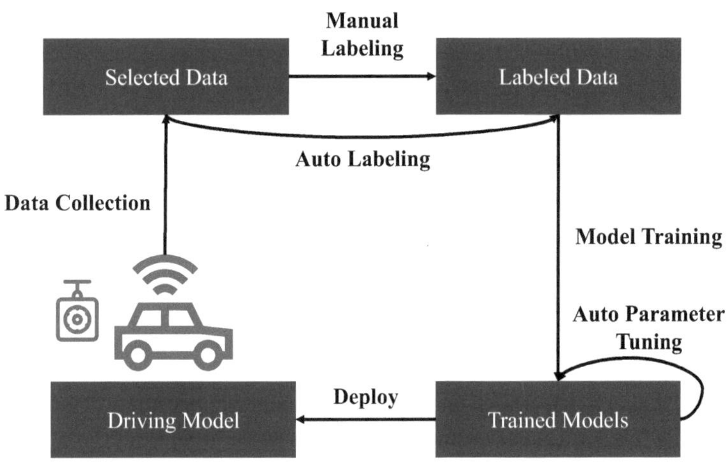

Fig. 1.8 Google Waymo's autonomous driving data closed loop framework

selectively transmitted, such as Tesla's "shadow mode" for selection, the application program deployed at the vehicle end finds that the autonomous driving algorithm running in "shadow mode" has significantly different decisions and behaviors from the driver; or using CAN data for selection, judging the abnormality of vehicle driving data (such as sudden steering wheel or braking actions); or like Tesla's trigger mode for selection, searching for similar scenes based on a known scene pattern.

The cloud/server end will further filter the collected data, such as removing duplicate scenes, mining new scenes, looking for safety-critical scenes and extreme situations, etc.

The data loop requires a cloud computing/edge computing platform and big data processing technology, which cannot be implemented in a single vehicle or machine. Big data and cloud computing have developed for many years, providing infrastructure support for the data loop in terms of data batch/stream processing, workflow management, distributed computing, status monitoring, and database storage.

Data annotation is also an important link in the loop. The efficiency of annotation is an important indicator of a platform. Manual annotation is a very common method, but semi-automatic annotation, or even Tesla's fully developed automatic annotation, requires more investment in algorithms, including the development of offline large models and the construction of human-in-the-loop 4D annotation algorithms.

Model training requires a powerful GPU computing platform, running multi-machine distributed deep learning training tasks in machine learning training frameworks such as PyTorch and TensorFlow. Distribution is mainly data parallel, with less model parallel.

Model testing and evaluation need to be carried out in simulation environments, closed environments, and open environments. Simulation testing generally defines a scene library, generates a variety of driving scenes to test the autonomous driving algorithm, including Software in the Loop (SIL), Hardware in the Loop (HL), and Vehicle in the Loop (VIL) methods.

When the model is deployed to the vehicle end, model compression and acceleration are also necessary steps. The focus of compression is to reduce the amount of network parameters, and the purpose of acceleration is to reduce computational complexity, enhance parallel capabilities, etc.

Especially the method of knowledge distillation, which compresses a network with large depth and width into a smaller network, where the compressed model simulates the function learned by the complex model. The main idea of the distillation method is to transfer knowledge from the teacher's large model to the student's small model by learning the class distribution obtained from the softmax output.

1.6 Large Scale Model

The development of foundational large language models (LLMs) [8], such as GPT, LLaMA, and PaLM, marks a major milestone in artificial intelligence (AI). As an example of foundational models [9], LLMs have enabled the emergence of new AI capabilities such as in-context learning (ICL), chain of thoughts (CoT), and instruction following. The emergence of LLMs marks the maturation of a new paradigm of pre-training + fine-tuning, while demonstrating the scaling law of large model pre-training.

The combination of visual and textual modalities has also led to the construction of large visual-language models (VLMs) [10], such as CLIP and GPT-4o, and extended to the field of multimodal language models (MLMs) [11]. Recent work has also seen these large-scale models work proficiently in various text generation, language translation, and perception tasks. They are trained on large-scale datasets and use advanced algorithms to excel in understanding and generating human-like responses. In particular, in challenging AI application scenarios such as embodied intelligence [12] and world models [13], generative AI models such as diffusion transformer (DiT) networks and special multimodal models, like vision-language-action (VLA), are producing impressive results. With the help of LLM, AI agents are also enhanced with capabilities such as tool/function calling, memory, and reflection [14], and agentic workflows enhance the reasoning and planning functionalities of AI entities.

Autonomous vehicles, as a special instance of robots, have also ushered in new research directions and performance improvement potential in the wave of large language models [15], especially in the face of decision-making and planning challenges in complex traffic environments. To complete a driving task with strict real-time running requirements, it is very important to balance the rapid response of driving maneuvers and the analytical reasoning of safety-critical scenarios.

Originally, it was only used for model alignment to eliminate bias and adverse reactions, namely human feedback-based reinforcement learning (HFRL). With the introduction of self-verifiers or evaluators, RL has changed the reasoning paradigm of LLM. The recent emergence of OpenAI o1/o3 and DeepSeek R1 has given a new law of scale for reasoning/testing-time, which is similar to the dual-system as human thinking system, requiring the implementation of a fast and intuitive system 1 and a slower and more cautious reasoning system 2 [16]. This dual-system design is expected to solve the intelligence bottleneck faced by embodied intelligence and autonomous driving.

1.7 Summary

This chapter is the beginning of the analysis of autonomous driving technology, first introducing its levels, two different paths of development, namely modular development structure, end-to-end development structure and landing scenarios, and then introducing the concept of data loop, and then introducing the large scale model applied in autonomous driving. Chapter 2 will further analyze the basic theories and tools needed for autonomous driving research and development.

References

1. PENDLETON S D, Perception, Planning, Control, and Coordination for Autonomous Vehicles, MDPI Machines, Feb. 2017, 5(1): 6–59.
2. HAN F, CSDN AV Levels. https://blog.csdn.net
3. BOJARSKI M et al., End to end learning for self-driving cars, arXiv preprint arXiv: 1604.07316, April, 2016
4. BOJARSKI M et al., Explaining How a Deep Neural Network Trained with End-to-End Learning Steers a Car, arXiv preprint arXiv:1704.07911, April, 2017
5. SANTANA E, HOTZ G, Learning a Driving Simulator, arXiv preprint arXiv:1608.01230, Aug., 2016
6. KAPRPATHY A, Building the Software 2.0 Stack, Spark + AI Summit, San Francisco, June, 2018
7. HUANG Y, How to Build a Data Closed Loop Platform for Autonomous Driving. https://medium.com
8. W X Zhao, K Zhou, J Li, et al., "A Survey of Large Language Models", arXiv preprint arXiv:2303.18223, 2023
9. R Bommasani et al., "On the Opportunities and Risks of Foundation Models", arXiv preprint arXiv:2108.07258, 2021
10. Z Li, X Wu, H Du, et al., "Benchmark Evaluations, Applications, and Challenges of Large Vision Language Models: A Survey", arXiv preprint arXiv:2501.02189, 2025
11. C Li, Z Gan, Z Yang, and et al. "Multimodal Foundation Models: From Specialists to General-Purpose Assistants" (survey), arXiv preprint arXiv:2309.10020, 2023
12. Y Liu, W Chen, Y Bai and et al. "Aligning Cyber Space with Physical World: A Comprehensive Survey on Embodied AI", arXiv preprint arXiv:2407.06886, 2024
13. J Ding, Y Zhang, Y Shang, et al., "Understanding World or Predicting Future? A Comprehensive Survey of World Models", arXiv preprint arXiv:2411.14499, 2024
14. L Wang et al., "A Survey on Large Language Model based Autonomous Agents", arXiv preprint arXiv:2308.11432, 2023
15. Y Huang, Y Chen, Z Li, "Applications of Large Scale Foundation Models for Autonomous Driving" (survey), arXiv preprint arXiv:2311.12144, 2023
16. Z Li, D Zhang, M Zhang, et al., "From System 1 to System 2: A Survey of Reasoning Large Language Models", arXiv preprint arXiv:2502.17419, 2025

Chapter 2
Basic Theory of Autonomous Driving

Developing an autonomous driving system requires some professional knowledge, the most fundamental of which are machine learning and deep learning, in addition to the basics of computer vision and image processing, and also involves some numerical computation and optimization algorithm knowledge.

This chapter cannot go into detail about these fields, but only briefly introduces related basic concepts and algorithms, and extracts key points for in-depth discussion. If you really want to fully learn and master this knowledge, you need to read specialized papers and works.

Section 2.1 introduces the framework of computer vision, focusing on a few key points for analysis, such as SIFT feature extraction, camera calibration, motion estimation and structure from motion (SFM), and other modules, such as stereo vision, shape from X, and object detection/recognition/tracking/segmentation, will be highlighted in Chap. 5, so they are not repeated here; Sect. 2.1 briefly introduces image processing theory, especially analyzing two image denoising algorithms, namely bilateral filtering (BLF) and non-local means (NLM) filtering; Sect. 2.3 is an overview of optimization theory, especially discussing two common nonlinear least squares methods: G-N method and L-M method; Sect. 2.4 outlines the theory of machine learning, focusing on commonly used support vector machines and random forests; Sect. 2.5 provides an overview of deep learning theory, with a focus on convolutional neural networks (CNN), recurrent neural networks (RNN), generative adversarial networks (GAN), and Transformers; Sect. 2.6 gives the overview of neural network's compression and acceleration techniques; and finally an introduction of efficient Transformers architecture applied in large-scale model theories and applications is given in Sect. 2.7.

Y. Huang and Z. Yang, *Autonomous Driving Handbook*,
https://doi.org/10.1007/978-981-95-3725-9_2

2.1 Computer Vision

What is computer/machine vision? Simply put, it is like the human eye, the computer understands the scene through images, understands what the scene contains, the position, size and shape of each thing, and the layout of the scene. For the scenarios encountered in autonomous driving, what computer vision needs to do is: Determine whether the vehicle is on the highway, in the street, or even on a suburban road, and what cars or pedestrians are on the road, how many lanes are on the road, what traffic signs (stop, speed limit or warning, etc.), whether it is at an intersection, whether there are traffic lights, and whether there are other obstacles on the road (such as construction sites, parked goods, or even unexpected garbage).

Professor J Malik of the University of California, Berkeley, once called the task of vision the three Rs: Recognition, Reconstruction, and Reorganization [1] (see Fig. 2.1).

- The recognition task is for object detection, tracking, and classification.
- The reconstruction task includes feature matching, multi-view geometry, localization, and mapping.
- The reorganization task is mainly object segmentation.

These three Rs are interrelated, recognition helps reconstruction and reorganization, reconstruction can also improve the performance of recognition and reorganization, and reorganization also collaborates with recognition and reconstruction tasks.

To be more specific, letting computers simulate human vision is not something a simple system can accomplish, but it is modular and hierarchical. Modularity means that the machine will implement some functions of human vision, such as depth perception, motion perception, and object separation. The hierarchy divides the visual functions into low-level 2D image domain, middle-level 2.5D motion domain and

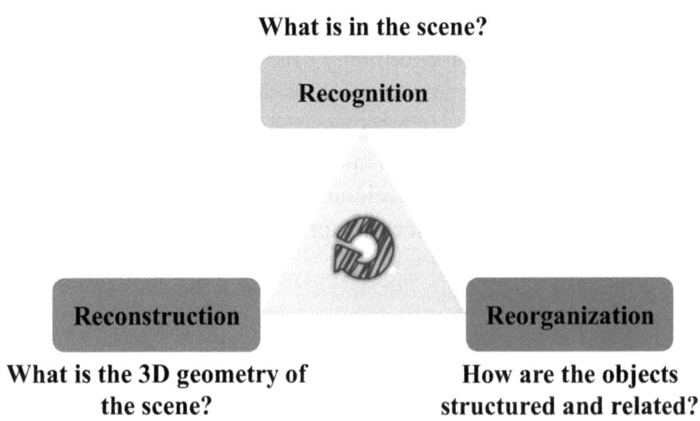

Fig. 2.1 Malik's 3R computer vision architecture

disparity (related to depth) domain and high-level 3D shape reconstruction and recognition. This section is based on the work of computer vision expert R.Szeliski (see reference [2]), and unfolds along the main line of photometry, image, and geometric shape, covering the basic theory and application of computer vision from the bottom to the top, as shown in Fig. 1.11 of [2].

2.1.1 Computer Vision Bottom Layer

The main task of the bottom layer of computer vision is image pixel-level processing, and there are also some low-level feature processing.

1. Image Domain Sampling Aliasing
The sampling and aliasing in the image domain is actually a digitization problem of the camera image; Fig. 2.25 [2] is an example of 2D signal aliasing, where Fig. 2.25(a) in [2] is the original full-resolution image; Fig. 2.25(b) in [2] is a 4x downsampling implemented using a 25% fill factor box filter; Fig. 2.25(c) in [2] is a 4x downsampling implemented using a 100% fill factor box filter; Fig. 2.25(d) in [2] is a 4x downsampling implemented using a 9-tap filter. It can be observed that low-quality filters will alias high frequencies into visible frequencies, while 9-tap filters completely eliminate high-frequency aliasing. Note: The sampling pitch is the physical interval between adjacent sensing units on the camera, and lenses with smaller sampling pitches have higher sampling densities. The fill factor is the size of the active sensing area of the camera (the product of the horizontal and vertical sampling pitches), and it is generally recommended to use a high fill factor, so as to capture more light and less aliasing.

2. Image Processing
The field of image processing can be independent of computer vision, and Sect. 2.2 will introduce it in detail.

3. Feature Extraction
Feature detection can find some significant image features in the image, they are generally sparse, for some visual tasks, it is a concise way of image description, such as corners and edges. It should be said that before deep learning "conquered" the field of computer vision, the choice of features was crucial to the performance of the visual system. The most successful and famous hand-crafted feature should be the Scale Invariant Feature Transform (SIFT), proposed by Professor Lowe of the University of British Columbia, Canada. SIFT is a feature that remains invariant to translation, scale, and rotation, and also has light intensity invariance, and is robust to local geometric distortions.
The steps to extract SIFT features are as follows.

1. Construct a multi-scale Gaussian pyramid structure of an image, and the feature points are also called key points, which can be obtained through the Difference of Gaussian (DoG) image. As shown in Fig. 4.11 of [2], scale-space feature

detection is performed using a sub-octave difference of Gaussian pyramid, where Fig. 4.11(a) of [2] is the Gaussian difference image produced by subtracting adjacent layers of the sub-octave Gaussian pyramid; Fig. 4.11(b) of [2] shows the extreme values (maximum and minimum) of the 3D volume obtained by comparing the pixel with its 26 neighbors (8 on the same layer and 9 on the upper and lower layers). These operations ensure the illumination intensity, translation, and scale invariance of SIFT.

2. After finding the extreme values in scale space, it is necessary to use the Taylor series expansion to obtain a more accurate position (sub-pixel). Because DoG has a high response to edges, it needs to be removed. For this, a method similar to Harris corner detection is used, defining a 2×2 Hessian matrix to calculate the principal curvature. For edges, one eigenvalue is much larger than the other (such as more than 10 times). In addition, extreme values less than the threshold need to be removed.
3. SIFT needs to estimate the dominant direction and achieve rotation invariance based on this. Figure 4.18(a) in [2] is the dominant direction estimation of SIFT, which requires creating a histogram of all gradient directions (amplitude weighted, or first remove small amplitude gradients), and then find the significant peak, which is the dominant direction estimation.
4. The representation of SIFT is also called a descriptor, which uses a 16×16 neighborhood around the key point and divides it into 16 4×4 sub-blocks. For each sub-block, create an 8-bin direction histogram (see Fig. 4.18(b) in [2]). SIFT is a 128-bin vector.

4. Feature Matching

Feature matching or feature tracking in the time domain is based on feature similarity to search for the same feature in other images. The constraints used for matching and tracking are different, the former is geometric constraints, and the latter is motion constraints; Fig. 4.28 in [2] is the famous feature tracking KLT (Kanade–Lucas–Tomasi) method. The principle of KLT is similar to the estimation of image motion optical flow (which belongs to the middle part).

5. Image Segmentation

Image segmentation emphasizes contour boundary information, but it is not completely corresponding to the semantic meaning of object boundaries; Sect. 5.11 will introduce semantic and instance segmentation methods in detail.

6. Feature-based Calibration

Feature-based alignment can be the calibration between 2D images, similar to the global motion estimation of images. It can also be the calibration of 2D features and 3D features, such as image stitching and camera calibration (internal parameters) in the middle part, or the calibration between 3D features, such as pose estimation (calibration of external parameters). The following will introduce the camera internal and external parameter calibration method based on the Zhang's calibration method proposed by Dr. Zhang Zhengyou [3]. Section 5.1 will introduce the calibration problem between different sensors [camera, lidar, millimeter wave radar, and GPS (Global localization System)/IMU (Inertial Measurement Unit)].

1. First, it is necessary to determine how many coordinate systems exist in the visual system, namely the scene coordinate system, the image coordinate system, and the camera coordinate system, and the task of camera calibration is to determine the conversion relationship between the three coordinate systems. The internal parameters are the conversion coefficients from the camera coordinate system to the image coordinate system, and the external parameters are the conversion coefficients from the scene coordinate system to the camera coordinate system.
2. Assuming the image point is represented as $\mathbf{m} = [u \quad v \quad 1]^T$, and the scene 3D point is represented as $\mathbf{M} = [X \quad Y \quad Z \quad 1]^T$, their relationship can be represented as

$$s\mathbf{m} = \mathbf{A}[\mathbf{Rt}]^T\mathbf{M} \tag{2.1}$$

   where s is a scale factor; $[\mathbf{Rt}]$ is the camera external parameter matrix including the rotation matrix and translation vector; $\mathbf{A}$ is the camera internal parameter matrix, in the form of

$$\mathbf{A} = \begin{bmatrix} \alpha \; \gamma \; u_0 \\ 0 \; \beta \; v_0 \\ 0 \; 0 \; 1 \end{bmatrix} \tag{2.2}$$

   where (u_0, v_0) are the coordinates of the image principal point; α and β are the scale factors of the image coordinate axes u and v; γ represents the skewness of the image coordinate axes.
3. Camera calibration requires a calibration field with a real value, and the flexibility of Zhang's calibration method compared to other methods is that this calibration field is a plane. Assuming the plane equation is $Z = 0$, it can be simplified to

$$s\begin{bmatrix} u \\ v \\ 1 \end{bmatrix} = \mathbf{A}\begin{bmatrix} \mathbf{r}_1 \; \mathbf{r}_2 \; \mathbf{r}_3 \; \mathbf{t} \end{bmatrix}\begin{bmatrix} X \\ Y \\ 0 \\ 1 \end{bmatrix} \tag{2.3}$$

$$= \mathbf{A}\begin{bmatrix} \mathbf{r}_1 \; \mathbf{r}_2 \; \mathbf{t} \end{bmatrix}\begin{bmatrix} X \\ Y \\ 1 \end{bmatrix} \tag{2.4}$$

   where $\mathbf{r}_i (i = 1 \sim 3)$ is the column vector of the rotation matrix. In this way, the relationship between the 3D point and the 2D image point is simplified to a homography matrix, $\mathbf{H} = \mathbf{A}[\mathbf{r}_1 \quad \mathbf{r}_2 \quad \mathbf{t}]$, it can be seen that $\mathbf{H}$ has a scale factor, which means there are only 8 unknown parameters.

4. Obviously the rotation matrix is orthogonal, and $\mathbf{A}^{-T}\mathbf{A}^{-1}$ actually represents an absolute conic, so the following formula can be derived. Define a 6D vector:

$$\mathbf{b} = [B_{11}\ B_{12}\ B_{22}\ B_{13}\ B_{23}\ B_{33}]^T \tag{2.5}$$

where

$$\mathbf{B} = \mathbf{A}^{-T}\mathbf{A}^{-1} \equiv \begin{bmatrix} B_{11} & B_{12} & B_{13} \\ B_{12} & B_{22} & B_{23} \\ B_{13} & B_{23} & B_{33} \end{bmatrix} \tag{2.6}$$

$$= \begin{bmatrix} \frac{1}{\alpha^2} & -\frac{\gamma}{\alpha^2\beta} & \frac{v_0\gamma - u_0\beta}{\alpha^2\beta} \\ -\frac{\gamma}{\alpha^2\beta} & \frac{\gamma^2}{\alpha^2\beta^2} + \frac{1}{\beta^2} & -\frac{\gamma(v_0\gamma - u_0\beta)}{\alpha^2\beta^2} - \frac{v_0}{\beta^2} \\ \frac{v_0\gamma - u_0\beta}{\alpha^2\beta} & \frac{\gamma(v_0\gamma - u_0\beta)}{\alpha^2\beta^2} - \frac{v_0}{\beta^2} & \frac{(v_0\gamma - u_0\beta)^2}{\alpha^2\beta^2} + \frac{v_0^2}{\beta^2} + 1 \end{bmatrix} \tag{2.7}$$

Define the column vector of $\mathbf{H}$ as $\mathbf{h}_i = (h_{i1}\ \ h_{i2}\ \ h_{i3})^T$, then a homogeneous system of equations can be obtained, that is,

$$\begin{bmatrix} \mathbf{v}_{12}^T \\ (\mathbf{v}_{11} - \mathbf{v}_{22})^T \end{bmatrix} \mathbf{b} = 0 \tag{2.8}$$

$$\mathbf{V}\mathbf{b} = \mathbf{0} \tag{2.9}$$

where

$$\mathbf{v}_{ij} = \begin{matrix} [h_{i1}h_{j1} & h_{i1}h_{j2} + h_{i2}h_{j1} & h_{i2}h_{j2} \\ h_{i3}h_{j1} + h_{i1}h_{j3} & h_{i3}h_{j2} + h_{i2}h_{j3} & h_{i3}h_{j3}]^T \end{matrix} \tag{2.10}$$

The solution $\mathbf{b}$ of the above system of equations is the eigenvector corresponding to the smallest eigenvalue of $\mathbf{V}^T\mathbf{V}$.

5. In this way, the solution of the internal and external parameters of the camera ($\mathbf{b} \to \mathbf{A} \to \mathbf{R}, \mathbf{t}$) can be obtained as follows:

$$v_0 = \frac{B_{12}B_{13} - B_{11}B_{23}}{B_{11}B_{22} - B_{12}^2} \tag{2.11}$$

$$\lambda = B_{33} - \frac{[B_{13}^2 + v_0(B_{12}B_{13} - B_{11}B_{23})]}{B_{11}} \tag{2.12}$$

$$\alpha = \sqrt{\frac{\lambda}{B_{11}}} \tag{2.13}$$

$$\beta = \sqrt{\frac{\lambda B_{11}}{B_{11}B_{22} - B_{12}^2}} \tag{2.14}$$

$$\gamma = -\frac{B_{12}\alpha^2\beta}{\lambda} \tag{2.15}$$

$$u_0 = \frac{\gamma v_0}{\beta} - \frac{B_{13}\alpha^2}{\lambda} \tag{2.16}$$

$$\mathbf{r}_1 = \lambda \mathbf{A}^{-1}\mathbf{h}_1 \tag{2.17}$$

$$\mathbf{r}_2 = \lambda \mathbf{A}^{-1}\mathbf{h}_1 \tag{2.18}$$

$$\mathbf{r}_3 = \mathbf{r}_1 \times \mathbf{r}_2 \tag{2.19}$$

$$\mathbf{t} = \lambda \mathbf{A}^{-1}\mathbf{h}_3 \tag{2.20}$$

where

$$\lambda = \frac{1}{\left\|\mathbf{A}^{-1}\mathbf{h}_1\right\|} = \frac{1}{\left\|\mathbf{A}^{-1}\mathbf{h}_2\right\|} \tag{2.21}$$

6. Of course, the accuracy of the above closed-form solution is not enough, and it cannot guarantee the orthogonality of **R**, so a global optimal maximum likelihood estimation (MLE) is needed as follows:

$$\sum_{i=1}^{n}\sum_{j=1}^{m}\left\|\mathbf{m}_{ij} - \hat{\mathbf{m}}(\mathbf{A}, \mathbf{R}_i, \mathbf{t}_i, \mathbf{M}_j)\right\|^2 \tag{2.22}$$

where $\hat{\mathbf{m}}(\mathbf{A}, \mathbf{R}_i, \mathbf{t}_i, \mathbf{M}_j)$ is the 2D image projection of the 3D point $\mathbf{M}_j$ ($i = 1 \sim n$ images, $j = 1 \sim m$ matched image features); this nonlinear least squares problem (see Sect. 2.3) can be solved using the Levenberg–Marquardt algorithm (with the closed-form solution as the initial value for iteration).
7. In addition, the camera has a nonlinear distortion problem, its model can be represented by a polynomial, and the solution parameters can also be put into the above MLE, which will not be elaborated here.

2.1.2 Middle Layer in Computer Vision

The middle layer will involve the processing of the motion domain and the disparity domain.

1. Motion Estimation in the Image Domain

Motion estimation in the image domain is the apparent motion discovered from the continuous or discontinuous image pairs taken, which can be dense pixel optic flow estimation, or feature tracking for the extracted sparse features, such as the previously mentioned KLT method; the generalized KLT method [2], i.e., parameterized motion model replacing the simple translation model, will be introduced below.

1. Based on the Taylor expansion formula, the motion estimation problem based on the sum of squared difference (SSD) is represented as an incremental motion method:

$$\mathbf{E}_{LK-PM} = \sum_i [\mathbf{I}_1(\mathbf{x}'(\mathbf{x}_i; \mathbf{p} + \Delta\mathbf{p})) - \mathbf{I}_0(\mathbf{x}_i)]^2 \tag{2.23}$$

$$\approx \sum_i [\mathbf{I}_1(\mathbf{x}_i') + \mathbf{J}_1(\mathbf{x}_i')\Delta\mathbf{p} - \mathbf{I}_0(\mathbf{x}_i)]^2 \tag{2.24}$$

$$= \sum_i [\mathbf{J}_1(\mathbf{x}_i')\Delta\mathbf{p} + \mathbf{e}_i]^2 \tag{2.25}$$

where $\mathbf{I}_1, \mathbf{I}_0$ are the two images where motion occurs; $\mathbf{x}_i = (x_i, y_i)$ is the i-th image feature point or pixel position; p is the parameterized motion vector replacing the translation vector (it can be affine transformation or homography); and $\Delta\mathbf{p}$ is its incremental step; the Jacobian matrix is

$$\mathbf{J}_1(\mathbf{x}_i') = \frac{\partial \mathbf{I}_1}{\partial \mathbf{p}} = \nabla \mathbf{I}_1(\mathbf{x}_i')\frac{\partial \mathbf{x}'}{\partial \mathbf{p}}(\mathbf{x}_i) \tag{2.26}$$

Here the image gradient is $\nabla \mathbf{I}_1$, and $J_{x'} = \frac{\partial \mathbf{x}'}{\partial \mathbf{p}}(\mathbf{x}_i)$.

2. Then, the linear equation of incremental motion estimation is

$$\mathbf{A}\Delta\mathbf{p} = \mathbf{b} \tag{2.27}$$

Where

$$\mathbf{A} = \sum_i \mathbf{J}_{x'}^T(x_i)[\nabla \mathbf{I}_1^T(\mathbf{x}_i')\nabla \mathbf{I}_1(\mathbf{x}_i')]\mathbf{J}_{x'}(\mathbf{x}_i) \tag{2.28}$$

$$\mathbf{b} = -\sum_i \mathbf{J}_{x'}^T(\mathbf{x}_i)[\mathbf{e}_i \nabla \mathbf{I}_1(\mathbf{x}_i')] \tag{2.29}$$

3. The entire motion estimation process is hierarchical: A Gaussian pyramid is constructed, starting from the top of the pyramid (the coarsest resolution) and stepping motion estimation (initial motion is zero), then passing the estimated results to the next layer (considering the scale change of the translation amount), and continuing to step estimation until the bottom of the pyramid (the finest resolution).

2. Classic motion reconstruction

Classic motion reconstruction (structure from motion, SfM [4]), also known as simultaneous localization and mapping (SLAM), the former concept is broader, while the latter requires the temporal relationship to be in an online execution mode. Theoretically, SFM/SLAM is an important task in computer vision that obtains image sequences and estimates the 3D structure of the scene through the motion of the camera under the basic assumption of a static scene. In the field of robotics, this task also estimates the posture and position of the on-site camera, i.e., the localization task. The basic principles of visual SLAM will be introduced below (Chapter 6 will discuss SLAM and lane line semantic maps in dynamic environments, and Chapter 7 will discuss lidar localization, lane line-based localization, and multi-sensor fusion localization).

1. Starting from the principle of stereo geometry, the epipolar constraint is defined: the optical centers of the two cameras are $\mathbf{c}_0$ and $\mathbf{c}_1$, respectively, the 3D space point $\mathbf{p}$ is projected onto the two image planes at points $\mathbf{x}_0$ and $\mathbf{x}_1$, respectively, then the intersection points of the line $\mathbf{c}_0\mathbf{c}_1$ and the two image planes are the **epipoles** $\mathbf{e}_0$ and $\mathbf{e}_1$, and the plane $\mathbf{p}\mathbf{c}_0\mathbf{c}_1$ is called the **epipolar plane**. The intersection lines of it and the two image planes $\mathbf{l}_0$ and $\mathbf{l}_1$ are the **epipolar lines**. As shown in Fig. 11.3 of [2], it can be seen that the transformation between the two camera coordinate systems satisfies $(\mathbf{R}, \mathbf{t})$, and it also shows that the image point $\mathbf{x}_0$ of the camera $\mathbf{c}_0$ must fall on its epipolar line $\mathbf{l}_1$ in the image corresponding point $\mathbf{x}_1$ of the camera $\mathbf{c}_1$, and vice versa.
2. Here we need to define an **essential matrix E** and a **fundamental matrix F**: $\mathbf{E} = [\mathbf{t}]_x \times \mathbf{R}$, where $[\mathbf{t}]_x$ is an antisymmetric matrix, i.e., defining a vector $\hat{\mathbf{n}} = (\hat{n}_x, \hat{n}_y, \hat{n}_z)$, then

$$[\hat{\mathbf{n}}]_x = \begin{bmatrix} 0 & -\hat{n}_z & \hat{n}_y \\ \hat{n}z & 0 & -\hat{n}_x \\ -\hat{n}_y & \hat{n}_x & 0 \end{bmatrix} \tag{2.30}$$

And $\mathbf{F} = \mathbf{A}^{-T}\mathbf{E}\mathbf{A}^{-1}$, where $\mathbf{A}$ is the camera's internal parameter matrix; for both matrix $\mathbf{E}$ and matrix $\mathbf{F}$, they satisfy the epipolar constraint equation:

$$\mathbf{x}_1^T\mathbf{E}\mathbf{x}_0 = 0, \quad \mathbf{x}_1^T\mathbf{F}\mathbf{x}_0 = 0 \tag{2.31}$$

The former is the correspondence of image feature points when the camera is calibrated, and the latter is the correspondence of image feature points when the camera is not calibrated.

Next, the epipolar line constraint (taking matrix $\mathbf{F}$ as an example, matrix $\mathbf{E}$ is the same) is as follows:

$$\mathbf{l}_1 = \mathbf{F}\mathbf{x}_0, \quad \mathbf{l}_0 = \mathbf{x}_1\mathbf{F} \tag{2.32}$$

The epipolar point constraint (taking matrix $\mathbf{F}$ as an example, matrix $\mathbf{E}$ is the same) is as follows:

$$\mathbf{F}\mathbf{e}_0 = 0, \quad \mathbf{F}^T\mathbf{e}_1 = 0 \tag{2.33}$$

3. According to the above epipolar constraint equation of image feature points, solve $\mathbf{F}$ and $\mathbf{E}$ by the 8-point algorithm, taking $\mathbf{F}$ as an example, given a set of matched feature points $(n > 7)$:

$$(x_1, y_1) \leftrightarrow (x'_1, y'_1) \tag{2.34}$$
$$(x_2, y_2) \leftrightarrow (x'_2, y'_2) \tag{2.35}$$
$$(x_3, y_3) \leftrightarrow (x'_3, y'_3) \tag{2.36}$$
$$\vdots$$
$$(x_n, y_n) \leftrightarrow (x'_n, y'_n) \tag{2.37}$$

Define the elements of matrix **F** as $f_{ij}(i = 1 \sim 3, j = 1 \sim 3)$, then the linear homogeneous equation group is

$$\begin{bmatrix} x_1x'_1 & x_1y'_1 & x_1x_1y'_1 & y_1y'_1 & y_1x'_1 & y'_1 & 1 \\ x_2x'_2 & x_2y'_2 & x_2x_2y'_2 & y_2y'_2 & y_2x'_2 & y'_2 & 1 \\ x_3x'_3 & x_3y'_3 & x_3x_3y'_3 & y_3y'_3 & y_3x'_3 & y'_3 & 1 \\ & & & \vdots & & & \\ x_nx'_n & x_ny'_n & x_nx_ny'_n & y_ny'_n & y_nx'_n & y'_n & 1 \end{bmatrix} \begin{bmatrix} f_{11} \\ f_{12} \\ f_{13} \\ f_{21} \\ f_{22} \\ f_{23} \\ f_{31} \\ f_{32} \\ f_{33} \end{bmatrix} = 0 \tag{2.38}$$

Simplified as

$$\mathbf{Qf} = 0 \tag{2.39}$$

Perform singular value decomposition (SVD) to get $\mathbf{Q} = \mathbf{USV}^T$, and the final solution of **F** is the singular vector on the far right of **V**.
In addition, since the **E** matrix is completely determined by the external parameters of the camera (**R** and **t** have a total of 6 unknown parameters, but there is an uncertain translation scale), there is a 5-point algorithm to solve the **E** matrix [5].

4. The external parameters of the camera can be obtained by decomposing **E**, the steps are as follows:
Similarly, perform SVD on the **E** matrix: $\mathbf{E} = \mathbf{U}diag(1, 1, 0)\mathbf{V}^T$.
Then the projection matrix of the first camera is simply set as $\mathbf{P}_1 = [\mathbf{I}|0]$, while the projection matrix $\mathbf{P}_2$ of the second camera has four possible situations, as shown in Fig. 9.12(a)~(d) of [4].

$$\mathbf{P}_2 = [\mathbf{UWV}^T + \mathbf{u}_3] \tag{2.40}$$
$$\mathbf{P}_2 = [\mathbf{UWV}^T - \mathbf{u}_3] \tag{2.41}$$
$$\mathbf{P}_2 = [\mathbf{UW}^T\mathbf{V}^T + \mathbf{u}_3] \tag{2.42}$$
$$\mathbf{P}_2 = [\mathbf{UW}^T\mathbf{V}^T - \mathbf{u}_3] \tag{2.43}$$

where

$$\mathbf{W} = \begin{bmatrix} 0 & -1 & 0 \\ 1 & 0 & 0 \\ 0 & 0 & 1 \end{bmatrix} \tag{2.44}$$

Observing the angle between the view direction and the direction from the camera center to the 3D point, it can be found that among the 4 possibilities, only the situation in Fig. 9.12(a) of [4] is a reasonable solution.

5. After determining the posture of the two viewpoints, the matched feature points $\mathbf{x}$ and $\mathbf{x}'$ can reconstruct their 3D coordinates $\mathbf{X}$, that is, the theory of triangulation. There is initially a linear solution. If the projection matrices of the two cameras are $\mathbf{P}$ and $\mathbf{P}'$, and their column vectors are $\mathbf{p}_i$ and $\mathbf{p}_i'$ ($i = 1 \sim 3$), then there is a set of equations:

$$\mathbf{AX} = 0 \tag{2.45}$$

where

$$\mathbf{A} = \begin{bmatrix} u\mathbf{p}_3^T - \mathbf{p}_1^T \\ v\mathbf{p}_3^T - \mathbf{p}_2^T \\ u\mathbf{p'}_3^{\prime T} - \mathbf{p'}_1^{\prime T} \\ v'\mathbf{p'}_3^{\prime T} - \mathbf{p'}_2^{\prime T} \end{bmatrix}, \ \mathbf{x} = \begin{bmatrix} u \\ v \\ 1 \end{bmatrix}, \ \mathbf{x}' = \begin{bmatrix} u' \\ v' \\ 1 \end{bmatrix} \tag{2.46}$$

However, due to the existence of some error interference, the above linear solution does not exist, so a nonlinear solution is needed. Here, the epipolar constraint equation defined by the $\mathbf{F}$ matrix, $\mathbf{x}^T\mathbf{F}\mathbf{x}' = 0$, is used to obtain the minimized error function as

$$C(\mathbf{X}) = d(\mathbf{x}, \hat{\mathbf{x}})^2 + d(\mathbf{x}', \hat{\mathbf{x}}')^2 \tag{2.47}$$

Next, the epipolar lines l and l' are used to define the error, as shown in Fig. 12.3 of [4], and the objective function is rewritten as

$$d(\mathbf{x}, \mathbf{l}(t))^2 + d(\mathbf{x}', \mathbf{l}'(t))^2 \tag{2.48}$$

6. In addition, given the known reconstructed 3D point set, how to match it with the 2D image feature points of the new viewpoint? The solution to this problem is called PnP (perspective-n-point), and the algorithm is as follows.

   - First, the 3D points are represented as the weighted sum of four control points.
   - The coordinates of the control points are the (12) unknowns to be solved.
   - The 3D points are projected onto the image, and a linear equation is established with the coordinates of the control points.
   - The coordinates of the control points are expressed as a linear combination of null eigenvectors.
   - The weights (β_i) of the above combination are new unknowns (< 4).
   - Add rigidity constraints to obtain the quadratic equation of β_i.

- Solve according to the number of β_i (whether linearized or re-linearized). **Note**: Sometimes the precision of 3D-2D matching is higher than that of 3D-3D matching.

7. Here, two concepts need to be supplemented: One is the random sample consensus method (RANSAC) of robust estimation, and the other is the bundle adjustment method (BA) of global optimization. The purpose of RANSAC is to robustly fit a model on a dataset containing outliers.

    a. Select a random (minimum) subset from the data points.
    b. Fit a model based on this random subset, that is, instantiate the model.
    c. Based on this instantiated model, all data points are divided into inliers that fit the model or outliers that do not fit the model.
    d. Iteratively repeat steps 1~3.
    e. Choose the largest set of inliers to re-estimate the final model.

The purpose of BA is to optimize the global parameter estimation. For SLAM, the 3D points X of the structure reconstruction and their corresponding 2D feature points x, as well as the estimated viewpoint transformation parameters (even including the camera internal parameters) P, are located under a framework of minimizing the reprojection error function D, that is,

$$E(\mathbf{P}, \mathbf{X}) = \sum_{i=1}^{m} \sum_{j=1}^{n} D(\mathbf{x}_{ij}, \mathbf{P}_i \mathbf{X}_j)^2 \tag{2.49}$$

Here, the bundle refers to the set of rays between 2D points and 3D points, and adjustment refers to the global optimization process. The solution is a nonlinear iterative gradient descent method, such as the Gauss–Newton method and the Levenberg–Marquardt method (Sect. 2.3 will discuss optimization theory). Because of the characteristics of the problem itself, the Jacobian matrix here is very sparse. In addition, if only the correction of the reconstructed 3D points is taken, it is called structure-only BA, and if only the viewpoint transformation correction is taken, it is called motion-only BA.

1. On the basis of the above, re-examine the SLAM process.

    a. First, we need to mention the concept of VO, which stands for Visual Odometry; VO is a part of SLAM, and VO mainly estimates the transformation between viewpoints, or motion parameters. It does not need to output the results of mapping, and BA is a motion-only mode.
    b. SLAM methods are divided into two approaches: One is the recursive filter method, and the other is the batch estimation method.

        - **Filter method**. Such as Kalman filter recursive estimation, it actually establishes an observation model and state transition (motion) model of the state

space. The observation model describes the probability of observing (landmarks) when the camera posture and landmark position (from the map) are known. The motion model is the probability distribution of system state (camera posture) transformation, that is, the Markov process. Then in recursive Bayesian estimation, the system state and the established map are updated simultaneously, where observations from different viewpoints are integrated to complete mapping, and the system state can be estimated to calculate the camera posture, that is, the localization problem.
- **Batch estimation method**. Also known as the "keyframe" method, its steps are as follows. a. First, by selecting the first two (key) frames, the initial 3D point cloud reconstruction is obtained by matching the feature points of binocular geometry. b. Normal mode: Assume that the 3D map is available and the incremental motion of the camera is estimated, track the feature points and use PnP to estimate the camera posture. c. Recovery mode: Assume that the 3D map is available, but tracking fails, so there is no incremental motion, and the camera posture can be relocalized relative to the previously reconstructed map. d. Keyframe BA: Maintain a "keyframe" subset, the state vector is the 3D landmarks of all keyframes and the corresponding camera posture, BA can adjust the state estimation in a thread parallel to the tracking module (note: The selection strategy of keyframes is a very important factor in algorithm performance).

2. The loop closure problem in SLAM. When the camera returns to its original position, it is called a loop closure, and it is necessary to detect the loop closure and optimize the structure and motion estimation under the BA framework. Loop closure detection and relocalization are similar, which can be achieved based on image feature matching, commonly known as image-based relocalization, when the "keyframe" subset is large, it is necessary to compress and accelerate feature matching, such as the bag of words method and KD-tree data structure.
3. The sensors in SLAM can be monocular, binocular, depth sensor (RGB-D), or even lidar, and can also be fused with IMU, called VINS (visual inertial navigation system).

3. Image Stitching Method

Image stitching is the process of merging multiple images of the same scene into one image, as long as there is a certain overlap area between the original images (Chap. 12 will introduce the application of image stitching).

4. Computational Photography

Computational photography is a new field combining computer vision and image processing in recent years, including the following works.

1. High dynamic range imaging is used to deal with the brightness saturation problem caused by changes in environmental lighting.
2. Super resolution is used to solve the problem of how to improve the low resolution of the captured image.
3. Stereo vision is a classic computer vision problem, similar to the human binocular system, its core is stereo correspondence and disparity estimation, the former is to find the corresponding pixels or features of a object in the left and right images in the scene, the latter is to calculate the disparity of the same object in the left and right eyes, this disparity is inversely proportional to the depth of the object in the scene; Sect. 5.12 will focus on this.
4. Multi-view stereo (MVS) is an extension of stereo vision, which can observe and capture images of the scene from multiple viewpoints (from outside to inside), and complete matching and depth estimation based on this, MVS and SfM/SLAM theoretically have a close relationship.

2.1.3 High Level Computer Vision

The high-level part of computer vision mainly deals with 3D processing.

1. 3D Shape Recovery
3D shape recovery (shape from X), here the clues can be texture, contour, shading, motion (the aforementioned SfM), multi-view (the aforementioned MVS), etc., Sect. 5.3 will focus on this direction.

2. Object Detection/Recognition
Object detection/recognition is an important part of computer vision, now the methods based on deep learning (see Sect. 2.5) have made this field develop rapidly, and the performance far exceeds the methods of traditional machine learning (see Sect. 2.4). Section 5.4 will discuss 3D object detection methods, Sects. 5.7–5.10 will discuss the special object detection and recognition tasks of autonomous driving, namely lane line detection, traffic sign detection and recognition, traffic signal light detection and recognition and general object detection, Sect. 5.13 will discuss the problem of human body posture estimation.

3. Intersection of Computer Vision and Computer Graphics

1. Image-based rendering (IBR) is a method of obtaining a representation of a scene model from image information using computer vision techniques, and then rendering a 3D model of the scene. Chapter 10 will cover the application of IBR technology, and Chap. 14 will discuss methods of solving IBR problems using neural networks.
2. Augmented reality (AR) adds virtual objects, rendered based on graphics, to a real scene analyzed by computer vision, creating a realistic "enhanced" scene.

Here, computer vision is a key technology for the 3D reconstruction of the real scene. In addition, the lighting of the virtual objects and the real scene need to be naturally integrated. Chapter 10 will introduce the application of AR in autonomous driving.

2.2 Image Processing

Image processing is the processing of image data on camera sensors for visual tasks, generally with the aim of improving image quality. However, it should be noted that some over-processing can affect the performance of visual algorithms, such as over-sharpening causing fake corner points or contouring and other distortions (artifacts).

2.2.1 Image Signal Processor

The Image Signal Processor (ISP) refers to the processing module at the front end of the camera. Because it is relatively mature, it is mostly implemented by chips. Figure 1 in [6] is a flowchart of an ISP implementation. The structure provided by some manufacturers may be slightly different, but it is basically like this, mainly including functions such as 3A [automatic focus, automatic exposure, automatic white balance], preprocessing, demosaicing, color transformation, post-processing. Preprocessing generally includes lens correction and bad pixel correction, some have noise reduction filtering before the "demosaicing" module. Post-processing generally includes tone mapping, sharpening, and gamma correction.

There are many types of image color spaces, in addition to RGB, YUV is commonly used in ISP. The conversion between RGB and YUV is linear and is very common in the field of cameras.

The "demosaicing" module is brought by the color filter array (CFA) of the camera sensor sensing light, such as the Bayer pattern (see Fig. 2.2), which is essentially a color interpolation process.

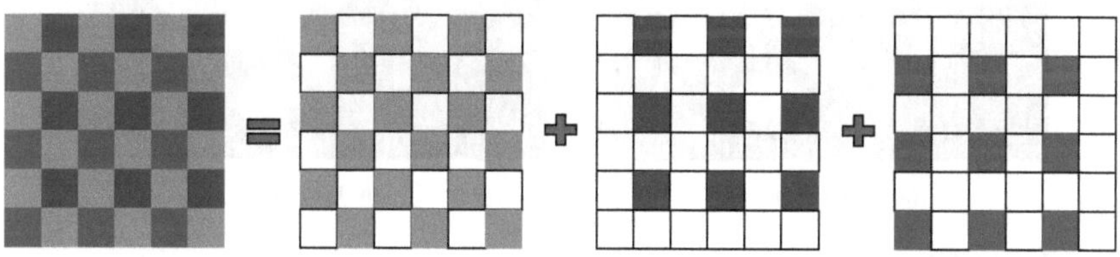

Fig. 2.2 Bayer pattern

2.2.2 *Image Filtering*

Filtering is a very important part of image processing, and noise removal is basically the most common filtering application. The simplest filtering methods include mean filtering, median filtering (methods derived from robust estimation theory), Gaussian filtering, and Wiener filtering.

Bilateral filtering (BLF) and non-local means (NLM) are methods that perform well and are easily accepted by the industry. The benchmark for image denoising algorithms is BM3D (block matching in 3D), but it has a large amount of computation and slow speed, so it has not been widely used.

The meaning of BLF is to perform Gaussian filtering in the spatial domain and brightness domain of the image at the same time, its form is

$$\mathbf{I}^{filtered}(\mathbf{x}) = \sum_{x_i \in \Omega} f_r(\|\mathbf{I}(\mathbf{x}_i) - \mathbf{I}(\mathbf{x})\|) g_s(|\mathbf{x}_i - \mathbf{x}|) \tag{2.50}$$

where $\mathbf{x}_i$ is the pixel in the filtering area of pixel $\mathbf{x}$, f_r is for brightness space filtering, and g_s is for spatial domain filtering. Figure 2.3 is an example of BLF filtering.

Unlike BLF, NLM is not a single-pixel filtering method, but a patch-based method that combines pixels with similar textures for mean filtering, its form is

$$NLM_u(\mathbf{p}) = \frac{1}{C(\mathbf{p})} f(d(B(\mathbf{p}), B(\mathbf{q})))_u(\mathbf{q}) d\mathbf{q} \tag{2.51}$$

where $d(B(\mathbf{p}), B(\mathbf{q}))$ is the Euclidean distance between two image patches located at $\mathbf{p}$ and $\mathbf{q}$, f is a decreasing function, and $C(\mathbf{p})$ is a normalization factor.

For a color image $\mathbf{u} = (u1, u2, u3)$ and a central pixel $\mathbf{p}$, its pixel-wise denoising implementation can be written as

(a) The Original Image (b) The Result

Fig. 2.3 BLF Filtering Example

(a) The Original Image (b) The Result

Fig. 2.4 NLM filtering example

$$\hat{u}_i(\mathbf{p}) = \frac{1}{C(\mathbf{p})} \sum_{\mathbf{q} \in B(\mathbf{p}, r)} u_i(\mathbf{q}) \omega(\mathbf{p}, \mathbf{q}), \ C(\mathbf{p}) = \sum_{\mathbf{q} \in B(\mathbf{p}, r)} \omega(\mathbf{p}, \mathbf{q}),$$
$$i = 1, 2, 3 \tag{2.52}$$

where $B(\mathbf{p}, r)$ refers to the $(2r + 1)(2r + 1)$ size neighborhood located at pixel $\mathbf{p}$, the weight $\omega(\mathbf{p}, \mathbf{q})$ depends on the Euclidean distance, and can be written in the form of an exponential function. And the patch-wise denoising implementation can be written as

$$\hat{B}_i(\mathbf{p}) = \frac{1}{C} \sum_{Q=Q(\mathbf{q}, f) \in B(\mathbf{p}, r)} u_i(Q) \omega(B, Q), \ C = \sum_{Q=Q(\mathbf{q}, f) \in B(\mathbf{p}, r)} \omega(B, Q), \ i = 1, 2, 3 \tag{2.53}$$

The weight function $\omega(B, Q)$ is similar to the weight calculation implemented by pixel-by-pixel denoising.

The difference between the two implementations lies in the different gains of PSNR (Peak Signal-to-Noise Ratio). Noise jitter near the boundary will be reduced in the latter's aggregation process. However, from the perspective of the quality of detail preservation, the patch-based NLM method has not shown significant improvement.

Figure 2.4 is an example of NLM filtering.

In addition, sparse coding and deep learning have also contributed some new filtering methods, which still need time to become widespread in the industry.

2.2.3 Image Enhancement

Image enhancement aims to improve the visual quality of an image or highlight specific features for subsequent analysis. Enhancement techniques can be broadly categorized into two approaches:

- **Global enhancement**: Methods that modify the overall appearance of the image by adjusting its intensity distribution, contrast, or brightness. A common example is histogram equalization, which redistributes pixel intensities to utilize the full dynamic range, thereby improving global contrast.
- **Detail enhancement**: Methods that emphasize local structures, edges, or textures to make fine details more visible. An example is unsharp masking, where a blurred version of the image is subtracted from the original to enhance high-frequency components, often guided by edge information.

Image enhancement can be performed either in the *spatial (pixel) domain*, where operations are directly applied to pixel values, or in the *frequency domain*, where image features are manipulated via their spectral components. A mixed space–frequency domain method decomposes the image into its low-frequency and high-frequency components: the low-frequency part is well-suited for global enhancement, while the high-frequency part is more appropriate for detail enhancement. Such hybrid approaches can balance overall contrast improvement with the preservation and sharpening of fine details.

2.3 Optimization Theory

Matrix theory is essential, numerical calculations are often used, and optimization algorithms are at the core. Computer vision and machine learning (deep learning) both use optimization theory to accomplish many tasks.

Optimization requires an objective function, whether it is to be minimized or maximized. This function has variables, and the task of optimization is to find the best value of the variables. Optimization problems can be divided into constrained and unconstrained types, as well as continuous optimization and discrete optimization.

To solve optimization problems, simple problems can have a closed solution, such as linear least squares (LLS); but complex problems cannot, and can only iteratively approximate the optimal solution, such as nonlinear least squares (NLS). For iterative approximation, we divide optimization methods into local methods and global methods, or into stochastic methods and deterministic methods.v

Each step of the iterative algorithm basically follows one of two strategies: line search or trust region. The former, which is the most common, first identifies a descent direction and then determines the next variable position that minimizes the objective function along this direction. The latter, less common strategy, constructs a local model that closely approximates the original objective function within a neighborhood around the current variable, called the trust region, and then solves a sub-problem to find the optimal position within this region.

Common line search methods include the steepest descent method [7], such as Newton's method, Gauss–Newton (G-N) method and its evolved L-M (Levenberg–Marquardt) method, coordinate descent method, conjugate gradient method, etc. Figure 2.5 in [7] is a direction diagram of the steepest descent method, where the optimal solution of the problem being solved is $\mathbf{x}^*$, and the gradient direction of the solution $\mathbf{x}_k$ during iteration is $\mathbf{p}_k$.

Here we will focus on the gradient descent method most commonly used in the field of computer vision: G-N method and its variant L-M method.

Suppose there is an observation vector $\mathbf{z}_i'$, its prediction model is $\mathbf{z}_i = z_i(\mathbf{x})$, where $\mathbf{x}$ is the model parameter, then the least squares (LS) method is to minimize the following cost function: weighted sum of squared error (SSE).

$$f(\mathbf{x}) = \frac{1}{2}\sum_i \Delta z_i(\mathbf{x})^T \mathbf{W}_i \Delta z_i(\mathbf{x}) \tag{2.54}$$

where $\mathbf{W}_i$ is an arbitrary symmetric positive definite (SPD) matrix, and the characteristic error function is

$$\Delta z_i(\mathbf{x}) = z_i'(\mathbf{x}) - z_i(\mathbf{x}) \tag{2.55}$$

The calculation of the Jacobian matrix $\mathbf{J}$ and the Hessian matrix $\mathbf{H}$ is

$$\mathbf{J} \equiv \frac{d\mathbf{z}}{d\mathbf{x}},\ g \equiv \frac{df}{d\mathbf{x}} = \Delta\mathbf{z}^T\mathbf{W}\mathbf{J} \tag{2.56}$$

$$\mathbf{H} \equiv \frac{d^2 f}{d\mathbf{x}^2} = \mathbf{J}^T\mathbf{W}\mathbf{J} + \sum_i (\Delta\mathbf{z}^T\mathbf{W})_i \frac{d^2\mathbf{z}_i}{d\mathbf{x}^2} \tag{2.57}$$

Then $\mathbf{H}$ is approximated as

$$\mathbf{H} \approx \mathbf{J}^T\mathbf{W}\mathbf{J} \tag{2.58}$$

As a gradient descent method, the G-N iteration step size, i.e., $\mathbf{z} \rightarrow \mathbf{z} + \Delta z$, is calculated by Eq. 2.59:

$$\mathbf{J}^T\mathbf{W}\mathbf{J} = -\mathbf{J}^T\mathbf{W}\Delta\mathbf{z} \tag{2.59}$$

For the above G-N iteration step size calculation, the matrix on the left may not be invertible. Therefore, an improved method is to add a damping factor λ to the diagonal elements, making the L-M iteration step size calculation into the following equation set:

$$(\mathbf{J}^T\mathbf{W}\mathbf{J} + \lambda diag(\mathbf{J}^T\mathbf{W}\mathbf{J})) = -\mathbf{J}^T\mathbf{W}\Delta\mathbf{z} \tag{2.60}$$

A more general optimization problem is the optimization of the objective function under constraints, which can be defined as follows:

Minimize $f(\mathbf{x})$
Subject to $g(\mathbf{x}) \leq 0, h(\mathbf{x}) = 0, \mathbf{x} \in \mathbf{X}$

where $\mathbf{x} = x_1, x_2, \cdots, x_{n_x}$, the function f is the objective function or standard function, and each constraint $g_i(\mathbf{x}) \leq 0, i = 1, 2, \cdots, n_g$ is an **inequality constraint**, and each constraint $h_i(\mathbf{x}) = 0, i = 1, 2, \cdots, n_h$ is called an **equality constraint**. A vector $\mathbf{x} \in \mathbf{X}$ that satisfies all constraints is called a **feasible solution** of the problem; the set of all these feasible solutions forms the **feasible region**. Therefore, this optimization problem is to find a feasible point $\mathbf{x}^*$ that satisfies $f(\mathbf{x}) \geq f(\mathbf{x}^*)$. Of course, this optimization problem can also be defined as a maximize $f(\mathbf{x})$ problem, and the inequality constraint can be written in the form of $g() \geq 0$.

It can be proved that the following first and second order derivative KKT (Karush–Kuhn–Tucker) necessary conditions hold: there exists a unique vector ν^*, λ^*, that satisfies

$$\nabla f(\mathbf{x}^*) + \nu^{*T} \nabla g(\mathbf{x}^*) + \lambda^{*T} \nabla h(\mathbf{x}^*) = 0 \tag{2.61}$$
$$\nu^* \geq 0 \tag{2.62}$$
$$g(\mathbf{x}^*) \leq 0 \tag{2.63}$$
$$h(\mathbf{x}^*) = 0 \tag{2.64}$$
$$\nu^{*T} \nabla g(\mathbf{x}^*) = 0 \tag{2.65}$$

and

$$\mathbf{y}^T (\nabla^2 f(\mathbf{x}^*) + \nu^{*T} \nabla^2 g(\mathbf{x}^*) + \lambda^{*T} \nabla^2 h(\mathbf{x}^*))\mathbf{y} \geq 0 \tag{2.66}$$

For all $\mathbf{y}$, satisfy $\nabla g(\mathbf{x}^*)\mathbf{y} = 0, i \in \mathcal{A}, \nabla h(\mathbf{x}^*)\mathbf{y} = 0$. Here $\mathcal{A}(\mathbf{x}^*)$ is called the active constraint or active set, defined as follows: for $i = 1, 2, \cdots, n_g$, if $g_i(\mathbf{x}) = 0$, the constraint g_i is active or binding at $\mathbf{x}$; if $g_i(\mathbf{x}) < 0$, it is inactive at $\mathbf{x}$; then the active constraint is defined as

$$\mathcal{A}(\mathbf{x}^*) \triangleq i{:}g_i(\mathbf{x}^*) = 0 \tag{2.67}$$

It can also be proved that the following second order derivative sufficient condition: if there exists a unique vector $\mathbf{x}^*, \nu^*, \lambda^*$ that satisfies the above KKT conditions, and

$$\mathbf{y}^T \nabla^2_{xx} \mathcal{L}(\mathbf{x}^*, \nu^*, \lambda^*)\mathbf{y} > 0 \tag{2.68}$$

For any $\mathbf{y} \neq 0$, satisfy:

$$\nabla g_i(\mathbf{x}^*)\mathbf{y} = 0, \ i \in \mathcal{A}(\mathbf{x}^*), \ \nu^* > 0 \tag{2.69}$$
$$\nabla g_i(\mathbf{x}^*)\mathbf{y} \leq 0, \ i \in \mathcal{A}(\mathbf{x}^*), \ \nu^* = 0 \tag{2.70}$$
$$\nabla h(\mathbf{x}^*)\mathbf{y} = 0 \tag{2.71}$$

where $\mathcal{L}(\mathbf{x}, \nu, \lambda) = f(\mathbf{x}) + \nu^T g(\mathbf{x}) + \lambda h(\mathbf{x})$, then $\mathbf{x}^*$ is a strict local minimum solution.

Thus, the KKT sufficient condition is obtained as follows: if $f(\mathbf{x})$ and $g_i(\mathbf{x})$ are convex differentiable functions, $h_i(\mathbf{x})$ is an affine function, for the above optimization problem, if $(\mathbf{x}^*, \nu^*, \lambda^*)$ satisfies the above KKT necessary conditions, then $\mathbf{x}^*$ is a global optimal solution of the objective function.

Constrained optimization problems can be transformed into unconstrained optimization objectives in certain ways, such as Lagrangian multipliers. The successive quadratic programming (SQP) method, also known as sequential or recursive quadratic programming, uses the Newton method (or similar to the Newton method) to directly solve the KKT conditions of the original problem. As a result, the accompanying sub-problems are proven to be a quadratic approximation minimization of the Lagrangian function, subject to a linear approximation of the constraints. Therefore, this process is also called a projected Lagrangian method or Newton–Lagrangian method. According to its properties, this method generates primal and dual (Lagrangian multipliers) solutions.

Assuming $f(\mathbf{x})$, $g_i(\mathbf{x})$ and $h_i(\mathbf{x})$ are all twice differentiable functions, given the iterative solution of $(\mathbf{x}^k, \lambda^k, \nu^k)$, where λ^k and ν^k are the Lagrangian multipliers of the equality and inequality constraints, respectively, consider the following $QP(\mathbf{x}^k, \lambda^k, \nu^k)$ sub-problem:

$$\min_{\mathbf{d}^k} f(\mathbf{x}^k) + \nabla f(\mathbf{x}^k)^T \mathbf{d}^k + \frac{1}{2}\mathbf{d}^{k^T} \nabla_{xx}^2 \mathcal{L}(\mathbf{x}^k, \lambda^k, \nu^k)\mathbf{d}^k \tag{2.72}$$

$$s.t.\ g_i(\mathbf{x}) + \nabla g_i(\mathbf{x}^k)^T \mathbf{d}^k = 0,\ i = 1, 2, \cdots, n_g \tag{2.73}$$

$$h_i(\mathbf{x}) + \nabla h_i(\mathbf{x}^k)^T \mathbf{d}^k = 0,\ i = 1, 2, \cdots, n_h \tag{2.74}$$

where $\mathcal{L}(\mathbf{x}, \lambda, \nu) \triangleq f(\mathbf{x}) + \nu^T g(\mathbf{x}) + \lambda^T h(\mathbf{x})$. Apart from the initial feasibility, the KKT conditions of $QP(\mathbf{x}^k, \lambda^k, \nu^k)$ require the Lagrangian multipliers λ^{k+1}, it can be found that ν^{k+1} satisfies

$$\nabla f(\mathbf{x}^k) + \nabla_{xx}^2 \mathcal{L}(\mathbf{x}^k, \lambda^k, \nu^k)\mathbf{d}^k + \nabla g(\mathbf{x}^k)^T \nu^{k+1} + \nabla h(\mathbf{x}^k)^T \lambda^{k+1} = 0 \tag{2.75}$$

$$[g(\mathbf{x}^k) + \nabla g(\mathbf{x}^k)^T \mathbf{d}^k]^T \nu^{k+1} = 0 \tag{2.76}$$

where $\nu^{k+1} \geq 0$, λ^{k+1} has no sign restriction. If $\mathbf{d}^k = 0$, $\mathbf{x}^k$ and λ^{k+1}, ν^{k+1} together produce a KKT solution of the original problem. Otherwise, set $\mathbf{x}^{k+1} \triangleq \mathbf{x}^k + \mathbf{d}^k$, step k by 1, and repeat the process. If the initial values $(\mathbf{x}^0, \lambda^0, \nu^0)$ and the optimal solution $(\mathbf{x}^*, \lambda^*, \nu^*)$ are sufficiently close, the iterative program can show quadratic convergence.

Furthermore, commonly used global optimization algorithms include the following.

1. Simulated Annealing Method: Simulates physical annealing, i.e., the heating process, isothermal process, and cooling process, to seek the optimal solution in the search space of statistics.

2. Genetic Algorithm: Uses simulated variable individuals to evolve to the optimal solution through mutation, crossover, and selection operations similar to gene reproduction.
3. Graph Cut Method: Finds a cut in the constructed vertices-edges graph to achieve max-flow/minimum cut, suitable for solving energy minimization problems.
4. Belief Propagation: Similar to dynamic programming, it is a simplified version of the Bayesian network. In the graphical model, messages/beliefs between adjacent nodes are continuously propagated, ultimately achieving a stable state of variable marginal probability.

2.4 Machine Learning

What is machine learning? It can be understood as the machine changing its structure, program, or data based on input samples or experience data to optimize its performance [8].

Machine learning can be divided into four types: supervised, unsupervised, semi-supervised, and reinforcement learning (RL).

The distinction between supervised, semi-supervised, and unsupervised learning is clear. It depends on whether the input data has labels. If it has labels, it is supervised learning; if it doesn't, it is unsupervised learning. If only a part of the data has labels, it is semi-supervised learning. Regardless, a machine learning method needs to define an objective function to calculate the prediction error of the model.

Semi-supervised learning uses unlabeled data to generate a prediction function with trainable model parameters, aiming to be more accurate than the prediction function obtained from labeled data. Since it combines supervised and unsupervised learning methods, the loss function of semi-supervised learning can have various shapes. A common method is to add a supervised learning loss term and an unsupervised learning loss term to the objective function of the training model.

There is a special method called **self-supervised learning** [9, 10], which uses pretext tasks to learn the representation of unlabeled data. Pretext tasks are unsupervised, but the learned representation usually cannot be directly used for downstream tasks, such as image classification, and must be fine-tuned first. Self-supervised learning can be interpreted as a type of unsupervised, semi-supervised, or a self-strategy. The methods of self-supervised learning are basically divided into **generative** learning, **contrastive** learning, and a hybrid of generative–contrastive learning, i.e., **adversarial** learning.

There is an iterative method called **active learning** [11], which aims to select the most useful samples from the unlabeled dataset and hand them over to the annotator to label, in order to reduce the labeling cost as much as possible while maintaining performance. Active learning is usually an iterative process, learning the model in each iteration, where some heuristic methods are used to select a set of data for labeling from the unlabeled data pool.

There is a method in supervised learning called **transfer learning** [12], which divides model training into two stages: The first stage is training on a large general dataset (such as ImageNet), regardless of whether it is supervised or not. The second stage is to use the trained weights to fine-tune on the target dataset. The model in the target data domain does not need to be trained from scratch, which can reduce the demand for training data and time in the target domain. **Domain adaptation** is a special case of transfer learning, which uses labeled data from one or more related source domains to perform new tasks in the target domain.

A typical example of transfer learning is zero-shot learning (ZSL) [13]. ZSL methods aim to recognize unseen classes that are not present during the training process. Recent approaches directly learn mappings from the image feature space to the semantic space, while others employ nonlinear multimodal embeddings. Although most ZSL methods rely on discriminative loss functions, a few generative models represent categories as probability distributions.

To learn from limited supervision information, a new direction in machine learning is called **few-shot learning (FSL)** [14]. Based on how to use prior knowledge, FSL can be divided into three categories: ① Enhance supervised experience with data prior knowledge; ② Constrain the hypothesis space with model prior knowledge; ③ Change the way of searching for the best parameters in the hypothesis space with algorithm prior knowledge. FSL can be considered as a type of supervised learning, semi-supervised learning, or reinforcement learning (RL), depending on what other data is available besides limited supervision information.

Continual learning [61], also known as **incremental learning** or **lifelong learning**, refers to the machine learning method that can continuously accumulate knowledge obtained from different tasks without having to retrain from scratch. The difficulty lies in how to overcome **catastrophic forgetting** and the **stability–plasticity dilemma**. Plasticity refers to the ability to integrate new knowledge, while stability refers to the preservation of original knowledge during encoding. Continual learning can be divided into three categories: experience replay, regularization, and parameter isolation.

A concept related to continuous learning is open set recognition (OSR), which refers to the existence of incomplete world knowledge during training, and the ability to submit unknown classes to the algorithm during testing. The classifier is required not only to accurately classify the seen classes, but also to effectively handle unseen classes. **Open world learning** [16] can be seen as a subtask of continuous learning.

Reinforcement learning is actually about how an autonomous agent perceives and acts optimally in an environment to achieve a given goal. When an agent is interacting with the environment, then we have the state S, the action A, the reward r, and the agent's goal is to learn the control strategy from state S to action A, maximizing the cumulative reward, where each reward also decays over time, and the decay factor γ satisfies $0 \leq \gamma < 1$.

The task of reinforcement learning is to learn how to choose actions to maximize the following objective:

$$r_0 + \gamma r_1 + \gamma^2 r_2 + \cdots$$

Furthermore, the utility of machine learning algorithms can also be divided into prediction, classification, clustering, and association.

Machine learning is a branch of artificial intelligence (AI), AI is to imitate human intelligent activities, learning is one of them. Data mining is for detecting the existence of patterns, while statistical analysis is based on probability models through data inference, and information retrieval is to retrieve desired information from a set of data.

Machine learning can be divided into two types: discriminative and generative. The former is bottom-up learning, such as the conditional random field (CRF) model; the latter is top-down learning, such as the Markov random field (MRF) model.

The following issues need to be noted in machine learning.

1. Data imbalance: The gap in the number of positive and negative samples is too large. In perceptual data, negative samples often greatly outnumber positive samples, which requires consideration of how to ensure the performance of the classifier.
2. Development dataset: How to handle newly emerging class data or unfamiliar class data.
3. Curse of dimensionality: Very sensitive to "noise."
4. Overfitting: Simply put, it is when the model obtained from training data performs poorly in test data; an intuitive feeling is that the model has learned the "noise" rather than the distribution characteristics of the real data.
5. Performance evaluation indicators: ROC (receiver operation characteristic curve), precision/recall, confusion matrix, mAP (mean average precision), etc.

There are several classic machine learning methods.

1. Decision tree.
2. Boosting.
3. Random forest (RF) method.
4. Support vector machine (SVM).
5. K-NN (K-nearest neighbor) classification.
6. Principal component analysis (PCA).
7. Spectral clustering (SC).
8. Locally linear embedding (LLE).
9. Neural network (multi-layer perceptron), MLP is generally not deep learning.

The following will focus on two very popular supervised machine learning algorithms: SVM and RF.

2.4.1 Support Vector Machine

In SVM, the support vector is its main feature [17]. However, let's start with a simple scenario.

Suppose the data belongs to two types and is separable, and the task of machine learning is to obtain a hyperplane that separates the two types of vectors in the data. As shown in Fig. 5 in [17], the margin represents a maximum isolation band parallel to the hyperplane that does not include data, and the support vector is actually the data point closest to the established separation hyperplane. Sometimes SVM is also referred to as a classifier that maximizes the margin.

The above is the simplest case. Now, the problem becomes more complex, turning into a scenario where different types of data cannot be completely separated. At this time, SVM defines a soft margin, which means that the hyperplane can separate many but not all data points. In more complex cases, some classification problems do not have a simple separation hyperplane, which leads to the kernel trick, i.e., there is a kernel-based nonlinear transform that transforms the data into a linear space where a separation hyperplane exists (generally, this linear space is definitely higher-dimensional than the original space), Common kernel functions are generally polynomial, radial basis functions, and sigmoid functions.

So how to formalize an SVM problem with inseparable data? As supervised learning, there is a set of labeled training data $\{\mathbf{x}_i, y_i\}, i = 1, 2, \cdots, l, y_i \in \{-1, 1\}, \mathbf{x}_i \in \mathbf{R}_d$, the conditions to be met are as follows:

$$\mathbf{x}_i \cdot \mathbf{w} + b \geq 1, \text{ for points where } y_i = 1 \tag{2.77}$$

$$\mathbf{x}_i \cdot \mathbf{w} + b \leq -1, \text{ for points where } y_i = -1 \tag{2.78}$$

This can be combined into

$$y_i(\mathbf{x}_i \cdot \mathbf{w} + b) - 1 \geq 0, \ \forall i \tag{2.79}$$

where $\mathbf{w} \cdot \mathbf{x} + b = 0$ is the hyperplane equation, $\mathbf{w}$ and b are classifier parameters, here the margin is $2/\|\mathbf{w}\|$. Introducing the Lagrange multiplier α_i, this training task can be represented as

$$L_p = \frac{1}{2}\|\mathbf{w}\|^2 - \sum_{i=1}^{l} \alpha_i y_i (\mathbf{x}_i \cdot \mathbf{w} + b) + \sum_{i=1}^{l} \alpha_i \tag{2.80}$$

where the support vectors are those data points where $\alpha_i > 0$.

Re-emphasizing an important concept, the Karush–Kuhn–Tucker (KKT) condition, is represented as follows:

$$\frac{\partial}{\partial \mathbf{w}_v} L_p = \mathbf{w}_v - \sum_i \alpha_i y_i x_{iv} = 0, \ v = 1, 2, \cdots, d \tag{2.81}$$

$$\frac{\partial}{\partial b} L_p = -\sum_i \alpha_i y_i = 0 \tag{2.82}$$

$$y_i(\mathbf{x}_i \cdot \mathbf{w} + b) - 1 \geq 0, \ i = 1, 2, \cdots, l \tag{2.83}$$

$$\alpha_i \geq 0, \ \forall i \tag{2.84}$$

$$\alpha_i (y_i(\mathbf{x}_i \cdot \mathbf{w} + b) - 1) = 0, \ \forall i \tag{2.85}$$

Substituting KKT, the above minimization objective function becomes an equivalent form of maximization objective function:

$$L_D = \sum_i \alpha_i - \frac{1}{2} \sum_{i,j} \alpha_i \alpha_j y_i y_j (\mathbf{x}_i \cdot \mathbf{x}_j) \tag{2.86}$$

For inseparable data, we need to introduce a positive slack variable ξ_i, represented as

$$\mathbf{x}_i \cdot \mathbf{w} + b \geq +1 - \xi_i, \ y_i = +1 \tag{2.87}$$

$$\mathbf{x}_i \cdot \mathbf{w} + b \leq -1 + \xi_i, \ y_i = -1 \tag{2.88}$$

$$\xi_i \geq 0, \ \forall i \tag{2.89}$$

Correspondingly, the objective function of the classification problem becomes:

$$L_p = \frac{1}{2} \|\mathbf{w}\|^2 + C \sum_i \xi_i - \sum_{i=1}^{l} \alpha_i \{y_i(\mathbf{x}_i \cdot \mathbf{w} + b) - 1 + \xi_i\} - \sum_{i=1}^{l} \mu_i \xi_i \tag{2.90}$$

Then the KKT condition will correct or supplement the part with slack vectors:

$$\frac{\partial}{\partial \xi_i} L_p = C - \alpha_i - \mu_i = 0 \tag{2.91}$$

$$y_i(\mathbf{x}_i \cdot \mathbf{w} + b) - 1 + \xi_i \geq 0 \tag{2.92}$$

$$\alpha_i \geq 0 \tag{2.93}$$

$$\xi_i \geq 0 \tag{2.94}$$

$$\mu_i \geq 0 \tag{2.95}$$

$$\alpha_i \{y_i(\mathbf{x}_i \cdot \mathbf{w} + b) - 1 + \xi_i\} = 0 \tag{2.96}$$

$$\mu_i \xi_i = 0 \tag{2.97}$$

For classifiers with nonlinear kernel methods, it can be described as

$$f(\mathbf{x}) = \sum_{i=1}^{N_s} \alpha_i y_i \Phi(\mathbf{s}_i) \cdot \Phi(\mathbf{x}) + b = \sum_{i=1}^{N_s} \alpha_i y_i K(\mathbf{s}_i, \mathbf{x}) + b \tag{2.98}$$

where N_s is the number of support vectors $\mathbf{s}_i$, and the kernel function is defined as

$$K(\mathbf{x}_i, \mathbf{x}_j) = \Phi(\mathbf{x}_i) \cdot \Phi(\mathbf{x}_j) \tag{2.99}$$

SVM was a very successful classifier before deep learning became popular. As a classic method of statistical machine learning, it provides a learning method that minimizes structural risk, rather than empirical risk.

The advantages of SVM are as follows.

1. SVM is a convex optimization problem, its solution is definitely globally optimal.
2. It is applicable to both linear and nonlinear problems (using kernel methods).
3. Due to the support vectors, it avoids the "curse of dimensionality" in a sense.
4. The theoretical foundation is relatively complete.

The disadvantages of SVM are as follows.

1. As a quadratic programming problem, the solution is not suitable for super large data sets (the SMO training algorithm can alleviate this problem).
2. It is basically a binary classification problem, but SVR (support vector regression) is also applicable to regression problems; multiple SVMs can be combined to solve multiclassification problems.

2.4.2 Random Forest

Random forest, also known as random decision forest (RDF), is an ensemble learning method used for classification, regression, and other tasks. It builds multiple decision trees during training and outputs the classification or average prediction (regression) of each tree during inference. RDF can correct the overfitting results of decision trees on the training set.

The ensemble learning method constructs multiple individual learners to complete tasks, which can be divided into two categories: One type of individual learner has strong dependence, which is a serialized method that works in series, represented by boosting. The other type of individual learner has weak dependence, which is a parallelized method that works simultaneously, represented by bagging. Ensemble learning often achieves significantly improved generalization performance than a single learner by combining multiple individual learners.

1. RDF is an evolved version of bagging

1. RDF uses decision trees as weak learners.
2. Based on the use of decision trees, RDF has improved the establishment of decision trees.

   a. For ordinary decision trees, one optimal feature is selected from all n sample features on the node to divide the left and right subtrees of the decision tree.
   b. RDF only randomly selects a part of the sample features on the node, this number is less than n, set as m.
   c. Among the selected m sample features, one optimal feature is selected to divide the left and right subtrees of the decision tree, thereby further enhancing the generalization ability of the model.
   d. If $m = n$, then there is no difference between the decision tree of RDF and the ordinary decision tree: the smaller the m, the more robust the model, but the fit to the training set will be worse. That is to say, the smaller the m, the smaller the variance of the model, but the bias will increase.

Apart from the above two points, there is no difference between RDF and ordinary bagging.

Decision trees are a popular machine learning method. Especially when the tree is very deep, it is more suitable for learning highly irregular patterns, but these patterns have low bias but very high variance, often causing overfitting on the training set. And RDF is a method of averaging multiple deep decision trees, training on different parts of the same training set, with the aim of reducing variance, but bias will increase slightly. Therefore, some reasonable losses greatly promote the performance of the final model.

2. Advantages and disadvantages of RDF

(1) Advantages of RDF

   a. Highly parallel training, advantageous for large sample training speed in big data.
   b. Because it can randomly select decision tree node division features, it can still efficiently train decision tree models when the sample feature dimension is very high.
   c. Can give judgments on the importance of each feature.
   d. The model trained by random sampling has small variance and strong generalization ability.
   e. Compared to the boosting method of Adaboost, RDF is relatively simple to implement.
   f. Insensitive to some missing features.

(2) Disadvantages of RDF

a. It is easy to overfit on sample sets with large noise.
b. Features with more divisions can have a greater impact on decisions, thereby affecting fitting performance.

2.5 Deep Learning

What is deep learning [18]? According to the relevant definition of deep learning, it should be a hierarchical concept structure that can teach computers to learn complex concepts from simple concepts. If a graph is used to represent how these concepts are constructed, it is a deep multi-layer graph.

In a sense, deep learning is feature/representation learning, or hierarchical feature/abstract learning. The most typical deep learning model is the feed-forward deep network or multi-layer perceptron (MLP).

So why didn't MLP succeed 20 years ago? The main reasons are: insufficient training data, insufficiently powerful computers, incorrect model parameter initialization, and incorrect model node response functions.

Deep model architectures are divided into the following 3 types.

1. Feedforward network: MLP, convolutional neural network (CNN).
2. Backward feedback network: stacked sparse coding, deconvolutional nets.
3. Bidirectional feedback network: deep Boltzmann machines, stacked auto-encoders.

CNN should be the most popular deep learning model, and it is also the most influential in computer vision. Here is a brief review of the development history of classic CNN models.

- AlexNet should be considered the first deep CNN (see Fig. 2 in [19]).
- ZFNet uses DeconvNet and visualization technology to monitor the learning process.
- VGGNet uses 3×3 small filters to replace 5×5 and 7×7 large filters, thereby reducing computational complexity.
- GoogleNet extends the idea of NIN to define the basic Inception (see Fig. 2 in [20]) module (using multi-scale transformations and combinations of different size filters, i.e., 1×1, 3×3, 5×5) to build the model.
- Highway Networks borrow the gating unit from RNN-LSTM.
- ResNet is a revolutionary work (see Fig. 2 in [21]), borrowing the idea of skip connections from Highway Networks, which can train deep models to improve performance and reduce computational complexity.
- Inception-V3/4 replaces 5×5 and 7×7 large filters with 1×7 and 1×5 small filters, and uses 1×1 filters to do the previous feature bottleneck, making the convolution operation like a cross-channel related operation.

- DenseNet mainly solves the vanishing gradient problem through cross-layer links (see Fig. 2 in [22]).
- SE-Net is designed for feature selection, and the gating mechanism is still adopted.
- The attention mechanism is adapted from LSTM to implement the context model for object perception.

The following focuses on the basic architecture analysis of CNN, the currently eye-catching GAN, RNN networks, and the currently high-performing Transformer network.

2.5.1 Convolutional Neural Network

As shown in Fig. 2 in [23], a CNN consists of two main parts: feature extraction and classifier. In the feature extraction layer, each layer of the network receives the output from its previous layer as its input and passes its output as input to the next layer.

This CNN architecture is composed of three types of layers: convolution, max-pooling, and classification. There are two types of layers at the bottom and middle of the network: convolution layers and max-pooling layers. Even layers are used for convolution, and odd layers are used for max-pooling operations. The output nodes of the convolution and max-pooling layers are combined into a 2D plane, called a feature map. Each plane of a layer is usually derived from one or more planes of the previous layer. The nodes of the plane are connected to a small area of each connected plane of the previous layer. Each node of the convolution layer extracts features from the input image through convolution operations. The overall architecture of CNN includes an input layer, multiple alternating convolution layers and max-pooling layers, a fully connected layer, and a classification layer.

Higher-level features come from features propagated from lower layers. As features propagate to the highest layer or high level, the dimensions of the features decrease depending on the kernel size of the convolution and max-pooling operations. However, to ensure the accuracy of classification, the number of feature maps is usually increased to better represent the input image features. The output of the last layer of CNN is used as the input of the fully connected network, which is called the classification layer. Feedforward neural networks have been used as classification layers.

The expected number of feature selections in the classification layer is used as input relative to the dimensions of the final neural network weight matrix. However, the fully connected layer is expensive in terms of the network or learning parameters. There are several techniques, including average pooling and global average pooling, as alternatives to the fully connected network. In the top classification layer, the softmax layer calculates the scores of related categories. The classifier selects the highest score and outputs the related category.

1. Basic Concepts

Computational neuroscience has conducted important research on the computational model of building artificial neurons. Artificial neurons trying to mimic human brain behavior are the basic components of building artificial neural networks. The basic computational element (neuron) is called a node (or unit), which receives input from external sources and has some internal parameters (including weights and biases learned during training) that produce output. This unit is called a perceptron.

The basic model of the perceptron is shown in Fig. 2 in [24], which shows the basic nonlinear model of the neuron, where $\mathbf{x}_1, \mathbf{x}_2, \cdots, \mathbf{x}_m$ are input signals; $\mathbf{w}_{k1}, \mathbf{w}_{k2}, \cdots, \mathbf{w}_{km}$ are synaptic weights; $\mathbf{v}_k$ is the linear combination of input signals; $\varphi(\cdot)$ is the activation function (such as sigmoid); $\mathbf{y}_k$ is the output. The offset b_k is added to the linear combiner of the output, which has the effect of applying an affine transformation, producing the output $\mathbf{y}_k$.

The function of the neuron can be expressed by the following formula:

$$\mathbf{v}_k = \sum_{j=1}^{m} \mathbf{w}_{kj}\mathbf{x}_j \tag{2.100}$$

$$\mathbf{y}_k = \varphi(\mathbf{v}_k + b_k) \tag{2.101}$$

Artificial neural networks or general neural networks are composed of multi-layer perceptrons (MLP), which contain one or more hidden layers, each layer containing multiple hidden units (neurons).

An NN model architecture with L-layer MLP is shown in Fig. 2 in [25], where $\mathbf{x}$ represents the input, $\mathbf{w}^k$ represents the weight of the k-th layer MLP, $\mathbf{b}^k$ represents the offset of the k-th layer MLP, $\mathbf{a}^k$ represents the output of the $k-1$ layer which is also the input of the k-th layer ($\mathbf{a}^0 = \mathbf{x}$).

So, the output of MLP is

$$\mathbf{y} = f(\mathbf{x}) = \varphi(\mathbf{w}^L \cdots \varphi(\mathbf{w}^2 \varphi(\mathbf{w}' x + \mathbf{b}^2) \cdots + \mathbf{b}^L) \tag{2.102}$$

The **learning rate** is an important part of training DNN. It is the step size considered during training, making the training process faster. However, choosing the learning rate is sensitive. If η is large, the network may start to diverge rather than converge; moreover, if η is chosen to be small, the network needs more time to converge, and it may easily fall into local minima.

There are three common methods to reduce the learning rate during training, namely constant, factor, and exponential decay. First, a constant ζ can be defined to manually reduce the learning rate based on the defined step function. Secondly, the learning rate can be adjusted during training according to the following equation:

$$\eta_t = \eta_0 \beta^{t/\varepsilon} \tag{2.103}$$

The format of the exponential decay step function is

$$\eta_t = \eta_0 \beta^{[\frac{t}{\varepsilon}]} \tag{2.104}$$

Momentum is a method that helps SGD (Stochastic Gradient Descent) accelerate training. The main idea behind it is to use the moving average of the gradient, rather than just the current actual value of the gradient. Its mathematical expression is as follows:

$$v_t = \gamma v_{t-1} - \eta \nabla \mathcal{F}(\gamma \theta_{t-1}) \tag{2.105}$$

$$\theta_t = \theta_{t-1} + v_t \tag{2.106}$$

where γ is the momentum; η is the learning rate of the t-th round of training. The main advantage of using momentum during training is to prevent the network from getting stuck in local minima. The momentum value is $\gamma \in (0, 1)$. Higher momentum values exceeding their minimum may make the network unstable. Usually, γ is set to 0.5 until the initial learning is stable, and then it is increased to 0.9 or higher.

Weight decay is used to train deep learning models as an L2 regularization method, which helps to prevent overfitting of the network and model generalization. The L2 regularization of $F(\theta, x)$ can be defined as

$$\Omega = \|\theta\|^2 \tag{2.107}$$

$$\hat{\varepsilon}(\mathcal{F}(\theta, x), y) = \varepsilon(\mathcal{F}(\theta, x), y) + \frac{1}{2}\lambda\Omega \tag{2.108}$$

the gradient of the weight θ is

$$\frac{\partial \frac{1}{2}\lambda\Omega}{\partial\theta} = \lambda \cdot \theta \tag{2.109}$$

where $\lambda = 0.0004$.

Batch normalization (BN) reduces internal covariance by shifting input samples, thereby accelerating the DL process. This means that the input is linearly transformed to zero mean and unit variance. For whitened inputs, the network converges faster and shows better regularization during training, which affects the overall accuracy. Since data whitening is performed outside the network, it does not produce the effect of whitening during model training. In deep neural networks, the input of the n-th layer is a combination of the $n - 1$ layer, not the original feature input. As training progresses, the effects of normalization or whitening decrease, respectively, leading to the vanishing gradient problem. This may slow down the entire training process and lead to saturation. For better training, BN is applied to the internal layers of deep neural networks. This method ensures faster convergence in theory and benchmark experiments. In BN, the features of a layer are independently normalized to mean 0 and variance 1. The following algorithm gives a pseudocode implementation of BN.

$$\textbf{Algorithm}: \ \textbf{Batch Normalization (BN)}$$

$$\textbf{Input}: \ \text{Mini-batch } x \text{ values}: \mathcal{B} = \{x_1, x_2, \cdots, x_m\} \tag{2.110}$$

$$\textbf{Output}: \ \{y_i = BN_{\gamma,\beta}(x_i)\} \tag{2.111}$$

$$\mu_{\mathcal{B}} \leftarrow \frac{1}{m}\sum_{i=1}^{m} x_i \qquad \text{// Mini-batch mean} \tag{2.112}$$

$$\sigma_{\mathcal{B}}^2 \leftarrow \frac{1}{m}\sum_{i=1}^{m}(x_i - \mu_{\mathcal{B}})^2 \qquad \text{// Mini-batch variance} \tag{2.113}$$

$$\hat{x}_i \leftarrow \frac{x_i - \mu_{\mathcal{B}}}{\sqrt{\sigma_{\mathcal{B}}^2 + \varepsilon}} \qquad \text{// Normalize} \tag{2.114}$$

$$y_i = \gamma \hat{x}_i + \beta \equiv BN_{\gamma,\beta}(x_i) \qquad \text{// Scale and shift} \tag{2.115}$$

2. Activation Functions

The common activation functions are defined as follows, and Fig. 2 in [26] shows the curve graph of common activation functions.

(1) Sigmoid

$$y = \frac{1}{1 + e^x} \tag{2.116}$$

Note: The softmax function is an extension of the sigmoid function, where each unit of the input is a vector, i.e., $y_i \in \mathbf{y}$, $i = 0, 1, \cdots, C - 1$, $\mathbf{y} = [y_0, y_1, \cdots, y_{C-1}]$, the output is a value in the range [0, 1], and its formula is as follows:

$$\text{softmax}(\mathbf{y})_i = \frac{e^{y_i}}{\sum_j e^{y_j}}$$

(2) Hyperbolic Tangent (Tanh)

$$y = \frac{e^x - e^{-x}}{e^x + e^{-x}} \tag{2.117}$$

(3) Rectified Linear Unit (ReLU)

$$y = \max(x, 0)$$

(4) Leaky ReLU takes a small positive number $a > 0$

$$y = \begin{cases} x, & x \geq 0 \\ ax, & x < 0 \end{cases} \tag{2.118}$$

(5) Exponential Linear Unit (ELU)

$$y = \begin{cases} x, & x \geq 0 \\ a(e^x - 1), & x < 0 \end{cases} \tag{2.119}$$

3. Convolutional Layer, Pooling Layer, Classification Layer

(1) Convolutional Layer
The feature maps from the previous layer and the learnable kernel convolve in the convolutional layer. The output of the kernel passes through a linear or nonlinear activation function, such as sigmoid, hyperbolic tangent, softmax, rectified linear, and identity function, to generate output feature maps. Each output feature map can be combined with multiple input feature maps. Overall, we have:

$$x_j^l = f(\sum_{i \in M_j} x_i^{l-1} \cdot k_{ij}^l + b_j^l) \tag{2.120}$$

where x_j^l is the current layer output; x_i^{l-1} is the previous layer output; k_{ij}^l is the current layer kernel; b_j^l is the current layer bias. M_j represents the selected input map. For each output map, an additional bias b is given. However, the input map will convolve with different kernels to generate corresponding output maps.

(2) Pooling Layer
The subsampling layer performs a downsampling operation on the input map, which is commonly referred to as the pooling layer. In this layer, the number of input and output feature maps does not change. For example, if there are N input maps, then there will be N output maps. Due to the downsampling operation, the size of each dimension of the output map decreases, which depends on the size of the downsampling mask. For example, if a 2×2 downsampling kernel is used, all image output dimensions will be half of the corresponding input image dimensions. This operation can be expressed as

$$x_j^l = \text{down}(x_j^{l-1}) \tag{2.121}$$

Two types of operations are mainly performed in the pooling layer: average pooling or max-pooling. In the average pooling operation, the function usually summarizes the $N \times N$ patches from the previous layer's feature map and selects the average value. In the max-pooling operation, the highest value is selected from the $N \times N$ patches of the feature map. Therefore, the size of the output map decreases. In special cases, the output map is multiplied by a scalar. There have been some alternative subsampling layers, such as fractional max-pooling layer and subsampling with convolution.

(3) Classification Layer

The classification layer is a fully connected layer that calculates the score of each class based on the features extracted from the previous convolutional layer. The final layer feature map is represented as a vector of scalar values, which are passed to the fully connected layer. The fully connected feed-forward neural layer serves as the softmax classification layer.

In the backpropagation of CNN, the fully connected layer is updated according to the method of the fully connected neural network (FCNN). A full convolution operation is performed on the feature map between the convolutional layer and its previous layer to update the filters of the convolutional layer. Figure 2 in [27] shows the basic operations of convolution and pooling of the input image.

(4) Network Parameters and Memory Requirements

The number of network parameters is an important indicator to measure the complexity of deep learning models. The size of the output feature map is

$$M = (N - F)/S + 1 \tag{2.122}$$

where N refers to the size of the input feature map; F refers to the size of the filter or receptive field; M refers to the size of the output feature map; S represents the stride. Padding is usually applied in the convolutional layer to ensure that the input and output feature maps have the same size. The padding size depends on the size of the kernel. The number of rows and columns of padding are as follows:

$$P = (F - 1)/2 \tag{2.123}$$

where P is the padding size; F is the kernel dimension. There are various metrics used for model comparison. However, in most cases, the number of network parameters and memory requirements will be considered. The formula for calculating the amount of parameters in the l-th layer is as follows:

$$\text{Parm}_l = (F \cdot F \cdot \text{FM}_{l-1}) \cdot \text{FM}_l \tag{2.124}$$

where the total number of parameters in the l-th layer is denoted as Parm_l; FM_l is the total number of output feature maps; FM_{l-1} is the total number of input feature maps or channels. If the offset parameters are added to the weight parameters, the above equation can be written as

$$\text{Parm}_l = (F \cdot (F + 1) \cdot \text{FM}_{l-1}) \cdot \text{FM}_l \tag{2.125}$$

For example, suppose the l-th layer has $\text{FM}_{l-1} = 32$ input feature maps and $\text{FM}_l = 64$ output feature maps, and the filter size is $F = 5$, in this case, the total number of offset parameters for this layer is

$$\text{Parm}_l = (5 \times 5 \times 33) \times 64 = 528\ 000 \tag{2.126}$$

Therefore, the memory size required for the lth layer operation (Meml) can be represented as

$$\mathrm{Mem}_l = (N_l \cdot N_l \cdot \mathrm{FM}_l) \tag{2.127}$$

4. Backpropagation

Deep NNs are trained using the popular backpropagation algorithm and SGD. In the case of MLP, the NN model can be easily represented with computational graphs such as directive acyclic graphs. The representation of DL allows a single-path network to effectively calculate the gradient from the top layer to the bottom layer using the chain rule. The pseudocode for the basic backpropagation algorithm is given below.

Algorithm: Backpropagation

Input: A network with $l$layers, activation function σ_l

Output of hidden layer: $\mathbf{h}_l = \sigma_l(\mathbf{W}_l^T \mathbf{h}_{l-1} + \mathbf{b}_l)$

Network output: $\tilde{\mathbf{y}} = \mathbf{h}_l$

Calculate gradient: $\delta \leftarrow \dfrac{\partial_\varepsilon(\mathbf{y}_i, \tilde{\mathbf{y}}_j)}{\partial \mathbf{y}}$ (2.128)

Proceed from $i \leftarrow l$ to 0

Calculate the gradient of the current layer:

$$\frac{\partial_\varepsilon(\mathbf{y}_i, \tilde{\mathbf{y}}_j)}{\partial \mathbf{W}_l} = \frac{\partial_\varepsilon(\mathbf{y}_i, \tilde{\mathbf{y}}_j)}{\partial \mathbf{h}_l} \frac{\partial \mathbf{h}_l}{\partial \mathbf{W}_l} = \delta \frac{\partial \mathbf{h}_l}{\partial \mathbf{W}_l} \tag{2.129}$$

$$\frac{\partial_\varepsilon(\mathbf{y}_i, \tilde{\mathbf{y}}_j)}{\partial \mathbf{b}_l} = \frac{\partial_\varepsilon(\mathbf{y}_i, \tilde{\mathbf{y}}_j)}{\partial \mathbf{h}_l} \frac{\partial \mathbf{h}_l}{\partial \mathbf{b}_l} = \frac{\partial \mathbf{h}_l}{\partial \mathbf{b}_l} \tag{2.130}$$

Apply $\dfrac{\partial_\varepsilon(\mathbf{y}_i, \tilde{\mathbf{y}}_j)}{\partial \mathbf{W}_l}$ and $\dfrac{\partial_\varepsilon(\mathbf{y}_i, \tilde{\mathbf{y}}_j)}{\partial \mathbf{b}_l}$ to gradient descent

Backpropagate the gradient to the next layer

Output of hidden layer: $\mathbf{h}_l = \sigma_l(\mathbf{W}_l^T \mathbf{h}_{l-1} + \mathbf{b}_l)$

$$\delta \leftarrow \frac{\partial_\varepsilon(\mathbf{y}, \tilde{\mathbf{y}})}{\partial \mathbf{h}_l} \frac{\partial \mathbf{h}_l}{\partial \mathbf{h}_{l-1}} = \delta \frac{\partial \mathbf{h}_l}{\partial \mathbf{h}_{l-1}} \tag{2.131}$$

End.

5. Gradient Descent and Stochastic Gradient Descent

(1) Gradient Descent

Gradient descent is a first-order optimization algorithm used to find the local minimum of an objective function. The following pseudocode illustrates the concept of gradient descent.

Algorithm : **Gradient Descent**

Input : Loss function ε, learning rate η, dataset $\mathbf{X}$,

: output $\mathbf{y}$, and model $\mathcal{F}(\theta, \mathbf{x})$

Output : Optimal θ that minimizes ε

Repeat until convergence:

$$\tilde{\mathbf{y}} = \mathcal{F}(\theta, \mathbf{x}) \tag{2.132}$$

$$\theta = \theta - \eta \cdot \frac{1}{N} \sum_{i=1}^{N} \frac{\partial_{\varepsilon}(\mathbf{y}, \tilde{\mathbf{y}})}{\partial \theta} \tag{2.133}$$

End

(2) Stochastic Gradient Descent

Due to the long training time being a major drawback of traditional gradient descent, stochastic gradient descent is used for training deep neural networks (DNN). The following pseudocode explains the concept of stochastic gradient descent in detail.

Algorithm : **Stochastic Gradient Descent**

Input : Loss function ε, learning rate η, dataset $\mathbf{X}$,

: output $\mathbf{y}$, and model $\mathcal{F}(\theta, \mathbf{x})$

Output : Optimal θthat minimizes ε

Repeat until convergence:

Randomize $(\mathbf{X}, \mathbf{y})$;

For each batch $(\mathbf{x}_i, \mathbf{y}_i)$ of $(\mathbf{X}, \mathbf{y})$ do

$$\tilde{\mathbf{y}}_i = \mathcal{F}(\theta, \mathbf{x}_i) \tag{2.134}$$

$$\theta = \theta - \eta \cdot \frac{1}{N} \sum_{i=1}^{N} \frac{\partial_{\varepsilon}(\mathbf{y}_i, \tilde{\mathbf{y}}_i)}{\partial \theta} \tag{2.135}$$

End

Stochastic gradient descent has many important applications beyond deep learning. It is the main method for training large linear models on large-scale data.

In the model training of CNN, the following issues should also be noted.

1. AdaGrad/AdaDelta: Adjust the learning rate based on gradient history information.
2. Dropout: A method to deal with overfitting, select nodes to turn off responses with a certain probability.
3. Data augmentation: Increase data variation to overcome overfitting.
4. Preprocessing: Normalization, Principal Component Analysis (PCA) whitening.
5. Initialization: Such as the Xavier method.

2.5.2 Generative Adversarial Networks

What is the principle of generative adversarial networks (GAN)? [28] It is easier to explain with the example of generating images. Suppose there are two networks, the generator network G and the discriminator network D, then:

1. G is a network that generates images, which inputs a random noise z and generates an image through it, denoted as $G(z)$.
2. D is a network that discriminates images, which confirms whether an image is "real." Its input parameter is x, representing an image, and the output $D(x)$ represents the probability that x is a real image. If it is 1, then it is 100% a real image; if the output is 0, then it is not a real image at all.

Figure 2 in [29] is a visual representation of GAN. In the training process, the generator network G tries to generate real images to deceive the discriminator network D, while the discriminator network D tries to distinguish the images generated by G from real images. In this way, G and D form a dynamic "gaming process." As a result of the game, G can generate images $G(z)$ that are "fake but look real," and for D, it is difficult to determine whether the images generated by G are real or not, so $D(G(z)) = 0.5$.

Mathematically, the objective function that GAN needs to optimize is as follows:

$$\min_G \max_D V(G, D) = \min_G \max_D \mathbb{E}_{x \sim p_{data}}[log D(x)] + \mathbb{E}_{x \sim p_z}[log(1 - D(G(z))] \tag{2.136}$$

where $p_{data}(x)$ and $p_z(z)$, respectively, represent the probability distribution of real data defined in data space $\mathbf{X}$ and the probability distribution of latent variable z defined on latent space $\mathbf{Z}$, and $V(G, D)$ is the binary cross entropy function, commonly used for binary classification problems.

Since D wants to classify real or fake samples, $V(G, D)$ is a natural choice for the objective function of the classification problem. From the perspective of D, if the sample comes from real data, D will maximize its output; if the sample comes from G, D will minimize its output. At the same time, G wants to deceive D, so when a fake sample is presented to D, it tries to maximize D's output. By deriving $V(G, D)$, the optimal discriminator can be calculated:

$$D^*(x) = \frac{p_g(x)}{p_g(x) + p_{data}(x)} \tag{2.137}$$

Substituting this optimal discriminator into the previous objective function, the equation becomes the JS divergence (Jensen Shannon divergence, JSD) between $p_{data}(x)$ and $p_g(x)$. Therefore, the optimal generator that minimizes $JSD(p_{data}\|p_g)$ is the data distribution $p_{data}(x)$, and substituting the optimal generator into the optimal discriminator, D becomes 0.5, meaning they are hard to distinguish.

The f-divergence $D_f(p_{data}\|p_g)$ is a means of measuring the difference between two distributions through a specific convex function f. Under the condition that f is a convex function and satisfies $f(1) = 0$, $Df(p_{data}\|p_g)$ can act as the divergence between two distributions. Due to the condition $f(1) = 0$, if the two distributions are equivalent, their ratio becomes 1, and their divergence becomes 0. f is generally called the generating function, and it is called the f-divergence function to avoid confusion with the generator G. KL divergence (Kullback–Leibler divergence, KLD), reverse KLD, JSD and other divergence functions can be derived from the f-GAN framework of a specific generating function f.

Wasserstein GAN (WGAN) proposed important research on the distance between $p_{data}(x)$ and $p_g(x)$. GAN learning transforms the latent variable z into the generating function g_θ of $p_g(x)$, rather than directly learning the probability distribution $p_{data}(x)$ itself. Therefore, a measure between $p_g(x)$ and $p_{data}(x)$ is needed to train g_θ. WGAN suggests using the earth mover distance (EMD), also known as the Wasserstein distance, as a measure of the difference between the two distributions. The definition of the Wasserstein distance is as follows:

$$W(p_{data}, p_g) = \inf_{\gamma \in (p_{data}, p_g)} \mathbb{E}_{(x,y)\sim\gamma}[\|x - y\|] \tag{2.138}$$

where $\prod(p_{data}, p_g)$ represents the set of all joint distributions, and the marginal distributions of $\gamma(x, y)$ are $p_{data}(x)$ and $p_g(x)$, respectively.

Deep Convolutional Generative Adversarial Network (DCGAN) has made significant contributions to GAN, and the Convolutional Neural Network (CNN) architecture it proposed has greatly stabilized GAN training. DCGAN proposed several useful architectural suggestions.

1. The discriminator D replaces pooling with strided convolution, and the generator G replaces pooling with fractional-strided convolution.
2. Use batch normalization in both D and G.
3. Remove fully connected hidden layers.
4. In G, use ReLU for all layers except the output layer, and use the Tanh function.
5. In D, all layers use the Leaky ReLU generator.

Since DCGAN has solved the instability of training GAN, it has become the benchmark for various GAN models.

If the generator and discriminator are both conditioned on some additional information y, such as class labels or data from other modalities, then GAN extends to conditional models, i.e., y enters the discriminator and generator as an additional input layer for adjustment. This network is called the Conditional Generative Adversarial Net (CGAN), and Fig. 2 in [30] shows a intuitive representation of CGAN.

GAN is now being used in many fields, such as image translation, super-resolution, object detection, object change image, image generation and video generation, as well as the generation of serial data, such as music, language, and speech.

The challenge of training GAN is greater than CNN, and some tips to note are as follows.

1. Normalize the input.
2. Pay attention to correcting the loss function.
3. Use spherical noise distribution (not uniform distribution, but Gaussian distribution).
4. Perform batch normalization (BN).
5. Avoid sparse gradients, such as using activation functions ReLU and max-pooling.
6. Use soft labels with added noise.
7. It is recommended to use DCGAN or hybrid models, such as the combination of KL divergence + GAN, or the combination of VAE (variational auto-encoder) + GAN.
8. Use stability methods based on reinforcement learning.
9. Use the Adam optimizer for training the generator G, and Stochastic Gradient Descent (SGD) for the discriminator D.
10. Track failures that occur early in training, check if the modulus of the gradient is too large or too small.
11. Unless there is sufficient reason, do not balance losses through statistics.
12. If there are true value annotations, it is best to use them, such as using the auxiliary GAN method.
13. Add noise interference to the input and gradually reduce the interference amplitude over time.
14. You can try training more discriminators (especially when the input contains noise).
15. You can try batch discrimination.
16. Use discrete variables in CGAN, use the dropout method in the generator G during the training/testing phase to avoid overfitting.

2.5.3 Recurrent Neural Network

The Recurrent Neural Network (RNN) is unique in that it allows operations on a series of vectors over time.

In the Elman network, the output of the hidden layer and the normal input of the hidden layer are used together as input. In addition, in the Jordan network, the output of the output unit is used as the input of the hidden layer. Conversely, the Jordan network uses the output of the output unit as the input for both itself and the hidden layer. This can be represented mathematically as follows.

(1) Elman network

$$\mathbf{h}_t = \sigma_h(\mathbf{w}_h\mathbf{x}_t + \mathbf{u}_h\mathbf{h}_{t-1} + \mathbf{b}_h) \tag{2.139}$$

$$\mathbf{y}_t = \sigma_y(\mathbf{w}_y\mathbf{h}_t + \mathbf{b}_y) \tag{2.140}$$

(2) Jordan network

$$\mathbf{h}_t = \sigma_h(\mathbf{w}_h\mathbf{x}_t + \mathbf{u}_h\mathbf{y}_{t-1} + \mathbf{b}_h) \tag{2.141}$$

$$\mathbf{y}_t = \sigma_y(\mathbf{w}_y\mathbf{h}_t + \mathbf{b}_y) \tag{2.142}$$

where $\mathbf{x}_t$ is the input vector; $\mathbf{h}_t$ is the hidden layer vector; $\mathbf{y}_t$ is the output vector, $\mathbf{w}$ and $\mathbf{u}$ are the weight matrices, and $\mathbf{b}$ is the offset vector.

The loop allows information to be passed from one step of the network to the next. The Recurrent Neural Network can be seen as multiple copies of the same network, each passing messages to its successor.

The main problem with the RNN method is gradient vanishing.

1. LSTM

The key idea of the Long Short Term Memory (LSTM) model is the cell state. As shown in Fig. 2 in [31], LSTM removes or adds information to the cell state through gates: input gate ($\mathbf{i}_t$), forget gate ($\mathbf{f}_t$), and output gate ($\mathbf{o}_t$), which can be defined as follows:

$$\mathbf{f}_t = \sigma(\mathbf{W}_f[\mathbf{h}_{t-1}, \mathbf{x}_t] + \mathbf{b}_f) \tag{2.143}$$

$$\mathbf{i}_t = \sigma(\mathbf{W}_i[\mathbf{h}_{t-1}, \mathbf{x}_t] + \mathbf{b}_i) \tag{2.144}$$

$$\tilde{C}_t = \tanh(\mathbf{W}_C[\mathbf{h}_{C-1}, \mathbf{x}_t] + \mathbf{b}_C) \tag{2.145}$$

$$C_t = \mathbf{f}_t C_{t-1} + \mathbf{i}_t \tilde{C}_t \tag{2.146}$$

$$\mathbf{o}_t = \sigma(\mathbf{W}_o[\mathbf{h}_{t-1}, \mathbf{x}_t] + \mathbf{b}_o) \tag{2.147}$$

$$\mathbf{h}_t = \mathbf{o}_t \tanh(C_t) \tag{2.148}$$

The LSTM model is popular in time information processing. Most papers containing the LSTM model have some minor differences.

2. GRU

The Gated Recurrent Unit (GRU) also comes from LSTM. The main reason for the popularity of GRU is its low computational cost and simple model, as shown in Fig. 2 in [32]. In terms of topology, computational cost, and complexity, GRU is a lighter version of the standard LSTM. This technique combines the forget gate and input gate into a single update gate, and merges the cell state, hidden state, and some other changes. The simpler GRU model is becoming increasingly popular. Mathematically, GRU can be represented as follows.

$$\mathbf{z}_t = \sigma(\mathbf{W}_z[\mathbf{h}_{t-1}, \mathbf{x}_t]) \tag{2.149}$$

$$\mathbf{r}_t = \sigma(\mathbf{W}_r[\mathbf{h}_{t-1}, \mathbf{x}_t]) \tag{2.150}$$

$$\tilde{\mathbf{h}}_t = \tanh(\mathbf{W}[\mathbf{r}_t\mathbf{h}_{t-1}, \mathbf{x}_t]) \tag{2.151}$$

$$\mathbf{h}_t = (1 - z_t)\mathbf{h}_{t-1} + z_t\tilde{\mathbf{h}}_t \tag{2.152}$$

GRU requires fewer network parameters, which makes model inference faster. In addition, given enough data and computational power, LSTM can provide better performance.

2.5.4 *Transformer Network*

The Transformer is a new type of neural network different from CNN [15, 33, 34], mainly using the self-attention mechanism to extract inherent features. The Transformer was initially applied to Natural Language Processing (NLP) tasks and brought significant improvements, such as BERT, which pre-trains unlabelled text's Transformer by jointly processing the left, right, top, and bottom contexts. For this reason, the Transformer has been introduced into computer vision and now seems to have become an alternative to CNN.

A Transformer consists of an encoder module and a decoder module, each containing several identical encoder/decoder architectures. Each encoder consists of a self-attention layer and a feed-forward neural network, while each decoder consists of a self-attention layer, an encoder–decoder attention layer, and a feed-forward neural network.

In the self-attention layer, the input vector is first converted into three different vectors, namely the Query vector **q**, Key vector **k**, and Value vector **v**, where the dimensions are $d_q = d_k = d_v = d_{model} = 512$. Then, vectors from different inputs are packed into three different matrices **Q**, **K**, and **V**. Afterward, the attention function between different input vectors is calculated through the following steps, as shown in Fig. 2 in [35](a):

- Step 1: Calculate the score $S = \mathbf{Q} \cdot \mathbf{K}^T$ between different input vectors.
- Step 2: Normalize the score $S_n = S/\sqrt{d_k}$ for gradient stability.

- Step 3: Convert the score into probabilities $P = \text{softmax}(S_n)$ using the softmax function.
- Step 4: Obtain the weighted value matrix $\mathbf{Z} = \mathbf{V} \cdot P$.

This entire process can be unified into one function:

$$\text{Attention}(\mathbf{Q}, \mathbf{K}, \mathbf{V}) = \text{softmax}(\frac{\mathbf{Q} \cdot \mathbf{K}^T}{\sqrt{d_k}}) \cdot \mathbf{V} \tag{2.153}$$

Step 1 calculates the score between two different vectors, which determines the attention to other words when encoding the current word. Step 2 normalizes the score to have a more stable gradient for better training. Step 3 converts the score into probabilities. Step 4 multiplies each $\mathbf{V}$ vector by the total probability, and vectors with larger probabilities will receive more attention in the following layers.

The encoder–decoder attention layer in the decoder module is almost the same as the self-attention layer in the encoder module, except that the Key matrix $\mathbf{K}$ and Value matrix $\mathbf{V}$ are exported from the encoder module, and the Query matrix $\mathbf{Q}$ is exported from the previous layer.

Note that the above process is independent of the position of each word, so the self-attention layer lacks the ability to capture the position information of words in the sentence. To solve this problem, a position encoding of dimension d_{model} is added to the original input embedding to obtain the final input vector of the word. Specifically, the position is encoded using the following formulas:

$$\text{PE}(pos, 2i) = \sin(\frac{\text{pos}}{10000^{\frac{2i}{d_{\text{model}}}}}) \tag{2.154}$$

$$\text{PE}(pos, 2i+1) = \cos(\frac{\text{pos}}{10000^{\frac{2i}{d_{\text{model}}}}}) \tag{2.155}$$

where pos represents the position of the word in the sentence; i represents the current dimension of the position encoding.

By adding a multi-head attention mechanism, the self-attention layer is further improved, enhancing the performance of the standard self-attention layer. Note that for a given reference word, several other words are usually focused on when reading through the sentence. Therefore, the single-head self-attention layer limits the ability to focus on a specific position (or several specific positions) without affecting attention to other equally important positions. This is achieved by giving the attention layer different representation subspaces. Specifically, different heads use different Query matrices, Key matrices, and Value matrices, which can project the input vector into different representation subspaces after training (due to random initialization).

Specifically, given the input vector and the number of heads h, the input vector is first converted into three different sets of vectors, namely the query set, key set, and value set. Each set contains h vectors, with dimensions $d_q' = d_k' = d_v' = d_{\text{model}}/h =$

64. Then, vectors from different inputs are packed into three different sets of matrices $\{\mathbf{Q}_i\}_{i=1}^h$, $\{\mathbf{K}_i\}_{i=1}^h$, $\{\mathbf{V}_i\}_{i=1}^h$. The process of multi-head attention is as follows:

$$\text{MultiHead}(\mathbf{Q}', \mathbf{K}', \mathbf{V}') = \text{Concat}(\text{head}_1, \cdots, \text{head}_h)\mathbf{W}^O \tag{2.156}$$

where $\mathbf{W}^O$ is the linear projection matrix.

$$\text{head}_i = \text{Attention}(\mathbf{Q}_i, \mathbf{K}_i, \mathbf{V}_i) \tag{2.157}$$

Figure 2 in [36] shows the detailed structure of the Transformer: A residual connection is added in each sub-layer of the encoder and decoder to strengthen the information flow and achieve better performance. Layer normalization is then performed. The output of the above operations can be described as

$$\text{LayerNorm}(\mathbf{X} + \text{Attention}(\mathbf{X})) \tag{2.158}$$

Note that here $\mathbf{X}$ is used as the input to the self-attention layer because the matrices $\mathbf{Q}$, $\mathbf{K}$, and $\mathbf{V}$ of Query, Key, and Value all come from the same input matrix $\mathbf{X}$.

A feed-forward neural network is applied after the self-attention layer of each encoder and decoder. Specifically, the feed-forward NN consists of two linear transformation layers and a ReLU activation function in between, which can be represented as the following function:

$$\text{FFNN}(\mathbf{X}) = \mathbf{W}_2\sigma(\mathbf{W}_1\mathbf{X}) \tag{2.159}$$

where $\mathbf{W}_1$ and $\mathbf{W}_2$ are the two parameter matrices of the two linear transformation layers; σ represents the ReLU activation function. The dimension of the hidden layer is $d_h = 2048$.

The last layer of the decoder aims to convert the vector stack back into a word. This is achieved through a linear layer and a softmax layer. The linear layer projects the vector into a logits vector with d_{word} dimensions, where d_{word} is the number of words in the vocabulary. Then, the softmax layer converts the logits vector into probabilities.

Most Transformers used in computer vision tasks utilize the encoder module of the original Transformer. In short, it can be seen as a new feature selector different from CNN and RNN. Compared to CNNs that only focus on local features, Transformers can capture long-distance features, meaning that Transformers can easily obtain global information. Compared to RNNs that must compute hidden states in sequence, Transformers are more efficient because the outputs of self-attention layers and fully connected layers can be computed in parallel and are easy to accelerate.

2.6 Neural Network Model Compression and Acceleration

Deep neural network models have high computational costs and large storage requirements, which can be a hindrance in low-memory resource devices or applications with strict latency requirements. Therefore, one idea is to perform model compression and acceleration in deep networks without significantly reducing model performance [37, 38].

Currently, there are roughly four schemes for compressing and accelerating CNN models.

1. Parameter pruning and sharing: Explore redundancy in model parameters and attempt to remove redundant and unimportant parameters.
2. Low-rank factorization: Use matrix/tensor decomposition to estimate the information parameters of deep CNN models.
3. Transferred/compact convolutional filter: Design special structure convolutional filters to reduce parameter space and save storage/computation.
4. Knowledge distillation: Learn to distill models and train more compact neural networks to reproduce the outputs of larger networks.

Typically, parameter pruning and sharing, low-rank factorization, and knowledge distillation methods can be used for deep neural network models with fully connected layers and convolutional layers; in addition, the transferred/compact convolutional filter method is only applicable to models with convolutional layers. Low-rank factorization and transferred/compact convolutional filter methods provide an end-to-end pipeline that can be easily implemented in a CPU/GPU environment. Parameter pruning and sharing will use different methods such as vector quantization, binary encoding, and sparse constraints. In summary, implementing compression and acceleration requires multiple steps.

As for training methods, parameters can be extracted from pre-trained training methods for pruning and sharing, low-rank factorization models, or training can be started from scratch. Transferred/compact convolutional filter and knowledge distillation models can only support training from scratch. These methods are independently designed and complement each other. For example, transferred network layers can be used together with parameter pruning and sharing, or model quantization and binarization can be used together with low-rank factorization approximation.

2.6.1 Parameter Pruning and Sharing

For a typical three-step model compression method, it includes pruning, quantization, and encoding. This method uses weight sharing to quantize connection weights and uses Huffman encoding to quantize weights and codebooks. First, learn its connectivity through normal network training; then, prune small weight connections;

finally, retrain the network to learn the final weights of the remaining sparse connections. Hessian weights can be used to measure the importance of network parameters, so one metric for compression is to minimize the Hessian-weighted average quantization error of clustering network parameters.

The extreme case of each weight 1-bit representation is a binary weight neural network. There are quite a few methods for training CNNs directly with binary weights, such as BinaryConnect, BinaryNet, and XNORNetworks. The main idea is to directly learn binary weights or activations during model training. Networks trained with backpropagation can tolerate special weight distortions, including binary weights.

In network architectures that include fully connected layers, eliminating parameter redundancy in the fully connected layer is crucial because it is often the largest part of memory consumption. These network layers use the nonlinear transformation $f(\mathbf{x}, \mathbf{M}) = \sigma(\mathbf{MX})$, where $\sigma(\cdot)$ is the element-wise nonlinear operator, $\mathbf{x}$ is the input vector, and $\mathbf{M}$ is the $m \times n$ parameter matrix. When $\mathbf{M}$ is a general dense matrix, the cost of storing $m \times n$ parameters and computing matrix–vector products is of the order $O(m \times n)$. Therefore, the most direct method of pruning parameters is to impose $\mathbf{x}$ as a parameterized structure matrix. An $m \times n$ matrix can be described with much fewer parameters than $m \times n$, called a structured matrix. Typically, this structure not only reduces storage costs but also speeds up network training and inference through fast algorithms for matrix–vector multiplication and gradient computation.

2.6.2 *Low-Rank Factorization*

Most of the computation in CNN models is convolutional operations, reducing convolutional layers will improve the compression rate and overall acceleration rate. The convolutional kernel can be viewed as a 4D tensor, and there is a lot of redundancy in the 4D tensor. Tensor decomposition is a particularly effective way to eliminate redundancy. The fully connected layer can be viewed as a 2D matrix, and the low-rank characteristics of the matrix are conducive to network compression. Low-rank approximation is done layer by layer. After completing one layer, fix its parameters and fine-tune the parameters of the upper layer based on the reconstruction error.

2.6.3 *Transferred/Compact Convolutional Filter*

The CNN model fully reflects the translational invariance of the input image, and its parameters are efficient, which is the key reason why deep CNNs do not overfit. Although there is no strong theory to verify this, a large amount of empirical data supports this conclusion, that is, translational invariance and convolutional parameter allocation play an important role in ensuring the predictive performance of CNNs. The idea of transfer/dense convolution filters comes from the theory of equivariant groups.

Let $\mathbf{x}$ be the input, $\Phi(\cdot)$ be the network or layer, and $\mathbf{T}(\cdot)$ be the transfer matrix. The concept of equivalence is defined as

$$\mathbf{T}'\Phi(\mathbf{x}) = \Phi(\mathbf{T}\mathbf{x}) \tag{2.160}$$

The meaning of Eq. 2.160 is that the input $\mathbf{x}$ is passed through the transfer matrix $\mathbf{T}(\cdot)$ and then passed through the network or layer $\Phi(\cdot)$, and the network output is transformed after the network maps $\mathbf{x}$, both give the same result. According to this equivalence theory, it is reasonable to apply the transfer to the layer or filter $\Phi(\cdot)$ to compress the entire network model. From an empirical observation, deep CNNs also benefit from the use of a large number of convolution filters, which apply the transfer matrix $\mathbf{T}(\cdot)$ to a small group of basic filters, because they are regularizers of the model itself.

2.6.4 Knowledge Distillation

Knowledge Distillation (KD) is a technique used to compress deep and wide neural networks into shallower and more lightweight models. In this process, the smaller compressed model, often referred to as the *student* model, is trained to simulate the function or behavior learned by a larger, more complex *teacher* model. The central idea of KD-based methods is to transfer knowledge by having the student model learn from the class probability distribution produced by the softmax output of the teacher model. This "soft target" distribution contains richer information than the one-hot ground truth labels, as it encodes the relative probabilities across all classes, thereby enabling the student to generalize more effectively.

The KD framework follows a "teacher–student" paradigm to facilitate the training of deep networks, where the student defines a loss function based on the teacher's soft output as a form of regularization. This framework enables the student network to be trained by mimicking not only the final predictions but also the output behavior of the teacher network. Although conceptually simple, KD has demonstrated strong performance in a wide range of image classification tasks. Building upon this idea, the FitNets model proposes a strategy for training thin and deep networks to replace wide and shallow ones. FitNets extends the basic KD framework by encouraging the student to learn from the intermediate feature representations of the teacher network, effectively aligning internal feature maps. However, this assumption of full feature mimicry may still be overly strict, as the representational capacity gap between teacher and student networks can make precise imitation difficult in practice.

2.6.5 MobileNets

Google's MobileNets [39, 40], based on a streamlined architecture, use depthwise separable convolutions to build lightweight deep neural networks. There are two simple global hyperparameters that allow for effective trade-offs between latency and accuracy. These hyperparameters allow the model to choose the right size model for its constraints when building.

Based on depthwise separable convolution, MobileNets model decomposes standard convolution into depthwise convolution and 1×1 convolution, also known as pointwise convolution. Depthwise convolution equips a filter for each input channel. Then, pointwise convolution applies 1×1 convolution to combine the output depthwise convolution. Standard convolution can filter and combine into a set of new outputs in one step. Depthwise separable convolution divides it into two layers, a filtering layer and a combination layer. This decomposition can significantly reduce computation and model size.

The MobileNets structure is built on the theory of depthwise separable convolution, and all layers except the first one are fully convolutional structures. One with regular convolution, batch normalization (BN) and ReLU layers, and the other with depthwise convolution, 1×1 pointwise convolution, and batch normalization (BN) and ReLU layers after convolution. In the depthwise convolution layer and the first layer, strided convolution is used to handle downsampling. Finally, the average pooling operation reduces the spatial resolution to 1 before the fully connected layer.

If depthwise convolution and pointwise convolution are considered as separate layers, MobileNets has 28 layers. The definition of the MobileNets architecture is shown in Table 2.1. Conv represents convolution, dw represents depthwise, s1/2 represents the stride of stride convolution, softmax is the activation function, and Classifier represents the classifier.

Subsequently, Google proposed an improved model called MobileNet V2, which improved its performance in multiple tasks and benchmarks, and across different model size ranges. It also defined a new framework called SSDLite, with the following characteristics.

1. Based on the inverted residual structure, where shortcut connections are set between thin bottleneck layers.
2. The intermediate expansion layer uses lightweight depth convolution to filter features as a source of nonlinearity.
3. Removing nonlinearity in narrow layers to maintain representational capacity is important.
4. Allows the input/output domain to be separated from the expression of the transformation.

Table 2.1 MobileNets architecture with layer configurations, filter shapes, and input sizes

Model/stride	Filter shape	Input size (pixels)
Conv/s2	$3 \times 3 \times 3 \times 32$	$224 \times 224 \times 3$
Conv dw/s1	$3 \times 3 \times 32$ dw	$112 \times 112 \times 32$
Conv/s1	$1 \times 1 \times 32 \times 64$	$112 \times 112 \times 32$
Conv dw/s2	$3 \times 3 \times 64$ dw	$112 \times 112 \times 64$
Conv/s1	$1 \times 1 \times 64 \times 128$	$56 \times 56 \times 64$
Conv dw/s1	$3 \times 3 \times 128$ dw	$56 \times 56 \times 128$
Conv/s1	$1 \times 1 \times 128 \times 128$	$56 \times 56 \times 128$
Conv dw/s2	$3 \times 3 \times 128$ dw	$56 \times 56 \times 128$
Conv/s1	$1 \times 1 \times 128 \times 256$	$28 \times 28 \times 128$
Conv dw/s1	$3 \times 3 \times 256$ dw	$28 \times 28 \times 256$
Conv/s1	$1 \times 1 \times 256 \times 256$	$28 \times 28 \times 256$
Conv dw/s2	$3 \times 3 \times 256$ dw	$28 \times 28 \times 256$
Conv/s1	$1 \times 1 \times 256 \times 512$	$14 \times 14 \times 256$
5× Conv dw/s1 Conv/s1	$3 \times 3 \times 512$ dw	$14 \times 14 \times 512$
	$1 \times 1 \times 512 \times 512$	$14 \times 14 \times 512$
Conv dw/s2	$3 \times 3 \times 512$ dw	$14 \times 14 \times 512$
Conv/s1	$1 \times 1 \times 512 \times 1024$	$7 \times 7 \times 512$
Conv dw/s2	$3 \times 3 \times 1024$ dw	$7 \times 7 \times 1024$
Conv/s1	$1 \times 1 \times 512 \times 1024$	$7 \times 7 \times 1024$
Conv dw/s2	$3 \times 3 \times 1024$ dw	$7 \times 7 \times 1024$
Conv/s1	$1 \times 1 \times 1024 \times 1024$	$7 \times 7 \times 1024$
Avg Pool/s1	Pool 7×7	$7 \times 7 \times 1024$
FC/s1	1024×1000	$1 \times 1 \times 1024$
Softmax/s1	Classifier	$1 \times 1 \times 1000$

ReLU will definitely cause information loss, and this loss is irreversible. There are two possible solutions.

1. Replace ReLU with a linear activation function.
2. If a larger number of channels can reduce information loss, then use more channels.

The bottleneck block looks similar to the residual block, where each block contains an input and several subsequent bottlenecks and their expansions. Assuming the manifold of interest is low-dimensional, it can be captured by inserting a linear bottleneck layer into the convolution block.

Ordinary convolution maps channel and spatial information simultaneously, with a large number of parameters; separable convolution blocks decouple channel and spatial information, changing multiplication to addition, saving a certain degree of parameter space; adding a bottleneck after the separable convolution block, mapping it to a low-dimensional space; starting from a low-dimensional space to perform separable convolution block, then expanding to a higher dimension, and then reducing dimension through 1 × 1 convolution.

For the structures of the residual block and the inverted residual block. The classic residual block has layers with a large number of channels, while the inverted residual block only connects to the bottleneck.

The 1×1 convolution in the architectures of MobileNets V1 and MobileNets V2 only changes the thickness of the feature map, without changing its width and length.

2.7 Architecture of Efficient Transformer and its Application

The development of deep learning, specifically with the emergence of Transformer [41–43], and particularly its application in large language models [44–49], has led to two urgent challenges at the fundamental level because the structure has become extremely large and computationally intensive. The first problem involves the intensive computation of training and reasoning of large Transformer-based models. The second problem is the exponential growth of memory overhead proportional to the size of the attention-based model.

Transformer modules are used in different ways. The main Transformer modes include [41]: (1) encoder only (e.g., for classification), (2) decoder only (e.g., for language modeling), and (3) encoder–decoder (e.g., for machine translation).

In the encoder–decoder mode, there are usually multiple multi-head self-attention modules, including standard self-attention in the encoder and decoder, and encoder–decoder cross-attention that allows the decoder to utilize information from the decoder. This affects the design of the self-attention mechanism.

In the encoder only mode, there is no restriction or constraint that the self-attention mechanism must be causal, that is, only depend on the present and past tokens.

In the encoder–decoder mode, the encoder and encoder–decoder cross-attention may be non-causal, but the decoder's self-attention must be causal. Designing effective self-attention mechanisms requires supporting causality for autoregressive (AR) decoding, which may be a common limitation factor.

Self-attention is a key defining feature of the Transformer model. The mechanism can be viewed as a graph-like inductive bias that associates all tokens in a sequence with a correlation-based pooling operation. A well-known problem with self-attention is its quadratic time and memory complexity, which hinders model scalability in many settings. As a result, a large number of variants have been proposed recently to address this problem, also known as efficient Transformers.

The quadratic time and space complexity of the attention module significantly slows down the pre-training and inference of Transformer-based models. To improve performance, many techniques have been proposed to reduce the burden of the attention module. These techniques can be roughly categorized into [44]: shared attention mechanisms, kernelized or low-rank attention mechanisms, fixed-pattern strategies, learnable-pattern strategies, and hardware-assisted attention mechanisms.

In traditional multi-headed attention (MHA), each head has a different linear transformation on the input matrices query (Q), key (K), and value (V), allowing different representations and attention mechanisms across different subspaces. In multi-query attention (MQA) [50], all heads share a set of key-value weights across all query heads. Therefore, MQA can speed up inference, but may hurt output quality. Group-Query Attention (GQA) [51] improves inference quality by interpolating MQA and MHA using a key-and-value header in each group query header.

Kernelization or low-rank techniques adopted by models such as Linformer and Performer exploit the low-rank representation of the self-attention matrix or employ attention kernelization techniques to improve efficiency. Low-rank methods focus on compressing the dimensions of the keys and values in the attention. For example, Linformer [52] proposed to split the scaled dot product attention into smaller units through linear projection. Kernelization is a variant of low-rank techniques that focuses on approximating the attention matrix. For example, Performer [53] uses positive orthogonal random features to compress the Softmax attention kernel.

Hardware-assisted attention mechanisms focus on developing hardware-specific techniques to improve the efficiency of attention. For example, FlashAttention [54] reduces the number of memory accesses between the GPU high bandwidth memory (HBM) and the GPU on-chip SRAM when computing the attention module in the large language model (LLM). Unlike the standard attention mechanism that transfers values and results multiple times between the HBM and SRAM, FlashAttention combines all attention operations into a single kernel and flattens the weight matrix into smaller patches to better fit in the small SRAM. Therefore, only one communication is required to process each attention block, which significantly improves the efficiency of processing the entire attention block.

Based on FlashAttention, FlashAttention-2 [55] minimizes the computational FLOPs of non-matrix multiplication, optimizes the online softmax technique, introduces parallelism of sequence length, and improves the workload division between warps within each thread block to reduce synchronization, thereby achieving optimized performance on modern GPU chips.

Nevertheless, FlashAttention-2 has not yet fully utilized the new features of the latest hardware. Compared with the optimized matrix multiplication library GEMM, its utilization on the H100 GPU is only 35%, while the latter has 80–90%. Therefore, FlashAttention-3 [56] adopts three techniques to accelerate the attention mechanism on the Hopper GPU, mainly by exploiting the asynchrony of the hardware-specific Tensor Cores and Tensor Memory Accelerator (TMA) for matrix multiplication: (1) overlapping the overall computation and data movement through a warp-specific pipeline; (2) hiding the softmax under asynchronous block-wise GEMM, where block-wise matrix multiplication and softmax operations are interleaved; and (3) leveraging hardware-accelerated FP8 low-precision GEMM, using block quantization and incoherent processing techniques to mitigate accuracy loss.

PagedAttention [57] is a commonly used dynamic memory allocation method in LLM service systems. It allows GPU memory to be allocated on demand, thereby alleviating the key-value cache fragmentation problem, which in previous systems severely affects batch size (and thus reduces throughput). However, when trying to allocate physical memory at runtime, PagedAttention ends up changing the virtual memory layout of the key-value cache from contiguous to non-contiguous. Such a design leads to cumbersome programming and performance overhead.

vAttention [58] proposes to store the key-value cache in contiguous virtual memory without committing physical memory in advance. It takes advantage of CUDA's support for low-level virtual memory APIs to avoid the software complexity of storing the key-value cache.

The K-V cache technique, as the name implies, involves storing and reusing previous key-value (K-V) pairs in a multi-head self-attention (MHSA) block [44]. This technique significantly optimizes the decoder latency of autoregressive implementations and has been widely used in reasoning systems deployed on large language models (LLMs). As the length of tokens generated during LLM reasoning increases, the required K-V cache size increases significantly, resulting in large memory consumption and long reasoning latency. Therefore, reducing the size of the K-V cache is key to improving reasoning efficiency.

The earliest implementations usually allocated storage space in advance based on the preset maximum length of each request. However, this approach can lead to a serious waste of storage resources when request generation is terminated prematurely. Currently, optimized service systems generally use paging to manage the K-V cache, thereby reducing the waste of redundant K-V cache memory.

Existing K-V cache optimization techniques can generally be divided into two categories [44]: The first category of techniques is to compress the K-V cache, such as [59], which empirically studies the impact of per-channel quantization and other types of quantization (such as quantization before rotation position embedding), and then proposes KVQuant, which combines these quantization methods to quantize LLaMA's K-V cache to 3 bits; the second category of techniques is to evict some K-V pairs from the cache, such as [60], which proposes a "significant persistence" hypothesis that only tokens that are critical in the early stages will have a significant impact on subsequent stages. Based on this hypothesis, Scissorhands is designed to significantly reduce the K-V cache without affecting the quality of the model.

2.8 Summary

This section mainly provides an overview of the basic theories involved in autonomous driving research and development, including the most important computer vision theory and deep learning theory, as well as other related image processing methods, optimization theory, and machine learning algorithms. Besides, it analyzes the efficient Transformer' architecture and implementation, which is involved in large-scale models. Chapter 3 will provide a brief explanation and discussion of the hardware platform for autonomous driving.

References

1. MALIK J et al., The three R's of computer vision: Recognition, reconstruction and reorganization, Pattern Recognition Letters, March, 2016, 72(3): 4–14.
2. SZELISKI R, Computer Vision: Algorithms and Applications, Springer, September, 2010.
3. ZHANG Z,A Flexible New Technique for Camera Calibration, IEEE Trans. on Pattern Analysis and Machine Intelligence, 2000, 22(11): 1330–1334.
4. HARTLEY R, ZISSERMAN A, Multiple View Geometry in Computer Vision, Cambridge Press, 2003.
5. NISTÉR D, An efficient solution to the five-point relative pose problem, IEEE Trans. on Pattern Analysis and Machine Intelligence, 2004, 26(6): 750–770.
6. RAMANATH R et al., Color image processing pipeline, Signal Processing Magazine, 2005, 22(1): 34–43.
7. NOCEDAL J, WRIGHT S J, Numerical Optimization, Springer, 1999.
8. MITCHELL T M, Machine Learning, WCB, McGraw-Hill, 1997.
9. OUALI Y, HUDELOT C, TAMI M, An Overview of Deep Semi-Supervised Learning, arXiv preprint arXiv:2006.05278,2020
10. LIU X, ZHANG F, HOU Z, MIAN L, WANG Z, ZHANG J, TANG J, Self-supervised Learning: Generative or Contrastive, arXiv preprint arXiv:2006.08218,2021
11. REN P,XIAO Y, CHANG X, HUANG P-Y, LI Z, GUPTA B B., CHEN X,WANG X, A Survey of Deep Active Learning, arXiv preprint arXiv:2009.00236,2020
12. ZHANG L, GAO X, Transfer Adaptation Learning: A Decade Survey, arXiv preprint arXiv:1903.04687,2020
13. XIAN Y, LAMPERT C H., SCHIELE B, AKATA Z,Zero-Shot Learning-A Comprehensive Evaluation of the Good, the Bad and the Ugly, arXiv preprint arXiv:1707.00600,2017
14. WANG Y, YAO Q, KWOK J, NI L M.,Generalizing from a Few Examples: A Survey on Few-Shot Learning, arXiv preprint arXiv:1904.05046,2020
15. VASWANI A., SHAZEER N., PARMAR N., USZKOREIT J., JONES L., GOMEZ A. N., KAISER L., and POLOSUKHIN I., Attention is all you need, Proceedings of the 31st Int. Conf. on Neural Information Processing Systems, Long Beach, California, December 2017: 6000–6010.
16. GENG C, HUANG S-J, CHEN S, Recent Advances in Open Set Recognition: A Survey, arXiv preprint arXiv:1811.08581,2020
17. BURGES C, A Tutorial on Support Vector Machines for Pattern Recognition, Data Mining and Knowledge Discovery, 6, 1998, 2(2): 121–167.
18. GOODFELLOW I et al., Deep Learning, MIT press, 2016.
19. Krizhevsky, Alex, Ilya Sutskever, and Geoffrey E. Hinton. "Imagenet classification with deep convolutional neural networks." Advances in neural information processing systems 25 (2012).
20. Szegedy, Christian, Wei Liu, Yangqing Jia, Pierre Sermanet, Scott Reed, Dragomir Anguelov, Dumitru Erhan, Vincent Vanhoucke, and Andrew Rabinovich. "Going deeper with convolutions." In Proceedings of the IEEE conference on computer vision and pattern recognition, pp. 1–9. 2015.
21. Sarwinda, Devvi, Radifa Hilya Paradisa, Alhadi Bustamam, and Pinkie Anggia. "Deep learning in image classification using residual network (ResNet) variants for detection of colorectal cancer." Procedia Computer Science 179 (2021).

22. Huang, Gao, et al. "Densely connected convolutional networks." Proceedings of the IEEE conference on computer vision and pattern recognition. 2017.
23. Sakib, Shadman, Nazib Ahmed, Ahmed Jawad Kabir, and Hridon Ahmed. "An overview of convolutional neural network: Its architecture and applications." (2019).
24. Gallant, Stephen I. "Perceptron-based learning algorithms." IEEE Transactions on neural networks 1, no. 2 (1990): 179–191.
25. Desai, Meha, and Manan Shah. "An anatomization on breast cancer detection and diagnosis employing multi-layer perceptron neural network (MLP) and Convolutional neural network (CNN)." Clinical eHealth 4 (2021): 1–11.
26. Wang, Yingying, Yibin Li, Yong Song, and Xuewen Rong. "The influence of the activation function in a convolution neural network model of facial expression recognition." Applied Sciences 10, no. 5 (2020): 1897.
27. Alom, Md Zahangir, Tarek M. Taha, Chris Yakopcic, Stefan Westberg, Paheding Sidike, Mst Shamima Nasrin, Mahmudul Hasan, Brian C. Van Essen, Abdul AS Awwal, and Vijayan K. Asari. "A state-of-the-art survey on deep learning theory and architectures." electronics 8, no. 3 (2019): 292.
28. HONG Y et al. How Generative Adversarial Networks and Their Variants Work: An Overview, arXiv preprint arXiv:1711.05914,2017
29. Hong, Yongjun, Uiwon Hwang, Jaeyoon Yoo, and Sungroh Yoon. "How generative adversarial networks and their variants work: An overview." ACM computing surveys (CSUR) 52, no. 1 (2019): 1–43.
30. Radford, Alec, Luke Metz, and Soumith Chintala. "Unsupervised representation learning with deep convolutional generative adversarial networks." arxiv preprint arxiv:1511.06434 (2015).
31. Jia, Tan, Li Zhiyi, Y. A. N. G. Huan, Z. H. A. O. Rongxiang, and J. U. **. "A Multi-Level Collaborative Load Forecasting Method for Distribution Networks Based on Distributed Optimization [J]." Journal of Shanghai Jiaotong University 55, no. 12 (2021): 1544–1553.
32. Zhang, J. Y., G. He, Z. Dai, and Y. Liu. "Named entity recognition of enterprise annual report integrated with BERT [J]." Journal of Shanghai Jiao Tong University 55, no. 02 (2021): 117–123.
33. HAN K, WANG Y, CHEN H, et al., A Survey on Visual Transformer, arXiv preprint arXiv:2012.11556, 2020.
34. KHAN S, NASEER M, HAYAT M, ZAMIR S W, KHAN F S, and SHAH M, Transformers in Vision: A Survey, arXiv preprint arXiv:2101.01169, 2021.
35. Chen, **g**g, Qirong Mao, and Dong Liu. "Dual-path transformer network: Direct context-aware modeling for end-to-end monaural speech separation." arxiv preprint arxiv:2007.13975 (2020).
36. Ghosh, Subhadip, Endalk Y. Gebru, Chandramouli V. Kashyap, Ramesh Harjani, and Sachin S. Sapatnekar. "Accelerating OTA Circuit Design: Transistor Sizing Based on a Transformer Model and Precomputed Lookup Tables." In 2025 Design, Automation & Test in Europe Conference (DATE), pp. 1–7. IEEE, 2025.
37. ZHANG Q et al, Recent advances in convolutional neural network acceleration, arXiv preprint arXiv:1807.08596,2018
38. CHENG Y et al, A Survey of Model Compression and Acceleration for Deep Neural Networks, arXiv preprint arXiv:1710.09282,2017
39. HOWARD A et al, MobileNets: Efficient Convolutional Neural Networks for Mobile Vision Applications, arXiv preprint arXiv:1704.04861,2017
40. SANDLER M et al, MobileNetV2: Inverted Residuals and Linear Bottlenecks, arXiv preprint arXiv:1801.04381,2018
41. Y Tay et al., "Efficient Transformers: A Survey", arXiv:2009.06732, 2020.
42. B Zhuang et al., "A Survey on Efficient Training of Transformer", arXiv:2302.01107, 2023.
43. S Kim et al., "Full Stack Optimization of Transformer Inference: a Survey", 2302.14017, 2023.
44. Z Wan et al., "Efficient Large Language Models: A Survey", arXiv:2312.03863, 2023.
45. X Miao et al., "Towards Efficient Generative Large Language Model Serving: A Survey from Algorithms to Systems", arXiv:2312.15234, 2023.

46. A Chavan et al., "Faster and Lighter LLMs: A Survey on Current Challenges and Way Forward", arXiv:2402.01799, 2024.
47. K Chang et al., "Efficient Prompting Methods for Large Language Models: A Survey", arXiv:2404.01077, 2024.
48. Z Zhou et al., "A Survey on Efficient Inference for Large Language Models", arXiv:2404.14294, 2024.
49. Y Jin et al., "Efficient Multimodal Large Language Models: A Survey", arXiv:2405.10739, 2024.
50. N Shazeer, "Fast Transformer Decoding: One Write-Head is All You Need" (Multi-query Attention, MQA), arXiv:1911.02150, 2019.
51. J Ainslie et al., "GQA: Training Generalized Multi-Query Transformer Models from Multi-Head Checkpoints", arXiv:2305.13245, 2023.
52. S Wang et al., "Linformer: Selfattention with linear complexity", arXiv:2006.04768, 2020.
53. K M Choromanski, V Likhosherstov, D Dohan, and et al. "Rethinking attention with Performers", arXiv:2009.14794, 2020.
54. T Dao et al., "FlashAttention: Fast and Memory-Efficient Exact Attention with IO-Awareness", arXiv:2205.14135, 2022.
55. T Dao, "FlashAttention-2: Faster Attention with Better Parallelism and Work Partitioning", arXiv:2307.08691, 2023.
56. J Shah et al., "FlashAttention-3: Fast and Accurate Attention with Asynchrony and Low-precision", arXiv:2407.08608, 2024.
57. W Kwon et al., "Efficient Memory Management for Large Language Model Serving with PagedAttention", arXiv:2309.06180, 2023.
58. R Prabhu, A Nayak, J Mohan, R Ramjee, and A Panwar. "vAttention: Dynamic memory management for serving LLMs without PagedAttention", arXiv:2405.04437, 2024.
59. C Hooper, S Kim, H Mohammadzadeh, and et al., "Towards 10 million context length LLM inference with KV cache quantization", arXiv:2401.18079, 2024.
60. Z Liu, A Desai, F Liao, and et al. "Scissorhands: Exploiting the persistence of importance hypothesis for LLM KV cache compression at test time". arXiv:2305.17118, 2023.
61. LANGE M D, ALJUNDI R, MASANA M, PARISOT S, JIA X, LEONARDIS A, SLABAUGH G, TUYTELAARS T, A continual learning survey: Defying forgetting in classification tasks, arXiv preprint arXiv:1909.08383,2021

Chapter 3
Hardware Platform for Autonomous Driving

The main difference between autonomous vehicles and traditional vehicles is that they are equipped with a large number of new sensors and powerful computing hardware that supports autonomous driving algorithms. The autonomous driving hardware platform discussed here mainly includes vehicle control parts, steer-by-wire chassis, sensors, and computing platforms.

In addition, with the surge of "Software Defined Vehicles" (SDV), the Electronic and Electrical Architecture (EEA) has also seen new evolutions and developments, including the emergence of skateboard chassis. The development of intelligent vehicles conforms to the current trend of the "new four modernizations" (electrification, networking, intelligence, and sharing) of automobiles, and also promotes the in-depth application of computer technology, communication network technology, artificial intelligence, and big data cloud computing in the automotive industry.

Section 3.1 introduces sensors, namely cameras, lidar, millimeter-wave radar, ultrasonic radar, inertial measurement units (IMU), Global Positioning System (GPS), and Vehicle-to-Everything (V2X); Sect. 3.2 briefly describes vehicle control, including VCU/ECU/MCU, CAN bus, ESP, and EPS; Sect. 3.3 mainly discusses current commercial computing platforms, such as Intel Mobileye's EQ series, NVIDIA's Drive Xavier and Orin, TI's TDA4VM, and Qualcomm's Ride; Sect. 3.4 introduces steer-by-wire chassis technology; Sect. 3.5 mainly discusses the technology of automotive electronic and electrical architecture.

3.1 Sensors

Sensors include cameras, lidar, millimeter-wave radar, ultrasonic radar, inertial measurement units, Global Positioning System (GPS), and Vehicle-to-Everything (V2X) systems.

Y. Huang and Z. Yang, *Autonomous Driving Handbook*,
https://doi.org/10.1007/978-981-95-3725-9_3

3.1.1 Camera

The camera is an optical sensor that converts the light signal of a scene into an electronic signal through photoelectric conversion. It is divided into CCD (charge-coupled device), as shown in Fig. 3.1 and CMOS (complementary metal oxide semiconductor) types, as shown in Fig. 3.2.

CCD cameras are generally global shutters, that is, they extract all pixels of the image at once. The global shutter is a scene exposed at the same time, all pixels collect light at the same time, and are exposed at the same time. At the beginning of the exposure, light is collected, and at the end of the exposure, the light circuit is closed. The data read at this time is a photo. Most CMOS cameras are rolling shutters, that is, they extract the pixels of the image line by line or column by column. It is a line-by-line exposure method. At the beginning of the exposure, it scans line by line for exposure until all pixels are exposed.

In addition, cameras can form a monocular system, such as the Mobileye vision system, or a multi-eye system (commonly binocular, such as Ambarella's CV-1/2 system and Tesla's 8-camera system. In addition, a special multi-eye system is the

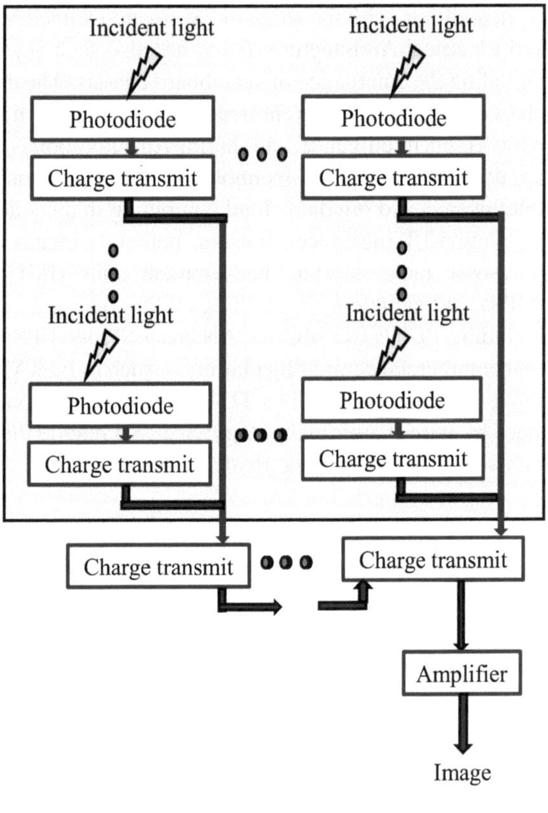

Fig. 3.1 Internal structure of CCD

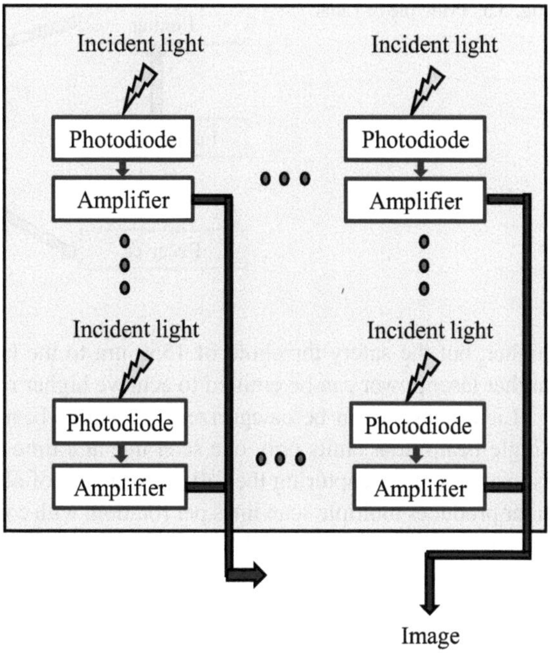

Fig. 3.2 Internal structure of CMOS

surround view system, which is a road stitching image composed of fisheye cameras, mainly used for automatic parking.

The main disadvantage of the camera system is that the illumination is poor at night or in tunnels, and the contrast also drops a lot. High Dynamic Range (HDR) imaging technology can overcome some problems of insufficient or saturated illumination. The Mobileye vision system uses an HDR camera. There are developed infrared thermal lenses, but cost and power consumption are still concerns.

3.1.2 *Lidar*

Lidar (Light Detection and Ranging) is an active optical ranging sensor. Its emitter emits a beam of laser light. After encountering an object, it is diffusely reflected and returns to the receiver, as shown in Fig. 3.3. The distance of the object can be calculated by multiplying the time difference between sending and receiving signals by the speed of light and then dividing by 2.

The wavelength of lidar is divided into two types: One is within 1000 nm, the typical value is 905 nm, silicon can be used as a receiver, the cost is low and the product is mature; the other is 1000–2000 nm, the typical value is 1550 nm. Silicon cannot detect this band, and Ge or InGaAs detectors are needed. The cost will be

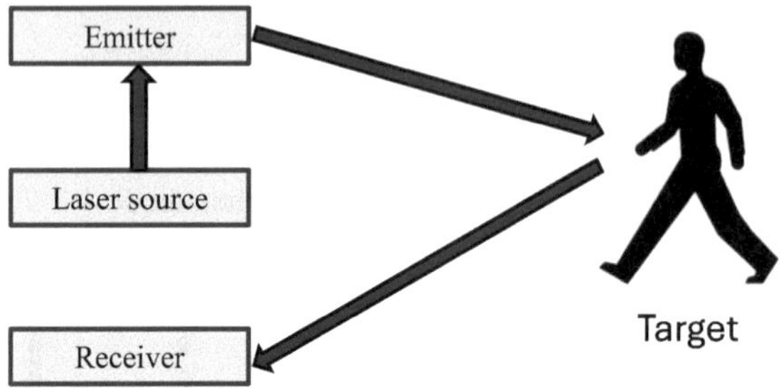

Fig. 3.3 Principle of Lidar

higher, but the safety threshold of 1550 nm to the human eye is also higher, so a higher laser power can be emitted to achieve higher ranging sensitivity.

Lidar sensors can be categorized into single-beam and multi-beam devices. A single-beam lidar emits only one scan line at a time and generates 2D data, which is insufficient for capturing the full 3D structure of objects. In contrast, multi-beam lidar produces multiple scan lines per rotation, with commercially available products ranging from 4, 8, 16, 32, 64, up to 128 beams. The number of beams is directly correlated with the cost of the device; thus, without considering beam count, it is impossible to estimate lidar cost. For example, Velodyne's latest mechanical lidar achieves 128 beams, whereas the lidar adopted in the Audi A8 provides only 4 beams, leading to significant cost differences. In practice, lidars with fewer than 8 beams (commonly 1 or 4) are typically mounted around the body of autonomous vehicles, while higher-resolution lidars with at least 16 beams (commonly 16, 32, 64, or even 128) are usually roof-mounted to provide richer 3D perception.

From the working mode, lidar is divided into scanning and non-scanning types. Scanning lidar measures distance by point-by-point scanning, and is further divided into mechanical, rotating mirror, micro-electro-mechanical system (MEMS), prism, and phased array types. Non-scanning lidar is flash lidar, which emits a planar array of light for distance measurement like a flashlight.

The mechanical lidar rotates the optical structure as a whole through the motor. The technical points include the number of system channels, measurement range, spatial angle resolution, system integration and reliability, etc. The advantages of mechanical lidar are fast scanning speed, small receiving field of view, strong light interference resistance, and high signal-to-noise ratio; the disadvantages are high price, complex optical path adjustment and assembly, long production cycle, and low reliability.

The rotating mirror lidar is different from the mechanical lidar. Its laser emission module and receiving module are stationary, only the scanning mirror is mechanically rotating. The laser unit emits a laser to the rotating scanning mirror, which is deflected and emitted forward (scanning angle is 145°), and the light reflected by the object is received by a detector on one side after the optical system.

The working principle of MEMS (Micro-Electro-Mechanical System) lidar is as follows: a laser source emits high-frequency pulsed light, which is collimated into a beam with a small divergence angle. The two-dimensional MEMS scanning mirror then adjusts its deflection angles to steer the outgoing beam across the X and Y axes, enabling point-by-point scanning of the scene. The echo signal reflected by the target is collected by the receiving optical system and directed to an APD (avalanche photodiode) array detector. Based on the ToF (Time of Flight) principle, the control system calculates the round-trip flight time of the laser pulse with high precision, thereby determining the distance to the target.

The prism lidar emits a laser from the pulse laser diode (PLD) of the transceiver module, which becomes parallel light through the reflection mirror and convex lens. The two rotating prisms of the scanning module change the optical path, causing the laser to be emitted from a certain angle. The laser hits the object, reflects back from the original optical path, and is received by the avalanche photodiode (APD). Compared with the MEMS scheme, it can achieve a large light aperture, and the distance will be measured further.

Optical phased array (OPA) technology can adjust the phase relationship of each phased unit by applying voltage, and use the coherent principle to achieve the deflection of the emitted beam, thereby completing the system's scanning measurement of a certain range of space.

Flash lidar mainly emits a large piece of laser covering the detection area in a short time, and then uses a highly sensitive array receiver to complete the drawing of the surrounding environment.

Solid-state lidar is widely regarded as the future development trend of lidar technology, owing to its fast scanning speed, high accuracy, good controllability, and compact form factor. However, the technology is still in the process of maturing and faces several challenges before large-scale deployment. While the advantages of lidar are evident–most notably its ability to generate high-precision 3D point cloud representations of objects–it also exhibits the following limitations.

1. The laser emitted by the lidar itself is not encoded, and it is impossible to identify whether this beam of light is emitted by the laser emitter or is an interference signal. Therefore, hackers can deliberately fabricate illusions by simulating vehicle and pedestrian signals, causing the vehicle to be forced to slow down or brake.
2. The lidar emits in a radiating shape, so the farther the place, the fewer laser points received, and the point cloud produced is sparser, and it is more difficult to recognize obstacles the farther away.
3. Some objects have weak surface reflection of lasers, which will cause "black holes;" while some strong reflections will interfere with the reflection of other objects, such as a large amount of water on the road surface.
4. The refresh rate of the lidar during rotation is relatively slow.
5. Lidar is easily affected by weather and atmosphere, such as heavy snow, heavy rain, and foggy weather.

3.1.3 Millimeter Wave Radar

Vehicle-mounted millimeter-wave radar primarily operates on the principle of frequency-modulated continuous wave (FMCW). It emits frequency-modulated millimeter waves and determines the position and relative velocity of a target by analyzing the frequency difference between the transmitted and reflected signals.

The structure of a millimeter wave radar receiver mainly includes a signal processing unit and a controller, including antennas, couplers, mixers, A/D converters, DSPs, filters, low noise amplifiers, etc. The applications of millimeter wave radar are divided into short-range SRR (short-range radar), mid-range MRR (mid-range radar), and long-range LRR (long-range radar). The main transmission waves of millimeter wave radar are 24 GHz and 77 GHz, and the latter has become a trend.

Millimeter wave radar is not affected by common environmental factors such as rain, fog, and dust, so it can be said that millimeter wave radar is the only all-weather sensor in the field of autonomous driving. But it has the following disadvantages.

1. Low angular resolution, high noise, and inability to determine the boundaries of the target.
2. Sensitive to metal, but weak in detecting the human body.
3. It can only obtain the target distance and radial speed on the 2D fan surface of the radar scan, without height information, and cannot accurately reconstruct the structure and shape of the scene.

Note: If you want to get the structure and shape of the scene, you need to use synthetic aperture radar imaging technology, but the cost of this radar is high.

The working range of the millimeter wave radar product Delphi ESR-25: switching between short-range wide-angle (60 m, 90°) and long-range narrow-angle (175 m, 22°).

Recently, 4D millimeter-wave radar has emerged as a research hotspot due to its ability to significantly enhance perception capabilities. Unlike traditional 3D radar tensors (RT), the 4D RT of 4D millimeter-wave radar incorporates power measurements across four dimensions: Doppler, range, azimuth, and elevation. By preserving full 3D spatial information, 4D millimeter-wave radar enables accurate 3D perception, supporting tasks such as 3D object detection that were previously dominated by lidar.

The emergence of the new generation of 4D radar has made up for the low resolution of traditional automotive radar, and provides height measurement, which is very suitable for high-level autonomous driving applications. In addition, it also provides some other low-level features, such as radar cross-section (RCS) or signal-to-noise ratio (SNR).

For the new generation of 4D imaging radar, the 4D point cloud will be the main output format, containing spatial and speed information.

Radar sensors have been a part of the automotive field for more than 20 years due to their robustness to harsh weather conditions and the ability to measure speed. The latest advances in high-definition (HD) imaging radar have reduced the angular resolution to less than 1°, thus approaching laser scanning performance. However, how to handle the increased data volume of HD imaging radar and how to estimate angular positions are still challenges for engineering implementation.

3.1.4 Ultrasonic Radar

Ultrasonic radar is similar to the previously mentioned lidar and millimeter wave radar, but the transmitted wave is sound wave, the ranging range is short, close to about 6 cm, far to about 3 m, and the band is 50 kHz. The structure of an ultrasonic receiver, including a transmission wave generator, a receiving amplifier, and a timer.

The general working frequency of ultrasonic radar 50 Hz, the cost is low, the ranging is short, and it is mostly used in parking situations.

3.1.5 Inertial Navigation

Based on the previous vehicle posture and known speedometer to calculate the current posture, it is called dead reckoning (DR), and inertial navigation is a sensor to implement DR.

The vehicle-mounted inertial navigation system (INS) uses motion and rotation sensors to calculate the vehicle's posture and position, and the inertial measurement unit (IMU) is the main member of the INS, generally including three orthogonal gyroscopes and three orthogonal accelerometers, used to measure angular velocity and linear acceleration respectively. The typical structure of an IMU. The most important calculation part is the integration module. The integration module is the final output link. Integration is equivalent to summation and will accumulate errors. Therefore, this is also the reason for the accumulation of IMU errors. Therefore, the general vehicle-mounted DR system bundles the IMU and GPS together. Otherwise, it needs lidar or a camera to provide a way for error correction.

3.1.6 GPS

The Global Navigation Satellite System (GNSS) refers to the provision of localization and timing signals through a globally covered satellite system. The basic operating idea is that the receiver measures the time of arrival of the satellite signal, while recording the satellite's transmission time, thus obtaining the propagation time of the signal in space, i.e., estimating the distance from the GNSS receiver to the satellite, known as the pseudo-range.

The GNSS receiver estimates its own position from the pseudo-range using multi-point MLAT (multi-point localization) technology, i.e., calculating from multiple satellite measurement data. The accuracy of position estimation depends on the accuracy of the pseudo-range and the geometric distribution of the satellites. The current GNSS includes the American GPS, Russian GLONASS, European Galileo system, and Chinese Beidou system.

Differential GPS (dGPS) improves GPS localization accuracy, mainly through a reference base station system at a known location. Each base station can broadcast the deviation between the GPS position and its internally calculated position, so that users receiving the signal can correct their own position accordingly.

RTK (real-time kinematics) is a type of differential GPS method that uses signal carrier phase signals. Depending on a single reference base station or interpolated virtual base station, it can correct the position in real time, achieving centimeter-level accuracy (note: a typical RTK base station covers around 10–20 km).

3.1.7 Vehicle Networking

Vehicle-to-Everything (V2X) [1] communication is the transmission of information from a vehicle to any entity that may affect the vehicle, and vice versa. It is a type of vehicular communication system that includes other more specific types of communication, such as V2I (Vehicle-to-Infrastructure), V2N (Vehicle-to-Network), V2V (Vehicle-to-Vehicle), V2P (Vehicle-to-Pedestrian), V2D (Vehicle-to-Device), and V2G (Vehicle-to-Grid).

The real-time communication of V2X can be applied in road safety and is an important auxiliary means for autonomous driving, such as forward collision warning, lane change warning/blind spot warning, emergency electronic brake light warning, intersection movement assistance, emergency vehicle approach, road work warning, queuing (automatic), etc.

Based on the underlying technology used, V2X communication technology is divided into two types: WLAN-based and cellular-based.

The standardization of WLAN-based V2X has replaced the standardization of cellular-based V2X systems. In 2012, IEEE first released the WLAN-based V2X (IEEE 802.11p) specification, supporting direct communication between vehicles (V2V) and between vehicles and infrastructure (V2I). This technology is known as Dedicated Short Range Communication (DSRC). DSRC uses the basic radio communication provided by IEEE 802.11p. In 2016, 3GPP released the V2X specification based on LTE (Long Term Evolution) as the underlying technology, commonly referred to as "Cellular V2X" (C-V2X), to distinguish it from V2X technology based on IEEE 802.11p. In addition to direct communication (V2V, V2I), C-V2X also supports wide-area communication on cellular networks (V2N). Compared with V2X systems based on IEEE 802.11p, C-V2X has two main advantages: additional communication modes and a native migration path to 5G (note: 5G, also known as the fifth-generation cellular communication technology, is a general term for solutions integrated with various new wireless access and optimized existing wireless technologies).

V2X on-board units or devices (OBU/E) include antennas, localization systems, processors, vehicle operating systems, and human-machine interfaces (HMI). V2X roadside units or devices (RSU/E) include antennas, localization systems, processors, vehicle infrastructure interfaces, and other interfaces.

3.2 Vehicle Control

Vehicle control mainly includes various control units and on-board data buses, as well as the Electronic Stability Program (ESP) and Electric Power Steering system (EPS).

3.2.1 VCU/ECU/MCU/HCU

The Vehicle Control Unit (VCU) mainly coordinates the operation of components such as the gearbox, power battery, motor, and engine. The VCU is the core of the whole vehicle control of electric vehicles, undertaking tasks such as data exchange, safety management, and power management.

The Electronic Control Unit (ECU), also known as the on-board computer or car computer, consists of a microprocessor (CPU), memory (ROM, RAM), input/output (I/O) interface, analog-to-digital converter (A/D), and shaping, driving, etc. The various functions that different ECUs are responsible for in the car, such as the rear blower, tracking system, overhead console, advanced forward lighting, radio navigation system, electronic adjustable steering wheel, safety sunroof shade, etc.

The Micro Control Unit (MCU) is responsible for the control between various peripheral devices and interface devices, and is also spread across dozens of secondary systems (sub-systems) such as suspension, airbags, door control, audio, etc.

The Hybrid Control Unit (HCU) is responsible for the power distribution and torque management of the system. The HCU generally consists of a boost valve (normally open valve), pressure reducing valve (normally closed valve), return pump, and energy storage device.

3.2.2 CAN Bus

The on-board bus is the data communication network that connects the underlying devices or instruments in the on-board network. There are currently four types of car buses: CAN bus, LIN bus, FlexRay bus, and MOST bus. Here, the CAN bus is used as an example for introduction.

The Controller Area Network (CAN), or CAN bus, is a serial data communication protocol. Its communication interface integrates the physical layer, object layer, and data link layer functions of the CAN protocol, which can complete the frame processing of communication data, including bit stuffing, data block encoding, cyclic redundancy check, priority discrimination, etc.

3.2.3 ESP

The Electronic Stability Program (ESP) is a set of comprehensive strategies for controlling vehicle stability, including the Anti-lock Brake System (ABS) and Anti-slip Regulation (ASR).

① ESP hydraulic device and integrated engine control unit.
② Wheel speed sensor.
③ Steering angle sensor.
④ Yaw rate and lateral acceleration sensor.
⑤ Engine control unit.

ESP is a functional extension based on the ABS and ASR systems. Its purpose is to effectively prevent loss of control under the dynamic limit state of the vehicle while improving vehicle handling performance, thus enhancing the safety and handling of the vehicle.

3.2.4 EPS

The Electric Power Steering (EPS) system is a power steering system that directly relies on the motor to provide auxiliary torque. Compared with the traditional Hydraulic Power Steering (HPS) system, the EPS system eliminates the power steering oil pump, hose, hydraulic oil, transmission belt, and pulley mounted on the engine of the HPS system, and has the advantages of simple adjustment, flexible assembly, and providing steering assistance under various conditions.

EPS mainly consists of a torque sensor, vehicle speed sensor, motor, reduction gear, steering angle sensor, electronic control unit (ECU), etc.

3.3 Computing Platform

There are various computing platforms for autonomous driving, such as CPU, GPU, DSP, ASIC, FPGA, etc. Here, the characteristics of each platform are introduced using the products of several autonomous driving hardware suppliers as examples, including NVIDIA's Drive AGX Xavier and Orin, Intel Mobileye's EQ series chips, TI's TDA4VM, and Qualcomm's Ride.

3.3.1 NVIDIA Drive AGX Xavier and Orin

NVIDIA launched the Xavier platform at CES 2018, which was then claimed to be the world's most powerful SoC system. Xavier can process autonomous driving data from vehicle radar, cameras, lidar, and ultrasonic systems, and is the earliest mass-produced AI chip. Xavier integrates 9 billion transistors, with the CPU using NVIDIA's self-developed 8-core ARM64 architecture (code-named Carmel), and the GPU using 512 CUDA cores of Volta, supporting FP32/FP16/INT8. At 20W power consumption, the single-precision floating-point performance is 1.3 TFLOPS, and the Tensor core performance is 20 TOPS. When unlocked to 30W, it can reach 30 TOPS.

In December 2019, NVIDIA launched the new generation SoC Orin for the ADAS and autonomous driving field [2]. Orin has 17 billion transistors, equipped with NVIDIA's next-generation GPU (i.e., GPU based on the Ampere architecture) and ARM Hercules. The CPU core, capable of providing 200TOPS of computing power, is seven times that of Xavier, with a power consumption of 45 W, mass-produced in 2022, and aimed at L2+ autonomous driving scenarios.

3.3.2 Mobileye's EQ-X

Mobileye's EQ [3] series mainly supported L2, i.e., assisted driving, in the previous EQ1–EQ3. Starting from EQ4, it can support autonomous driving functions above L3. EQ4 can support 8-channel camera data and can receive lidar and millimeter-wave radar data. EQ5 is designed to receive 16-channel camera data and other sensor data, and can provide a software SDK to implement differentiated algorithms.

3.3.3 TI's TDA4VM

The TDA4VM chip [4] is based on the evolved Jacinto 7 architecture, aimed at ADAS and autonomous driving applications. TI has a strong background in signal processing chip development and market experience. TDA4VM combines high-performance computing, deep learning engines, signal and image processing dedicated accelerators, and meets functional safety requirements. The ADAS and AV platforms built with TDA4VM support various sensor modes in centralized ECUs or independent sensors.

The core of the TDA4VM chip includes a next-generation DSP with scalar and vector cores, dedicated deep learning and traditional algorithm accelerators. These devices also include ARM and GPU processors for general computing, integrated ISPs, video codecs, Ethernet hubs, and independent MCU islands.

TDA4VM supports an 8 million pixel (MP) camera, can simultaneously operate 4–6 3 million pixel cameras, and can also perform deep fusion (post-fusion) of various other sensor data such as millimeter-wave radar, ultrasonic radar, and lidar on a single chip platform. For running deep learning models, it can support 8TOPS computing power. The power consumption of TDA4VM is about 5–20 W.

3.3.4 Qualcomm's Snapdragon Ride

Qualcomm, known for its mobile chip products, has designed a Snapdragon Ride platform [5], which is a vehicle-cloud ADAS infrastructure, including:

- Low-power SoC platform, including high-performance computing and AI engines with built-in functional safety.
- Out-of-the-box visual solutions that support front and surround cameras.
- A rich toolset that provides a simulation and continuous learning framework, where the DNN (Deep Neural Network) used can be trained faster in the cloud, and the newly trained network can be redeployed from the cloud to the device.

A key component of the Snapdragon Ride hardware platform is the Snapdragon Ride SoC, which has hardware modules that support ADAS applications, including:

- Machine learning and visual processing blocks for neural network perception.
- Visual accelerators for image preprocessing and post-processing applications.
- Graphics accelerator, support for visualization and general (GP) GPU parallel processing.
- Video processor that supports streaming.
- System safety manager compliant with ISO 26262.
- Multi-core ARM CPU for running general driving strategy applications.

The Snapdragon Ride Vision SoC supports multiple cameras with resolutions up to 8MP, including front-view and surround-view cameras, provides Ethernet and CAN communication, can process radar and lidar data; this vision system can also integrate various other components, including parking assistance, driver monitoring, map crowdsourcing, etc.; the safety processor provides a safety island for system safety design, quick start-up, and safety diagnosis, and also ensures vehicle network security.

3.4 Steer-By-Wire Chassis

The chassis system accounts for about 10% of the total vehicle cost, and the steer-by-wire chassis is a key component of autonomous driving, because if it does not get its support, the control signal output by autonomous driving may not be correctly executed.

Steer-by-wire (drive-by-wire or X-by-wire) replaces mechanical [6], hydraulic, or pneumatic connections with electrical wires (electrical signals), so it does not need to rely on the driver's force or torque input.

The by-wire chassis mainly includes the braking system, steering system, drive system, and suspension system. It is characterized by fast response speed, high control precision, and strong energy recovery, making it an indispensable component for autonomous driving.

The safety of by-wire chassis technology is the most basic and core element for autonomous driving. Although the efficiency of pure mechanical control was low in the past, its reliability was high. By-wire technology is suitable for autonomous driving, but it also faces the risks brought by electronic software failures. Only by achieving dual or even multiple redundancies in functionality can it ensure that its basic functions can still be realized in the event of a failure.

The most mainstream test development car for global L4 autonomous driving startups is the Lincoln MKZ, due to its high-performance, high-precision by-wire capabilities, ease of use for reverse engineering modifications, and the presence of mature by-wire modification service providers AS and DataSpeed. Together, they provide a stable and easy-to-use platform for autonomous driving startups.

The electric vehicle chassis has a three-electric system (battery, motor, electric control), energy recovery and thermal management system, by-wire steering and braking system, suspension system, etc. The skateboard chassis is a modular arrangement of steering, braking, three-electric, and suspension modules installed on the chassis, which can change the required modules according to the vehicle model, thereby shortening the development cycle. Because of its skateboard-like shape, it is named the skateboard chassis. The skateboard chassis has high flexibility and can meet the needs of the autonomous driving system.

The core advantage of the skateboard chassis is software definition and software-hardware decoupling. It can simplify the mechanical structure, reduce the boundaries and limitations brought by components and hardware, and achieve safer chassis functions through the upgrade of the distributed drive algorithm of the hub motor. The chassis can truly be completely defined by software, specifically reflected in the distributed drive algorithm. Because the distributed drive algorithm can completely liberate the chassis, turning it from a traditional car chassis into a real wheeled robot. Behind this, a set of algorithm-driven design and flexible manufacturing system is needed, and the implementation of distributed manufacturing of multiple categories and small batches, in order to fully release the modularity and flexibility advantages of the skateboard chassis.

Startups such as Arrival, Rivian, Canoo, and REE in Europe and America, as well as Chinese startups Upower (Youpao Technology) and PIX Moving, have all announced the adoption of the skateboard chassis. Meanwhile, car companies such as Toyota, Hyundai, and Citroen, as well as Tier 1 companies like Schaeffler and ZF, have all begun to develop skateboard chassis.

3.5 Electronic and Electrical Architecture

With the intelligent development driven by the trends of networking, intelligence, sharing, and electrification (CASE) in the automotive industry, the distributed architecture of cars is shifting towards a centralized architecture. E2A is the overall layout plan integrating various car sensors, processors, electronic and electrical distribution systems, and software and hardware (including data center platforms and high-performance computing platforms).

Through E2A [7], powertrain, driving information, and entertainment information can be converted into actual physical layouts for power distribution, signal networks, data networks, diagnostics, fault tolerance, power consumption management, and other electronic and electrical solutions. The basic division of automotive E2A can be divided into three eras: distributed multi-MCU network architecture, functional cluster domain controller, zone connection domain controller, and central platform computer (CPC).

Autonomous driving cars need to use a large number of sensors, and the in-car wiring harness is also growing rapidly. The amount of data that needs to be transmitted in the car is surging, and the wiring harness not only carries more signals, but also requires faster data transmission rates.

Under the new generation E2A platform, autonomous driving has truly decoupled software and hardware through standardized APIs, can obtain stronger computing power support, and the bandwidth of data communication has also been enhanced. Resource allocation and task scheduling are more flexible, and it is also convenient for OTA (over-the-air) [8, 9].

For intelligent car E2A, Aptiv proposes a solution combining "brain" and "nerve," including three parts: central computing cluster, standard power and data backbone network, and power data center. This intelligent car architecture focuses on three characteristics: flexibility, continuous updateability throughout the life cycle, and relative fault tolerance and robustness of the system architecture.

The E2A of Tesla Model 3 [10] is divided into domain control architecture and power distribution architecture. The driving assistance and entertainment system AICM (Autonomous Driving and Entertainment Control Module) control is merged into the CCM (Central Processing Module) central computing module, and the power distribution architecture considers the power redundancy requirements needed by the autonomous driving system.

3.6 Summary

This chapter introduces the hardware platform needed for autonomous driving, including sensors, vehicle control, computing platform, by-wire chassis, electronic and electrical architecture, etc. Chapter 4 will introduce the software platform for autonomous driving.

References

1. Wikipedia, Vehicle to Everything. https://en.wikipedia.org/wiki/Vehicle-to-everything
2. KARUMBUNATHAN L S., NVIDIA Jetson AGX Orin Series: A Giant Leap Forward for Robotics and Edge AI Applications, Nvidia Technical Brief, 2022
3. Mobileye EQ. https://www.mobileye.com
4. TI TDA4. https://www.mouser.com
5. Qualcomm Ride. https://developer.qualcomm.com
6. Pico Technology Inc, Electronic Throttle Control Drive by Wire or Fly by Wire. https://www.picoauto.com
7. MORRIS C, History of Electric Cars using Skateboard Platforms. https://cleantechnica.com
8. Ecotron AI's blog, Evolution of the Automotive Electrical Electronics Architecture. https://ecotron.ai
9. Aptiv, Smart Vehicle Architecture. https://www.aptiv.com
10. LEOPOLD G, Inside Tesla's Model 3. https://www.eetimes.com/inside-teslas-model-3/

Chapter 4
The Software Platform for Autonomous Driving

With the development of autonomous driving technology, more software companies are investing in the development of autonomous driving software platforms, such as NVIDIA's commercial product DriveWorks, and Baidu's open-source Apollo.

In the current trend of automobile intelligence, "software-defined cars" have become an industry consensus. Compared with traditional cars, the goal of intelligent cars is to create a brand new driving experience, which is also the key to differentiation. Therefore, software and algorithms are gradually becoming the core competitive elements of the automotive industry, and the threshold for vehicle manufacturing has evolved from the ability to integrate tens of thousands of parts into the ability to combine and run billions of lines of code.

With the evolution of the automotive electronic and electrical architecture (EEA), the software architecture of intelligent cars is gradually transitioning from signal-oriented architecture to service-oriented architecture. As a software design method and concept, service-oriented architecture has features such as standardized interfaces, loose coupling, and easy expansion.

Autonomous driving software platforms can basically be divided into the following three levels.

- System software, consisting of a hardware abstraction layer, OS kernel, and middleware.
- Functional software, including a general framework for autonomous driving, neural network modules, sensor modules, and related middleware.
- Application software, including algorithms and applications, supports intelligent cockpits and autonomous driving (including perception, localization, decision-making, planning, and control execution).

Currently, in the development of automotive software platforms, the most famous is the AUTOSAR alliance, which has proposed an open-source software development standard. Based on the traditional software development V-model, the automotive

Y. Huang and Z. Yang, *Autonomous Driving Handbook*,
https://doi.org/10.1007/978-981-95-3725-9_4

software design standard aSPICE and the famous functional safety standard ISO 26262 have proposed their own software development processes.

The following will discuss the issues of the autonomous driving software platform in detail. Section 4.1 introduces the AUTOSAR software architecture standard in the automotive industry; Sect. 4.2 introduces the famous V-model in software development; Sect. 4.3 briefly describes an automotive software design process standard aSPICE; Sect. 4.4 introduces the software development process defined by the famous safety standard ISO 26262; Sect. 4.5 introduces the autonomous driving software framework DriveWorks developed by NVIDIA; Sect. 4.6 analyzes the operating system platform used by autonomous driving; Sect. 4.7 introduces the autonomous driving cloud platform; Sect. 4.8 introduces DevOps and MLOps.

4.1 AUTOSAR

Software architecture (SA) was proposed in the 1990s. It is used to structure complex software systems and provide high-level system descriptions. The AUTOSAR (automotive open system architecture) alliance [1], led by BMW and some Tier-1 companies, aims to develop a standardized solution for distributed, function-driven automotive electronic software development. AUTOSAR provides a three-layer software structure.

1. Basic Software Layer (BSW): Standardized software modules, without any functional tasks, only provide necessary services for running the upper functional part.
2. Runtime Environment (RTE): Middleware, extracted from the network topology, supports information exchange within/between ECU between application software components or between application software components and the basic software layer.
3. Application layer: Application software components interact with RTE.

The basic software layer is abstracted into four levels.

1. Services layer.
2. ECU abstraction layer.
3. Microcontroller abstraction layer.
4. Complex driver program.

Adaptive AUTOSAR is a standardized interface for ARA (AUTOSAR runtime for adaptive applications). It includes two parts: the interface of operating system functions and communication middleware. The latter handles data exchange between adaptive AUTOSAR services and various local/remote applications. These interfaces allow OEMs to implement autonomous driving, online OTA software upgrades, Internet of Things (IoT), streaming services, etc.

Compared with classic AUTOSAR, this adaptive AUTOSAR platform can achieve dynamic linking of service-clients at ECU runtime, and provide necessary traffic infrastructure, cloud servers, microprocessors (such as ARM) and HPC (GPU) hardware, etc.

The execution management function group is responsible for the start and stop of the ECU and maintains the application program status. Together with platform health management, it provides necessary resources for the maintenance of applications that start degradation or similar strategies, so that the ECU can respond in a crisis according to a predefined strategy.

The communication middleware enables service-oriented communication between local applications and between local applications and applications in other ECUs, which includes the service interface of AUTOSAR Adaptive.

The biggest difference between classic AUTOSAR platform and the adaptive AUTOSAR platform is that the adaptive AUTOSAR adopts a service-oriented communication mode, uses virtual address space in memory, and the scheduling mode is also dynamic.

4.2 Software Development V-Model

The software development V-model flowchart [3, 4], this model has been widely adopted in the automotive industry, is the reference model for the industrial standard MISRA (Motor Industry Software Reliability Association) C programming specification (with a history of 20 years), and is also the reference model for building the ISO 26262 standard.

The V-model represents a systematic process of software construction plus traceability & verification/validation; the left side of the V-model is the path from requirements-design-implementation, and each step can be divided into multiple subsystems, which can be viewed in parallel (there is only one set of system requirements, but each subsystem can have different designs); the right side of the V-model is the evaluation from small components to system level, iteratively traceable verification (trace & verify/validate) of increasingly larger parts of the system.

According to the V-model, the testing of autonomous driving faces challenges in the following situations.

- The driver is not in the loop.
- Overly complex requirements.
- Use of non-deterministic algorithms.
- Use of inductive learning algorithms.
- Need for fault-safe/fault-operable implementation.

To solve these problems, there are several available methods.

- Through the phased deployment method of sequentially relaxing operating scenarios.
- Adopt a "monitor-driven" structure, separating the most complex autonomous functions from the simpler safety functions.
- Fault injection to get more effective edge case testing.

4.3 aSPICE Software Development Process

There are two recognized software design process standards in the automotive industry: one is aSPICE (automotive software performance improvement and capability determination), and the other is the software part of ISO 26262.

aSPICE originates from the general SPICE (ISO/IEC 15504) standard [2], it is still built on top of the V-model, which means that every process from requirements to source code has corresponding tests.

According to the V-model, the development process on the left side of aSPICE is as follows.

- Analyze customer requirements.
- Map customer requirements to system requirements.
- Decompose system requirements into logical services, including design decisions.
- For each software service, get software requirements from system requirements.
- Break down software requirements into units, manage available resources (memory, CPU time, etc.).
- Design and implement each software unit. The corresponding testing process on the right side of aSPICE is as follows.
- Unit test of the design: Does the code conform to the design? Does it meet non-functional requirements (such as not crashing)?
- Integration test of software architecture: Is the unit composition as a service still valid?
- Software certification test software requirements: Does the service meet its requirements?
- System integration test system architecture: When all services are combined into the whole system, is it effective?
- System certification test system requirements: Does the entire system/car meet the requirements?
- Acceptance test completed by the customer.

In addition to these processes, aSPICE also supports and manages the entire process, including archiving and planning. Agile development can be achieved with aSPICE, flexible methods start with fewer requirements, but still execute the entire V-model process.

aSPICE evaluation includes a rating from "not implemented" (N) to "fully implemented" (F) 0-5 levels.

- Level 0: The process can at most "partially" implement the products defined by aSPICE (source code, requirements, architecture description, test reports, etc.).
- Level 1: "Largely" (L) capable of producing the specified products.
- Level 2: Fully capable of producing working products, and can respond by achieving goals, checking progress, and reacting when goals are missed.
- Level 3: Possessing central standards, all of which must be followed in work and projects.
- Level 4~5: In fact, it's not important, the goal is to establish organizational processes, more "predictable" and "innovative."

4.4 ISO 26262 Software Development Process

ISO 26262 [3] (Part 6) also provides a revised software development process diagram according to the V-model.

The software development process diagram of ISO 26262 (Part 6) stipulates the requirements for automotive application software product development, including:

- General topics of software product development.
- Software requirements specifications.
- Software architecture design.
- Software unit design and implementation.
- Software unit verification.
- Software integration and verification.
- Test requirements verification.

It also stipulates the requirements related to the use of configurable software.

The guidelines for ISO 26262 software architecture design are shown in Table 4.1.

The error detection mechanism at the ISO 26262 software architecture level is shown in Table 4.2.

4.5 NVIDIA DriveWorks Software Platform

NVIDIA's DriveWorks [5] is a software framework based on its autonomous driving hardware platform, including some core algorithms (sensor data processing, such as image processing and point cloud processing, as well as the vehicle's own interface algorithms), deep learning training models (obstacle detection, traffic light detection, traffic sign recognition, lane line detection, intersection detection, drivable road segmentation, map localization) and system calibration tools. On top of DriveWorks,

Table 4.1 ISO 26262 software architecture design guidelines method

Method		ASIL			
		A	B	C	D
a	Graded structure of software components	++	++	++	++
b	Restricted use of software components	++	++	++	++
c	Restricted use of interfaces	+	+	+	+
d	High cohesion within each software component	+	++	++	++
e	Restricted coupling between software components	++	++	++	++
f	Appropriate scheduling characteristics	++	++	++	++
g	Restricted use of interrupts	+	+	+	++

Table 4.2 ISO 26262 software architecture level error detection mechanism

Method		ASIL			
		A	B	C	D
a	Range check of input/output data	++	++	++	++
b	Reasonableness check	+	+	+	++
c	Detection of data errors	+	+	+	+
d	External monitoring facilities	·	+	+	++
e	Control flow monitoring	·	+	++	++
f	Various software designs	·	·	+	++

NVIDIA provides some autonomous driving software SDKs, such as the Drive AV module with perception + planning + control and the Drive IX module for human-machine interaction.

4.6 Vehicle Operating System

The underlying layer of NVIDIA's autonomous driving software platform is its operating system Drive OS. The vehicle operating system must be a real-time operating system (RTOS), especially for the support of autonomous driving (other aspects include in-vehicle entertainment, vehicle networking, and central control display).

RTOS is designed to provide services for real-time data processing applications and usually operates without buffer delay. The processing time (including any OS delay) is required to be 0.1 s or shorter. Real-time systems require processing to be completed within a specified time, otherwise the system will fail. These processes are event-driven or time-shared. Event-driven systems switch between tasks based on priority, while time-sharing systems switch tasks based on clock interrupts. Most RTOS use preemptive scheduling algorithms.

RTOS has advanced algorithms for scheduling, and the flexibility of the scheduler can achieve a wider range of process orchestration. The key factors of a real-time operating system are the minimum interrupt delay and the minimum thread switching delay; the importance of a real-time operating system lies in the response speed, not the amount of work performed.

4.6.1 ROS

The Robot Operating System (ROS) [6] is a mature and flexible control programming framework in the field of robotics. The entire ROS can be a distributed computing structure, that is, each computer executes various parts of the control process, but as an entity, it is the robot in action.

ROS is suitable for autonomous vehicles because there is already a large amount of related open-source code, complete visualization tools, so it is easy to start an autonomous driving project in the ROS environment. The disadvantages of ROS are as follows.

1. Single point of failure. All ROS applications are completely dependent on a software component roscore (a Python script), which handles all coordination between different parts of ROS.
2. Insecurity. Currently, ROS does not support security mechanisms to prevent third parties from entering the ROS network, nor does it read communications between nodes.

The entire system is composed of several nodes, and a node communicates with other nodes through a publish/subscribe method. For example, a sensor acts as a node of ROS, it can publish the information it obtains in the form of an information stream, and the published information can be obtained by other nodes (such as perception, planning or control units).

Based on the problems existing in ROS, ROS2 was born (the previous version was called ROS1). Compared with ROS1, ROS2 adopts (or plans to adopt) the following strategies.

- Support for multiple robots.
- Support for small embedded devices and microcontrollers.
- Real-time system: Support for real-time control, including real-time communication between processes and machines.

- Support for non-ideal network environments: low-quality, high-latency network environments.
- Support for large-scale construction, development, and deployment based on ROS.

ROS2 uses DDS (data-distribution service) based on RTSP (real-time publish-subscribe) protocol as the middle layer. DDS is an industrial standard for publish/subscribe communication in real-time and embedded systems. This point-to-point communication mode is similar to the middle layer of ROS1, but DDS does not need to rely on the master node to complete the communication between two nodes, making the system more fault-tolerant and flexible.

The architecture of ROS2 includes the following two layers:

1. rmw (ROS middleware interface): The interface between the ROS2 software stack and the underlying middleware implementation. Here, the underlying middleware of ROS2 is implemented by DDS or RTPS, responsible for discovery, publish and subscribe mechanisms, service request-reply mechanisms, and message type serialization.
2. rcl (ROS client library): A higher-level API used to implement client libraries, which does not directly touch the middleware implementation, but is abstractly implemented through the ROS middleware interface rmw.

There is also a ros_to_dds component, mainly providing an interface for users to directly access the DDS layer. One of the goals of this abstract interface is to completely isolate ROS user space code from the middleware in use, so that the impact on user code is minimized when changing DDS suppliers or even middleware technologies.

ROS2 introduces Quality of Service (QoS) policies to configure communication between nodes, which can enhance the adaptability of ROS2 in different application scenarios. ROS1 only supports TCP-based communication, while ROS2 can show the reliability of TCP (Transmission Control Protocol) and the high real-time performance of UDP (User Datagram Protocol). In ROS2, users can achieve different communications by selecting QoS configuration files, which include:

- QoS settings for publish/subscribe mechanisms.
- QoS settings for services.
- QoS settings for sensor data.
- QoS settings for parameters.
- Default QoS settings for the DDS middleware layer.

4.6.2 QNX

QNX Neutrino [7] is a commercial real-time operating system (RTOS), developed by a company that has been acquired by the mobile company BlackBerry. QNX uses a microkernel design and modular architecture, which is the architecture diagram of QNX.

The microkernel is a streamlined version of the OS core that provides core OS functions. RTOS is an OS that can process at a sufficiently fast speed when external events occur and data is generated, and its results can be obtained within a specified time to control the generation process or quickly respond to processing, schedule all resources to complete real-time tasks, and control all tasks to run in a coordinated manner.

The QNX kernel provides four services: process scheduling, inter-process communication, low-level network communication, and interrupt handling. Therefore, the QNX kernel is more reliable than traditional macrokernel (Linux) systems.

4.6.3 BOSCH Iceoryx

BOSCH's open-source software Iceoryx is a middleware for inter-process communication (IPC) in various operating systems. The different processes in the autonomous driving system need to transmit a large amount of data, and the common communication mode is publish/subscribe. Typical middleware solutions for IPC copy messages when passing messages to or from the middleware.

More copies can be made within the middleware stack, or the message payload can be serialized. Therefore, if n users subscribe to a publisher, at least n+1 message copies can be easily obtained. When the communication speed reaches Gb/s, each message copy created in the middleware adds a significant cost in terms of runtime and latency.

Iceoryx [8] is a shared memory based IPC technology that combines publish subscribe architecture, service discovery, C++, and lock-free algorithms. By adding an API that avoids copying, it can achieve true zero-copy—an end-to-end method from publisher to subscriber, without the need to create a single copy.

Using the Iceoryx API, the publisher writes messages directly into the memory block previously requested from the middleware. When the message is passed, the subscriber receives references to these memory blocks, while maintaining its own queue with a configurable capacity.

Each subscriber can see which messages are still being processed and which messages can be discarded. The Iceoryx middleware calculates references in the background and can release memory blocks when the reader completes operations.

Iceoryx is a data-agnostic shared memory transmission mechanism that provides a fairly low-level API. Assuming the API is not directly accessed by users, but is integrated into a larger framework that provides a high-level API and some tools, such as AUTOSAR and ROS; in this case, Iceoryx can still ensure support for zero-copy API. If a framework is also based on a publish/subscribe architecture, the integration of Iceoryx is very simple.

4.6.4 DORA

DORA [9] is a data flow computing platform for autonomous driving developed based on the Rust programming language. Compared with ROS and AUTOSAR, it also introduces the separation of data flow control plane and data plane, can define the design architecture of the data plane, and is an open-source middleware project for the field of autonomous driving and robotics. Figure 4.1 shows the reference architecture of DORA, its core is a cross-end edge cloud data flow computing framework.

Figure 4.2 shows a more detailed software architecture of DORA, mainly divided into control plane and data plane. The components of the control plane mainly include: processing of data flow YAML declaration, management and telemetry of data flow; responsible for parsing YAML blueprint, deploying, managing and monitoring the operation of the computing graph according to the blueprint, while providing command line and API. On top of it, more tools can be built using the cloud-native ecosystem, such as DORA Hub, Dashboard, and DevOps development tools, etc. The components of the data plane include nodes, operators, and DORA SDK library control plane. A node is a separate program process that can run one or more operators. A computer can run one or more nodes, so node deployment can span a single machine. Operators implement an algorithm or function processing. Communication between multiple nodes and operators is implemented by communication middleware.

DORA supports different communication middleware implementations such as publish/subscribe, RPC (Remote Procedure Call), shared memory, and point-to-point channels. Nodes and operators are based on a unified data structure to implement data input and output; DORA defines the standard node API and operator API (SDK) of the data flow graph through the declarative language YAML, as well as the data

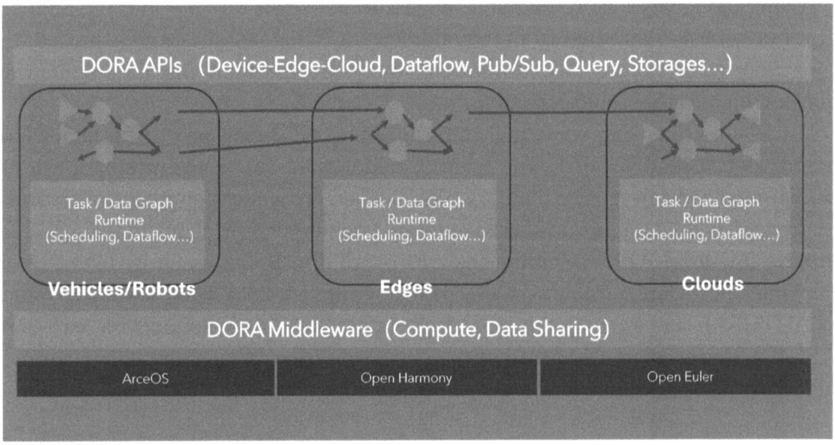

Fig. 4.1 DORA reference architecture

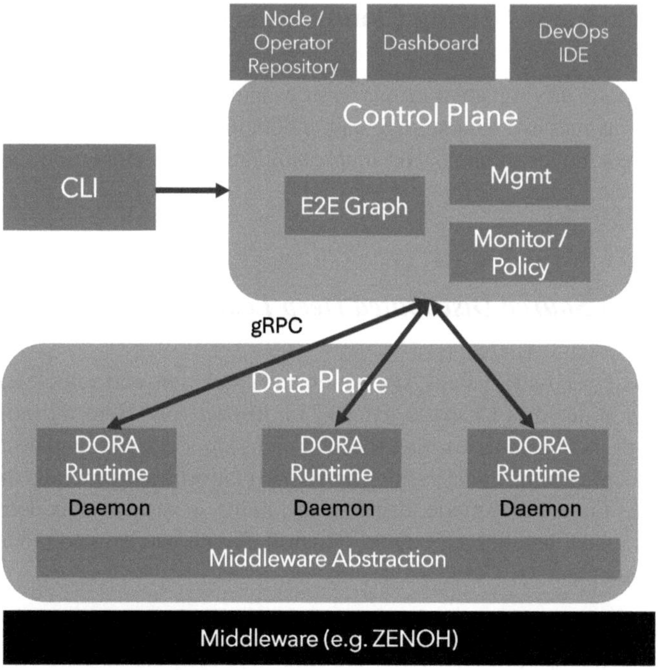

Fig. 4.2 DORA software architecture

plane topology relationship and real-time constraints of the end-to-end data flow graph within the node (such as the pipeline between tasks) and across nodes (for example, specific task scheduling algorithms in autonomous driving systems: soft real-time, deadline timestamp, and water level, etc.). The deployment requirements described by YAML are handled by the control plane, which meets the deployment requirements of the YAML blueprint by choosing different communication middleware implementations.

Providing developers with a transparent environment for development and production is also a core value of cloud-native. DORA integrates with the cloud through the control plane and cloud-native ecosystem tools, making full use of cloud-native technology to manage and telemetry data flow components (nodes-operators-sources-sinks); as well as the lifecycle management of software components. These tools allow developers to effectively debug component anomalies or unsafe behaviors, and provide support for data record preservation and offline reproduction of running scenarios.

4.7 Autonomous Driving Cloud Platform

The autonomous driving software platform cannot be separated from cloud computing, such as massive data storage and annotation, machine learning (including deep learning) model training, HD maps, simulation and emulation, and vehicle networking, etc.

4.7.1 Open Source Distributed Deep Learning Framework

This book will discuss HD maps, simulation and emulation, and vehicle networking technologies in detail in Chaps. 6, 10, and 13, respectively. Here we introduce the technology of the machine learning (deep learning) model training platform. When deploying deep learning model training on a cloud platform, it is generally distributed. According to the parallel mode, distributed training is basically divided into data parallel and model parallel. Of course, a hybrid of data parallel and model parallel can also be adopted.

- Model parallel: Different GPUs are responsible for different parts of the network model. For example, different network layers are assigned to different GPUs, or different parameters of the same layer are assigned to different GPUs.
- Data parallel: Different GPUs have multiple copies of the model, each GPU is assigned different data, and the results of all GPU calculations are combined in some way.

Model parallel is not commonly used, while data parallel involves how to synchronize model parameters between various GPUs, divided into synchronous updates and asynchronous updates. Synchronous update waits for all GPU gradient calculations to be completed, then calculates new weights, synchronizes new values, and then performs the next round of calculations.

Asynchronous update is that each GPU gradient calculation does not need to wait, immediately updates the weight, then synchronizes the new value for the next round of calculation.

The distributed training system includes two architectures: the parameter server architecture and the ring-allreduce architecture.

In terms of physical architecture, the PS is basically consistent with the master-worker architecture of Spark (an open-source unified analysis engine for large-scale data processing). It is basically divided into two major parts: the server cluster and multiple worker clusters, and the resource manager is responsible for the overall resource allocation and scheduling. The server cluster contains multiple server nodes, each of which is responsible for maintaining some parameters, and the server manager is responsible for maintaining and allocating server resources. Each worker cluster corresponds to an application (i.e., a task), and there is no communication between worker clusters or between worker nodes within a worker cluster. Worker nodes can only communicate with the server.

The architecture diagram of Ring-AllReduce, which forms a ring communication with multiple GPUs. Each card has a left-hand card and a right-hand card; the network structure of each card is fixed, so the parameter structure inside is the same; each communication only sends parameters to the card on the right, and then receives data from the card on the left; through continuous iteration, the entire parameter synchronization, that is, reduce, is achieved.

PyTorch [10] is a deep learning framework released by the Facebook team in January 2017. Its popularity on GitHub has surpassed frameworks like Theano, Caffe, and MXNet. PyTorch has now established partnerships with multiple cloud platforms and can be installed and used. For example, on AWS, PyTorch models can be trained in AWS Deep Learning AMIs, AWS Deep Learning Containers, and Amazon SageMaker, and finally deployed using TorchServe. PyTorch has two methods to split models and data across multiple GPUs:

- DataParallel
- DistributeDataParallel.

DataParallel is easier to use. However, communication is the bottleneck, GPU utilization is usually low, and it does not support distribution. DistributedDataParallel supports model parallelism and multi-process, both single-machine/multi-machine, and is distributed training.

PyTorch itself provides several techniques to accelerate distributed data parallelism, such as bucketing gradients, overlapping computation with communication, and skipping gradient synchronization, etc.

Google's TensorFlow [11] can be said to be the most popular open-source deep learning framework today, which can be used in various tasks related to deep learning. The use of TensorFlow in model design and training is also very convenient. You can use the high-level Keras API; for large-scale machine learning training tasks, you can also use the Distribution Strategy API to perform distributed training on different hardware configurations without changing the model definition.

The Estimator API allows you to write distributed training code, allowing you to customize the model structure, loss function, optimization method, and how to train, evaluate, and export, etc., while shielding details related to underlying hardware devices, distributed network data transmission, etc.

tf.distribute.MirroredStrategy supports synchronous distributed training on multiple GPUs on a single machine. This strategy creates a replica for each GPU device. Each variable in the model will be mirrored between all replicas. These variables will together form a single conceptual variable called MirroredVariable. These variables will keep each other synchronized by applying the same updates.

tf.distribute.experimental.MultiWorkerMirroredStrategy is very similar to MirroredStrategy. It implements synchronous distributed training across multiple worker processes, each of which may have multiple GPUs. Like MirroredStrategy, it also creates replicas of all variables in the model across all worker processes on each device.

tf.distribute.experimental. ParameterServerStrategy supports parameter server training on multiple machines. In this setting, some machines are designated as worker processes and some are designated as parameter servers. Each variable in the model will be placed on the parameter server. Computations are replicated to all GPUs in all worker processes (Note: This strategy only applies to the Estimator API).

Kubeflow [14] is an open-source machine learning platform that allows machine learning pipelines to orchestrate workflows running on Kubernetes. This solution is based on Google's method of deploying TensorFlow models, namely TensorFlow Extended. Kubeflow consists of the following logical components.

1. Kubeflow Pipelines: Build and deploy portable, scalable machine learning workflows based on Docker containers, including a user interface, an engine for scheduling machine learning workflows, an SDK for defining pipelines, and Notebooks for the SDK to interact with the system.
2. KFServing: Enable serverless inference on Kubernetes, providing high performance and highly abstract interfaces for ML frameworks such as PyTorch, TensorFlow, scikit-learn, and XGBoost.
3. Multi-user: Simplify user operations, allowing users to only view and edit the Kubeflow components and models in their configuration.
4. Training operators: Train machine learning models through operators, provide TFJob for running TensorFlow model training on Kubernetes, and PyTorchJob for PyTorch model training, etc.
5. Notebooks: Deploy services for managing and generating Jupyter Notebooks.

The pipeline architecture of Kubeflow is mainly divided into eight parts: ① Python SDK, a domain-specific language (DSL) for creating Kubeflow pipelines; ② DSL compiler, which converts Python code into static configuration YAML files; ③ Pipeline web server, the front-end service for the pipeline; ④ Pipeline Service, the back-end service for the pipeline; ⑤ Kubernetes resources, which create custom resource definitions (CRDs) to run the pipeline; ⑥ Machine learning metadata service, which stores data exchanges (input/output) between task flow containers; ⑦ Artifact storage, which stores Metadata and Pipeline packages, views; ⑧ Orchestration controllers, which handle task orchestration, such as Argo Workflow.

4.7.2 Example Architecture of Autonomous Driving Cloud

The following are examples of the basic architecture and functions of some commercialized autonomous driving cloud platforms.

The framework of the data integration solution provided by Baidu Cloud includes defect data (noise and unseen data) mining, AI model iteration pipeline, and data management.

The framework of Baidu Cloud's [18] data training platform integrated training platform connects data, models, and resources, provides professional model evaluation services, helps to identify model defects; its pipeline operation mode can simplify the model training process; it is compatible with multiple model frameworks such as Apollo, Paddle, Caffe, and provides tens of thousands of extreme cases to identify model defects; it also provides millions of kilometers of road test mileage and nearly a thousand types of scene categories.

The architecture of Huawei Cloud's autonomous driving solution [17] consists of modules such as safety compliance, road test data management, algorithm training, scenario simulation, scenario library trading, and platform management. These modules are loosely coupled and can meet the following three types of needs.

- Module construction on demand: Quickly connect the module to the research and development platform based on reference code.
- E2E rapid construction: Build an end-to-end autonomous driving research and development platform based on reference code.
- User-selected integration: Provide standard APIs for quick integration of selected steps such as annotation and simulation.

Lingjun is Alibaba Cloud's intelligent computing product [20], which has various product forms such as public cloud and private cloud, and provides heterogeneous computing fusion cluster instances; it supports CPU+GPU and other heterogeneous computing power fusion deployment, unified scheduling, providing diversified intelligent computing services for different application computing loads; it supports the full-process application fusion deployment of autonomous driving model development and simulation testing, carrying the full-cycle development of autonomous driving, and eliminating unnecessary data migration costs.

Lingjun [19] calls other basic resources of Alibaba Cloud, such as object storage service (OSS), CPFS (cloud parallel file storage), and machine learning platform (platform of artificial intelligence, PAI). PAI provides a full-chain AI development service including data annotation, model construction, model training, model deployment, and inference optimization. It has built-in 140+ optimized algorithms, various scenario plugins, and provides low-threshold, high-performance cloud-native AI engineering capabilities. The machine learning platform PAI provides an end-to-end large model platform, which can realize the complete process from large model subscription, fine-tuning training, model distillation, to offline inference, online deployment, and service experience.

The solution framework of Tencent's autonomous driving cloud [21]: ①It provides end-to-end data trust and privacy computing services from data collection, processing, and application, covering data collection, data delivery, data injection, and data and algorithm training after injection into the cloud. ②It provides data storage capabilities, adopts a combination of software and hardware storage solutions to support special data storage scenarios, effectively reducing storage costs. ③It provides a process-oriented configuration platform, which connects data import, scene extraction, data annotation, perception training, rule control simulation and evaluation functions, and creates research and development tasks in one stop. ④It provides a series of functions and services including data collection, data annotation tools and services, simulation simulation, simulation scene library, simulation evaluation system construction, and algorithm iteration.

4.8 DevOps and MLOps

DevOps, simply put, is to better optimize the processes of development (DEV), testing (QA), and operations (OPS), integrate development and operations, through highly automated tools and processes, making software building, testing, and release faster, more frequent, and reliable.

DevOps is a complete workflow oriented to IT operations, IT Automation is based on Continuous Integration (CI)/Continuous Deployment (CD) to optimize all aspects of program development, testing, and system operations. Trunk development is a prerequisite for CI, and automation and centralized code management are necessary conditions for implementing CI. DevOps is an extension of CI thinking, and CI/CD is the technical core of DevOps.

The core goal of MLOps is to enable AI models to run stably and efficiently in production environments from training to deployment, meeting the end-to-end business needs of customers.

To achieve this goal, it also puts forward corresponding requirements for the core technology of AI systems. Such as deployment automation, which will put forward clear requirements for the front-end design of the AI framework. If the front-end design of the AI framework is not conducive to exporting complete model files, it will cause a large number of downstream to have to introduce "patches" for their business scenario requirements in the deployment link.

The demand for deployment automation will also give birth to some software components around the AI core system, such as model inference deployment optimization, reproducibility of model training prediction results, and scalability of AI production systems.

4.9 Summary

This chapter continues to introduce the platform for autonomous driving—the software part. It includes the software standard AUTOSAR in the automotive industry, the widely recognized software engineering development process V-model, the automotive software process standard aSPICE, and the software development process provided by the functional safety standard ISO 26262. It also introduces the autonomous driving software platform DriveWorks developed by NVIDIA, as well as the commonly used operating systems for autonomous driving, cloud platforms for autonomous driving, and DevOps and MLOps. From Chap. 5 onwards, the development process of the autonomous driving module will be analyzed, including perception, mapping, localization, planning, control, simulation, and vehicle networking.

References

1. S. Fuerst and M. Bechter, "AUTOSAR for Connected and Autonomous Vehicles: The AUTOSAR Adaptive Platform," *IEEE Int. Conf. on Dependable Systems and Networks Workshop (DSN-W)*, June 2016.
2. Wikipedia, "SPICE," https://de.wikipedia.org/wiki/ISO/IEC_15504
3. ISO, "ISO 26262," https://www.iso.org/standard/51362.html
4. Wikipedia, "V-Model," https://en.wikipedia.org/wiki/V-Model_(software_development)
5. Nvidia, "DriveWorks," https://developer.nvidia.com/drive/driveworks
6. ROS, "ROS Documentation," http://wiki.ros.org/
7. QNX, "Neutrino RTOS," https://blackberry.qnx.com/en/products/foundation-software/qnx-rtos
8. Iceoryx, https://iceoryx.io/v1.0.0/
9. Dora Framework, https://github.com/dora-rs
10. PyTorch, https://github.com/pytorch/pytorch
11. TensorFlow, https://github.com/tensorflow/tensorflow
12. Apache MXNet, https://github.com/apache/mxnet
13. PaddlePaddle, https://github.com/PaddlePaddle/Paddle/
14. Kubeflow, https://github.com/kubeflow
15. Horovod, https://github.com/horovod/horovod
16. BytePS, https://github.com/bytedance/byteps
17. Huawei Cloud, "Autonomous Driving Solution," https://www.huaweicloud.com
18. Baidu Cloud, "Autopilot Data Services," https://cloud.baidu.com
19. Alibaba Cloud Lingjun AI, https://www.aliyun.com/product/ai/lingjun
20. Alibaba Cloud Machine Learning Platform, https://www.aliyun.com/product/bigdata/learn
21. Tencent Cloud, "Autonomous Driving Solution," https://cloud.tencent.com/solution/auto

Chapter 5
Perception Module of Autonomous Driving

Autonomous driving needs to answer the following questions:

- Localization—Where am I?
- Perception—What's around me?
- Prediction—What will happen next?
- Planning—What should I do?
- Control—How to do it?

The perception module is like the "eyes" of an autonomous vehicle. It directly connects with sensors, obtains a large amount of input data from sensors, and then obtains an understanding of the surrounding environment after analysis and processing, including obstacle detection, tracking and distance estimation, lane line detection and segmentation, traffic sign detection, and traffic signal light recognition detection. Since current autonomous driving systems are equipped with multiple/different types of sensors, sensor fusion is a necessary and important part of perception, and some systems even list it separately, independent of the perception module.

Section 5.1 introduces the calibration of various on-board sensors; Sect. 5.2 introduces various methods of distance measurement with a single camera; Sect. 5.3 introduces the depth estimation algorithm of a single image; Sect. 5.4 analyzes the current 3D obstacle detection method of a single camera, that is, the projection of the 3D target cuboid frame in the image, not just the 2D rectangle frame; Sect. 5.5 discusses obstacle tracking methods, including single-target and multiple-target situations; Sensor fusion is an important issue; Sect. 5.6 analyzes the fusion methods from the two levels of data and task; Sects. 5.7 to 5.9 mainly discuss three issues of concern to the on-board vision system: lane line detection, traffic sign detection, and traffic signal light recognition detection; Sect. 5.10 focuses on the issue of the drivable area seen by the visual camera, that is, semantic road segmentation; Sect. 5.11 is an analysis of the perception module based on stereo vision. Due to the need for pedestrian behavior analysis, human posture estimation is discussed in Sect. 5.12;

Y. Huang and Z. Yang, *Autonomous Driving Handbook*,
https://doi.org/10.1007/978-981-95-3725-9_5

Sect. 5.13 specifically mentions the implementation algorithm of the Driver Monitoring System (DMS); Sect. 5.14 deeply discusses the current latest BEV (bird's-eye view) perception system; and Sect. 5.15 analyzes the extension of the BEV method: occupancy network.

5.1 Sensor Calibration

Sensor calibration is a basic requirement for autonomous driving. A car is equipped with multiple/different types of sensors, and the coordinate relationship between them needs to be determined.

This work can be divided into two parts: internal parameter calibration and external parameter calibration. Internal parameters determine the mapping relationship inside the sensor, such as the focal length, eccentricity, and pixel aspect ratio of the camera (as well as distortion coefficients), while external parameters determine the transformation relationship between the sensor and some external coordinate system, such as attitude parameters (rotation and translation six degrees of freedom).

The biggest problem with calibrating different sensors is how to measure the best, because the types of data obtained are different:

(1) The camera is an array of RGB image pixels.
(2) Lidar is 3D point cloud distance information (possibly with grayscale values of reflection).
(3) GPS-IMU provides the position and attitude information of the vehicle body.
(4) Radar is a 2D reflection map.

In this case, the objective function to minimize calibration error will vary depending on the pairing of different sensors.

Furthermore, calibration methods are divided into targetless and target methods. The so-called target is a custom-made object with real values. For these two methods, the former is carried out in a natural environment with fewer constraints; the latter requires a dedicated control field, such as a typical checkerboard flat panel.

Currently, there are some open-source projects related to sensor calibration, such as Kalibr and Autoware, and even more specific types of calibration open-source projects, such as between lidar and lidar-IMU calibration. OpenCV also provides some calibration tools. Recently, SenseTime has provided a complete calibration toolbox for different application scenarios of autonomous driving: OpenCalib.

5.1.1 Camera Calibration

Camera calibration has long been a fundamental prerequisite for 3D reconstruction in computer vision. One of the most widely used methods is Zhang's calibration technique, which leverages the invariance properties of the absolute conic to formulate a plane-based calibration algorithm. This method simplifies the calibration process by eliminating the need for a controlled 3D environment and instead relying on a planar calibration target observed from multiple viewpoints (see Sect. 2.1 for a detailed explanation of the algorithm).

In a typical setup, a checkerboard pattern is printed on a flat panel with a known grid size, and a set of at least five images is captured from different orientations. From each image, the subpixel locations of the checkerboard corner points are extracted using standard feature detection techniques. These image correspondences, along with the known geometry of the checkerboard, allow Zhang's method to compute both the intrinsic parameters (e.g., focal length, principal point, skew) and the extrinsic parameters (camera pose relative to the checkerboard), while also accounting for lens distortion.

This calibration approach remains popular due to its simplicity, accuracy, and minimal setup requirements, and it forms the basis of many classical and modern computer vision pipelines.

5.1.2 Lidar Calibration

Take Ref. [3] as an example to introduce a self-calibration of an assembled lidar.

This LiDAR system is constructed using three SICK LMS-151 2D laser scanners, each with an angular resolution of 0.5°. The scanners are mounted on a rotating plate operating at a frequency of 1–2 Hz, enabling 360° azimuthal coverage and near-complete elevation coverage. The only unobservable region is the cylindrical volume along the vertical rotation axis, which remains occluded during operation.

The spatial configuration of each laser scanner L_i $(i = 1, 2, 3)$ on the rotating plate is defined by three parameters:

- τ_i: The radial distance from the origin of the laser beam to the center of the rotating plate.
- α_i: The angle between the laser's scanning plane and the tangent vector of the plate at the mounting point.
- λ_i: The angular position of the beam origin around the radial axis of the plate.

For simplicity and without loss of generality, we define $\lambda_1 = 0$. These parameters uniquely determine the position and orientation of each laser scanner in the rotating frame, allowing for precise modeling of the aggregated 3D point cloud generated during rotation.

Reference [3] defines the crispness of the point cloud as a quality measure and minimizes a function called Rényi quadratic entropy (RQE) as the optimization goal for online calibration of lidar (note: the authors of Ref. [3] also discussed the solution to the clock bias problem of lidar).

The point cloud measurement $\hat{\boldsymbol{X}} = \{\hat{\boldsymbol{x}}_1, \hat{\boldsymbol{x}}_2, \ldots, \hat{\boldsymbol{x}}_N\}$ conforms to a distribution $p(\boldsymbol{x})$, that is, the probability of obtaining measurements from a given position $\boldsymbol{x}$. To get an approximation of $p(\boldsymbol{x})$, the Parzen window density estimation method is used. Applying a Gaussian kernel at each real data point can obtain a Gaussian mixture model (GMM) of $p(\boldsymbol{x})$, that is:

$$p(\boldsymbol{x}) = \frac{1}{N}\sum_{i=1}^{N} G(\boldsymbol{x} - \hat{\boldsymbol{x}}_i, \sigma^2 \boldsymbol{I}) \tag{5.1}$$

where $G(\boldsymbol{\mu}, \boldsymbol{\Sigma})$ is a Gaussian function with mean $\boldsymbol{\mu}$ and variance $\boldsymbol{\Sigma}$. $\boldsymbol{\Sigma} = \sigma^2 \boldsymbol{I}$, where σ can be adjusted.

The measure of crispness can be linked to the entropy of $p(\boldsymbol{x})$. According to the definition of entropy, RQE is chosen as the following measure:

$$H_{RQE}[\boldsymbol{X}] = -log \int p(\boldsymbol{x})^2 d\boldsymbol{x} \tag{5.2}$$

Substituting GMM into it, we get the following form:

$$H_{RQE}[\hat{\boldsymbol{X}}] = -log\left(\frac{1}{N^2}\sum_{i=1}^{N}\sum_{i=1}^{N} G(\hat{\boldsymbol{x}}_i - \hat{\boldsymbol{x}}_j, 2\sigma^2 \boldsymbol{I})\right) \tag{5.3}$$

It can be considered as a measure of the compactness of points in $\hat{\boldsymbol{X}}$, and after simplification, we get the cost function of optimization:

$$E(\hat{\boldsymbol{X}}) = \sum_{i=1}^{N}\sum_{i=1}^{N} G(\hat{\boldsymbol{x}}_i - \hat{\boldsymbol{x}}_j, 2\sigma^2 \boldsymbol{I}) \tag{5.4}$$

Define the external parameters of the lidar $\Theta = [\Theta_1^T, \Theta_2^T, \Theta_3^T]^T$, and finally the point cloud $\hat{\boldsymbol{X}} = \{\hat{\boldsymbol{X}}_1, \hat{\boldsymbol{X}}_2, \hat{\boldsymbol{X}}_3\}$, where $\hat{\boldsymbol{X}}_i = \{\hat{\boldsymbol{x}}_1, \hat{\boldsymbol{x}}_2, \ldots, \hat{\boldsymbol{x}}_m\}$; measurement $\boldsymbol{Z} = \{\boldsymbol{Z}_1, \boldsymbol{Z}_2, \boldsymbol{Z}_3\}$, where $\boldsymbol{Z}_i = \{z_1, z_2, \ldots, z_m\}$, $z_j = h_i(\boldsymbol{x}_j; \Theta_i)$ is the sensor model.

Ignoring the time-domain delay problem, the target function of calibration can be obtained as

$$E(\Theta \mid \boldsymbol{Z}) = \sum_{i=1}^{N}\sum_{j=1}^{N} G(h^{-1}(z_i; \Theta) - h^{-1}(z_j; \Theta), 2\sigma^2 \boldsymbol{I}) \tag{5.5}$$

where $h^{-1}(z_x; \Theta)$ is the inverse sensor model $x = i, j$.

5.1.3 Hand–Eye Calibration

Hand–eye calibration is a commonly researched method, which can be considered as an optimization problem under certain constraints. In a broad sense, a "hand" (such as GPS/IMU) and an "eye" (such as a LiDAR/camera) are both fixed on a machine. After the machine moves, the posture changes of the "hand" and "eye" must satisfy certain constraint relationships. By solving an equation, the coordinate transformation relationship between the "hand" and "eye" can be obtained, usually in the form of the equation $\mathbf{AX} = \mathbf{XB}$.

Hand–eye systems are divided into two types: eye-in-hand and eye-to-hand. Here we refer to the latter, that is, the situation where both the hand and the eye are moving. Hand–eye calibration is divided into two-step method and one-step method. The most famous paper for the latter is *Hand Eye Calibration Using Dual Quaternion*, which uses the dual quaternion method. Generally, the accuracy of the one-step method is higher than that of the two-step method. The former estimates rotation first and then translation.

5.1.4 Camera-GPS/IMU Calibration

Take Ref. [4] as an example to introduce the calibration method between the camera and GPS/IMU.

Although the method proposed in Ref. [4] is for drone calibration, it is also suitable for vehicles.

In fact, IMU–camera calibration and LiDAR–camera calibration are similar. First, solve a hand–eye calibration task, and then optimize the calibration result. IMU has no feedback information available, only attitude data, so it does a pose graph-based optimization. Just as the flowchart of IMU–camera optimization, divided into two paths. The camera data on the left is the image, and the inertial data on the right gets the INS pose. The pose can help the image do SFM reconstruction. The initial system calibration and inertial attitude data together determine the initial pose of the camera; then get the initial 3D point coordinates, and enter the graph optimization with the camera calibration and camera pose to get the refined 3D points, camera pose, and system calibration.

5.1.5 LiDAR–Camera Calibration

Take Ref. [1] as an example to analyze the calibration algorithm between the **LiDAR and the camera**.

Obviously, this is to solve an extended problem of hand–eye calibration, that is, 2D–3D calibration.

The iterative closest point (ICP) algorithm is used to solve the posture change of the LiDAR, and feature matching is used for the camera movement. The latter has a scale problem of monocular SFM. Reference [1] proposes a solution based on sensor fusion: the initial estimate comes from the scale-free camera motion and the scaled LiDAR motion; then the scaled camera motion will be added to the re-estimated LiDAR point cloud data. Finally, their extrinsic parameters can be obtained through hand–eye calibration. In the initialization module, after image feature point detection and matching, estimate camera motion, depth data do ICP calibration, and the two together achieve extrinsic calibration; in the iterative state, re-estimate camera motion, including 3D–2D projection, tracking and camera motion estimation, etc., finally enter the extrinsic calibration module output.

The typical solution of hand–eye calibration is the two-step method: first solve the rotation matrix $\boldsymbol{R}$, and then estimate the translation vector $\boldsymbol{t}$, the formulas are as follows:

$$\boldsymbol{R}_a^i \boldsymbol{R} = \boldsymbol{R}\boldsymbol{R}_b^i \tag{5.6}$$

$$\boldsymbol{R}_a^i \boldsymbol{t} + \boldsymbol{t}_a^i = \boldsymbol{R}\boldsymbol{t}_b^i + \boldsymbol{t} \tag{5.7}$$

$$\boldsymbol{k}_a^i = \boldsymbol{R}\boldsymbol{k}_b^i \tag{5.8}$$

$$\boldsymbol{R}_a^i + s^i \boldsymbol{t}_a^i = \boldsymbol{R}\boldsymbol{t}_b^i + \boldsymbol{t} \tag{5.9}$$

$$\boldsymbol{R}^* = \underset{\boldsymbol{R}}{\operatorname{argmin}} \sum_i \left| \boldsymbol{R}_a^i \boldsymbol{R} - \boldsymbol{R}_b^i \right| \tag{5.10}$$

Now because of the scale problem, the above solution is unstable, so we need to use the data of the LiDAR to improve it.

$$\boldsymbol{v} = \operatorname{Proj}(\boldsymbol{R}\boldsymbol{p} + \boldsymbol{t}) \tag{5.11}$$

And the problem to be solved becomes

$$\mathbf{R}_a^{i*},\ \mathbf{t}_a^{i*} = \underset{\mathbf{R}_a,\ \mathbf{t}_a}{\operatorname{argmin}} \sum_i \left| \mathbf{v}'_j \cdot \operatorname{Proj}\left(\mathbf{R}_a(\mathbf{R}\mathbf{p}_j + \mathbf{t}) + t_a\right) \right| \tag{5.12}$$

The initial solution to the above optimization problem is obtained through the classic perspective-3-point (P3P) algorithm.

After obtaining the motion parameters of the camera, six parameters including rotation and translation can be obtained in the two-step hand–eye calibration method, among which the translation estimation is as follows:

$$\boldsymbol{t}^* = \underset{\boldsymbol{t}}{\operatorname{argmin}} \sum_i \left| (\boldsymbol{R}_a^i \boldsymbol{t} + \boldsymbol{t}_a^i) - (\boldsymbol{R}_b^i \boldsymbol{t} + \boldsymbol{t}_b^i) \right| \tag{5.13}$$

Note: *Here, the estimation of camera motion and hand–eye calibration is alternately performed to improve estimation accuracy. In addition, some strategies that affect the calibration accuracy of camera motion have been found, and the analysis is as follows.*

It can be seen that:

(1) The smaller the overall motion (rotation and translation) of the camera a, the smaller the projection error.
(2) The smaller $(\alpha - \beta)$, the smaller the projection error.

Point (1) indicates that the camera motion should be small during calibration, and point (2) indicates that the depth change of the surrounding environment during calibration should be small, such as walls. In addition, it was found that when the rotation angle of the camera motion is increased, the error propagation from the camera motion estimation to the hand–eye calibration will be small.

Note: *This method cannot be used in outdoor natural environments because it is difficult to determine the image points of point cloud projection.*

How to optimize the **calibration of the lidar-camera**? It is not to estimate the calibration parameters through the matching error of 3D point clouds and image points, but to directly calculate the depth map formed by the point cloud on the image plane, and there is a global matching measure with the image obtained by the camera. The problem is that this kind of method requires a lot of iterations, and the best practice is to generate initial values based on hand–eye calibration.

The following analyzes three methods: Stanford University's method only uses lidar point clouds, and directly uses point clouds and image matching to achieve calibration; the University of Michigan also uses lidar reflection values; the University of Sydney improves on this basis. It should be pointed out that the latter two methods are not as simple as Stanford University's method.

Stanford University's method is to correct the "drift" of calibration online. Accurate calibration should make the depth discontinuity points in the figure match the edges of the inverse distance transform (IDT).

The target function of calibration is defined as follows:

$$J_C = \sum_{f=n-w}^{n} \sum_{p=1}^{|X^f|} \boldsymbol{X}_p^f \cdot \boldsymbol{D}_{i,j}^f \tag{5.14}$$

where w is the video window size, f is the frame number, (i, j) is the pixel position in the image, $\boldsymbol{p}$ is the 3D point of the point cloud, $\boldsymbol{X}$ represents lidar point cloud data, and $\boldsymbol{D}$ is the result of the image after IDT.

Reference [5] defines the task of calibration as solving the transformation relationship between two sensors. It is the coordinate transformation between 3D space points in the lidar and camera coordinate systems, and the solution is $\boldsymbol{R}$ and $\boldsymbol{t}$. It is assumed that the intrinsic matrix $\boldsymbol{K}$ of the camera is known.

The defined mutual information (MI) function is an entropy value, and its calculation formula is as follows:

$$\mathrm{MI}(X, Y) = H(X) + H(Y) - H(X, Y) \tag{5.15}$$

where

$$H(X) = -\sum_{x \in X} \sum_{y \in Y} p_X(x) \log p_X(x) \tag{5.16}$$

$$H(Y) = -\sum_{x \in X} \sum_{y \in Y} p_Y(y) \log p_Y(y) \tag{5.17}$$

$$H(X, Y) = -\sum_{x \in X} \sum_{y \in Y} p_{X,Y}(x, y) \log p_{X,Y}(x, y) \tag{5.18}$$

The calculation of the middle probability is as follows:

$$\hat{p}(X = k_x, Y = k_y) = \frac{1}{n} \sum_{i=1}^{n} K_\Omega \left(\begin{bmatrix} X \\ Y \end{bmatrix} - \begin{bmatrix} X_i \\ Y_i \end{bmatrix} \right); (k_x, k_y) \in ([0, 255] \times [0, 255]) \tag{5.19}$$

$$\hat{p}(X = k) = \frac{1}{n} \sum_{i=1}^{n} K_\omega (X - X_i); k \in [0, 255] \tag{5.20}$$

The solution algorithm is the gradient method, as given in the following pseudocode (the calculations involved are seen in the above formulas). The input is the point cloud, reflection map, image, and initial external parameters, and the output is the estimated external parameters. The iterations in the loop are the calculation of the joint histogram, the kernel density estimation of the joint distribution, the MI estimation, and the parameter update of the gradient F based on MI.

Algorithm: Calibration method based on mutual information

Algorithm 1: Automatic external calibration maximizing MI

Input: 3D point cloud $\{P_i, i = 1, 2, \cdots, n\}$, reflection map $\{X_i, i = 1, 2, \cdots, n\}$, image $\{1\}$, initial estimate $\{\Theta_0\}$

Output: Estimated parameters $\hat{\Theta}$

while $||\Theta_{k+1} - \Theta_k|| >$ THRESHOLD **do**
  $\Theta_k \rightarrow |\mathbf{R}|\mathbf{t}$
  Initialize joint histogram: $\mathrm{Hist}(X, Y) = 0$
  for $i = 1 \rightarrow n$ **do**
    $\tilde{p}_i = K|\mathbf{R}|\mathbf{t}|\tilde{P}_i$
    $Y_i = I(p_i)$
    Update joint histogram: $\mathrm{Hist}(X_i, Y_i) = \mathrm{Hist}(X_i, Y_i) + 1$
  end
  Calculate the kernel density estimation (KDE) of the joint distribution: $p(X, Y; \Theta_k)$
  Calculate MI: $\mathrm{MI}(X, Y'; \Theta_k)$
  Calculate the current estimate: $\Theta_{k+1} = \Theta_k + \lambda F(\mathrm{MI}(X, Y; \Theta_k))$, where F is either a gradient function or some heuristic, λ is the adjustment parameter
end

Reference [6] is a further improvement of the University of Michigan method. First, its sensor configuration includes GPS, lidar, and camera.

A new measure, the gradient orientation measure (GOM), is defined as follows:

$$\mathrm{GOM} = \frac{\sum_{j=1}^{n} |\mathbf{g}(1, j) \cdot \mathbf{g}(2, j)|}{\sum_{j=1}^{n} \|\mathbf{g}(1, j)\| \cdot \|\mathbf{g}(2, j)\|} \tag{5.21}$$

where $\mathbf{g}(1, \cdot)$, $\mathbf{g}(2, \cdot)$ are the gradients of the image and lidar point cloud, and GOM is actually a measure of their correlation.

The calibration flowchart of the camera–lidar is mainly divided into feature computation and gradient optimization. The former includes the orientation and amplitude calculation of the image and point cloud; the latter is GOM calculation and optimization based on particle swarm.

When matching point cloud data with image data, the point cloud must be projected onto the cylindrical image to establish spatial correspondence between the two modalities.

Its projection formula is as follows:

$$x_{\mathrm{cam}} = x_0 - \frac{cx}{z} + \Delta x_{\mathrm{cam}}, \quad y_{\mathrm{cam}} = y_0 - \frac{cy}{z} + \delta y_{\mathrm{cam}} \tag{5.22}$$

$$x_{\mathrm{cam}} = x_0 - c \cdot \arctan(\frac{-y}{x}) + \Delta x_{\mathrm{cam}}, \quad y_{\mathrm{cam}} = y_0 - \frac{cz}{\sqrt{x^2 + y^2}} + \Delta y_{\mathrm{cam}} \tag{5.23}$$

where x_{cam}, y_{cam} are the X, Y positions of the image point; x, y, z are the point coordinates of the environment; c is the principal distance of the model; x_0, y_0 are the principal point positions of the image; and Δ_x, Δ_y are the correction items of the camera.

Before calculating the gradient of the point cloud, the point cloud needs to be projected onto the sphere, the formula is as follows:

$$x_{\mathrm{sphere}} = \arccos\left(\frac{z}{\sqrt{x^2 + y^2 + z^2}}\right) \tag{5.24}$$

$$y_{\mathrm{sphere}} = \arctan\left(\frac{y}{x}\right) \tag{5.25}$$

Finally, the gradient calculation method of the point cloud is: first define p_x, p_y, and p_v as the x position, y position, and grayscale value of the current point; n_x, n_y, and n_v are the x position, y position, and grayscale value of the neighboring point; g_x, g_y, g_{mag}, and g_{or} are the x direction value, y direction value, gradient amplitude, and orientation of the gradient, then the gradient orientation angle is calculated by comparing with the neighboring point as shown below. The final output is the gradient orientation g_{or}.

Algorithm 2: Calculate 3D lidar point cloud gradient

$g_x = 0$
$g_y = 0$
$g_{\text{mag}} = 0$ **for** *neighbor point* $\boldsymbol{n}$ **do**
  $x = p_x - n_x$
  $y = p_y - n_y$
  $v = p_v - n_v$
  $g_{mag} = g_{mag} + |\frac{v}{8}|$
  $g_x = g_x + \frac{vx}{8}$
  $g_y = g_y + \frac{vy}{8}$
end
$g_{or} = (g_y, g_x)$

The task of calibration becomes to solve for the maximum GOM, and the method used is the Monte Carlo method, similar to the particle filter.

In addition, there are two recent calibration methods based on deep learning: RegNet and CalibNet.

Reference [7] is probably the first deep convolutional neural network (CNN) to infer the six-degree-of-freedom (DoF) external parameter calibration of multiple sensors, i.e., lidar and monocular camera. RegNet maps the three steps of calibration (feature extraction, feature matching, and global regression) into a single real-time CNN model. During training, the system is randomly readjusted to train RegNet to infer the correspondence between the depth measurement of the lidar projected onto the camera and the RGB image, and finally regress the calibration external parameters. In addition, multiple CNNs are executed iteratively, trained on data with varying degrees of decalibration.

Firstly, a point $\boldsymbol{x}$ given in the sensor coordinate system is transformed into a point $\boldsymbol{y}$ in the world coordinate system, defined as an affine transformation matrix $\boldsymbol{H}$, i.e., $\boldsymbol{y} = \boldsymbol{Hx}$. The task of estimating the transformation matrix $\boldsymbol{H}$ is called external calibration. Applying deep learning, the problem of external calibration needs to be redefined. Given the initial calibration $\boldsymbol{H}_{init}$ and the basic factual calibration H_{gt}, the decalibration matrix $\boldsymbol{\phi}_{\text{decalib}}$ is determined, which is defined as follows:

$$\boldsymbol{\phi}_{\text{decalib}} = \begin{bmatrix} \mathbf{0} \ \boldsymbol{R} \ \mathbf{0} \ \boldsymbol{t} \\ \mathbf{0} \ \mathbf{0} \ \mathbf{0} \ \mathbf{1} \end{bmatrix} \tag{5.26}$$

Then, $\boldsymbol{H}_{\text{init}}$ can be randomly changed to obtain a large amount of training data. In order to establish an observable correspondence in the calibration process, the lidar points are projected onto the camera image plane using $\boldsymbol{H}_{\text{init}}$ and the camera's internal parameter matrix $\boldsymbol{P}$, i.e.:

$$z_c \begin{bmatrix} u \\ v \\ 1 \end{bmatrix} = \boldsymbol{P}\boldsymbol{H}_{\text{init}}\boldsymbol{x} \tag{5.27}$$

For each pixel (u, v), if there is no projected lidar point, the inverse depth value (camera coordinate) z_c of the projected point is stored or set to zero. Compared to the number of image pixels, most common lidar sensors only provide a small amount of measurement data, so the depth image is very sparse. To deal with this sparsity, max pooling is used in the input depth image to sample the projected lidar depth points.

The structure of RegNet implements calibration between depth and RGB images. The depth points are projected onto the RGB image using the initial calibration $\boldsymbol{H}_{\text{init}}$. In the first and second parts of the CNN network, NiN blocks are used to extract rich features for matching, where the index shows the kernel size k of the first convolution layer of the NiN block. The number of feature channels is shown in the upper right corner of each module. The last part of the CNN network regresses the decalibration by collecting global information using two fully connected layers (note: a NiN block consists of a $k \times k$ convolution, followed by several 1×1 convolutions). During training, the decalibration matrix is randomly permuted, forming different image plane projection data of depth points.

The projection of depth points strongly varies with the given initial calibration value. When the initial calibration deviates from the calibrated basic fact (GT) by a rotation of 20° and a translation of 1.5m, it may cause most points of the projected lidar point cloud to be outside the image area, making it difficult to establish a correspondence with the RGB image.

Even so, the trained CNN network can still improve the calibration. Using the new estimated calibration parameters $\hat{\boldsymbol{H}} = \boldsymbol{H}_{\text{init}}\hat{\boldsymbol{\phi}}_{\text{decalib}}^{-1}$, the depth points can be projected again, thus producing more depth points for related calculations. Then, this step can be iterated multiple times.

Reference [8] proposes CalibNet, a self-supervised deep learning framework designed to estimate the six-degree-of-freedom (6-DoF) rigid transformation between a LiDAR sensor and a 2D camera in real time. Unlike traditional calibration approaches that rely on hand-crafted correspondences or supervised learning with ground-truth transformations, CalibNet does not require direct supervision during training. Instead, it learns to predict calibration parameters by optimizing for geometric and photometric consistency between the projected point cloud and the corresponding RGB image.

Given an input pair consisting of an RGB image from a camera and a raw point cloud from a LiDAR sensor, CalibNet infers the rigid transformation $\boldsymbol{T} \in SE(3)$ that best aligns the two modalities. The network is trained to minimize losses based on image reconstruction (photometric error) and point cloud alignment (geometric error), enabling it to self-correct misalignments between modalities without needing explicitly labeled calibration data.

The framework effectively maps the input point cloud to the image coordinate frame using the predicted transformation and evaluates alignment quality through differentiable loss functions. When initialized with incorrect extrinsic parameters, the misaligned point cloud projection yields distorted or semantically inconsistent visual outputs. After calibration with CalibNet, the transformed point cloud aligns accurately with image structures, demonstrating successful multimodal registration.

This approach allows CalibNet to serve as a plug-and-play calibration module that adapts in real time, which is particularly useful for dynamic or changing sensor setups in robotics and autonomous systems.

The calibration framework of CalibNet includes the following components: (a) an RGB image captured from the calibrated camera; (b) a module that takes the original lidar point cloud as input and estimates a six-degree-of-freedom rigid transformation $\boldsymbol{T}$ to optimally align the image and point cloud; (c) a colorized point cloud resulting from an incorrect calibration setting; and (d) the corrected colorized point cloud obtained after applying the transformation estimated by CalibNet.

CalibNet takes as input an RGB image, a corresponding mis-calibrated lidar point cloud, and the camera calibration matrix $\boldsymbol{K}$. As a pre-processing step, the point cloud is projected onto the image plane to generate a sparse depth map. However, due to the inaccuracy of the initial mis-calibration, the projection results in a sparse depth map that is often severely misaligned with the RGB image. Both the RGB image and the sparse depth map are normalized to the range of ± 1. Subsequently, the sparse depth map is transformed into a semi-dense representation using a 5×5 max pooling operation.

CalibNet takes as input an RGB image, a corresponding mis-calibrated LiDAR point cloud, and the camera's intrinsic calibration matrix $\boldsymbol{K}$. As a pre-processing step, the LiDAR point cloud is projected onto the image plane using the initial (possibly inaccurate) extrinsic transformation. This projection yields a sparse depth map in the camera frame, constructed by assigning depth values to pixels corresponding to projected 3D points.

Due to the mis-calibration, the projected point cloud typically fails to align properly with image structures, resulting in a sparse depth map that is geometrically inconsistent with the RGB image. This misalignment is visually evident, as shown in qualitative examples where the depth map does not conform to object boundaries or scene layout.

To prepare the inputs for the network, both the RGB image and the sparse depth map are normalized to a value range of ± 1. To reduce the sparsity and enhance local structure, the depth map is then processed using a 5×5 max pooling operation, yielding a semi-dense depth map. This operation fills in gaps in the sparse depth map and provides a more structured geometric signal to guide calibration learning.

These normalized and enhanced representations are then fed into the network, which uses them to estimate the 6-DoF rigid transformation that corrects the initial misalignment between the LiDAR point cloud and the camera image.

Although the network's output can directly predict translation, the output rotation vector in SO(3) needs to be converted into its corresponding rotation matrix. An element $\boldsymbol{\omega}$ $\in$SO(3) can be converted into SO(3) using an exponential map.

Once the calibration parameters predicted by the network are converted into a rigid transformation $\boldsymbol{T} \in \mathrm{SE}(3)$, a 3D spatial transformer layer is applied to transform the input depth map using $\boldsymbol{T}$. In this framework, the original 3D spatial transformer is extended to accommodate sparse or semi-dense input depth maps, enabling effective transformation despite incomplete depth information.

Once the network predicts the calibration parameters, they are converted into a rigid body transformation $\boldsymbol{T} \in \mathrm{SE}(3)$, representing a 6-DoF transformation between the LiDAR and camera coordinate frames. This transformation is then applied to the input depth map using a 3D spatial transformer layer. The spatial transformer warps the 3D coordinates of the projected LiDAR points according to $\boldsymbol{T}$ and reprojects them onto the image plane using the known camera intrinsics $\boldsymbol{K}$.

To accommodate the nature of LiDAR data, which is typically sparse or semi-dense, the standard 3D spatial transformer layer is extended to support sparse data inputs. This extension ensures that the transformation can be applied effectively without introducing interpolation artifacts or density assumptions that may not hold in practice.

The transformed depth map, now aligned according to the predicted transformation, is expected to be geometrically consistent with the RGB image. This alignment is critical for computing the photometric and geometric losses used to guide the self-supervised training of the network. By directly incorporating the transformation operation into the learning process, the network is able to refine its predictions through differentiable warping, improving calibration accuracy over time.

The network structure diagram of CalibNet mainly consists of two asymmetric branches, RGB and "depth," each of which performs a series, so the filters of the "depth" stream are reduced at each stage. The outputs of the two branches are then connected along the channel dimension and passed through a series of additional full convolution and batch normalization layers for global feature aggregation. The output streams of rotation and translation are separated to capture any modal differences that may exist between rotation and translation. The output of the network is a 1×6 vector $\boldsymbol{\xi} = (\boldsymbol{v}, \boldsymbol{\omega}) \in \mathrm{SE}(3)$, where $\boldsymbol{v}$ is the translation velocity vector and $\boldsymbol{\omega}$ is the rotation velocity vector.

The following is the definition of the loss function:

(1) Photometric Loss: After passing through the predicted $\boldsymbol{T}$-transform depth map, check the dense pixel error between the predicted depth map and the correct depth map.
(2) Point Cloud Distance Loss: The 3D space transformer layer allows point cloud transformation after back-projection. At this stage, try to minimize the 3D–3D point distance between the uncalibrated transformed points and the target point cloud. Distance measurements include:

- Chamfer distance.
- Earth mover's distance.
- Centroid ICP distance.

Finally, the entire loss function is defined as the sum of the photometric loss and the point cloud distance loss.

5.1.6 Camera–Radar Calibration

Taking Ref. [2] as an example, this method introduces a calibration procedure for aligning millimeter-wave radar with a monocular camera. The objective is to estimate the rigid transformation between the radar coordinate system X_r–Y_r–Z_r and the camera coordinate system X_c–Y_c–Z_c.

To establish this relationship, a set of corresponding target points is independently observed by both sensors. The radar provides 2D or 3D measurements of targets in its local radar plane, while the camera captures the same targets as image plane points using known intrinsic parameters. By associating these radar plane and image plane correspondences, it becomes possible to derive the extrinsic parameters—rotation and translation—that define the transformation between the radar and camera coordinate frames.

This calibration process is essential for sensor fusion tasks, enabling accurate projection of radar measurements into the image space and vice versa. It also lays the foundation for jointly interpreting spatial information from the two modalities in applications such as object detection, tracking, and localization.

Calibration is actually the calculation of the homography matrix parameters between the image plane and the radar reflection plane, as follows:

$$\begin{pmatrix} u \\ v \\ 1 \end{pmatrix} = \boldsymbol{T}_{\mathrm{I}}^{\mathrm{R}} \begin{pmatrix} x_r \\ y_r \\ 1 \end{pmatrix} = \begin{bmatrix} t_{11} & t_{12} & t_{13} \\ t_{21} & t_{22} & t_{23} \\ t_{31} & t_{32} & t_{33} \end{bmatrix} \begin{pmatrix} x_r \\ y_r \\ 1 \end{pmatrix} \tag{5.28}$$

5.1.7 Lidar–IMU Calibration

The calibration between the lidar and the IMU is also one of the commonly used calibration methods in autonomous driving, and its calibration accuracy has a great impact on the lidar mapping and localization module. Usually, when the external parameters of the lidar are calibrated to the IMU, the calibration accuracy is judged by whether the local map of the lidar is good. The calibration process is to solve the external parameters from the lidar to the IMU through the local map constructed by the sliding window.

The algorithm proposed in Ref. [9], by minimizing the eigenvalues of the covariance matrix to minimize the distance sum from the feature points to the feature plane, and optimize, to achieve the purpose of calibrating the external parameters from the lidar to the IMU.

The global optimization BA algorithm minimizes the distance from each plane feature point to the plane, and its formula is as follows:

$$(\boldsymbol{T}^*, \boldsymbol{n}^*, \boldsymbol{q}^*) = \underset{\boldsymbol{T},\boldsymbol{n},\boldsymbol{q}}{\operatorname{argmin}} \frac{1}{N} \sum_{i=1}^{N} (\boldsymbol{n}^T (\boldsymbol{p}_i - \boldsymbol{q}))^2 \tag{5.29}$$

$$= \underset{\boldsymbol{T}}{\operatorname{argmin}} (\min_{\boldsymbol{n},\boldsymbol{q}} \sum_{i=1}^{N} (\boldsymbol{n}^T (\boldsymbol{p}_i - \boldsymbol{q}))^2) \tag{5.30}$$

where $\boldsymbol{n}$ is the normal; $\boldsymbol{T}$ is the pose matrix; $\boldsymbol{q}$ is a plane feature point; and $\boldsymbol{p}_i$ are all points on the plane ($i = 1, \cdots, N$). After optimization, the position of the feature points and the feature normal (direction vector) can be written as a function of the pose $\boldsymbol{T}$, so only the pose $\boldsymbol{T}$ needs to be optimized.

Adaptive voxelization is a hierarchical spatial partitioning technique that accelerates feature correspondence search in point cloud registration tasks. Assuming an approximate initial pose for each scan—such as one obtained from odometry in LiDAR SLAM algorithms like LOAM—adaptive voxelization proceeds by recursively subdividing the 3D space into voxels of varying resolution, based on local geometric properties.

The process begins by voxelizing the 3D space using a default voxel size (e.g., 1 m). For each voxel, the feature points contained within (aggregated across all scans) are analyzed by computing the eigenvalues of their covariance matrix. If the points in the voxel approximately lie on a plane (i.e., their covariance matrix has one dominant eigenvalue and two smaller ones), the voxel is considered planar and stored along with its enclosed feature points. If the planar condition is not met, the voxel is recursively subdivided into eight smaller voxels. This subdivision continues until a minimum voxel size is reached (e.g., 0.125 m). As a result, the voxel map becomes spatially adaptive, allocating finer resolution only where needed based on local structure.

This adaptive voxel map is used to represent the 3D environment, where different regions may have varying voxel resolutions tailored to the complexity of the local geometry. For example, in an area such as a tree crown, fine-scale voxels (e.g., 0.125 m) may be generated to capture detailed structure. From both front and side views, these voxels reveal the localized planar surface distribution, enhancing spatial representation and alignment accuracy.

To support efficient storage, indexing, and retrieval, the resulting voxel map is organized using an octree structure, with all voxel entries indexed via a hash table. This design enables rapid access during matching and registration processes, making it well suited for large-scale mapping and localization tasks in complex environments.

The method in Ref. [9] can obtain the pose of the lidar in the world coordinate system at any time t. Since the impact of the offset on the system gradually increases with the delay of time, local mapping optimization is used to further correct the offset.

Based on this, a local map is constructed, and then further search for the correspondence between each frame of laser point cloud and the local map between the current frame and the next frame.

It is recommended that the autonomous driving car to be calibrated should drive in an eight-shaped trajectory. The calibration scene has several requirements: ① ensure that the ground is flat enough; ② ensure that there are enough features around, such as walls, lane lines, utility poles, and stationary vehicles; ③ the calibrated vehicle circulates three times in a figure-eight trajectory, maintaining a speed of 10km/h; and ④ try to avoid dynamic targets in the surroundings, such as vehicles.

5.2 Obstacle Ranging of Monocular Vision

There are two different methods for a single-lens autonomous driving system to estimate the distance of obstacle targets. One is to directly calculate the distance between the target and the vehicle based on the target's detection box information, camera calibration parameters (i.e., camera pose), and road information. The second is to first estimate the depth map from the monocular image, and then obtain the target distance information in combination with other geometric constraints.

Here we first introduce the first method, and Sect. 5.3 will introduce the depth map estimation method.

Reference [10] summarizes the methods for estimating vehicle speed based on vision, which also includes the part of ranging.

Monocular ranging is necessarily based on certain assumptions, such as

- Based on homography matrix/transformation (assuming the road is flat).
- Add virtual lines, patterns, or areas.
- Some known data, such as the actual size of certain targets (vehicles and license plates).

The methods to improve the accuracy of monocular ranging are as follows:

- Use a telephoto lens (large focal length), because the error decreases with the increase of the focal length.
- Use as large a known size target as possible as a reference.
- Avoid measuring vehicles far from the camera. The closer the target, the smaller the error in distance estimation.

Reference [11] presents a detailed description of the imaging geometry used in vehicle-mounted monocular vision systems, establishing the spatial relationship between the on-board camera, the vehicle, and surrounding objects in the observed scene. The analysis includes the orientation and height of the camera relative to the road surface, as well as the geometric projection model that maps 3D world coordinates onto the 2D image plane.

In this system, let vehicle A be equipped with a monocular camera of focal length f, mounted at a height H above the road. The camera observes preceding vehicles B and C, located at distances Z_1 and Z_2 from the camera, respectively. In the image plane, the vertical pixel coordinates of the projections of the base points (e.g., the bottom of the bounding boxes) of these objects are denoted as y_1 and y_2.

According to the pinhole camera model, the relationship between the object distance Z_i and its image coordinate y_i is given by

$$y_i = \frac{fH}{Z_i} \quad \Rightarrow \quad Z_i = \frac{fH}{y_i}$$

This equation allows for depth estimation of objects using known camera parameters and observed image coordinates.

Empirical evaluations report that the relative depth estimation error increases with distance. Specifically, the system yields approximately a 5% ranging error at a distance of 44 m, and around a 10% error at 90 meters. These results highlight the trade-off between simplicity and accuracy in monocular-based distance estimation, especially in long-range scenarios where pixel-level errors can lead to larger distance uncertainties.

Reference [12] builds a forward collision warning system (FCW) based on monocular vision. Assuming the width W of the front vehicle and its width w in the image are known, then its vehicle distance is $d = FW/w$.

The distance of the front vehicle can be calculated as

$$d = \frac{1}{\cos^2\theta} \cdot \frac{F_c \cdot H_c}{y_b - y_h} - H_c \tan\theta \tag{5.31}$$

The Forward Collision Warning (FCW) system operates through a two-stage algorithmic workflow. The first stage focuses on target detection and tracking. It begins by generating potential object hypotheses based on sensor data, which are then verified to eliminate false positives. Once verified, the system continuously tracks the positions and movements of surrounding vehicles or obstacles to ensure reliable perception in real time.

The second stage involves collision risk assessment and warning generation. In this stage, the system identifies relevant targets, estimates their distance from the ego vehicle, and calculates the time-to-collision (TTC) by analyzing the relative speed and trajectory of each target. Based on this information, the system evaluates the risk of an imminent collision. If the TTC falls below a predefined safety threshold, a warning is issued to alert the driver or automated control system to take preventive action.

Together, these two stages enable the FCW system to detect potential collision risks early and respond appropriately under dynamic traffic conditions.

Reference [12] defines a concept of a virtual horizontal line, the formula for estimating its position is $Y_h = Y_b - H_c W_a / w_a$, where H_c is the camera height and Y_b is the vertical position of the bottom of the vehicle frame. Then the vehicle distance can be calculated as follows:

$$d = \frac{F_c \cdot H_c}{y_b - y_h}$$

Reference [13] first calculates the motion of feature points, obtains the motion of the camera itself from this, and infers the pitch angle from its translation vector.

From the geometric relationship between the pitch angle and the distance of the front car, the distance of the front car can be calculated as follows:

$$D = \frac{H_c}{\tan(\alpha + \beta)} \tag{5.32}$$

First detect Harris corners, then calculate optic flow, implement motion reconstruction (SFM) based on the famous eight-point method, verify motion parameters, and optimize motion estimation based on the Gauss–Newton method; then, it is divided into two branches, one side estimates the pitch angle through the translation vector, and the other side estimates the pitch rate, i.e., angular velocity, through the rotation matrix. Finally, they are combined to estimate the average pitch angle.

In a Structure from Motion (SFM) system, the depth of scene points is inferred by leveraging the geometric relationship between camera motion and scene structure. This is achieved through the epipolar geometry principle, commonly used in stereo vision. When the camera captures images from two different viewpoints, the relative displacement between the two camera poses—especially the translation vector—encodes valuable information about the spatial layout of the scene. By analyzing how scene points shift across the image sequence and applying the epipolar plane constraint, the system can triangulate the 3D positions of those points. As a result, SFM enables three-dimensional reconstruction from monocular image sequences without the need for depth sensors.

After estimating the rotation matrix $\boldsymbol{R}$ and translation vector $\boldsymbol{t}$ of the camera motion, the calculation formula for distance is as follows:

$$Z = \frac{|\boldsymbol{t}|}{\frac{|\boldsymbol{p}_1| \sin \omega}{\cos \gamma}} = \frac{\boldsymbol{t}(((\boldsymbol{p}_1 + \boldsymbol{R}\boldsymbol{p}_2) \times \boldsymbol{t}) \times \boldsymbol{R}\boldsymbol{p}_2)}{\boldsymbol{p}_1(((\boldsymbol{p}_1 + \boldsymbol{R}\boldsymbol{p}_2) \times \boldsymbol{t}) \times \boldsymbol{R}\boldsymbol{p}_2)} \tag{5.33}$$

The relationship between the translation vector and the pitch angle of a vehicle's (or camera's) motion on the road surface can be described geometrically: the vehicle's trajectory follows a curved or linear path along the road, while the translation vector represents the direction and magnitude of motion in 3D space. The pitch angle, defined as the angle between the horizontal road surface and the forward motion direction, directly influences the orientation of the translation vector relative to the ground plane.

The calculation formula for the pitch angle is

$$\beta = \arctan\left(-\frac{t_y}{\sqrt{t_x^2 + t_z^2}}\right) \tag{5.34}$$

Recently, some work has used data-driven, end-to-end, deep learning methods to estimate target distance. This method is essentially seen as a regression problem of the detection model, and the following introduces examples from Refs. [14, 15].

Reference [14] developed the first end-to-end learning model to directly predict the distance of a given target in an image. In addition to introducing a learning-based model, it further designed an enhanced model with a keypoint regressor, which defines a projection loss to enforce better distance estimation, especially for targets close to the camera.

The basic model framework consists of three main components. First, a feature extractor processes the entire image to generate a comprehensive feature map. Second, a distance regressor operates on object-specific ROI features to directly predict the distance to each detected object. Third, a multi-class classifier uses the same ROI features to determine the object category. This modular design allows for simultaneous spatial localization and semantic classification of objects within the scene.

Although the basic model can predict promising target-specific distances from the ROI feature map, it still cannot meet the accuracy requirements of autonomous driving, especially for targets close to the camera.

To improve the accuracy of distance prediction, an enhanced method is introduced by incorporating a keypoint regressor into the model. This component applies a projection constraint that guides the network to better align 2D image features with 3D geometric cues, thereby refining the output of the distance regressor. The complete pipeline of the enhanced model includes four components: a feature extractor that generates a global feature map from the input image; a keypoint regressor that predicts object keypoints to support geometric reasoning; a distance regressor that estimates the depth of objects based on feature and keypoint information; and a multi-class classifier that determines the object category using ROI features.

One of the main challenges in training deep neural networks for estimating the distance to specific targets lies in the lack of datasets that include explicit distance annotations for each target in RGB images. While existing object detection datasets typically provide only bounding boxes and target class labels, and dense depth prediction datasets offer pixel-level depth maps, neither provides direct distance measurements tied to individual detected objects.

To address this gap, a common method for generating ground-truth distance annotations involves leveraging synchronized LiDAR data. Specifically, the 3D bounding box parameters of a target are first used to extract the subset of LiDAR points corresponding to that target. These points are then sorted by their depth values (i.e., distances from the sensor), and a representative depth value is selected. To mitigate the influence of outliers and noisy points, a quantile-based approach is used: for example, selecting the depth value at the 10th percentile of the sorted list. This value is then assigned as the ground-truth distance for the target in the RGB image.

To generate ground-truth distance values for targets in RGB images, a multi-step process is employed. First, 3D bounding box parameters are used to segment the target points from the corresponding LiDAR point cloud. Next, the segmented LiDAR points are sorted according to their depth values. To reduce the influence of noise and outliers, the ground-truth distance is not taken from the minimum depth value, but instead from a representative point within the distribution. Specifically, the depth value at the nth position in the sorted list is selected, where $n = 0.1\times$ (number of segmented points). This strategy ensures a more robust and reliable distance label for each target.

In addition, the LiDAR points are projected onto the corresponding RGB image plane to obtain image coordinates as keypoint ground truth. The ground truth of target-specific distance and keypoints is added to the KITTI/nuScenes (mini version) object detection dataset labels along with the RGB image, to build a specific distance dataset.

Reference [15] is a study conducted by Google on remote target distance estimation, proposing R4D, a framework that accurately estimates the distance of remote targets by known distance references in the scene. R4D builds a graph for all known reference targets. The edges of the graph encode the relative distance information between the target and the reference target, using an attention module to measure the importance of the reference target, and combining it into a distance prediction for the target.

Two datasets were introduced for this purpose, the pseudo-remote KITTI dataset and the Waymo Open Dataset—remote labels. Since the KITTI dataset does not provide the true distance of remote targets, the pseudo-remote KITTI dataset is a derived dataset that deletes LiDAR points beyond 40 m and considers all targets beyond 40 m as remote targets. More importantly, a new large-scale dataset was built based on the Waymo Open Dataset, which includes annotated real remote targets (distances of 80–300 m). In summary, although these two datasets differ in their definition of "remote" (40 m or 80 m and above), they both include LiDAR points, camera images, and distance labels for remote targets.

R4D trains a model to locate remote targets with known distances, referred to as designated targets, by representing both the designated target and its references in a graph structure. In this framework, each target is modeled as a node, and edges connect the designated target to its reference targets. The model extracts target-reference (Tar-Ref) embeddings to propagate spatial and contextual information from the references to the designated target. These embeddings are then fed into an attention module, which dynamically assesses the relative importance of each reference. By aggregating the weighted reference information, the model produces a fused distance prediction for the designated target, enabling accurate localization based on multiple spatial cues.

The R4D architecture is designed to model the relationship between a designated target and its reference targets by capturing both visual and geometric cues. To achieve this, the model extracts two types of embeddings: a joint embedding that encodes visual features shared between the target and its references, and a geographic-distance embedding that captures their relative spatial relationships. Recognizing that different references may contribute unequally to the final prediction, an attention module is employed to selectively aggregate these pairwise relationships based on their contextual relevance. The model is trained using auxiliary supervision that minimizes the error in predicting the relative distances between the target and its reference points, thereby enhancing the accuracy of the final distance estimation.

To guide the model in focusing on the most informative reference targets, an attention-based module is introduced. Inspired by Google's VectorNet, the model constructs embeddings that capture both local and global contextual information. Specifically, local pairwise relationships between the designated target and each reference are encoded individually, while global embeddings are generated by applying a multi-layer perceptron (MLP) followed by average pooling across all reference features. This combination allows the model to weigh reference targets according to their overall contribution to accurate distance prediction.

Then, the global embedding is connected with the original designated target-reference embedding. Considering these global-local embeddings, the attention module predicts and normalizes the importance weights of each reference with a fully connected layer and softmax layer. Finally, the original designated target-reference embedding is weighted averaged with these weights to obtain a final embedding.

Predicting the designated target distance from the designated target embedding is simpler than predicting the designated target distance from other indirect clues (such as reference embedding). To encourage the model to learn the pair relationship between the designated target and the reference target, rather than short-circuiting clues, additional supervision is provided in Ref. [15]. The design of this additional supervision is similar to residual representation, which is widely used in computer vision to assist the optimization process.

Specifically, during the training phase, a relative (or residual) distance head is added to each designated target-reference target embedding. The relative distance Δd between the designated target and the reference target, i.e., $\Delta d = d_\mathrm{t} - d_\mathrm{r}$, where d_t is the distance of the designated target and d_r is the distance of the reference target.

5.3 Depth Map Estimation of Monocular Vision

This section is about directly estimating the depth map of a monocular image, i.e., the depth value of each pixel. In this way, it is similar to a virtual depth sensor, like Microsoft's Kinect depth sensor. If it is inversely projected into 3D space, it forms a virtual point cloud, similar to the work of a lidar. Section 5.4 directly detects 3D obstacles from monocular images, obtains 3D frames, and actually implicitly obtains the depth data of 3D obstacles, just ignoring the depth of the background area.

The depth estimation problem in the field of computer vision belongs to 3D reconstruction, i.e., shape recovery (shape from X). This X includes stereo vision, multiple-view stereo, silhouette, motion (including SfM and SLAM), focusing, hazing, shading, occlusion, texture, vanishing points, etc. The first five are multi-image inputs, deducing depth distance from spatial geometry, temporal transformation and focal length change, and the rest are monocular inputs.

5.3.1 Traditional Methods

If image blur is explicitly modeled, the response of image edge blur and its first and second derivatives can be used to infer depth information from a single image. This approach, known as depth from defocusing, leverages the relationship between the amount of blur and the distance of objects from the camera. By analyzing the spatial variation of edge sharpness and its derivative patterns, it becomes possible to estimate scene depth even in the absence of stereo or motion cues.

Another phenomenon that reveals depth information is atmospheric scattering, which causes haze in images. This effect enables depth recovery from haze by leveraging the physics of light propagation through the atmosphere. In general, when an image is captured in hazy conditions, the depth of each pixel can be inferred using a scattering model. The observed image brightness C is related to the scene depth z through an exponential attenuation function. Specifically, the brightness can be modeled as follows:

$$\tilde{C} = C_0 e^{\beta z} + S(1 - e^{-\beta z}) \tag{5.35}$$

where C_0 is the intrinsic brightness of the scene without atmospheric scattering, S is the brightness of the sky or ambient light, and β is the scattering coefficient that characterizes the density of the haze. By inverting this model, it is possible to estimate depth from single hazy images.

The above two tasks can be considered as two different processing methods of image restoration (deconvolution) algorithm: non-blind image restoration algorithm and blind image restoration algorithm.

Variations in shading across an object's surface can reveal depth information, giving rise to the technique known as shape-from-shading (SfS). This approach estimates the 3D shape of a surface by analyzing the relationship between image brightness and surface geometry. Similar to Structure from Motion (SfM), SfS is an ill-posed problem that requires additional constraints to yield a unique and stable solution. One common constraint is the assumption about the reflectance properties of the object surface. Typically, SfS methods are built upon one of four surface reflectance models: purely Lambertian surfaces, purely specular (mirror-like) surfaces, mixed reflectance surfaces, or more complex hybrid models. Among these, the Lambertian model is most widely adopted, where the surface reflects light uniformly in all directions, resulting in brightness that depends only on the angle between the surface normal and the light source. Solving the SfS objective function, often formulated as an energy minimization or integral equation, is computationally demanding and typically requires advanced numerical methods similar to those used in finite element analysis.

An example application of SfS is the reconstruction of facial structure, where variations in skin shading across a 2D image are used to infer the underlying 3D geometry of the face. This technique demonstrates how subtle changes in image brightness can encode detailed surface curvature, enabling depth estimation from a single image.

The deformation of texture in an image provides important cues about the underlying shape of the textured surface. By analyzing the way texture patterns distort due to surface curvature and perspective, it is possible to infer the surface geometry. From such analysis, a surface normal map can be estimated to capture the orientation of the surface at each point, and a corresponding depth map can be derived to represent the relative distance of each point from the camera.

Texture segmentation is a necessary and challenging foundation for solving the optimization problem of recovering 3D shape from texture, also known as *shape from texture*. To address this problem, several constraints on the texture elements (textels) are typically imposed, including **homogeneity** (uniformity of texture across a region), **isotropy** (uniformity in all directions), and **stationarity** (invariance of statistical properties across the image). These constraints help guide the estimation of surface orientation and depth from texture distortion.

Another important cue for depth estimation is *occlusion*, with *curvature* being a key manifestation. This leads to the approach of *shape from curvature*, where *isophotes*—closed contours of equal image intensity—are analyzed. These isophotes can be extracted by applying thresholds to grayscale images, and their curvature provides clues to the underlying 3D shape. By varying the grayscale threshold across the full intensity range (e.g., 0 to 255), multiple isophotes can be obtained. Estimating the depth associated with each isophote and averaging the results produces an approximate depth map of the scene.

Despite the utility of these approaches, a central challenge remains: *segmentation*. Accurate segmentation is often required as a basis for both texture- and occlusion-based depth inference. However, segmentation itself is a difficult problem in computer vision, often described as the "chicken-and-egg" problem—since reliable segmentation depends on depth cues, and accurate depth estimation often relies on prior segmentation.

Occlusion provides valuable cues for depth estimation, and one of its geometric manifestations is curvature—leading to the approach known as shape recovery from curvature. A key concept in this context is the isophote, which refers to a contour line of constant brightness within an image. Isophotes can be extracted by applying grayscale thresholds, and their curvature can be used to infer depth corresponding to each threshold level. By sweeping the threshold over the full grayscale range, typically $[0, 255]$, and computing depth estimates at each level, a superimposed average depth map can be generated to capture overall scene structure. However, accurate occlusion reasoning often depends on reliable image segmentation, which remains a persistent challenge in computer vision—frequently described as the "chicken-and-egg" problem, since good segmentation requires depth information, and vice versa.

Finally, consider the concept of the vanishing point, which arises when a set of parallel lines on a given plane appear to converge under perspective projection. This geometric phenomenon allows for depth recovery from geometrical perspective. Specifically, when the vanishing point is identified, the corresponding plane can be analyzed to infer relative depth: regions closer to the vanishing point are assigned larger depth values, reflecting their greater distance from the camera. This approach is particularly effective in artificial environments—such as indoor scenes—where parallel structural elements (e.g., walls, floors, and ceilings) are common and well defined. By leveraging the regularity of these geometric cues, depth maps can be generated without relying on stereo or motion-based techniques.

One of the earliest approaches to estimating depth maps using machine learning was based on Markov Random Fields (MRFs). In this framework, depth estimation is formulated as an energy minimization problem, where the energy function typically consists of a data term and a connectivity (or smoothness) term. The data term enforces consistency between the observed image features and the estimated depth values, while the connectivity term imposes spatial smoothness by encouraging neighboring pixels to have similar depths, unless strong evidence suggests otherwise. By optimizing this energy function, MRF-based methods produce depth maps that balance local accuracy with global coherence.

Using super-pixels obtained from image segmentation can lead to smoother and more coherent depth estimation results. This approach groups pixels with similar visual properties into regions, allowing the model to enforce spatial consistency more effectively. A well-known system that employs this technique is Make3D, which demonstrated that leveraging super-pixels can significantly enhance the quality of single-image depth prediction by reducing noise and preserving structural boundaries across the scene.

Another method, as described in Ref. [16], applies machine learning to perform semantic segmentation following image segmentation. In this approach, pixels are categorized into coarse semantic classes such as "ground," "sky," and "vertical" surfaces (e.g., buildings or trees). Based on these labels, a simplified 3D representation is constructed using billboard geometry, where vertical surfaces are represented as flat planes textured with image patches. This technique creates a "pop-up" 3D scene that gives an impression of depth and spatial layout, offering a lightweight and visually intuitive form of 3D reconstruction from a single image.

Another approach frames depth map estimation as a search problem, based on the assumption that visually similar images tend to share similar depth structures. In this method, a database of images with known depth maps is used to retrieve the most similar images to a given input. The depth maps of these retrieved images are then fused to generate an initial estimate for the input image. This preliminary depth map is further refined through smoothing techniques to enhance spatial consistency. Finally, image-based rendering (IBR) techniques—specifically depth-based image rendering (DBIR)—are applied to synthesize stereo image pairs, enabling 3D scene visualization. For a detailed description of the algorithmic framework, see Ref. [17].

For video sequences, optical flow can be leveraged to perform motion segmentation, which provides valuable temporal information for depth inference. Building upon the previously described image-based methods, this approach incorporates motion cues by segmenting moving objects and background regions based on their distinct motion patterns. The temporal consistency across frames helps to refine depth estimates, especially in dynamic scenes. By integrating motion segmentation with depth fusion and smoothing techniques, more accurate and stable depth maps can be generated over time. For a detailed explanation of this algorithmic framework, see Ref. [18].

Another improvement involves using dictionary learning to optimize the depth estimation process. In this method, a clustering-based dictionary is learned during the training phase by associating RGB image patches with their corresponding depth patches. During inference, the input RGB image is decomposed into local patches, which are matched against the trained RGB-depth dictionary pair to predict depth values. This allows for efficient and accurate depth reconstruction by leveraging learned correlations between appearance and geometry. The dictionary learning framework significantly accelerates the search process while improving the quality of the estimated depth map. For further details, see Ref. [19].

5.3.2 Deep Learning Methods

Inferring depth information from a single image (monocular depth estimation) is an ill-posed problem. With the rapid development of neural networks, depth estimation based on deep learning has been widely researched in recent years and has achieved good accuracy. At the same time, deep neural networks estimate dense depth maps from a single image in an end-to-end manner.

Reference [20] categorizes depth estimation from the perspective of training ground truth into supervised methods, unsupervised methods, and semi-supervised methods. Although the training process of unsupervised and semi-supervised methods depends on monocular videos or stereo image pairs, the network only predicts depth maps from a single image during the testing period.

Unsupervised methods mainly use the same video's previous and next frames as stereo vision pairs (with an additional pose estimation task) to construct a depth estimation framework. Compared with supervised and semi-supervised methods, unsupervised methods learn from geometric constraints rather than learning depth from ground truth. Therefore, the training process depends on the monocular sequence captured by the camera. However, since the monocular sequence does not contain absolute scale information, unsupervised methods have problems such as scale ambiguity, scale inconsistency, and occlusion.

If real stereo vision sequences are used to train the depth estimation model, the learning of this model is called a semi-supervised method. Similarly, using sparse lidar point clouds as supervision signals is also similar.

Furthermore, incorporating camera parameters into the depth estimation framework and using domain adaptation techniques during the training process will significantly improve the transferability of the depth network, which is a hot topic in recent years.

First, the spatial constraints of binocular stereo vision are used for unsupervised learning of monocular depth estimation, see Refs. [21–23].

The algorithm proposed in Ref. [22] involves a depth estimation framework that outputs left and right disparity maps, denoted as d^l and d^r, through a multi-scale prediction pipeline. The training process minimizes a composite loss function that includes three key components: an appearance matching loss (C_{ap}), which compares the original images $\boldsymbol{I}^l/\boldsymbol{I}^r$ with their reconstructed counterparts $\tilde{\boldsymbol{I}}^l/\tilde{\boldsymbol{I}}^r$; a disparity smoothness loss (C_{ds}), which penalizes high gradients in the disparity maps to encourage spatial smoothness; and a left-right consistency loss (C_{lr}), which enforces agreement between the predicted disparity and its corresponding reprojected counterpart, i.e., between d^l and $d^l(d^r)$, and between d^r and $d^r(d^l)$. These loss terms are computed at four different output scales to improve the network's robustness and multi-resolution accuracy. The architecture uses convolutional layers (C), up-convolution layers (UC), bilinear interpolation (S), and upsampling (US), while skip connections (SCC) are employed to preserve spatial features across scales.

The model architecture presented in Ref. [21] replaces the fully connected layers of AlexNet with fully convolutional layers, each comprising 2048 filters of size 5×5. This architectural change significantly reduces the number of parameters and enables the network to handle variable-sized inputs during testing. More importantly, the fully convolutional design preserves the spatial structure of the image, which is crucial for depth estimation tasks. After the seventh layer (L7), the network performs stepwise upsampling of the predictions, aligning with the requirements of a stereo vision autoencoder.

A key challenge in deep convolutional networks is that image details tend to be lost in the deeper layers. To address this, the model employs a "skip-line structure" that fuses coarse depth predictions from deeper layers with fine-grained local information from earlier layers. This fusion leads to sharper and more accurate depth maps. Starting from the L9 output, which generates a 22×76 depth map, the network uses a series of simple 4×4 convolutional layers. Each layer progressively refines the coarse low-resolution depth map into a higher-resolution estimate, ultimately producing a detailed depth prediction that better captures object boundaries and spatial variation.

The blocks C, P, L, F, and D in the figure correspond to convolution, pooling, local response normalization, FCN (fully convolutional network), and upsampling layer. FCN blocks F1 and F2 upsample predictions from layers (L7, L8) and combine them with the inputs of pooling layers P3 and P2, respectively.

Reference [23] presents an algorithm for depth estimation along with a carefully designed loss function. The algorithm follows a structured pipeline that typically includes feature extraction, depth prediction, and refinement stages. Its loss function integrates multiple components—such as reconstruction loss, smoothness regularization, and consistency constraints—to guide the network toward producing accurate and spatially coherent depth maps. This combination ensures robust performance across diverse visual scenes.

(1) Using the principle of binocular geometry, depth prediction is learned directly from stereo camera images in an unsupervised manner; this is achieved by directly calibrating one stereo image to another; this process only requires known camera

calibration parameters and the depth map predicted by CNN. This semi-supervised method can learn from both supervised and unsupervised cues.

(2) Supervised, unsupervised, and regularization terms are seamlessly combined in the semi-supervised loss function formula; compared to previous methods, this method treats both cameras equally in a stereo vision setting; all three loss components define the cameras in a symmetrical way, which implicitly enforces the consistency of the predicted depth maps between the cameras.

(3) A deep residual network architecture is used in the encoder–decoder scheme, which includes long skip connections for fine detail prediction at the output resolution; the input to the network is an RGB image, and the encoder is similar to the ResNet-50 architecture (without fully connected layers), continuously extracting low-resolution high-dimensional features from the input image; the encoder subsamples the input image at five levels, the first level convolves the image to half the input resolution, and each subsequent level stacks multiple residual blocks; the decoder uses residual blocks to extract the encoder's output; long skip connections are added between corresponding layers in the encoder and decoder, slightly improving the performance of all measures without affecting convergence; in addition, the network can predict more detailed depth maps than without skip connections; convolution outputs are generally only activated by the ReLU function, except at one point, i.e., at the input point of the sum of the residual blocks, where ReLU is used after the sum operation.

In the algorithm described in Ref. [23], several key variables are used to define the depth estimation framework. Let $\boldsymbol{I}$ denote the input image, and let $\rho(\boldsymbol{x})$ represent the inverse depth estimate at pixel location $\boldsymbol{x}$. The corresponding true depth at that pixel is denoted by $Z(\boldsymbol{x})$. The function $\omega(\boldsymbol{x}, \rho)$ models the deformation of the image at pixel $\boldsymbol{x}$ as a function of the inverse depth ρ, capturing how the pixel shifts under changes in viewpoint. Additionally, ϕ is a smoothing function applied to enforce spatial regularity, often defined based on the image gradient to preserve edges while promoting consistency in flat regions. Together, these components contribute to the formulation of a loss function that balances depth accuracy, deformation consistency, and smoothness in the predicted depth map.

A natural approach to improving depth estimation in monocular videos is to leverage inter-frame motion information, allowing the model to infer scene geometry through temporal cues. This idea is implemented as a form of dual-task joint training, as described in Ref. [24], where depth estimation and camera motion (or pose) estimation are learned simultaneously. By jointly optimizing these two tasks, the model can enforce geometric consistency across frames, leading to more accurate and temporally stable depth predictions.

(1) The end-to-end method allows the model to directly map from input pixels to self-motion estimation (parameterized as a 6-DoF transformation matrix) and scene structure (parameterized as a pixel depth map under the reference viewpoint); view synthesis is used as a measure, in a completely unsupervised mode, simply using unlabeled image sequences or even camera motion information for training. The basis of this method is that the geometric view synthesis system can only consistently perform when its scene geometry and camera pose intermediate predictions

correspond to physical basic facts (GT). Although imperfect geometric shapes and/or pose estimates are acceptable for some scenes (such as textureless) in terms of view synthesis performance, the same model will fail miserably in another set of scenes with more diverse structures. Therefore, the goal here is to formulate the entire view synthesis pipeline as the inference process of a convolutional neural network, so that in the "meta" task of training view synthesis on large video data, the network is forced to learn about the intermediate tasks of depth and camera pose estimation in order to arrive at a consistent interpretation of the visual world.

(2) Is a supervised flowchart based on view synthesis: The depth network only takes the target view as input and outputs the pixel depth map $\hat{D}_t$. First, the pose network takes the target view ($\boldsymbol{I}_t$) and nearby/source views (such as $\boldsymbol{I}_{t-1}$ and $\boldsymbol{I}_{t+1}$) as input and outputs the relative camera poses ($\boldsymbol{T}_{t\to t-1}$, $\boldsymbol{T}_{t\to t+1}$). Then, the outputs of the two networks are used to warp the source view to reconstruct the target view, and its photometric reconstruction loss value is used to train the CNN. Finally, the entire framework is trained from videos in an unsupervised manner using view synthesis as a supervisory measure.

(3) Is the network architecture of the depth/pose/explainability prediction module: The width and height of each rectangular block represent the feature map output channels and spatial dimensions of the corresponding layer, respectively, and each decrease/increase in size represents a change by a factor of 2. Among them, (a) uses the DispNet architecture with multi-scale edge prediction for monocular depth, except for the first 7 convolutional layers with kernel sizes of 7, 7, 5, 5, all other layers are 3. The number of output channels of the first convolutional layer is 32; the pose and explainability networks share the first few convolutional layers, then one branch predicts the 6-DoF relative pose, and the other branch gets the multi-scale explainability mask. The number of output channels of the first convolutional layer is 16, except for the first two convolutions and the last two deconvolutions/prediction layers with kernel sizes of 7, 5, 5, 7, all other layers are 3.

The method proposed in Ref. [25] introduces joint training of depth maps and surface normal maps within a unified framework called GeoNet, a geometric neural network. By jointly estimating both depth and surface normals, the network leverages the intrinsic geometric relationship between them to enhance prediction accuracy. This coupled learning strategy allows GeoNet to enforce consistency between depth gradients and surface orientations, leading to more precise and coherent 3D scene reconstruction from monocular input.

(1) On the one hand, surface normals are determined by the local surface tangent plane of the 3D point cloud, estimated directly from the depth. On the other hand, the depth is constrained by the local surface tangent plane determined by the surface normals. Although it seems simple, designing a neural network to correctly utilize these geometric conditions is not easy. One possibility is to build a convolutional neural network (CNN) to directly learn this geometric relationship from the data, but it is also impossible to obtain from the depth, and even cannot approach reasonable normal results.

(2) Due to noise and other issues, the least squares module occasionally produces noisy estimates of surface normals. To improve accuracy, a residual module is used, which includes a three-layer CNN with a skip connection and a 1×1 convolution layer, aiming to smooth noise and combine the initial guess of surface normals to further improve estimation accuracy. Specifically, before being sent to the 1×1 convolution, the output of the CNN is connected with the initial estimate of the surface normals, which can be the output of another network. The structure of the depth-to-normal network is shown at the bottom of the figure, explicitly utilizing the geometric relationship between depth and surface normals, the network avoids the difficulty of learning geometrically consistent surface normals. More importantly, using GT depth as the input to the pre-trained network allows it to be connected and jointly fine-tuned with other networks, thus predicting depth maps from original images.

(3) Through the above two networks, a complete model can be built. First, a two-stream CNN is used to predict the initial depth and initial normal map. The basic structure used is VGG-16 and ResNet-50, based on an initial depth map predicted by a CNN, the depth-to-normal network is used to refine the normal map. Similarly, given the surface normal estimate, the normal-to-depth network is used to refine the depth map. The depth-to-normal network is pre-trained with GT depth maps, and for the normal-to-depth network, there is no need to learn any weights.

The unsupervised learning approach proposed in Ref. [26] jointly estimates depth, optical flow, and camera pose within a unified framework. The method operates in two stages: the rigid structure inference stage and the non-rigid motion refinement stage. In the first stage, the model infers the static layout of the scene using two dedicated sub-networks—DepthNet and PoseNet—which, respectively, regress the depth map and the relative camera pose. These outputs are fused to compute the rigid optical flow, capturing motion induced by the camera in a static environment.

To account for dynamic elements in the scene, the second stage employs ResFlowNet, which estimates the residual non-rigid flow that cannot be explained by rigid motion alone. This residual flow is then combined with the rigid flow to produce the final optical flow prediction. By assigning each sub-network a specialized subtask, the overall problem of understanding complex scene geometry is decomposed into simpler and more manageable components.

The framework relies on view synthesis as its primary supervisory signal: synthesized views from predicted depth and pose are compared against actual views to guide learning, despite the absence of ground truth. Additionally, continuity checks are applied during training to enforce spatial and temporal consistency across predictions, further enhancing the model's robustness in unconstrained environments.

Reference [27] proposes an end-to-end self-supervised framework, Dyna-DM, which uses continuous monocular video data to learn depth and motion. It introduces the Dice coefficient to detect real dynamic targets.

To self-supervise depth estimation, consider the continuous RGB frames $(\boldsymbol{I}_1, \boldsymbol{I}_2)$ in the video sequence, and estimate the depth $(\boldsymbol{D}_1, \boldsymbol{D}_2)$ of these two images with the reciprocal output by the disparity network (DispNet). To estimate the motion in the scene, two trainable pose networks (PoseNet) with different weights are used, one for optimizing self-motion and the other for handling target motion.

First, corresponding to the images ($\boldsymbol{I}_1, \boldsymbol{I}_2$), erase all potential dynamic target masks ($\boldsymbol{M}_1^i, \boldsymbol{M}_2^1$), and the **initial background images can be calculated**. These masks are obtained with the off-the-shelf instance segmentation model R-CNN.

The next step is to estimate self-motion ($\boldsymbol{P}_{1\to2}^{i=0}, \boldsymbol{P}_{2\to1}^{i=0}$) with these two background images. First, define backward and forward warping:

$$\mathcal{F}_{\text{fw}}(\boldsymbol{I}_1, \boldsymbol{D}_1, \boldsymbol{P}_{1\to2}, \boldsymbol{K}) \to \hat{\boldsymbol{I}}_{1\to2}^{\text{fw}}), \qquad \mathcal{F}_{\text{iw}}(\boldsymbol{I}_1, \boldsymbol{D}_1, \boldsymbol{P}_{2\to1}, \boldsymbol{K}) \to \hat{\boldsymbol{I}}_{1\to2}^{\text{iw}}) \quad (5.36)$$

where $\boldsymbol{K}$ is the camera parameter, fw is forward deformation, and iw is backward deformation. With the calculated self-motion, forward warp a pair of image-masks to produce forward warped image-mask pairs ($\hat{\boldsymbol{I}}_{1\to2}^{\text{fw}}, \hat{\boldsymbol{I}}_{1\to2}^{\text{fw},i=n}$). Feed the pixel product of the forward warped mask-image and the target mask-image ($\boldsymbol{M}_2^{i=n}, \boldsymbol{I}_2$) into the object motion network to generate the target motion estimation ($\boldsymbol{P}_{1\to2}^{i=n}, \boldsymbol{P}_{2\to1}^{i=n}$) for each potential dynamic target, $n = 1, 2, 3, \cdots$.

Through these target motion estimations, each target $\hat{\boldsymbol{I}}_{1\to2}^{\text{fw}\to\text{iw},i=n}$, $n = 1, 2, 3, \cdots$, and the background image $\hat{\boldsymbol{I}}_{1\to2}^{\text{iw},i=0}$ can be backward warped. The final warped image is a combination of the background warped image and all warped potential dynamic targets:

$$\hat{\boldsymbol{I}}_{1\to2} = \hat{\boldsymbol{I}}_{1\to2}^{\text{iw},i=0} + \sum_{n\in(1,2,\cdots)} \hat{\boldsymbol{I}}_{1\to2}^{\text{fw}\to\text{iw},i=n} \quad (5.37)$$

To detect truly dynamic targets, first define an initial **self-motion network**, which uses the pre-trained weights in the Insta-DM self-motion network. The masks of Insta-DM will also be used as the initial masks containing all potential dynamic targets.

$$\boldsymbol{M}_{1,2}^{\text{initial}} = (1 - \cup_{n\in(1,2,\cdots)} \boldsymbol{M}_1^{i=n} \cap (1 - \cup_{n\in(1,2,\cdots)} \boldsymbol{M}_2^{i=n} \quad (5.38)$$

Note: *Insta-DM is an end-to-end joint training framework designed to unsupervisedly model the six-degree-of-freedom motion, self-motion, and depth of multiple dynamic targets using only monocular camera input. To handle the challenges posed by dynamic scenes, the framework applies forward warping to dynamic targets—preserving their geometric consistency—and backward warping to static background regions, thereby avoiding distortion or stretching of moving objects. The architecture consists of a DepthNet and a PoseNet, both built upon a ResNet18-based encoder–decoder structure. These components are jointly trained to estimate scene depth and relative camera pose while accommodating object-level motion, enabling accurate depth and motion prediction in complex, unconstrained environments.*

With the pre-trained weights of the self-motion network, the background image is found using pixel multiplication, and the initial self-motion estimation between frames $\boldsymbol{I}_1$ and $\boldsymbol{I}_2$ is determined:

$$\boldsymbol{P}_{1\to 2}^{\text{initial}} = \text{Ego}(\boldsymbol{M}_{1,2}^{\text{initial}} \odot \boldsymbol{I}_1, \boldsymbol{M}_{1,2}^{\text{initial}} \odot \boldsymbol{I}_2) \tag{5.39}$$

Then, forward warp all these masks of potential dynamic targets with this initial self-motion.

Assuming the accuracy of pose and depth is acceptable, it means that if a target is represented by self-motion, i.e., if the target is static, the target mask will warp to the target mask in the target image. In other words, if a target is static, there will be a significant overlap between the warped mask (performing initial self-motion) and the mask of the target image. Conversely, if the object is truly dynamic, the warped mask and the mask of the target image will not match.

This type of overlap or difference can be captured using the Dice coefficient or Jaccard index (also known as the IoU index):

$$\text{Dice}(\hat{\boldsymbol{M}}_{1\to 2}^{i=n}, \hat{\boldsymbol{M}}_{2}^{i=n}) = \frac{2|\hat{\boldsymbol{M}}_{1\to 2}^{i=n} \cap \hat{\boldsymbol{M}}_{2}^{i=n}|}{|\hat{\boldsymbol{M}}_{1\to 2}^{i=n}| + |\hat{\boldsymbol{M}}_{2}^{i=n}|} < \theta \tag{5.40}$$

$$\text{Jaccard}(\hat{\boldsymbol{M}}_{1\to 2}^{i=n}, \hat{\boldsymbol{M}}_{2}^{i=n}) = \frac{|\hat{\boldsymbol{M}}_{1\to 2}^{i=n} \cap \hat{\boldsymbol{M}}_{2}^{i=n}|}{|\hat{\boldsymbol{M}}_{1\to 2}^{i=n} \cup \hat{\boldsymbol{M}}_{2}^{i=n}|} < \theta \tag{5.41}$$

where θ is the threshold.

The Dice value of a stationary target is greater than that of a dynamic target. Potential dynamic targets with Dice values lower than the selected θ value will be classified as true dynamic targets. By testing IoU and Dice, it was found that the Dice coefficient can more accurately detect dynamic targets, so the recommended solution is to use the Dice coefficient. The optimal θ value is found to be within the range of [0.8, 1].

However, the above reasoning is based on the assumption of accurate depth and pose estimation. In most cases, this is unrealistic, so it can be expected that the Dice value of dynamic targets will be larger. To mitigate this challenge, a larger frame interval can be used between the source frame and the target frame. Only the reconstruction between frames t and $t+2$ is calculated, not the reconstruction between frames t and $t+1$. The additional distance between frames provides more time for dynamic targets to deviate from their self-motion distorted position, which is beneficial for slowly moving targets.

With the frame rate of modern cameras, the additional distance will not cause jitter, but will make depth estimation more consistent and pose estimation more accurate. Since these targets may move very slowly, calculating the reconstruction between larger frames helps to determine imperceptible target motion.

To remove static targets, all potential dynamic targets are sorted in descending order according to their Dice values, and the top 20 targets are selected. These targets are most likely to be static and are often larger targets. If θ is less than 1, these targets with larger Dice values are removed first.

In summary, filtering with Dice numbers removes all static targets from the initial mask, providing an updated mask that only contains true dynamic targets. This filtering process is called ϑ filtering, and each true dynamic target can be used to calculate updated self-motion estimation and target motion estimation.

Further improvements are made in two aspects: removing small targets and reconsidering invariant pose.

Small targets are defined as objects occupying a very limited number of pixels in an image—typically less than 1% of the total image pixels. These targets are often located far from the camera, making their motion estimation unreliable or visually insignificant. In the Dyna-DM framework for dynamic target extraction, both the input image and an initial background mask are used to estimate self-motion. Prior to removing static targets, small targets—specifically those occupying less than 0.75% of the image pixels—are filtered out to refine the background mask. This updated mask is then further refined using a θ filter, which excludes targets whose Dice coefficient exceeds 0.9, as such high overlap suggests static behavior. Ultimately, only targets identified as truly dynamic are retained for updating the background mask, ensuring accurate separation of moving objects from the static scene.

Second, when calculating the pose in 3D space, some methods use the relationship between forward pose and backward pose as a loss function, called invariant pose:

$$\mathcal{L}_1 = |\boldsymbol{P}_{1\to 2} + \boldsymbol{P}_{2\to 1}| \tag{5.42}$$

The photometric consistency loss $\mathcal{L}_{pe}$ will be the main self-supervised loss function. According to Insta-DM, the consistency is applied to the photometric loss function with the weighted effective mask $\boldsymbol{V}_{1\to 2}(p)$, dealing with invalid instance areas, exiting viewpoints, and occlusions:

$$\mathcal{L}_{pe} = \sum_{p\in P} \boldsymbol{V}_{1\to 2}(p) \cdot [(1-\alpha)\cdot |\boldsymbol{I}_2(p) - \hat{\boldsymbol{I}}_{1\to 2}(p)| + \frac{\alpha}{2}(1 - \mathrm{SSIM}(\boldsymbol{I}_2(p), \hat{\boldsymbol{I}}_{1\to 2}(p)))] \tag{5.43}$$

The geometric consistency loss $\mathcal{L}_g$ is a combination of the depth inconsistency map and the effective mask instance:

$$\mathcal{L}_g = \sum_{p\in P} \hat{\boldsymbol{M}}_{1\to 2}(p) \cdot \boldsymbol{D}^{\mathrm{diff}}_{1\to 2}(p) \tag{5.44}$$

To smooth textures and noise while maintaining sharp edges, an edge-aware smoothing loss is used:

$$\mathcal{L}_s = \sum_{p\in P} (\nabla \boldsymbol{D}_1(p) \cdot e^{-\nabla \boldsymbol{I}_1(p)})^2 \tag{5.45}$$

Finally, due to the rare cases of moving targets moving at the same speed as the camera, a target height constraint loss $\mathcal{L}_h$ is used:

$$\mathcal{L}_h = \sum_{n \in \{1,2,\cdots\}} \frac{1}{\bar{D}} \cdot \left| D \odot M^{i=n} - \frac{f_y \cdot p_h}{h^{i=n}} \right|_1 \tag{5.46}$$

Finally, all the losses are added together:

$$\mathcal{L} = \lambda_1 \mathcal{L}_1 + \lambda_{pe} \mathcal{L}_{pe} + \lambda_g \mathcal{L}_g + \lambda_s \mathcal{L}_s + \lambda_h \mathcal{L}_h \tag{5.47}$$

Reference [28] is a method for estimating depth maps from multiple frames of a video. Multi-frame depth estimation methods have successfully improved depth accuracy by directly using the geometric constraints of multi-view stereo (MVS). Unfortunately, MVS is often affected by textureless areas, non-Lambertian surfaces, and moving targets, especially in real-world video sequences with unknown camera motion and depth supervision. Reference [28] proposes a framework, MOVEDepth, that uses monocular cues and motion guidance to improve multi-frame depth learning.

MOVEDepth directly addresses the inherent issues of MVS, enhancing multi-frame depth learning. The key to this method is to use monocular depth as a metric priority to construct the MVS cost volume, and adjust the depth candidates of the cost volume under the guidance of predicted camera speed. By learning the uncertainty in the cost volume, the monocular depth and MVS depth are further fused to obtain a robust depth estimation against ambiguity in multi-view geometry.

Given a video sequence, the framework begins by using PoseNet to estimate the camera's self-motion and speed between two consecutive frames, specifically between frame $T-1$ and frame T. Following this, DepthNet is employed to predict the monocular depth of the current frame. These two outputs—camera self-motion and monocular depth—serve as priors for performing homographic warping between encoded frame features, allowing the network to align image features across time. The resulting cost volume, which captures the pixel-wise similarity across warped views, is then decoded into both a depth map and an uncertainty map. These outputs provide essential guidance for the subsequent fusion of monocular depth with multi-view stereo (MVS) depth information. A notable aspect of this framework is its adaptive adjustment of the depth sampling range during homographic warping, based on the predicted camera speed. This mechanism mitigates issues caused by slow or static camera motion, which can otherwise degrade the quality of depth estimation in video sequences.

Multi-view warps the source image to the reference camera frustum to form a cost volume, and the depth is estimated as the highest output value of the cost volume.

MOVEDepth uses clues from monocular methods to supplement the limitations of Multi-View Stereo (MVS). Typically, learning-based MVS samples depth candidates within a fixed range, which is calculated by the SfM open-source software COLMAP or learned by the network. However, dense search for depth range across the entire scene is computationally expensive and cannot produce accurate depth. Coarse-to-fine sampling or confidence-based sampling can reduce the depth range. However, in the self-supervised multi-frame depth learning framework, these samplings are limited because they ignore the triangulation prior of nearby frames.

To this end, MOVEDepth proposes a speed-guided depth sampling strategy. The key point is to associate the triangulation prior with the camera motion speed. That is, when the camera moves at high speed, the viewpoint changes significantly, providing sufficient triangulation prior knowledge for multi-view geometry. In contrast, slow/static video frames have similar viewpoints, so the contribution of triangulation prior is limited. For video frames with sufficient triangulation prior, the depth range is expanded to infer accurate depth, while for video frames with insufficient triangulation prior, the depth range is narrowed to a more reliable monocular priority, i.e., a deterministic sampling strategy, rather than probabilistic sampling.

The depth estimated by MVS methods is often degraded in textureless regions, on non-Lambertian surfaces, or in the presence of moving targets—challenges that are inherent to multi-view geometry. To address these limitations, MOVEDepth introduces a fusion strategy based on uncertainty estimation. In this approach, unreliable MVS depth values are selectively replaced with more trustworthy monocular depth predictions from the current frame.

The framework employs an uncertainty decoder that learns an uncertainty map directly from the depth probability distribution. This process involves two branches: the upper branch decodes the depth probability into a depth estimate using a Local-max operation, while the lower branch encodes the same depth probability into an uncertainty map. This uncertainty map serves as a reliability indicator, guiding the fusion process to balance MVS and monocular depth according to their respective confidences. By doing so, MOVEDepth enhances overall depth accuracy and robustness in challenging visual scenes.

5.4 Monocular Vision 3D Obstacle Detection

Before monocular 3D object detection, the field of autonomous driving mainly used 2D bounding box detection, which can be divided into two types: single-step and two-step methods. Generally, the single-step method is fast but low in accuracy, while the two-step method is slow but high in accuracy. Two-step methods, such as R-CNN, fast R-CNN, and faster R-CNN, first require region proposals (RP), which avoids searching the entire image; single-step methods do not have RP or RPN, like YOLO1/2/3, SSD, and RetinaNet, and use anchors (reference boxes) to improve accuracy. During training, a set of anchors of different scales and positions can be preset, covering almost all scales and positions. Methods that eliminate anchors also perform well, with the advantage of simple architecture and easy hardware implementation, such as CornerNet, ExtremeNet, and FoveaNet.

The 2D object ranging mentioned in Sect. 5.3 traditionally estimates the distance on the road based on the detection box landing point and camera posture. This section introduces a method for estimating 3D object detection boxes based on monocular images, which directly estimates the position, size, and orientation of the 3D object's rectangular box. The result is still projected on the image plane for display, and the calculation of distance is relatively simple, determined by the ratio of the object's size in the image to the actual 3D corresponding size and the focal length.

As a 3D object detection task, it is necessary to compare with the methods of lidar. According to Ref. [29], a simple review of lidar object detection is conducted.

3D object detection of lidar point cloud data can be divided into: ① point-based (input permutation invariant) methods; ② voxel-based (grid-based representation) methods; ③ distance view-based methods; and ④ multi-view-based methods. Structurally, it is similar to 2D image object detection, also divided into single-step and two-step methods.

The PointPillar architecture is a method based on points and voxels. First, the point cloud is divided into a grid in the $x - y$ coordinates, creating a set of voxel units (pillars). Each point in the cloud $(x, y, z$, reflectivity) is converted from a 4D vector to a 9D vector. These points are extended by the distance to the arithmetic mean of the pillar and the distance to the center of the pillar. Internally, PointNet (a 3D point cloud feature extraction network) is used to extract dense robust features from sparse pillars. Each 2D pillar is characterized by a fixed-size feature vector. The encoder learns point cloud representations from the pillars. Finally, a 2D CNN (improved SSD) is applied for joint bounding box and classification prediction.

CenterPoint is a two-stage detection method. In the first stage, it locates the center of an object, from which other attributes such as size, orientation, and velocity are estimated through separate regression heads. The network models 3D objects as rotation-invariant points, making it more flexible and accurate in handling rotating objects compared to anchor-based detectors. In the second stage, features from points within the initial bounding box estimate are aggregated into a vector, which is then fed into an MLP to predict confidence scores and further refine the box attributes.

The detection head is implemented as a heatmap in the bird's-eye view (BEV). Each local peak in the heatmap corresponds to the center of a detected object. After feature extraction from the input point cloud, the representation is passed to a 2D CNN detection head that produces a K-channel heatmap, with each channel corresponding to a semantic class. The peak value indicates the object center. During training, the true bounding box centers are projected onto the heatmap, and the detection head is optimized to fit a 2D Gaussian distribution around these projections using a focal loss.

Image-based 3D object detection methods can generally be divided into two categories: ① 2D feature-based methods first estimate the 2D attributes of an object on the image plane (e.g., direction, depth, or position) from 2D features, and then lift the 2D detections into 3D space. Since these approaches typically adopt architectures similar to those of conventional 2D detection models, they can be further classified according to standard 2D detection paradigms, such as region-based and single-stage methods. ② 3D feature-based methods directly predict objects in 3D space using 3D features. Depending on how the 3D features are constructed, these methods can be further divided into feature elevation-based and data elevation-based approaches.

Detecting 3D objects from RGB images is a challenging task due to the lack of depth information in the input data. To more accurately estimate 3D bounding boxes, auxiliary data is used and supplementary features are extracted from RGB images, such as automatic labeling and data augmentation with CAD models; supervision and depth correction with lidar signals; and additional data such as depth estimation, temporal sequences, and stereo vision to improve performance.

It is worth noting that 3D visual object detection based on BEV space has received industry attention and will be introduced separately in Sect. 5.14.

The following analyzes monocular depth learning-based 3D object detection methods through some references.

The method described in Ref. [30] estimates the full 3D pose and position of an object in three main steps. First, it predicts the 3D orientation of the object. Second, it regresses the object's physical dimensions and 3D center. Finally, these components are combined to recover the complete six-degree-of-freedom (6-DoF) pose. A central challenge in this pipeline is establishing accurate correspondences between the 2D bounding box detected in the image and the projected 3D bounding box in space.

Each side of a 2D bounding box could, in theory, correspond to any of the eight corner points of a 3D bounding box, leading to a total of $8^4 = 4096$ possible configurations. However, practical assumptions can significantly reduce this number. For instance, in many driving scenarios, objects such as cars can be assumed to remain upright, which restricts the top and bottom edges of the 2D bounding box to match only the top and bottom faces of the 3D box. This reduces the number of valid correspondences to 1024.

Further constraints apply when the object's tilt angle relative to the camera is small (close to 0°). In such cases, the vertical edges of the 2D bounding box (defined by x_{min} and x_{max}) can only correspond to projections of the vertical edges of the 3D box, typically parameterized as $[\pm d_x/2, \pm d_z/2]$. Similarly, the horizontal edges of the 2D box (defined by y_{min} and y_{max}) correspond to projections from the horizontal edges of the 3D box, represented as $[\pm d_y/2, \pm d_z/2]$. These geometric constraints reduce the number of valid configurations further to $4^4 = 256$.

In datasets such as KITTI, objects are typically assumed to have zero pitch and roll angles, meaning they remain perfectly upright. Under this assumption, the configuration space can be further reduced to just 64 plausible 2D–3D projection correspondences. This simplification greatly facilitates accurate estimation of an object's 3D pose from 2D image data.

The relationship between the local and global orientations of a detected vehicle can be characterized by two angular components. Let θ denote the global orientation of the obstacle vehicle with respect to a fixed world coordinate frame, and let θ_{ray} represent the viewing direction of the ego vehicle, defined as the line of sight from the ego vehicle to the target. The local orientation θ_1 is then defined as the relative angle between the target vehicle's heading direction and the viewing ray.

In practice, the vehicle's physical dimensions are known, such as its height, denoted by d_y. The local orientation θ_1 is calculated with respect to the viewing ray that intersects the 2D bounding box but may not precisely pass through the actual

3D center of the vehicle. The global orientation θ of the vehicle is then reconstructed by combining the ego vehicle's viewing angle with the locally estimated orientation: $\theta = \theta_{ray} + \theta_1$.

To achieve this, the network is trained to estimate the local orientation angle θ_1, which is more stable and less sensitive to global scene variation. This design allows for accurate recovery of the obstacle vehicle's global pose even when the viewing geometry is imperfect or when the vehicle is off-center within the image frame.

To address the challenge of object orientation estimation, the MultiBin method is proposed. This approach discretizes the orientation space into multiple overlapping bins, allowing neighboring bins to share angular regions for smoother prediction. During inference, the bin with the highest predicted confidence is selected, and the associated angle offset $\Delta\theta$ is used to compute the final orientation by combining it with the bin center.

The underlying CNN architecture is composed of three branches. The first branch regresses the physical dimensions of the object, including width, height, and depth. The remaining two branches are dedicated to orientation estimation: one predicts the confidence score for each discretized orientation bin, while the other regresses the trigonometric components—$\cos(\Delta\theta)$ and $\sin(\Delta\theta)$—for each bin. These components allow the angular offset to be recovered in a continuous manner, yielding improved accuracy and robustness compared to direct angle regression.

By combining classification and regression within overlapping bins, the MultiBin method enhances the precision of orientation estimation, especially in scenarios where objects appear in ambiguous or edge-aligned viewpoints.

In summary, this method estimates 3D bounding boxes from 2D bounding boxes using geometric constraints, where orientation estimation is crucial.

The method proposed in Ref. [31] extends the approach of Ref. [32] by integrating a tracking module into the detection pipeline. This module enhances the temporal stability of object detection and pose estimation across consecutive video frames. By preserving consistent object identities and smoothing estimated trajectories over time, it reduces frame-to-frame jitter and significantly improves the robustness of the system in dynamic environments.

The learning algorithm predicts the 2D projection of the 3D center of an object using a Region Proposal Network (RPN) as part of the detection pipeline. To improve spatial accuracy, the architecture adopts ROIalign from Mask R-CNN rather than the ROIpooling used in Faster R-CNN. This substitution reduces spatial quantization errors and improves alignment between feature maps and object regions. For each Region of Interest (ROI) used in estimating the 3D bounding box, the model extracts and aggregates the following elements: the 2D bounding box, the projected location of the 3D center, a confidence score indicating prediction reliability, and a feature vector representing the object's appearance and geometry. These components collectively enable accurate 3D localization and pose estimation from monocular images.

A key challenge in monocular 3D object detection is the discrepancy between the 2D bounding box center and the true 3D center of the object. The projected 2D center—visible in the image—may not accurately correspond to the actual 3D center, and can even appear above or below the ground plane when reprojected into 3D space.

This discrepancy becomes more severe in scenarios involving occlusion or truncation, where parts of the object are missing from the image. Such misalignments between the estimated center and the actual ground-truth (GT) center can introduce significant errors in localization, degrade the predicted trajectory accuracy, and ultimately harm overall detection and tracking performance.

Reference [33] introduces the Orthographic Feature Transform (OFT) to address the challenge of inferring 3D bounding boxes from monocular 2D images. The core idea of OFT is to bridge the gap between 2D image features and 3D spatial reasoning by transforming image-aligned features into a bird's-eye-view representation. Specifically, voxel-based features $g(x, y, z)$ are constructed by aggregating 2D image features $f(u, v)$ over regions that project into corresponding 3D voxels. These voxel features are then collapsed along the vertical (y) axis to produce orthographic features $h(x, z)$, which represent a top-down view of the scene. This transformation enables efficient reasoning about object position and size in the 3D space using only monocular input, facilitating accurate 3D bounding box prediction in a structured and spatially coherent manner.

Within a deep learning framework, the OFT module enables the mapping of image feature maps into an orthogonal BEV representation, facilitating structured spatial reasoning. The process begins with a ResNet-based front-end feature extractor that generates image-aligned features. These features are then projected into the BEV space using the OFT, effectively transforming the perspective view into a top-down representation aligned with the ground plane.

A dedicated top-down network operates on this BEV representation to perform 3D object detection. At each position on the ground plane, the network predicts several key outputs: a confidence score S indicating object presence, position offsets Δ_{pos} to refine localization, size offsets Δ_{dim} to estimate object dimensions, and angle vectors Δ_{ang} to determine orientation. This architecture enables precise and efficient monocular 3D detection by leveraging spatially organized BEV features derived from image inputs.

The method proposed in Ref. [34] follows a two-stage pipeline for 3D object detection using monocular input. In the first stage, a dedicated network is used to estimate a dense depth map from the input image. This estimated depth information is then passed into a Region Proposal Network (RPN) to generate candidate object regions, providing geometric cues that guide region selection. Although effective, this approach is computationally intensive due to the separate depth estimation step.

In the second stage, the predicted disparity map is transformed into a 3D point cloud representation, which is fed into the detection module. This module performs several tasks: 3D position regression to localize objects in space, multi-category classification to identify object types, 2D bounding box regression to refine object boundaries in the image plane, 3D pose regression to estimate orientation, and 3D scale regression to determine object dimensions.

Reference [35], a work by Microsoft, proposes a method called Instance Depth Estimation (IDE), which avoids computing a dense full-image depth map. Instead, IDE directly estimates the depth of each object's 3D bounding box, providing a more efficient and instance-centric solution for monocular 3D object detection. To improve spatial alignment between feature maps and object regions, the framework replaces the traditional ROIpool operation with Mask R-CNN's ROIalign.

The overall architecture consists of four key modules. First, a 2D detection module identifies object proposals within the image. Second, the IDE module estimates the depth of each detected object instance without relying on pixel-level depth maps. Third, a 3D position estimation module predicts the spatial location of each object in world coordinates. Finally, a local corner point regression module refines the 3D bounding box by estimating the positions of its eight corners in the object's local coordinate system. This modular design allows for accurate and efficient estimation of object depth and pose using only monocular image input.

The algorithm framework of IDE includes a specialized DE subnet that predicts the depth of the 3D bounding box center, denoted as Z_c in the ABBox-3D representation. The DE subnet processes image features to generate two feature maps at different spatial resolutions, which are then passed through separate depth encoding branches.

The high-resolution feature map is combined with the 2D bounding box information and passed through a fully connected layer to estimate object-specific depth. In parallel, the low-resolution (coarse) feature map is combined with grid-based spatial information and processed through another fully connected layer to estimate coarse-resolution depth across the scene. These two depth predictions—fine and coarse—are then fused to generate the final depth output, balancing global spatial context with object-specific precision. This multi-scale fusion approach allows IDE to produce accurate and robust depth estimates for each detected object instance without requiring a full-scene depth map.

In monocular 3D object detection, the objective is to localize and characterize objects of interest in 3D space using only a single RGB image. Each detected object is represented by a class label and a 3D bounding box (BBox-3D). The BBox-3D encloses the full spatial extent of the object, including portions that may be occluded or truncated in the image.

The BBox-3D is defined by two components: the global 3D center of the object, denoted as $C = (X_c, Y_c, Z_c)$, and a set of eight corner points $O = \{O_k\}$ $(k = 1, \ldots, 8)$. The center C is expressed in the camera coordinate system and represents the object's absolute position in the scene. In contrast, the corner points O are defined in the object's local coordinate system, encoding the object's shape and orientation relative to its own center. This separation of global and local context allows for consistent and accurate 3D bounding box estimation even when parts of the object are not visible in the image due to occlusion or truncation.

The concept of the Instance Depth Map (IDM) is designed to encode localized depth information at the grid level within a feature map. Given a grid G divided from the backbone feature map, each grid unit $g \in G$ is responsible for predicting the depth of the 3D center of the nearest object instance, provided that the instance lies within a specified distance threshold σ. This localized representation allows the network to incorporate instance-aware spatial geometry directly into the depth estimation process.

To enforce accurate associations between grid units and instances, each unit is assigned to the nearest instance within the σ range, ensuring that overlapping or adjacent objects are correctly disambiguated. The predicted depth values correspond to the estimated distances from the camera to the object centers, making this representation especially effective in scenes with multiple objects. By providing fine-grained supervision and supporting dense spatial reasoning, the grid-based approach enables the model to achieve accurate and consistent 3D localization across the entire image.

Reference [36], a collaboration between Face++ and UCLA, presents a monocular 3D object detection method that integrates depth estimation with geometric constraints. In addition to predicting depth, the method incorporates a road-surface assumption, which acts as a prior to constrain the vertical position of detected objects, helping to resolve scale ambiguities and improve localization accuracy.

The model also utilizes a morphable wireframe representation to describe object structure. However, rather than relying on explicit 3D landmarks—which are often sensitive to occlusion and appearance variations—the method adopts an ambiguous 3D bounding box representation to enhance robustness.

A key component of the framework is the 3D–2D consistency module, which ensures coherence between 3D predictions and 2D image observations. This module operates at two levels: the 3D hypothesis level, where candidate object poses are generated in 3D space, and the 2D pseudo-observation level, where projections of these hypotheses are compared to image features. This multi-scale consistency check enforces alignment between predicted 3D structures and the visual evidence present in the image, leading to more accurate and stable 3D pose estimation.

The 3D–2D continuity module is designed to bridge 3D geometric hypotheses and 2D image observations, enabling robust monocular 3D pose and shape estimation. The module incorporates both input-level and hypothesis-level information. On the 2D side, it utilizes image frames and detected 2D landmarks as pseudo-observations. On the 3D side, it generates dual-scale hypotheses, which include rigid 3D bounding box frames and deformable wireframe models that encode variations in object rotation and scale.

During inference, these 3D hypotheses are projected into the 2D image space and compared with the observed 2D frames and landmarks. This comparison is formalized in a loss function that penalizes misalignment between predicted and observed features. The optimization of this loss leads to the final estimation of the object's 3D pose and shape, ensuring consistency between the predicted 3D structure and the visual evidence in the image. This continuity mechanism allows the model to maintain geometric coherence across representations and improves the accuracy of object localization under monocular settings.

Reference [37], a master's thesis project, proposes a method for joint training of 2D and 3D bounding frames using a Fully Convolutional Network (FCN). The model is designed to predict both 2D object locations in the image and their corresponding 3D structural information. For the 3D bounding frame, the method defines three principal directions: front-bottom, left-bottom, and front-left, which help establish the spatial orientation of the object in 3D space.

The 3D bounding frame itself is composed of eight corner points and six planar faces, forming a cuboid representation of the object. This structured geometric definition provides a clear correspondence between the object's appearance in the image and its underlying 3D layout. By training on these representations, the FCN is able to learn to infer both 2D object locations and their full 3D spatial configurations from monocular imagery.

A comparison between 2D and 3D frame annotations highlights the difference in spatial representation: 2D annotations involve drawing rectangular bounding boxes around objects in the image plane, while 3D annotations define cubic bounding boxes that capture the object's orientation, depth, and full extent in three-dimensional space. Unlike 2D boxes, which provide only position and size in the image, 3D frames offer richer geometric information necessary for tasks such as pose estimation, object reconstruction, and spatial reasoning.

The ground assumption is still used, which is key to obtaining the 3D frame. Define the vertex **X** and the camera position C in the 3D world as 3×1 vectors, and project the point X to the camera $\mathbf{x} = [u, v, 1]^T$, with the center at **C**, rotation **R**, and calibration **K**, then the image projection formula is

$$\lambda\mathbf{x} = \mathbf{KR}(\mathbf{X} - \mathbf{C}) = [\mathbf{KR}| - \mathbf{KRC}]\begin{bmatrix}\mathbf{X}\\1\end{bmatrix} \tag{5.48}$$

For inverse projection, only the direction of the light can be reconstructed, and additional information is needed to determine the exact position of the vertex **X**, that is, λ. It is known that

$$\lambda\mathbf{x} = \mathbf{KRX} - \mathbf{KRC} \tag{5.49}$$

$$\mathbf{KRC} + \lambda\mathbf{x} = \mathbf{KRX} \tag{5.50}$$

$$\mathbf{C} + \lambda(\mathbf{KR})^{-1}\mathbf{x} = \mathbf{X} \tag{5.51}$$

This means that to get the direction of the light from **X**, you need to calculate the inverse of KR and the position of the camera **C**. This can be easily achieved from the image projection matrix $\mathbf{P} = \mathbf{KR}[\mathbf{I}| - \mathbf{C}]$. The inverse of KR is easy to calculate, and the camera center C can be calculated as follows:

$$\mathbf{C} = -(\mathbf{KR})^{-1}\mathbf{p}_4 \tag{5.52}$$

where $\mathbf{p}_4$ is the fourth column vector of matrix **P**. According to the inverse projection and ground assumption (if the plane equation $ax + by + cz + d = 0$ is known, then the normal vector $n = [a \quad b \quad c]^T$), the position of the 3D frame on the ground can be obtained first, and λ is solved as follows:

$$\mathbf{n} \cdot (\mathbf{C} + \lambda\mathbf{l}_\mathbf{X}) + d = 0 \tag{5.53}$$

$$\mathbf{n} \cdot \mathbf{C} + \lambda\mathbf{n} \cdot \mathbf{l}_\mathbf{X} + d = 0 \qquad \lambda = -\frac{\mathbf{n} \cdot \mathbf{C} + d}{\mathbf{n} \cdot \lambda\mathbf{l}_\mathbf{X}} \tag{5.54}$$

When a 3D bounding frame is projected onto the ground plane in an image, the result appears as a parallelogram due to perspective distortion. This projected shape provides cues about the object's orientation and footprint in 3D space. By analyzing the geometry of the parallelogram, it is possible to infer the actual square base of the 3D bounding box, enabling recovery of the object's position and alignment relative to the ground. This projection relationship plays a crucial role in estimating accurate 3D poses from monocular images.

To recover a 3D bounding frame from a monocular image, one can begin by analyzing the projection of the object's base onto the ground plane. This projection typically appears as a parallelogram in the image due to perspective distortion. From this parallelogram, the true square base of the object can be inferred by applying geometric constraints and correcting for perspective. Once the square base is determined, the full 3D bounding box can be reconstructed by extending the square upward according to the estimated object height.

The estimation of the ground plane is a critical step in this process. It serves as the reference surface for interpreting the object's position and orientation in 3D space. The algorithm for estimating the ground plane proceeds as follows:

Given N, $\boldsymbol{G} = \{[x_1, y_1, z_1]^T, \cdots, [x_m, y_m, z_m]^T\}$, initialize $\text{dst}_{\text{min}} = \infty$. For $i = 1, \cdots, N$, the steps are as follows:

(1) Randomly draw three vertices from $\boldsymbol{G}$.
(2) Extract the coefficients of the plane $ax + by + cz + d = 0$ defined by the sampled vertices.
(3) Calculate the sum of the square distances of all vertices in $\boldsymbol{G}$ to the plane, dst.
(4) If dst $<$ dst_{min}, then $\text{dst}_{min} = \text{dst}$, store a, b, c, d.
(5) Output the best fit $ax + by + cz + d = 0$ plane coefficients a, b, c, d.

Using the previously mentioned, the bottom-left line as the normal vector of the frontal plane, and then calculate the frontal plane with the point of front-bottom-left. Find its intersection with the inverse projection to get the vertex position, so the height is estimated.

Reference [38] presents the original pseudo-LiDAR approach, which bridges monocular or stereo depth estimation with 3D object detection. The core idea is to convert image-based depth predictions into a 3D point cloud representation that mimics LiDAR data. Specifically, each pixel in the estimated depth map is back-projected into 3D space using the camera's intrinsic parameters, producing a dense pseudo-LiDAR point cloud. Once this point cloud is generated, existing LiDAR-based 3D object detection methods can be applied directly, allowing image-based systems to leverage mature LiDAR detection pipelines without requiring actual LiDAR sensors.

Initially, this method assumes that stereo vision estimates the depth map, but in fact, the depth map estimation of monocular vision is also acceptable. In order to be maximally compatible with the existing LiDAR detection pipeline, some post-processing is performed on the pseudo-LiDAR data. Since the real LiDAR signal only exists within a specific height range, pseudo-LiDAR points outside this range are ignored. All points above 1 m from the virtual LiDAR source (located on top of the autonomous vehicle) are deleted. In addition to depth, the reflectance of each pseudo-LiDAR point is set to 1.0.

The method proposed in Ref. [39] first investigates how the ground plane can provide valuable cues for depth inference in 3D object detection. Building on this observation, it enhances the representation of 3D anchor boxes and introduces a dedicated neural network module to effectively incorporate this prior knowledge into the deep learning framework.

The first step is anchor box filtering. For each anchor box, its prior distance to the camera is used to back-project it into 3D space. Because objects of interest are assumed to lie near the ground plane, anchor boxes that project far from the ground plane are discarded during both training and testing. This operation constrains the search space and guides the network to focus on locations where objects of interest are likely to appear.

The second step is the Ground-Aware Convolution (GAC) module. In anchor box-based detectors, features at the anchor center are typically used to estimate all 3D parameters. However, accurate depth inference requires leveraging information from ground pixels, similar to how humans perceive depth. To achieve this, the network must propagate feature extraction outward from the object center toward the surrounding ground region:

1. Identify the contact points between the object and the ground.
2. Apply perspective geometry to estimate the 3D positions of these contact points.
3. Employ a downward-focused receptive field to gather information from the identified contact points.

In the 3D object detection network architecture described, the system incorporates both geometric reasoning and semantic classification. GAC is used to generate the regression tensor R, which encodes information such as object location, size, and orientation. In parallel, a standard convolutional branch produces the classification tensor C, which assigns category scores to the detected objects.

To improve efficiency and accuracy, an anchor box filtering mechanism is applied to remove detections that are unlikely to correspond to valid road-surface objects. This filtering process discards anchors that are too far from the estimated ground plane, effectively reducing false positives from elevated or irrelevant regions of the scene. Together, these modules allow the network to perform precise 3D object detection with improved spatial alignment and category discrimination.

Standard object detection or depth prediction networks are typically constructed with uniform receptive fields and do not incorporate perspective geometry priors or camera parameters. The GAC module addresses this limitation by guiding the network to integrate ground-based reasoning into its inference process. Specifically, the prior depth value of each pixel is encoded as an additional feature map, which directs each pixel in the feature map to aggregate information from the pixels below it.

The GAC module leverages perspective geometry to embed spatial priors into 3D object detection. In computing the vertical offset, it assumes that the processed pixel corresponds to the center of a foreground object. Under this assumption, the network can infer the object's vertical position relative to the camera by exploiting geometric cues.

In contrast, when computing the depth prior, the pixel is assumed to lie on the ground plane and serves as a query point for estimating spatial depth. This dual-assumption strategy enables GAC to adaptively encode different geometric contexts depending on the task—whether it's regressing object offsets or inferring depth from the scene. By leveraging these perspective-based assumptions, the GAC module enhances localization accuracy and spatial awareness in monocular 3D detection tasks.

The architecture of the GAC module integrates perspective geometry and road plane assumptions to enhance 3D object detection accuracy. In this framework, the network first predicts the vertical offset for each pixel, which represents the estimated distance between the pixel and the base of the object in the vertical direction. Using this predicted offset, the module samples relevant features and depth priors from pixels located beneath the current pixel position, following the vertical axis in the image.

The depth prior at each sampled location is computed based on perspective projection principles, assuming that the corresponding pixel lies on a flat road surface. This allows the model to estimate the 3D position of objects even in the absence of explicit depth input. By combining learned features with geometrically grounded depth cues, the GAC module improves spatial reasoning and object localization in monocular 3D detection pipelines.

Another component of the GAC module predicts the depth residual between the prior estimate and the object. This component is differentiable across the entire network and is optimized in an end-to-end manner.

5.5 Object Tracking

Object tracking is a long-standing yet unsolved problem in computer vision, and it plays a critical role in autonomous driving systems, particularly in multi-object tracking (MOT). Research in this field has a rich history, involving diverse objects, such as faces, human bodies, hands, rigid objects, and deformable objects, represented in forms including points, contours, regions, and skeletons. Tracking approaches are also strongly influenced by the sensing modality; for example, depth sensors such as Kinect and traditional cameras (monocular or binocular) affect both the algorithms employed and the quality of the results. Even today, active research topics include deep learning-based methods, sensor fusion, multi-object tracking, and closed-loop frameworks that integrate detection and tracking.

5.5.1 Single Object Tracking

Following the discussion in [40], we first review the fundamental methods and categories of single object tracking (SOT) in monocular vision systems. A central question in this context is: how should the object be represented? Object representation is a critical factor influencing the accuracy, robustness, and efficiency of tracking algorithms, with different strategies adopted depending on the complexity of the object and the requirements of the application.

For instance, in the case of tracking a human body, several representations are commonly used. A simple approach is to employ bounding boxes that enclose the entire object. More detailed alternatives include contour-based models that capture the object's outline; keypoint-based skeletons that encode the positions of body joints; and part-based models that decompose the body into semantic regions such as the head, torso, and limbs. Each representation provides a distinct balance between computational cost, robustness to occlusion, and resilience to non-rigid deformations. Selecting an appropriate representation is therefore essential for achieving accurate and stable tracking in dynamic and unconstrained environments.

- Centroid.
- Multiple-point representation.
- Rectangular patch (bounding box).
- Elliptical patch.
- Part-based multiple patches.
- Object skeleton.
- Complete object contour.
- Control points on object contour.
- Object silhouette.

Objects in monocular visual tracking can be characterized using various feature spaces, each capturing different aspects of appearance or motion. Common features include color histograms, which capture chromatic distributions; edge-based descriptors, such as Histogram of Oriented Gradients (HOG), which represent shape and contour information; motion features, such as Motion HOG (MHOG), which extend HOG with temporal dynamics; and texture descriptors, such as Local Binary Patterns (LBP), which capture fine-grained surface details. The choice of feature space directly affects the tracker's robustness to illumination changes, deformation, occlusion, and background clutter.

Tracking methods can be broadly classified into three major categories according to how they represent and follow the object over time. The first category, point tracking, represents the object by a set of interest points (e.g., corners or keypoints) and tracks their motion across frames, often using techniques such as Kalman filtering or optical flow. The second category, kernel-based tracking, treats the object as a region characterized by statistical properties such as color or intensity distributions and applies iterative algorithms like mean-shift to locate the best match in each new frame. The third category, silhouette tracking, is designed for complex or non-rigid objects and involves matching shape contours or segmentation masks across frames, frequently using methods such as level sets or shape matching. This taxonomy provides a structured overview of the diverse strategies employed in visual tracking under monocular vision settings.

Point tracking is divided into two technical paths: deterministic methods and statistical methods. Kernel-based tracking is divided into two paths: multi-view-based and template-based. The former can be further divided into two directions, namely, view subspace methods and classifier-based approaches (i.e., discriminative methods). Silhouette tracking is also divided into two types: contour evolution methods and shape matching methods. The former is further divided into state-space approaches and direct minimization approaches. Direct optimization can be further classified into variational methods and heuristic methods.

The appearance model, i.e., the representation strategy of tracking, is the cornerstone of visual tracking, and the idea of "tracking by detection" has become both popular and widely accepted. Referencing [41], we now introduce the tracking methods in this regard.

Modeling the appearance of an object is a fundamental component of visual tracking, as the visual characteristics of an object can change significantly and unpredictably over time. Even with a simple rectangular representation, numerous real-world factors can complicate the appearance of the tracked object. These include variations in illumination that affect color and brightness, partial or full occlusion by other objects, non-rigid deformations (particularly for humans or animals), image noise, perspective distortion arising from viewpoint changes, and motion blur caused by rapid movement or camera shake.

These variations can cause tracking drift or even complete failure if not properly addressed. To mitigate this, robust appearance modeling seeks to extract invariant or adaptive features that can withstand such challenges, thereby enabling the tracker to maintain accurate and consistent localization of the object throughout the video sequence.

A typical object tracking pipeline in a monocular vision system consists of several key stages: initialization, appearance modeling, motion estimation, and object localization. Initialization may be performed manually by specifying the object in the first frame or automatically through object detection methods. Appearance modeling then becomes a central component, encompassing both visual representations (e.g., color, texture, and edges) and statistical models that describe the object's characteristics over time.

Within this pipeline, statistical modeling is particularly critical, as it governs how the tracker learns and updates its representation of the object. Three main approaches are commonly adopted: generative, discriminative, and hybrid models. Generative methods explicitly model the object's appearance and attempt to match it to candidate regions, using techniques such as mixture models, kernel density estimation, or subspace learning (e.g., PCA-based models). In contrast, discriminative methods formulate tracking as a classification problem that separates the object from the background, employing approaches such as boosting, support vector machines (SVMs), dictionary learning, discriminant component analysis (DCA), and random forests.

Hybrid models combine both generative and discriminative strengths, and can be implemented as single-layer or multi-layer mixtures. These models aim to enhance robustness by fusing likelihood-based matching with classification-based decision-making, which is particularly effective in challenging scenarios involving occlusion, background clutter, or appearance variation.

5.5.2 Multi-object Tracking

The key difference between multi-object tracking (MOT) and single object tracking (SOT) is the presence of mutual influence and even interactive behavior among multiple objects being tracked simultaneously. Multi-object tracking has long been studied outside of computer vision, most notably in radar-based systems. Reference [42], which focuses on radar signals, summarizes several foundational algorithms that have also been widely adopted in visual MOT systems: the Joint Probabilistic Data Association Filter (JPDAF), Multiple Hypothesis Tracking (MHT), and Random Finite Set (RFS)-based approaches.

In radar-based tracking systems, most detected objects are point targets, lacking rich appearance cues such as texture or shape. Consequently, radar tracking emphasizes accurate modeling of motion and position states, while the appearance model remains relatively simple or even negligible. A typical radar MOT system can be described using a system model composed of two main components: the observation space and the state space.

The observation space contains measurable quantities obtained from radar returns, such as range, radial velocity (via Doppler shift), angle of arrival, and radar cross-section (RCS). The state space, on the other hand, represents the internal variables used to model and predict each object's trajectory over time—typically including position, velocity, acceleration, and identity. By linking observations to latent states through probabilistic frameworks, the system performs core MOT tasks such as data association, state estimation, and track management.

In contrast, vision-based MOT relies heavily on appearance information in addition to motion cues. A comprehensive review (see [43]) introduces MOT approaches specific to visual systems. Several key challenges and modeling strategies are commonly considered:

- **Appearance model**. Largely shared with SOT, where features such as color histograms, texture descriptors, and deep feature embeddings are used to re-identify and discriminate objects across frames.
- **Motion model**. Also inherited from SOT, but often extended to account for group dynamics and scene constraints.
- **Interaction model**. Interaction among multiple objects can be represented using *social force models* that capture pairwise influences, or *crowd motion pattern models* that describe collective dynamics.
- **Exclusion model**. Applied at two levels: (i) the detection level, where two detections cannot be assigned to the same object simultaneously and (ii) the trajectory level, where distinct tracks are constrained not to converge arbitrarily.

- **Occlusion model**. While occlusion handling exists in SOT, MOT requires more sophisticated mechanisms to resolve mutual occlusions, including part-to-whole reasoning, hypothesize-and-test strategies, and buffer-and-recover schemes.

A central problem in MOT is how to associate multiple detections across frames in a principled way. This is often referred to as the *reasoning* step. Reasoning methods can be divided into two broad categories:

- **Probabilistic statistical methods**, which include Kalman filtering, particle filtering, Multiple Hypothesis Tracking (MHT), and Joint Probabilistic Data Association Filtering (JPDAF). These approaches explicitly model uncertainty and data association likelihoods.
- **Deterministic optimization methods**, which cast MOT as an optimization problem solved using tools such as Bayesian networks, bipartite graph matching, min-cost max-flow network flow formulations, Conditional Random Fields (CRFs), Markov Random Fields (MRFs), or maximum-weight independent set (MWIS) formulations.

5.5.3 Deep Learning-Based Object Tracking

Deep learning-based object tracking has seen significant advancements, and a comprehensive review by the Daimler research and development team (Ref. [44]) outlines the foundational structure of modern multi-object tracking (MOT) systems. Despite variations in implementation, the overall framework across most MOT systems remains consistent and can be broken down into four core components: data association, state prediction, state update, and track management.

Data association is responsible for matching current observations with existing tracks, often relying on motion models, appearance features, or learned similarity metrics. State prediction uses models—such as Kalman filters or neural networks—to estimate the future positions of tracked objects. The state update step then refines these predictions based on new observations, improving accuracy. Finally, track management handles the creation of new tracks, maintenance of ongoing ones, and termination of those that no longer correspond to any detections. This modular architecture provides a scalable and interpretable foundation for integrating traditional filtering methods with deep learning-based perception models in real-world tracking systems.

The application of deep learning in object tracking can generally be categorized into four main approaches. First, deep learning can be used for feature extraction, where convolutional neural networks (CNNs) learn robust appearance representations to improve target discrimination. Second, deep models can be employed for motion modeling, enabling more accurate prediction of object trajectories by capturing complex temporal dynamics. Third, deep learning supports end-to-end tracking frameworks, where the entire tracking pipeline—including detection, association, and state estimation—is learned jointly. Finally, deep learning can enhance data association through learned similarity metrics, attention mechanisms, or graph-based matching strategies, improving the robustness of multi-object tracking in crowded or occluded scenes. These four approaches reflect the diverse ways deep learning has transformed traditional tracking pipelines by improving adaptability, accuracy, and generalization.

1. **Feature learning (appearance model part)**, such as the classic CNN.
2. **Data association part**, such as Siamese network plus Hungarian/LSTM.
3. End-to-end learning method (more challenging), such as RNN/LSTM.
4. **State prediction (motion/trajectory)**, such as Behavior-CNN, Social-LSTM, and Occupancy Grid Map.

Reference [45] introduces a proposal-based, learnable framework for multi-object tracking (MOT), which formulates the tracking task as a combination of proposal generation, proposal scoring, and trajectory inference on an association graph. This design closely resembles the architecture of the two-stage object detector Faster R-CNN and enables a fully data-driven solution to the MOT problem.

Given a sequence of image frames and a set of object detections, the framework first constructs a graph to formulate the data association problem. In this graph, nodes represent either individual detections or partial trajectories, while edges encode the likelihood of linking two nodes based on spatio-temporal consistency and appearance similarity. Each node corresponds to a specific target, and different node colors can be conceptually assigned to distinguish between targets.

In the first stage, multiple candidate trajectories (proposals) are generated from the association graph, each representing a plausible tracking hypothesis. A trainable Graph Convolutional Network (GCN) is then used to evaluate these proposals by assigning quality scores, capturing both local and global structural information in the graph. In the final stage, a simple yet effective de-duplication mechanism is applied to eliminate redundant or overlapping trajectories, yielding the final tracking output.

By learning the proposal scoring and selection process directly from data, this framework improves robustness to occlusions and identity switches while offering a unified and flexible structure for learning multi-object tracking from annotated sequences.

Reference [46] introduces a unified model called TADAM, which tightly integrates location prediction and embedding-based association for robust multi-object tracking. Unlike traditional approaches that treat these two tasks independently, TADAM establishes synergy between them using three key components: time-aware target attention, distraction attention, and an identity-aware storage aggregation module.

The attention mechanism enables the model to prioritize relevant targets while suppressing distractions, especially in occluded or crowded scenes. This results in more reliable and discriminative embeddings for identity association. In turn, these high-quality embeddings are aggregated through the identity-aware storage module, reinforcing the model's understanding of consistent object identities over time. This feedback loop improves both attention and localization, effectively reducing drift and enhancing stability during occlusion.

In contrast to existing methods where location prediction and embedding association fail to support each other under occlusion—leading to localization errors and noisy embeddings—TADAM's occlusion-aware design allows the two components to collaborate. This cross-task interaction strengthens tracking robustness by jointly leveraging spatial prediction and temporal identity information, resulting in improved accuracy and resilience in challenging scenarios.

The model architecture of TADAM is a unified model, bringing mutual benefits between location prediction and data association, thereby enhancing robustness to occlusion and improving tracking performance. To achieve this goal, time-aware target attention and distraction attention are introduced to better focus on the target and suppress interference from distractors, while also introducing an identity-aware memory aggregation scheme to achieve more robust attention generation. Regarding the naming of the TADAM model, TA and DA, respectively, represent target attention and distraction attention, while M represents memory aggregation. All components are trained using the same data source in the unified model.

Reference [47] proposes a Transformer-based DL tracker MT3v2, evaluates its performance based on the model settings, and compares it with Bayesian methods in various different tasks.

In the architecture of MT3v2, the input measurement sequence is processed by the encoder, generating embeddings, and processed by the selection mechanism, generating initial estimates, target queries, and position encoding for the decoder. The embeddings from the encoder and the output from the selection mechanism are used by the decoder for an output, which describes a multi-Bernoulli density with k components.

In the selection mechanism of MT3v2, the embeddings are first fed into an FFN (feed-forward network), and fed into a sigmoid layer, generating scores; then the measurements with the top k scores are embedded into two FFNs, generating target queries and their corresponding position encodings.

To further improve the performance of MT3v2, the concept of iterative refinement is adopted in the decoder. As previously mentioned, the decoder outputs a sequence representing the k components of the MB density, each of which contains the mean, covariance, and existence probability of the Bernoulli distribution. Instead of calculating the predicted state mean sequence directly from the output of the last layer of the decoder, it starts from the initial state estimate calculated by the selection mechanism, and each layer of the decoder generates an adjustment to it. All adjustments to the initial estimate are summed to obtain the output of the decoder.

Another improvement during the training process is an auxiliary task, which attempts to predict which measurements come from which targets (and which are interference). Adding simpler auxiliary tasks usually improves the initial part of the training process (when the main task is still difficult to solve and may not provide much gradient information) and the generalization performance of the final model.

To achieve this, a supervised contrastive learning idea is used, where the model is trained to generate similar predictions for samples of the same class, but different predictions for samples of different classes. In the data generation process, each measurement is labeled with an integer indicating its source (target ID), or -1 if it is interference.

A solution called TransSTAM was proposed in Ref. [48], which uses Transformer to effectively model the appearance features of each target and the spatio-temporal relationship between targets. TransSTAM consists of two main parts: ① the encoder uses the powerful self-attention mechanism of Transformer to learn the distinguishing features of each trajectory and ② the decoder adopts a standard cross-attention mechanism, considering both spatio-temporal and appearance features to model the correlation between trajectories and detections.

Online MOT's TranSTAM is based on the tracking-by-detection framework and performs multi-object tracking (MOT) in online mode. TranSTAM follows a typical Transformer encoder–decoder architecture. Given a set of N trajectories from frame $t - T$ to frame $t - 1$, first, the CNN model and Absolute Space Position Encoding (ASPE) model are used to extract the appearance and spatial features of each detection, respectively; secondly, the encoder learns the distinguishing features of each trajectory with the help of the Relative Spatio-Temporal Position Encoding (RSTPE) module, thus, all features are connected to form the memory in the decoder; thirdly, an Allocation Position Encoding (APE) generates position encoding for queries, and the decoder uses a standard cross-attention mechanism to calculate the allocation matrix; finally, a simple Hungarian algorithm generates association results.

The Transformer encoder aims to learn more distinguishable feature representations for each trajectory. Most existing Transformer-based trackers heavily rely on appearance features and are unable to model long-term spatio-temporal correlations. TranSTAM uses a Transformer model to model both appearance and spatio-temporal features simultaneously. The input to the Transformer encoder is a set of trajectories from the past T frames.

The appearance and spatial features of each detection are first embedded through CNN and MLP networks, respectively. Then, a simple "add" fusion operator is used to fuse the appearance and spatial features of each detection. These detection features are arranged into a feature tensor, which is further passed to the Transformer encoder module. The Transformer encoder module consists of L Transformer layers, each of which has two parts, namely, the self-attention module and the feed-forward network (FFN).

The Transformer decoder generates an allocation matrix based on the output of the Transformer encoder and M detection features. Then, the detection features are used as queries and passed to the Transformer decoder module.

Since the decoder is permutation invariant, the queries must be different to produce different results. As mentioned above, when estimating the elements in the same column of the allocation matrix, their corresponding query features are exactly the same. To solve this problem, Allocation Position Encoding (APE) is proposed. For the (i, j) query used to calculate the correlation between the i-th trajectory and the j-th detection, APE extracts features from the i-th trajectory as position encoding, and then fuses it with the feature embedding of the j-th detection to form the final feature.

The Transformer decoder architecture implements APE and completes data association using a cross-attention mechanism. The decoder uses the detection features of the current frame as queries, and the encoded features of all trajectories as Key and Value. The decoder follows the standard architecture of Transformer, transforming into embeddings with multi-head self-attention and cross-attention mechanisms. The difference from the original Transformer is that the model uses RSTPE to encode the relative spatio-temporal distance between query detection and trajectory detection, and uses it as a bias term in the cross-attention module.

Through RSTPE, each query in a single Transformer layer can adaptively focus on all trajectory detections according to the learned attention weights. For each query, the model may pay more attention to the trajectory detections near it and less attention to the trajectory detections far from it. One advantage of calculating the allocation matrix with the Transformer decoder is that the Transformer layer can provide a global receptive field. Therefore, when calculating the correlation of a specific detection-trajectory pair, all trajectory features can be utilized, which is crucial for avoiding ID switches in MOT.

5.6 Sensor Fusion

In the perception module of autonomous driving, **sensor fusion** has become a standard feature, but the level of fusion varies. It can be at the hardware level (such as the products of domestic Hesai Technology and American Innovusion), or at the data level, or at the task level, such as obstacle detection, lane detection, segmentation, tracking, and vehicle self-localization.

Some sensors are difficult to fuse at the base level, such as cameras or lidar and millimeter-wave radar, because the target resolution of millimeter-wave radar is very low (unable to determine the size and contour of the target), but fusion can be explored at a higher level, such as target speed estimation and tracking trajectory.

Reference [49] divides the fusion of cameras and lidar into two major categories: strong fusion and weak fusion. According to the different combination stages of lidar and camera data representation, strong fusion is further divided into pre-fusion, deep fusion, post-fusion, and asymmetric fusion.

Data fusion is a method that directly fuses each modality data through spatial alignment and projection at the raw data level. In contrast, **pre-fusion** methods fuse lidar data at the data level and camera data at the data level or feature level (such as segmentation tasks).

Deep fusion methods perform cross-modal data fusion on the feature level of the lidar branch and the data level and feature level of the image branch. For example, some methods use feature extractors to obtain the embedded representation of lidar point clouds and camera images separately, and fuse the features of the two modalities through a series of downstream modules. However, unlike other strong fusion methods, deep fusion sometimes fuses features in a cascading manner, using both raw and high-level semantic information.

Post-fusion, also known as target-level fusion, refers to the fusion of the results of each modality in the pipeline. For example, some post-fusion methods use the output of the lidar point cloud branch and the camera image branch, and make final predictions based on the results of the two modes.

Reference [49] specifically points out: In addition to pre-fusion, deep fusion, and post-fusion, some methods handle cross-modal branches with different permissions, so they fuse the target-level information of one branch and the data-level or feature-level information of other branches, defined as asymmetric fusion. In asymmetric fusion, at least one branch dominates, while other branches provide auxiliary information to perform the final task.

The weak fusion method separately divided out in Ref. [49] does not directly fuse data/features/targets from branches in various ways, but operates data in other ways. It usually uses a rule-based method to use one modality data as a supervisory signal to guide the interaction of another modality.

The main problems faced by the fusion model of sensors (lidar and camera) are as follows:

(1) Misalignment and information loss. Traditional pre-fusion and deep fusion methods use external calibration matrices to directly project all lidar points to the corresponding pixels, and vice versa. However, due to sensor noise, this pixel-by-pixel alignment is not accurate enough. Therefore, taking the surrounding information as a supplement will produce better performance. In addition, there are some other information losses in the process of input feature space conversion. Usually, the projection of dimension reduction operations will inevitably lead to a large amount of information loss, for example, mapping 3D lidar point clouds to 2D BEV images. One solution is to map the two modal data to another high-dimensional representation specifically designed for fusion, which can effectively use the original data and reduce information loss.

(2) Unreasonable fusion operations. Simple operations such as cascading and element multiplication may not be able to fuse data with large distribution differences, and it is difficult to bridge the semantic gap between the two modalities. One idea is to try to use more complex cascading structures to fuse data and improve performance.

The forward view single-frame image is a typical scenario of autonomous driving perception tasks. However, most frameworks use limited information and do not design auxiliary tasks in detail to further understand the driving scene. Two directions to strive for are as follows:

(1) Use more latent information. Existing methods lack effective use of multidimensional and source information, most of which focus on forward view single-frame multimodal data, and other meaningful information, such as semantic, spatial, and scene context information, is not given attention. Some models try to use the results of image semantic segmentation tasks as additional features, while other models may use the features of the middle layer of the neural network backbone. In autonomous driving scenarios, many downstream tasks with explicit semantic information may greatly improve the performance of target detection tasks, such as lane detection and semantic segmentation. Therefore, future research can build a complete city scene cognition framework through various downstream tasks (such as detecting lanes,

traffic lights, and signs), to help the performance of perception tasks. In addition, the current perception tasks mainly rely on a single framework that ignores time information. Time series information contains serialized monitoring signals, which can provide more robust results compared to single-frame methods.

(2) Self-supervised representation learning. Mutual supervision signals naturally exist in cross-modal data sampled from the same scene but different angles. However, due to a lack of deep understanding of the data, it is currently impossible to mine the synergistic relationship between different modalities. Therefore, multimodal data should be used for self-supervised learning, including pre-training, fine-tuning, or contrastive learning. Through these mechanisms, the fusion model will deepen its understanding of the data and achieve better results.

Domain bias and data resolution are highly related to real scenarios and sensors. These shortcomings hinder the large-scale training and implementation of deep learning models for autonomous driving.

(1) Domain bias. In autonomous driving perception scenarios, raw data extracted by different sensors is accompanied by domain-related features. Different camera systems have their optical characteristics, and lidar may vary due to mechanical lidar and solid-state lidar. More importantly, the data itself may be domain-biased, such as weather, season, or geographical location. Therefore, the detection model cannot smoothly adapt to new scenes. Due to generalization failure, these shortcomings hinder the collection of large-scale datasets and the reusability of original training data.

(2) Conflict with data resolution. Sensors from different modalities usually have different resolutions. For example, the spatial density of lidar is significantly lower than the spatial density of images. Regardless of the projection method used, some information is eliminated because no corresponding relationship can be found. Whether it is the different resolution of feature vectors or the imbalance of original information, it may cause the model to be dominated by data from a specific modality. What is currently needed is a data representation system compatible with different sensor resolutions.

Note: *BEV space is suitable for models that do sensor fusion with cameras, usually at the feature level, which will be discussed specifically in* Sect. 5.14.

5.6.1 Data Level

Taking the data fusion of lidar and cameras as an example, it is actually the combination of the depth formed by the projection of lidar point clouds on the camera image plane and the depth estimated by the image. In theory, the depth estimated by the image can be back-projected into 3D space to form a point cloud and fuse with the lidar point cloud, but it is rarely used. The reason is that the error of the depth map will be magnified in 3D space, and the means of analyzing point clouds in 3D space are not as mature as the depth maps of images. After all, the research history of 2.5D space is longer, such as previous RGB-D sensors, Kinect, or RealSense.

The idea of this fusion is very clear. On the one hand, the cost of image sensors is low and the resolution is high (it can easily reach 2–4 K); on the other hand, the cost of lidar is high, the resolution is low, and the depth detection distance is short. However, the ranging accuracy of lidar point clouds is very high, far greater than those of infrared/Time-of-Flight (TOF) depth sensors, and its anti-interference ability in outdoor environments is strong. At the same time, as the main sensor of the passive vision system, the image has poor depth estimation accuracy, and what is more troublesome is poor stability and robustness. So, can the sparse depth data of lidar and the dense depth data of images be combined to form a complement?

In addition, how to upsample the sparse depth map to become dense is also a topic that has been researched, similar to image-based depth upsampling work. Also, when the point cloud obtained by the lidar is projected onto the image plane of the camera, it will be found that some non-reflective object surfaces have caused "black holes," and there is basically no data display in the far distance street or sky area. This involves another research topic—image-guided depth filling/completion.

The premise to solve this problem is that the calibration and synchronization of the lidar and camera are completed, so the point cloud of the lidar can be calibrated and projected onto the image plane of the camera, forming a relatively sparse depth map.

1. Traditional methods

First, consider the task as a depth map interpolation problem, then the method is similar to super-resolution (SR) and upsampling, but it requires image-guided.

To achieve this combination of images and depth, what is needed is the correlation between image features and depth map features. This assumption has been mentioned when the lidar and camera sensors are calibrated. Here, it is to apply it at the pixel/depth element/surface element/volume element level.

Basically, technically, there are two ways: local methods and global methods. It is similar to several other classic computer vision problems, such as optical flow calculation, stereo vision matching, and image segmentation.

Let's briefly review the history of image filtering. Mean filtering → Gaussian filtering → Median filtering → Anisotropic diffusion → Bilateral filtering (equivalent to the former) → Non-local mean filtering → BM3D, these are all local methods (bilateral filtering and non-local mean filtering have been introduced in Sect. 2.2). So what about joint bilateral filtering? Bilateral filtering and the well-known guided image filtering can both play a role here.

Recently, a method (see Ref. [50]) using traditional morphological filtering has been adopted, which performs no worse than deep learning CNN.

The global method naturally includes MRF, CRF, total variation (TV), dictionary learning, and sparse coding. As proposed in Ref. [51], using TV, although the sensor is TOF, it is also applicable to lidar. As proposed in Ref. [52], semantic segmentation is introduced, similar to depth ordering.

If the sparse depth map is seen as a problem that needs to be filled, then it belongs to another topic, namely, image-guided depth inpainting/completion. The techniques in this area are basically global methods, such as Ref. [53].

There is a method that uses the points projected from the lidar point cloud to the image plane as prior knowledge or "seeds" to correct the depth estimation process of the image. This is like a grid composed of sparse points projected from the lidar point cloud to the image, to guide/constrain stereo image matching. As proposed in Ref. [54], it reduces the disparity range of the disparity space image (DSI), calculates the SRI (spherical range image), maximum/minimum distance filtering, predicts maximum/minimum disparity, and finally constructs the DSI. The solution method is dynamic programming (DP).

Note: *An SRI image is the result of the distance function output of the spherical azimuth and elevation angles.*

2. Deep Learning Methods

Recently, papers on depth map fusion using deep learning have been continuously published. Firstly, the emphasis on sensor fusion in autonomous driving provides motivation. Secondly, the widespread application of deep learning in depth map estimation/segmentation/optical flow estimation and other fields has led researchers to start planning and implementing deep learning solutions for multi-sensor fusion.

Reference [55] studied various methods for **depth map completion** from extremely sparse data (such as lidar) from the design aspects of network architecture, loss function, benchmark dataset, and learning strategy, and basically divided the existing methods into two categories: one is only sparse input without guidance from different modal data, called non-guided depth map completion; the other is sparse depth map and its corresponding RGB image is used for input, called RGB-guided depth map completion.

The first type of completion method can be subdivided into three types: ① sparse-aware CNN, ② normalized CNN, and ③ auxiliary image training method.

Non-guided methods usually perform worse than RGB-guided methods and are affected by blurred consequences and target boundary distortion. This can be attributed to the lack of prior information about natural scenes. The depth map of a natural scene can be decomposed into smooth surfaces and discontinuities between surfaces, the latter forming step edges in the depth map. However, when the depth map is very sparse, prior information such as adjacent targets and sharp edges is obviously missing, so it is difficult to recover the complete depth map with only CNN.

Therefore, using RGB information as additional input is simple and reasonable. RGB images provide information about the scene structure, including textures, lines, and edges, which can supplement the missing clues of the sparse depth map and encourage depth continuity within smooth areas and discontinuity at boundaries. In addition, there are some monocular cues, such as vanishing points, that can promote depth map estimation.

RGB-guided depth map completion models can be mainly divided into five types: ① early fusion models; ② late fusion models; ③ explicit 3D representation models; ④ residual depth map models; and ⑤ models based on Spatial Propagation Network (SPN).

Early fusion models are quite simple, as the fusion of multimodal data only takes place at the input layer, and feature extraction is entirely dependent on the black-box CNN. Early fusion models usually do not perform as well as **late fusion models**, which can learn domain-specific knowledge and related features.

Multimodal feature fusion methods have evolved from direct connection or summation to the application of semantic relevance, attention mechanism, and spatially variant kernels, and from single spatial scale to more common multi-spatial scales.

Explicit 3D representation models can be divided into: ① 3D-aware convolution (3DAC); ② Intermediate Surface Normal Representation (ISNR); and ③ Learning geometric representation from point clouds (LfPC) method.

The 3D perception convolution method is superior in accuracy and efficiency to the intermediate surface normal representation and the method of learning geometric representation from point clouds. This is not surprising, as 3D perception convolution only acts on spatially close effective depth points, thus less affected by data loss, helping to reduce redundancy. However, a clear advantage of the method of learning geometric representation from point clouds is that it exhibits higher generalization accuracy under different weather and lighting conditions.

The **residual depth map model** pays particular attention to improving the geometric realism of the depth map. The residual model applies residual learning to predict a residual map and uses it as a compensation for a dense map. The advantages of residual learning are usually reflected in two aspects: first, it is considered a structural adjustment that can improve perceptual quality; secondly, since distant areas usually produce larger depth errors, and the depth errors in close areas are smaller, residual learning tends to supplement pixels with larger errors and keep near-zero values for pixels in close areas or perfect estimates.

The model **based on the Spatial Propagation Network (SPN)** is significantly superior to other types of methods in terms of accuracy. In addition, depth map refinement based on affinity not only makes the results more accurate due to the adoption of geometric constraints in depth map space, but also applies to any existing model. However, this model inevitably requires additional convolution modules to implement SPN, thereby increasing the complexity of the model. In addition, multiple iterations in the optimization process are very time-consuming.

The following are examples of specific depth map fusion methods in some paper works.

The first public depth map fusion work report is a paper by MIT doctoral student F.Ma. Almost at the same time, a paper from the German Andreas Geiger research group was also published (see Ref. [37]). Their pioneering work led to the launch of the 2018 depth map completion and prediction competition in the Kitti vision benchmark suite, but MIT won the depth completion championship that year.

The input is a depth map and the corresponding mask map, the latter is a binary map of the projection of the lidar point cloud onto the image plane (the mask pixel value is 1 for projection, and 0 otherwise), for which a model called Sparse CNN is designed. The network architecture of the sparse convolution layer, the input is the sparse depth map and binary observation mask, it goes through several sparse convolution layers (dashed line), the kernel size decreases from 11×11 to 3×3.

The method of MIT doctoral student F.Ma (see Ref. [56]), initially adopted a "brute-force" training method, where different models were designed for NYU Depth (indoor) and KITTI, the cube is the feature map, the size is represented as the number of features @ height $\times$ width; the encoding layer is composed of ResNet and 3×3 convolution. The decoding layer consists of four upsampling layers, followed by bilinear upsampling.

They then adopted a self-supervised learning CNN method from RGB to depth map, and used the continuity constraint of adjacent frame motion for self-learning. The encoder consists of a series of convolutions, with added filter groups for downsampling the feature space resolution. In addition, the decoder has a reverse architecture, and the transposed convolution upsamples the spatial resolution.

The self-supervised learning framework only requires synchronous sequences of color/gray images from a monocular camera and sparse depth images from LiDAR. Therefore, the self-supervised framework does not depend on any additional sensors, annotations, or other learning-based algorithms as building blocks. In addition, this framework does not depend on any specific neural network architecture choice, during training, both the current data frame $RGBd_1$ and the nearby data frame RGB_2 are used to provide supervision signals. However, during inference, only the current frame $RGBd_1$ is needed as input to produce depth prediction $pred_1$.

The structure of INRIA's work (see Ref. [57]) is still an encoder–decoder framework which does not use mask input (the article analyzes the reason is due to layer-to-layer transmission causing failure) and appends another training goal, i.e. semantic segmentation.

The encoder part of the network is a modified NASNet, which removes batch normalization (BN) after the first strided convolution layer when the input is sparse, because the zero values of missing pixels falsify the average calculation of the BN layer.

The authors of Ref. [57] found that the performance of CNN methods that directly merge RGB and depth maps at early layers is not as good as merging them later (compared to the fusion with the task layer, this is still considered early fusion). This is their second finding, which is consistent with the view in Ref. [58].

In a paper co-authored by ETH Zurich, Princeton University, and Microsoft (see Ref. [59]), it is proposed that a binary mask is still needed as input. They also introduced a surface normal map to enhance depth prediction, as well as a confidence mask, and specifically added a target-driven attention mechanism.

This end-to-end neural network structure for depth map fusion uses surface normals as an intermediate representation. The entire network mainly consists of two paths: the color path and the surface normal path. The color path takes the color image and sparse depth as input and outputs the complete depth. The surface normal path first predicts the surface normal image from the input color image and sparse depth, and then combines it with the sparse depth and the confidence mask learned from the color path to produce the complete depth. Each of these two paths is implemented with a set of deep completion units (DCU), and the final complete depth is produced by the learned weighted sum of the depths from the two paths.

In the end-to-end fusion system of DCU model, the two encoders used for RGB images/normals and sparse inputs consist of a series of ResNet blocks, followed by convolution, and finally reducing the feature resolution to 1/16 of the input. The decoder consists of four up-projection units, gradually increasing the feature resolution and integrating the features of the two encoders to produce dense output. Since the input sparse depth is strongly related to the decoder output (such as surface normals or depth), the features from the sparse depth should contribute more in the decoder.

Finally, to simply summarize how deep learning works in the field of depth map fusion, starting from brute-force training models, gradually adding geometric constraints, the idea of joint training is widely accepted. Delaying the merging of RGB images and depth maps is a consensus, training feature maps separately is better, whether to input mask maps is still up for discussion.

Reference [60] proposes a dynamic method based on attention to learn the correlation between adjacent pixels in a model based on spatial propagation networks. Specifically, a Dynamic Spatial Propagation Network (DySPN) uses a Nonlinear Propagation Model (NLPM), decomposes the neighborhood into parts at different distances, and recursively generates independent attention maps, which are then refined in an adaptive affinity matrix. In addition, a diffusion suppression (DS) operation is used to make the model converge early and prevent over-smoothing of the dense depth map. DySPN includes a simple nonlinear propagation model, a diffusion suppression operation, and three modules that greatly reduce complexity. Its pipeline framework uses a ResNet34-UNet model to parallelly output attention maps, initial depth maps, affinity matrices, and confidence predictions of sparse depth measurements; the attention maps are divided into six parts and applied to the DySPN module in a six-step propagation process.

NLPM is similar to a diffusion process described by a partial differential equation (PDE), which stops only after N iterations, but this does not apply to all pixels. If N is too high, the output will be over-smoothed, which is not conducive to depth completion. To solve this problem, a diffusion suppression operation is used here, which estimates the correlation of the fine depth map in each iteration, so that the model can converge in the early stage.

The basic idea of the N-step NLPM is to use global attention, with a complexity of $O(m^2n^2N)$, where the size of the depth map is (m, n). When attention is allocated to each sample (k) neighbor individually, the complexity drops to $O(k^2N)$. At the same time, the base network needs to estimate N times more feature maps than the linear propagation model, accompanied by a large amount of additional computation.

Analyzing the correlation matrix of the 7×7 convolutional spatial propagation network (CSPN), it can be found that the correlation weights of neighbors at the same distance are related. DySPN adopts three variations, decomposing the neighborhood into parts at different distances. The complexity of this decoupled attention mechanism is only $O(kN)$, independent of the image size (m, n).

There are three variations of DySPN. The first variation employs a 7×7 DySPN that partitions the neighborhood into four regions based on distance. The second is a dilated DySPN, which aggregates information from two distinct 3×3 neighborhoods, one of which uses a dilation rate of 3. The third variation is a deformable DySPN, which captures varying receptive fields by applying two deformable 3×3 kernels. These designs enable DySPN to reduce the number of required neighbors while maintaining expressive spatial aggregation.

A comparison between the 7×7 CSPN and the 7×7 DySPN reveals key differences in how correlation matrices are constructed. In CSPN, a fixed correlation matrix is used, with weights predominantly distributed along the outermost layer of the kernel. In contrast, DySPN employs an adaptive correlation matrix, dynamically adjusted by an attention mechanism to better capture local context.

A class UNet encoder–decoder network with ResNet34 as the backbone is used as the base network for a fair comparison between DySPN and other SPNs. The head of the base network is modified to simultaneously output attention maps, initial depth maps, correlation matrices, and confidence predictions of sparse depth measurements.

Reference [61] proposes a three-branch backbone, including an RGB-guided branch, a semantic-guided branch, and a depth map-guided branch. Specifically, the RGB-guided branch takes the sparse depth map and RGB image as input, generating color depth that includes scene color clues (such as target boundaries). The color-guided branch's predicted dense depth map, semantic image, and sparse depth map are passed as input to the semantic-guided branch for estimating semantic depth. The depth map-guided branch uses sparse depth, color depth, and semantic depth to generate a dense depth map. Color depth, semantic depth, and guided depth are adaptively fused to produce the output of the three-branch backbone. In addition, a Semantic-Aware Multimodal Attention Fusion Block (SAMMAFB) is applied to fuse the features between all three branches. A further improvement is to refine the dense depth map generated by the three-branch backbone with CSPN++ with dilated convolution.

The block diagram of SemAttNet consists of two main stages. In the first stage, multimodal inputs—comprising RGB images, semantic maps, and sparse depth maps from LiDAR—are processed through a three-branch backbone. This backbone generates an initial dense depth map by adaptively fusing features across the RGB, semantic, and sparse depth branches using the SAMMAFB module. In the second stage, the output is further refined using CSPN++ and dilated convolution to enhance depth estimation quality.

SemAttNet introduces a novel architecture that differs from conventional image-guided approaches by incorporating a dedicated branch for learning scene semantics. A Semantic-Aware Multi-Modal Attention Fusion Block (SAMMAFB) enables semantic-aware fusion across RGB, depth, and semantic modalities. Each branch produces both a depth map and a corresponding confidence map, which are then adaptively fused to generate a unified depth map. This fused result is subsequently refined using the CSPN++ module, leveraging dilated convolution to enhance depth estimation accuracy.

The architecture of the Semantic-Aware Multi-Modal Attention Fusion Block (SAMMAFB) is designed to effectively integrate information from multiple modalities. It takes as input the concatenated features from the color, semantic, and depth branches. Attention mechanisms are applied along both the channel and spatial dimensions to these combined features, resulting in a finely fused feature map that captures complementary information across modalities.

Specifically, the purpose of channel attention is to assign a weight to each channel relative to its contribution to learn important channels, thereby improving overall performance. Similarly, the spatial attention module focuses on learning the locations of these important channels. These two modules complement each other and are crucial for generating a fine fusion feature map.

The Color-Guided (CG) branch aims to learn important color clues for dense depth completion. The CG branch follows an encoder–decoder network architecture with skip connections. The encoder consists of 10 ResNet blocks, while the decoder consists of a convolution and 5 transposed convolution layers for upsampling. The synthesized dense depth map of this branch is a noisy depth estimation, but it provides a baseline for learning the structural information of the scene in other branches.

Semantic clues help understand the scene and are crucial for the depth completion task. However, the CG branch alone is not sufficient to learn semantic information. Therefore, we propose a Semantic-Guided (SG) branch to encourage the learning of effective semantic clues and supplement the RGB image for depth completion. The SG branch takes color depth, semantic image, and sparse depth map as input, and outputs a dense depth map composed of color and semantic clues. The decoder features of the CG branch are fused into the corresponding encoder features of the SG branch. Features from CG and SG are sent to SAMMAFB, which outputs a fine fusion depth feature map.

The Depth Guidance (DG) branch focuses on learning depth-dominant features, which helps generate accurate dense depth maps. It takes CG, SG, and sparse depth cascades as input and outputs dense depth maps. Similar to the feature fusion strategy in the CG and SG branches, the decoder features from the CG and DG branches are fused into the corresponding encoder features of the DG branch. However, in the DG branch, SAMMAFB takes feature maps of three modalities as input and outputs refined fusion feature maps. The refined fusion feature maps consist of useful information from the CG and SG branches, guiding the DG branch to learn effective depth feature representations. The network architecture is similar to the CG and SG branches, but two layers are added in the encoder and decoder architectures to adapt to the characteristics of the decoders of the CG and CG branches.

5.6.2 *Task Level*

High-level fusion is task level, such as obstacle detection/classification, tracking, segmentation, and localization. Sometimes fusion may occur in the feature space of a certain level, which also varies with the task.

1. Traditional Fusion Methods

Traditional fusion theories generally include algebraic methods (such as weighted sum), Kalman filtering, Bayesian theory, and evidence theory, among which Kalman filtering is the most common method.

Reference [62] uses a binocular camera and lidar fusion for obstacle detection in an intersection safe driving system.

The software framework of the binocular vision and lidar fusion system includes sensor fusion at the detection level and model-based tracking, as well as risk assessment systems, etc. Binocular vision processing and lidar processing are combined as a sensor fusion system for obstacle detection and then enter the tracking module. The fusion is done using obstacle marking and location information, unified for target tracking.

The binocular vision algorithm used in the system follows a structured processing pipeline that can be understood at both low and high levels of abstraction. At the low level, the system begins with image acquisition from stereo cameras, followed by several pre-processing steps such as brightness adjustment, downsampling, camera calibration, and distortion correction. Next, sparse optical flow estimation is performed to analyze pixel motion between frames, which is then used to estimate ego-motion. Based on these computations, the system generates a depth map and subsequently reconstructs a 3D point cloud of the scene.

At the high level, the system distinguishes between structured and unstructured environments and applies specialized modules accordingly. In structured environments, lane line detection is used to assist with road understanding. In both structured and unstructured scenarios, modules are deployed for obstacle segmentation, target detection and tracking, and obstacle classification. Additional components handle high-definition map detection, occupancy grid generation, and polygon extraction to model the navigable space. The outputs of the occupancy grid and polygon extraction processes are fused to produce the final environmental representation used for downstream decision-making and navigation tasks.

Reference [63] fuses lidar, camera, and radar data at the detection task level, all of which are forward-configured lidar, camera, and millimeter-wave radar.

The system framework for the perception task in Ref. [63] includes two parts, SLAM and detection and tracking of moving objects (DATMO).

The perception system includes lidar and cameras, using an evidence theory-based fusion mechanism, with fusion occurring at both the detection and tracking levels.

The schematic diagram of the entire multi-sensor perception fusion system includes two modules, frontal object perception (FOP) and moving object detection (MOC).

The fusion method adopted in the CMU paper (see Ref. [64]) is somewhat different, with the sensor configuration, including the CMU autonomous vehicle, lidar sensor, and horizontal field of view (HFOV) display.

The framework of the tracking system in Ref. [64] includes two layers, the sensor layer and the fusion layer. The sensor layer includes radar, lidar, and cameras, each sensor independently completes target feature extraction and detection, while the fusion layer includes fusion based on the Kalman filter.

The tracking methods of data association vary depending on the type of sensor data. First, the predicted motion target hypothesis is projected into the image space and associated with a set of detected visual targets. Next, a set of possible edge targets is generated from the predicted motion target hypothesis and matched with the extracted edge targets. Similarly, a set of possible point targets is generated and associated with the detected point targets. Finally, all information is fused on the EKF platform for tracking.

Reference [30] is about the detection fusion of millimeter-wave radar and monocular camera, which requires calibration of millimeter-wave radar and camera, and calculation of a homography plane transformation.

The fusion strategy is based on the point alignment module and region search, including point alignment, region search of target detection, and determination of real obstacles.

The point targets detected by the radar are considered as potential focus points, so the first level of fusion is radar-visual point alignment. The radar point serves as the reference center, and the image is segmented into small sections in order. This will help reduce or even eliminate the impact of complex environments, ultimately speeding up obstacle detection and further determining the outline of the obstacle.

The configuration of the on-board sensors is shown in Fig. 3 of [11], all of which are forward-configured cameras and millimeter-wave radars.

Based on the analysis of the driver's visual awareness on the road, detection should be limited to close range, which helps reduce the number of potential obstacles and thus reduce computational load. Then, the left and right boundaries of the real obstacle are the driver's visual focus, so the edges of the obstacle should be determined.

First, detection should be limited to the road area near the vehicle. Second, not too many obstacles will be detected at once. Then, its boundaries should be determined for further precise navigation. Assuming the road boundary is known, effective obstacle detection should be limited within this range. Therefore, radar-visual point alignment is carried out within this range, and visual search should also be limited within this area. The detected significant obstacles should not exceed two.

The region search strategy is based on obstacle boundary detection (assuming there are two potential targets, i.e., region 1 and region 2, with centers at (u_1, v_1) and (u_2, v_2), respectively).

The fusion framework for the entire target detection level consists of three main parts. The upper part corresponds to the implementation of point alignment, while the middle section focuses on target detection and boundary determination, where shadow detection and edge detection serve as the core components of the algorithm. For example, the shadow beneath a vehicle can be used to estimate its left and right boundaries. This feature is simple and continuous but may sometimes become invalid, such as when shadows from overpasses interfere with detection or even cause detection to fail. For pedestrians, their edges are significant, so pedestrians are like shadows in the image. From an optical perspective, it is known that targets like plants may diffuse, absorb and block light, and the light intensity received by the camera sensor will be relatively low, forming darker parts at the corresponding positions in the grayscale image. Here, the darker part can be regarded as the shadow of the obstacle, which provides an indication for detecting potential obstacles under special circumstances.

Reference [65] integrates visual SLAM work on the basis of LOAM (laser odometry and mapping). It first uses the motion estimated by visual odometry (VO) to register the lidar point cloud, then uses the odometry of the point cloud to re-correct the motion estimation of the camera, and then step by step to achieve map making.

The visual odometry (VO) module and the lidar odometry (LO) module are connected in series, but the input of the lidar provides "depth map calibration" to help the visual odometry module. In the lidar odometry module, you can see that the lidar scan is calibrated in the map at a speed 1 Hz update, while the visual frame is calibrated in the map at a depth update 60 Hz. At the same time, in this module, the motion estimation of the visual and lidar parts is calibrated and integrated (transform integration).

2. Deep Learning Methods

The most common is the obstacle detection task, and many deep learning detection models that combine lidar and camera information have been introduced.

Reference [66] generates a forward view and a bird's-eye view based on point clouds, defining a sparse non-homogeneous pooling layer that transitions between the two views. This way, a CNN model combines the bird's-eye view and forward image of the point cloud to achieve object detection, where the right dashed box is the architecture diagram of the sparse non-homogeneous pooling layer.

Suppose the size of the forward view is $H_\mathrm{f} \times W_\mathrm{f}$, and the size of the bird's-eye view is $L_\mathrm{b} \times W_\mathrm{b}$. The sparse non-homogeneous pooling layer uses a feature map (such as an image) as input, with (u_i, v_i) and (x_i, y_i) pairs as parameters. This layer constructs a $L_\mathrm{b} \times W_\mathrm{b} \times H_\mathrm{f} \times W_\mathrm{f}$ sparse matrix $\boldsymbol{M}$; the $H_\mathrm{f} \times W_\mathrm{f} \times C$ feature map is flattened into a dense $H_\mathrm{f} \times W_\mathrm{f} \times C$ matrix $\boldsymbol{F}$; then, the pooled feature map is $\boldsymbol{B} = \boldsymbol{M} \times \boldsymbol{F}$, which can be merged with the target feature map for further processing. Before merging, batch normalization (BN) is applied to the two feature maps. It is not recommended to fuse the feature maps of the bird's-eye view and the forward view in the early stages, as the resolution of the point cloud is lower than that of the camera image, so the pooling of the original input will result in very low utilization of the image data. Sparse pooling is best applied to deeper layers.

The detection CNN model is based on fusion, with the sparse non-homogeneous pooling layer between the two CNNs, finally outputting classification and bounding box regression results.

The RoarNet proposed in Ref. [67] is still based on the PointNet model to analyze point cloud data to obtain 3D region proposals, which are then merged with the 2D bounding boxes and poses estimated by the image CNN model. This model first predicts 2D bounding boxes and 3D poses of targets from 2D images. For each 2D target detection, a geometric agreement search is applied to predict the target position in 3D space.

Centered on each position prediction, a region proposal with a standing cylinder shape is set. Considering the prediction error of the bounding box and pose, multiple region proposals can be taken for a single target. Each region proposal is responsible for detecting a single target. The point cloud sampled from each region proposal is used as input, and the model predicts the position of the target relative to the center of the region proposal, which recursively sets a new region proposal for the next step. The model also predicts the objectness score, which reflects the probability of the target in the region proposal. Only those proposals with high objectness scores will be considered in the next step. This part is called RPN.

Finally, the model sets a new region proposal at the previously predicted position. The model predicts all the coordinates needed for 3D bounding box regression, including target position, rotation, and size. This part is the bounding box regression network (BRN).

The final model structure adopts the detection pipeline, compared with the single-step/two-step detection pipeline, it removes the redundant prediction module.

In Ref. [68], the authors first use a 2D CNN target detector to obtain 2D region proposals and classify them; then elevate the 2D region proposals to 3D to become frustum proposals; given the point cloud in the frustum (n points, c channels), instance segmentation is achieved through binary classification of each point; based on the segmented target point cloud (including c channels, m points), a lightweight regression network PointNet (T-Net) attempts to align points by translation, so that the centroid is close to the amodal box center; finally, the box estimation network estimates the target's amodal 3D bounding box.

The 3D object detection process begins with RGB-D input, where an RGB image and corresponding depth map are provided. A convolutional neural network (CNN) is first applied to the RGB image to generate 2D region proposals that indicate potential object locations. Each 2D proposal is then geometrically back-projected into 3D space using the depth information, forming a 3D viewing frustum that extends from the camera through the 2D bounding box.

Within this frustum, the depth map is used to extract the corresponding point cloud, isolating the 3D structure of the proposed object. These points are then passed into a Frustum PointNet, a neural network specifically designed to process unordered point cloud data. The Frustum PointNet predicts a 3D bounding box for the object within the frustum, estimating its position, orientation, and dimensions. This approach allows the system to accurately perform 3D object detection in an end-to-end manner, leveraging both image cues and geometric depth information to localize targets in space, even in the presence of partial occlusions or non-canonical object orientations.

The PointNet-based model used for 3D object detection follows a structured architecture. On the left, the network resembles an hourglass-shaped design, where point cloud features are extracted through successive layers of shared multi-layer perceptrons (MLPs), aggregated via a symmetric function such as max pooling, and then refined through additional layers to capture both local and global geometric patterns. This design enables the network to handle unordered point sets while preserving permutation invariance.

A key component within the architecture is the T-Net module, whose structure is shown on the right side of the model. T-Net predicts a spatial transformation matrix to align input point clouds into a canonical space, improving robustness to rotation and translation. The final output of the PointNet model includes a set of 3D bounding box parameters, such as the object's center coordinates, dimensions (height, width, length), and orientation. This output provides an accurate and compact representation of the detected object in 3D space, suitable for downstream tasks such as tracking, mapping, or scene understanding.

The MV3D method, described in Ref. [69], introduces a multi-view fusion approach for 3D object detection by combining information from LiDAR point clouds and RGB images. Specifically, it projects the LiDAR point cloud into both a bird's-eye view (BEV) and a front-view representation, which are then used alongside RGB images to form a rich, multimodal input.

The detection framework follows a two-stage pipeline. In the first stage, instead of relying solely on generic region proposal mechanisms, MV3D generates proposals from both the BEV and front-view representations of the point cloud, as well as from 2D object proposals extracted from the RGB image. These proposals serve as candidate regions for potential 3D objects. In the second stage, features from the multi-view representations are fused and refined to perform precise 3D bounding box regression and object classification. This fusion across views and modalities allows MV3D to capture complementary geometric and appearance cues, leading to more accurate and robust 3D detection performance.

Reference [70] presents an example of deep learning-based tracking layer fusion within a "tracking-by-detection" framework. The method formulates multi-object tracking as inference in a Deep Structured Model (DSM), where the goal is to jointly optimize detection and association using both camera and LiDAR data. A set of feed-forward neural networks is used to compute detection scores and matching scores between objects across frames. These networks take multimodal inputs to capture complementary appearance and geometric cues.

The inference process in the DSM is efficiently carried out through a feed-forward pipeline, followed by solving a linear program that determines the optimal associations between detections across time. Crucially, the entire system is trained end to end using a structured hinge loss, which enables simultaneous optimization of both the object detector and the data association (tracking) module. This joint training improves the consistency and accuracy of the tracking pipeline, especially in challenging conditions such as occlusions or dense object interactions.

The system takes sequences of RGB frames (camera images) and LiDAR point clouds as external inputs. Based on these inputs, the system generates discrete trajectories of the target. An end-to-end trainable architecture is used, which includes three learnable networks: Detection Net, Matching Net, and Scoring Net. The previous MV3D detector is directly used to generate 3D target proposals.

DeepFusion, as introduced in Ref. [71], presents two key innovations for improving the alignment and integration of LiDAR and image features in multimodal 3D object detection. The first is the InverseAug module, which addresses the challenge of preserving geometric consistency between LiDAR points and image pixels after data augmentation. Many geometric augmentations, such as rotation or scaling, distort the spatial alignment between LiDAR and camera views. To counteract this, InverseAug reverses the applied geometric transformations on keypoints—such as LiDAR points or voxel centers—by mapping them back from the augmented 3D space to their original coordinate system. Once restored, these keypoints can be accurately projected onto the 2D camera plane using the known LiDAR–camera calibration parameters. This reverse-mapping process ensures precise cross-modal correspondence, which is crucial for high-fidelity feature fusion.

The second innovation is the LearnableAlign module, which employs a cross-attention mechanism to dynamically model the interaction between image and LiDAR features. Rather than relying on rigid projection-based fusion, LearnableAlign enables the network to learn soft and adaptive correspondences, capturing contextual relationships between modalities during training. This enhances the flexibility and accuracy of the fusion process, allowing the model to better exploit complementary strengths of both LiDAR (geometric precision) and image data (texture and semantics). Together, these modules enable DeepFusion to achieve improved geometric alignment and robust multimodal feature integration in 3D detection tasks.

The Voxel Field Fusion (VFF) method is designed to enhance cross-modal consistency during LiDAR–image fusion by preserving spatial context and alignment throughout the feature projection and enhancement process. Unlike traditional approaches that treat fusion as a point-to-point mapping from image features to voxels, VFF models the interaction between modalities as ray-based projection and fusion within the voxel space, improving geometric alignment and feature integrity.

To achieve this, VFF introduces a learnable sampler that identifies important features on the 2D image plane and projects them onto the 3D voxel grid in a point-to-ray fashion. This means that instead of assigning a single voxel to a specific pixel location, the method traces a ray through the voxel field along the projection path, sampling and aligning image features with 3D spatial context. This approach significantly improves the consistency between image-based and LiDAR-based representations.

Furthermore, the ray-by-ray fusion module aggregates image and LiDAR features along each ray, incorporating richer contextual information from both modalities. To address inconsistencies introduced by data augmentation, VFF also includes a mixed enhancement mechanism, which learns to align the feature transformations across modalities. This helps bridge the modal gap that arises from applying different or asynchronous geometric augmentations to LiDAR and image inputs.

In contrast to previous methods that project image features to voxels before enhancement and do so in a direct point-to-point manner, VFF performs feature enhancement first and then executes a geometrically aware point-to-ray fusion. This preserves alignment even under strong augmentations and improves the fidelity and expressiveness of the fused voxel representation.

The 3D object detection pipeline of the Voxel Field Fusion (VFF) framework consists of several coordinated stages that integrate multimodal information for robust detection. The process begins with a hybrid enhancement module, which applies geometric and appearance augmentations to the input data from different modalities (e.g., LiDAR and images). This module is used only during training to improve the model's generalization and cross-modal robustness.

Next, modality-specific feature encoders extract high-level representations independently from LiDAR and image inputs. To align these features spatially, a ray constructor establishes correspondence by projecting image space features into 3D space along rays determined by the camera geometry.

Within the VFF module, a learnable sampler selects the most informative image features based on their relevance to the voxel field. These selected features are then passed through a ray fusion process, which aggregates information along each ray where high-response features reside. This step captures detailed cross-modal context while preserving geometric consistency.

Finally, the fused features—now embedded within the 3D voxel space—are processed by a detection backbone and prediction head, which generate the final 3D object proposals. This pipeline ensures that spatially accurate and semantically rich features from both modalities are effectively utilized for high-quality 3D object detection.

Here we mainly introduce the fusion methods of the data layer and the task layer, and for the deep learning framework, there is currently a fusion method of the feature layer that is valued by the industry and academia, namely, the BEV network, which will be discussed separately in Sect. 5.14.

5.7 Lane Line Detection

Lane lines are important semantic information on the road, and lane line detection is very important for autonomous driving systems. Lane lines can not only provide relevant road condition information to prevent lane deviation, but also assist in vehicle localization and detection of vehicles ahead. However, lane line detection faces many challenges, including extreme illumination, missing lane lines, and obstacles. In recent years, lane line detection algorithms based on deep learning have attracted attention due to their excellent performance.

1. Traditional methods

General lane line detection systems are mostly based on monocular images, and early algorithms are mostly traditional methods. The work in Ref. [72] is basically divided into several modules: image cleaning (obstacle occlusion detection, shadow elimination, exposure correction), feature extraction (road/lane line detection), road/lane line model fitting (lateral/longitudinal), temporal integration (front and back frames), and image-physical space (road coordinate system) correspondence, etc.

Here we focus on a stereo vision-based method for lane line detection, which leverages the advantage of binocular systems in capturing disparity information. The approach begins with the extraction and matching of feature points from the left and right stereo images. Using these matched features, a road model is estimated to represent the geometric layout of the scene. Based on this model, a disparity map is computed, and vanishing point estimation is performed using the principle that lane lines in the real world appear as parallel lines in 3D space but converge in the image due to perspective projection.

Once the disparity map and vanishing point are obtained, the system proceeds to extract the lane lines. A critical parameter in disparity computation is the disparity range, which defines the expected pixel offset between corresponding points in the stereo images. This range is highly sensitive and directly affects the accuracy and efficiency of the algorithm. To address this, the method estimates the road plane geometry and uses it to derive a reasonable disparity range. This estimated range is then fed back into the disparity calculation module to constrain and optimize the disparity search, improving both speed and precision in the stereo matching process. This feedback loop enhances the robustness of lane detection in various driving environments.

Disparity calculation is divided into two steps: the first step is sparse disparity based on feature point matching, where the feature points use SURF (speeded-up robust features) and SIFT (scale invariant feature transform) and the second step is to interpolate the sparse disparity into a dense disparity map based on dynamic programming (DP).

Lane line extraction in this method also utilizes a dynamic programming algorithm, but with distinct guiding cues—primarily the estimated vanishing point and image-derived edge features. The vanishing point estimation itself is performed using a dynamic programming strategy. In particular, the horizon line is detected through a horizontal band analysis that incorporates road disparity data and edge information. The vanishing point is then defined as the intersection between the horizon line and the normal direction of the lane lines. A global optimization step refines this estimation to obtain the most accurate vanishing point.

Once the vanishing point is determined, it serves as a crucial constraint for lane line detection, providing directional and curvature cues. The position of the lane line is inferred by estimating the lateral offset of its centerline relative to the vanishing point. To facilitate this process, a one-dimensional (1D) accumulator is designed. In this setup, the edge features extracted from each image row are used to compute a likelihood function. Leveraging the characteristic "dark–bright–dark" intensity profile of typical lane markings, the projected signal along the image rows tends to form distinct peak-pair structures (referred to as "add–subtract" peaks). These are effectively captured by the accumulator, which supports robust localization of lane positions under varying road and lighting conditions.

A more detailed visualization of the method is presented in Ref. [73], which outlines an intuitive pipeline for stereo image processing. The process begins with the application of bilateral filtering to the stereo image pair. This filtering technique is designed to reduce noise while preserving edge details, ensuring that important structural information is retained. Following this, edge extraction is performed to emphasize the image's geometric structure, which is critical for enhancing the reliability of subsequent disparity computation and depth estimation stages. This structured preprocessing approach enables more robust and accurate performance in stereo vision tasks.

Next, disparity estimation is performed to obtain dense depth information across the stereo image pair. Based on this disparity map, road plane detection is conducted by leveraging geometric priors, such as the assumption of a flat road surface and consistent disparity gradients. This allows for reliable isolation of the road region in the 3D scene.

Following road plane estimation, the algorithm proceeds with lane line detection. It integrates cues from the disparity map and edge features, as well as constraints from the estimated vanishing point and horizon line, to robustly extract lane markings. Additionally, an obstacle erasure module is employed to eliminate the influence of foreground objects (e.g., vehicles or pedestrians) that may occlude lane lines, thereby improving the accuracy and continuity of the detected lanes.

2. Deep Learning Methods

Next, we will focus on the more popular deep learning methods.

Reference [74] summarizes and reviews various methods from the perspectives of network structure and optimization objectives. The objective function is analyzed from three aspects: classification, regression, and unsupervised learning (deep generative models). Among them, the objective function of unsupervised training can be divided into generative adversarial and autoregressive (AR) types.

In applying deep learning methods for lane line detection, several issues need to be considered.

(1) Unlike ordinary semantic segmentation targets, lane lines have unique colors (white/yellow) and shapes (thin lines); The design of the neural network should reflect these unique characteristics.

(2) There are various types of lane lines, such as single solid/dashed lines and double white/yellow solid lines. Sometimes it is necessary to detect not only the position of the lane lines, but also the type of lane lines.

(3) For assisted driving, real-time performance is an important consideration.

Lane line detection methods based on segmentation have a common flaw: lane lines only occupy a small part of the entire image, and slight losses in segmentation can significantly reduce detection performance. Regression networks can directly predict lane line coordinates and separate different categories of lane lines. Regression-based lane line detection is easy to classify, the network is easy to converge, and segmentation-based lane line detection can flexibly respond to lane line variations.

Improvement strategies that can be adopted for lane line detection based on semantic segmentation networks include the following:

(1) Utilize spatial structure. Utilize the spatial structure of lane lines during network learning, such as dilated convolution, graph convolution networks, and attention mechanisms.

(2) Utilize scene priors. Information in the scene can assist in the segmentation of the target, such as vanishing points.

(3) Utilize positional regularity. Lane lines are different from other segmentation targets, and their distribution in the scene has a certain regularity. For example, lane lines are distributed on the road surface, they are parallel in BEV, and they have continuity in the scene.

In Ref. [75], VPGNet is a multi-task neural network trained end to end based on the principle of vanishing points, which can handle lane lines and road markings simultaneously. For this, the authors of Ref. [75] established a database of road and lane line markings, including various traffic conditions, such as rain/no rain/heavy rain/night.

VPGNet designs a data layer to introduce grid-level annotations, so that lanes and road markings can be trained simultaneously. In addition, it adopts a regression method of grid-level masks, where points on the grid are regressed to the nearest grid unit, and the multilabel classification task combines them into targets. This integrates two independent tasks with different features and shapes, namely, lanes and road markings. In the post-processing stage, the lane class only uses the output of the multilabel task, while the road marking class uses both the grid box regression and multilabel task outputs. In addition, a vanishing point detection task is added to infer the global geometric context relationship during the training of lanes and road markings.

VPGNet is a multi-task deep learning architecture designed for comprehensive road scene understanding. It integrates four functional modules: grid box regression, object detection, multilabel classification, and vanishing point prediction. These modules work in a coordinated manner, enabling the system to detect and classify lane markings and road symbols, while also estimating the vanishing region of the road scene. All tasks are accomplished in a single forward pass, which ensures both efficiency and consistency across outputs. The network architecture is carefully designed with shared feature extraction layers and task-specific branches, allowing for real-time performance and high accuracy in complex driving environments.

In Ref. [76], although the CNN model can easily extract semantic information from pixels, the spatial relationship of pixels in rows and columns has not been fully exploited. To learn semantic targets with strong shape prior knowledge but weak appearance coherence, such as traffic lane lines on the road that are obscured or even missing, the Spatial CNN (SCNN) model extends the layer-by-layer convolution in the feature map to slice-by-slice convolution, so that information passing between pixels is carried out in rows and columns of each layer. SCNN is particularly suitable for long continuous shape structures or large targets with strong spatial relationships but lack of appearance clues, such as lane lines, walls, and poles/columns.

The SCNN model adopts a structured training and prediction strategy tailored for lane detection. During training, the number of output channels in the `fc7` layer is set to 128, and the dilated convolution in the `fc6` layer is configured with a dilation rate of 4. Batch normalization (BN) is applied before each ReLU activation to improve convergence. Additionally, a lightweight sub-network is used to predict the existence probability of lane lines, with the target line width defined as 16 pixels. Both the input and target images are resized to 800×288 pixels to standardize processing. To address the class imbalance between lane lines and background, the background loss is down-weighted by a factor of 0.4.

During inference, the model outputs a probability map indicating lane locations. For each lane marker with an existence probability exceeding 0.5, the corresponding probability map is scanned at 20-row intervals to identify the pixel with the highest response. These peak points are then linked using a cubic spline to construct a smooth lane curve as the final prediction result.

A key distinction between the MRF/CRF model and the SCNN model lies in how spatial dependencies are captured within feature representations. While traditional MRF/CRF approaches are well suited to scenarios where the number of channels matches the number of classification categories, SCNN extends applicability to intermediate layers of the network that encode richer and more abstract features. Specifically, SCNN operates on a 3D tensor of dimensions $C \times H \times W$, where C is the number of channels, H is the height, and W is the width of the feature map. This tensor is conceptually sliced along the height dimension into H horizontal slices. The process begins by passing the first slice through a convolutional layer with C kernels, each having a kernel width of w. Rather than feeding the output directly to the next layer as in conventional CNNs, the output is added element-wise to the next slice to form an updated slice. This updated slice is then passed through the convolutional layer, and the process repeats iteratively until all slices are processed. This method enables efficient spatial message passing across the vertical axis of the feature map.

In Ref. [77], a neural network named LaneNet is proposed for lane line detection, which is divided into two main stages. The first stage, the lane edge proposal network, adopts an encoder–decoder architecture: the encoder consists of convolutional layers, while the decoder is composed of deconvolutional layers, and includes skip connections to preserve spatial detail. This network is responsible for extracting candidate lane line edges from the input image. The second stage, the lane line localization network, also follows an encoder–decoder structure. It takes the lane line edge map generated from the first stage as input and applies a series of 1D convolutional encoders to transform the input into a suitable representation. Then, a set of LSTM (long short-term memory) decoders—commonly used in recurrent neural networks—progressively decodes the features to produce the final lane line parameter predictions.

The first step input is an image through inverse perspective mapping (IPM), and the output is pixel-level lane edge classification. A pixel-level edge probability map is generated as the credibility of the lane lines through a high-resolution feature map. The second step is to detect the lane lines based on the lane edge proposal as output. The challenge is how to reduce the mis-detection caused by lane line-like markings on the road, such as arrows and characters. This method does not need to assume the number of types of lane lines, and it has small computational cost and high speed.

In Ref. [78], it is mentioned that some approaches train neural networks to perform pixel-level segmentation for lane line detection, enabling detection even when explicit lane markings are absent. However, these approaches are typically limited to handling a fixed number of lane lines and struggle with dynamic scenarios such as lane changes. To overcome this limitation, an instance segmentation-based method is proposed in which each lane line is treated as a separate instance. This enables the model to handle varying numbers of lanes and adapt to lane changes more effectively, while still allowing end-to-end training.

In the proposed approach, the input is a road image, and the neural network (LaneNet) outputs a lane line instance map in which each pixel corresponding to a lane line is assigned a unique instance ID. Before performing lane fitting, a learned perspective transformation—parameterized by a homography matrix (via a component called H-Net)—is applied to the segmented instances. Each detected lane is then fitted using a third-order polynomial in the transformed (top-down) view and subsequently projected back to the original image space. This method enhances the model's robustness to changes in lane topology and provides accurate lane localization in complex driving environments.

The work in Ref. [78] builds upon the previously introduced LaneNet architecture, utilizing an ENet-based encoder–decoder structure and modifying it into a dual-branch neural network design. The first branch is a segmentation branch that produces a binary lane mask, identifying which pixels belong to lane markings. The second branch is a manifold embedding branch, which assigns each lane pixel an N-dimensional embedding vector. These embeddings are trained such that pixels belonging to the same lane line have embedding vectors that are close in the learned space, while those belonging to different lane lines are well separated.

After inference, the binary segmentation mask from the first branch filters out the background pixels, leaving only the lane-relevant pixels with associated embeddings. These remaining embeddings are then clustered, with each cluster representing one detected lane line. The center of each cluster corresponds to a distinct lane, enabling accurate instance-level lane detection through this combination of segmentation and embedding-based clustering.

The method proposed in Ref. [79] is particularly noteworthy in that it enables the direct prediction of 3D lane lines from a single monocular image of a road scene. The 3D LaneNet architecture introduces two key innovations: a built-in inverse perspective mapping (IPM) mechanism and an anchor-based lane line representation. The IPM mechanism allows the network to simultaneously learn from both the front-view and bird's-eye-view representations of the scene, facilitating improved spatial understanding and geometric reasoning. The anchor-based lane representation enables a column-wise prediction scheme across the road width, which promotes efficient end-to-end learning and prediction.

The network ultimately outputs, for each longitudinal road slice, a confidence score indicating the presence of lane lines, along with the corresponding 3D curve coordinates of the detected lane lines that intersect the slice. This anchor-based approach is conceptually similar to single-stage object detectors, such as SSD and YOLO, where predefined anchors are used to detect objects in a dense prediction manner. One of the main advantages of this method is its ability to explicitly model and handle complex road structures, such as lane merges and splits, in a principled and scalable way.

The relationship between the camera coordinate system and the road coordinate system can be intuitively understood by considering how the camera, typically mounted on the vehicle, captures the road scene. The camera coordinate system is usually defined with its origin at the camera's optical center, the z-axis pointing forward along the viewing direction, the x-axis pointing to the right, and the y-axis pointing downward. In contrast, the road coordinate system is often defined with its origin on the road surface, the x-axis aligned with the vehicle's lateral direction, the y-axis pointing vertically upward, and the z-axis aligned with the forward driving direction. Understanding the transformation between these two systems is essential for accurate projection of 3D lane lines or objects from image space to real-world road geometry.

A dual context module can be constructed to process both front-view and top-view representations of a road scene simultaneously. The core component of this module is the projective transformation layer, which is a variant of the spatial transformer module. This layer enables differential sampling of the input feature map, such that the sampled output spatially aligns with a "virtual" top-down (bird's-eye) view, while the input remains aligned with the front-view image plane. The differential sampling is typically implemented through a grid-based method. By leveraging this projective transformation layer, the dual context module constructs highly descriptive feature maps. These feature maps preserve the spatial correspondence and semantic consistency between the original front view and the transformed top view, allowing for effective information fusion and enhanced understanding of the scene.

The structure of 3D LaneNet is composed of two parallel information flows, referred to as a dual-core design. One flow processes the front view of the scene, retaining rich image information, while the other flow operates in the top-view domain, enabling translation-invariant feature extraction and supporting the prediction of 3D lane line detection results. The front-view flow is built upon the VGG-16 architecture, and the top-view flow adopts a similar structure. This dual-branch design allows the model to leverage both perspective and bird's-eye representations, which enhances its capacity to localize and model lane lines in three-dimensional space.

The lane lines in a road scene are described using two sets: the set of lane centerlines $\{C_i\}_{i=1}^{N_C}$ and the set of lane delimiters $\{D_i\}_{i=1}^{N_D}$. Each element in these sets is a 3D curve expressed in the camera coordinate system. The goal of 3D LaneNet is to detect these 3D curves, identifying the lane centerlines and/or lane delimiters from a single input image.

To represent the 3D structure of a lane line, the method uses an anchor-based formulation. Anchors are defined as a series of fixed, vertically equidistant longitudinal positions $\{X_A^i\}_{i=1}^{N}$ along the x-axis. For each anchor X_A^i, the 3D lane line is parameterized by $2K$ neural network outputs $(\boldsymbol{x}^i, \boldsymbol{z}^i) = \{(x_j^i, z_j^i)\}_{j=1}^{K}$ and a fixed set of K predefined y-coordinates $\boldsymbol{y} = \{y_j\}_{j=1}^{K}$. Each point on the 3D lane line is therefore represented as $(x_j^i + X_A^i, y_j, z_j^i)$ in the road coordinate system, where x_j^i is a horizontal offset relative to the anchor.

In addition to the geometric representation, each anchor X_A^i is associated with a confidence score p^i, indicating the likelihood of a lane line passing through that anchor. Anchor association is determined based on a reference longitudinal coordinate Y_{ref}: the anchor X_A^i is selected for a given lane line if its x-coordinate is closest to the x-position of the lane line at $y = Y_{\text{ref}}$.

For each anchor, the network outputs three lane descriptions denoted by (t), which include both confidence and geometric information. Specifically, the first two outputs, (c_1, c_2), correspond to the centerlines of two possible lanes associated with the anchor, while the third output (d) represents the lane divider. The use of two centerlines per anchor enables the network to model scenarios where lane lines merge or split. At the reference longitudinal position Y_{ref}, it is common for the two centerlines to overlap, with their trajectories diverging at other positions along the road. This modeling approach allows for flexible representation of lane topologies involving merging and splitting behaviors.

However, the representation of lane dividers is inherently more complex than that of centerlines. While the centerlines are assumed to intersect the $y = Y_{\text{ref}}$ plane, lane dividers may not always follow this assumption. As a result, this anchor-based representation has limitations and may fail to capture certain configurations of lane dividers that do not pass through the predefined reference position.

The prediction head of 3D LaneNet produces the described output, through a series of convolutions with no padding in the y dimension, the feature map is reduced in dimension, and the final prediction layer size is $3 \times (2K + 1) \times 1 \times N$, where each column i corresponds to a single anchor X_A^i. For each anchor X_A^i and output $t \in \{c_1, c_2, d\}$, the network output is represented by $(\boldsymbol{x}_t^i, \boldsymbol{z}_t^i, \boldsymbol{p}_t^i)$. The final prediction performs 1D non-maximum suppression, only retaining the lane line with the local maximum confidence (compared with the left and right adjacent anchors). The remaining lanes represented by a small number (K) of 3D points are converted into smooth curves using spline interpolation.

A Global Association Network (GANet) was proposed in Ref. [80], offering a novel perspective on the lane line detection task. Unlike traditional point-to-point lane expansion methods, GANet introduces a keypoint-based regression approach, in which each key point directly predicts its offset to the starting point of the corresponding lane line. This decouples key points from each other, allowing the association process to be parallelized and thereby significantly improving computational efficiency.

To support accurate key point estimation, GANet incorporates a Lane Perception Feature Aggregator (LFA), which adaptively captures the local relationships between adjacent key points, enriching the local contextual understanding within the broader global association framework. The overall system takes a front-facing image as input and extracts multi-scale visual features through a CNN backbone, a self-attention (SA) layer, and a Feature Pyramid Network (FPN) bottleneck. In the decoding stage, two parallel heads predict key point confidence maps and offset maps. These are subsequently fused to group the key points into distinct sets, each corresponding to an individual lane line instance. The LFA module is applied prior to the key point prediction to improve contextual sensitivity and local structure modeling.

The process of lane line construction in the GANet follows a systematic approach based on keypoint regression. First, valid keypoints are selected from the predicted confidence map. For each keypoint, denoted by its position (x, y), the network also predicts an offset vector $(\delta x, \delta y)$, which points from the current keypoint toward the estimated starting point of the corresponding lane line. Using this offset, the estimated coordinates of the lane line's starting point are computed as $(sx', sy') = (x, y) + (\delta x, \delta y)$. Keypoints whose predicted starting points fall within a small neighborhood around a common location (sx, sy) are then grouped together, forming a complete lane line instance. This grouping process relies on the proximity and consistency of offset predictions, enabling efficient and accurate lane association across the image.

To estimate the starting point of the lane line, key points with values less than 1 on the offset map are selected as candidate starting points. Since there may be multiple key points in the same local area that match the above criteria, the geometric center point of this area is chosen to ensure uniqueness. In this way, all instances of lanes are initially determined by their starting points.

Afterward, the remaining key points are associated with their respective lanes based on the offset estimation between the key points and the corresponding starting points, so that each key point estimates the coordinates of the starting point of the lane line.

The LFA module enhances the representation of each keypoint by capturing its local geometric context. Specifically, the module first predicts the offsets between a given observed keypoint and M neighboring keypoints that lie on the same lane line. This is achieved using a convolutional layer. Subsequently, the features of these neighboring keypoints are aggregated using deformable convolution, allowing the network to flexibly sample and incorporate spatially adaptive contextual information. To improve the module's ability to learn the local geometric shape of lane lines, an auxiliary supervision signal is introduced by applying a loss on the predicted offsets. The Hungarian algorithm is employed to associate predicted offsets with ground-truth correspondences, enabling effective and accurate offset learning during training.

A method called SALAD, proposed in Ref. [81], enables 3D lane line detection without relying on external parameters or anchor settings. Unlike many previous methods, SALAD operates directly in the image view and regresses the 3D coordinates of lane lines without converting feature maps into the bird's-eye view (BEV). To support its development, a real-world autonomous driving dataset named ONCE-3DLanes was established, which includes precise 3D lane layout annotations. This dataset is constructed using the explicit correspondence between LiDAR point clouds and image pixels. The annotation pipeline follows a five-step procedure: first, road segmentation is performed to identify drivable regions; second, point cloud data is projected into the image space; third, lane annotations are obtained either manually or via automatic labeling algorithms; fourth, adaptive blending techniques are applied to smoothly integrate lane lines across different spatial perspectives; and finally, the point cloud data is used to recover the accurate 3D geometry of each lane line. This approach enables the automatic generation of high-quality 3D lane annotations for 211,000 road scenes.

SALAD is a spatially aware monocular lane line detection method capable of directly performing 3D lane line detection from a single camera image. Unlike conventional methods that rely on manual anchor point design or require external calibration parameters, SALAD automatically projects the input image into a top-view representation and employs a set of predefined anchor points to regress the 3D coordinates of the lane lines.

The SALAD model is designed with a dual-branch architecture following a convolutional backbone. The backbone first encodes the input image into high-level deep features. These features are then decoded by two separate branches: a semantic awareness branch, which produces a lane segmentation mask, and a spatial context branch, which extracts spatial cues essential for geometric reasoning. The outputs of these branches are subsequently fused to perform spatial reconstruction, yielding accurate 3D lane line positions in real-world coordinates. Furthermore, the model incorporates an enhanced joint 3D lane extension strategy, which improves generalization across diverse road scenes by leveraging both local and global lane continuity during training and inference.

Due to the use of downsampling and lack of global information, the predicted position of the lane points is not accurate enough. The spatial context branch receives image features and outputs a pixel-level offset map, which predicts the spatial position offset δu and δv of the lane line points on the image plane along the u-axis and v-axis, based on which the rough estimation of the position of the lane line points will be modified according to the global spatial context.

In order to recover 3D lane line information, the spatial context branch also generates a dense depth map to regress the depth offset δ_z of each pixel of the lane line. Considering that the ground depth of the image plane increases along the row, a predefined offset α_r and scale β_r are assigned to each row of the depth map for residual regression. A depth completion is applied on the sparse depth map to obtain a dense depth map, providing sufficient training signals for the spatial context branch.

5.8 Traffic Sign Detection and Recognition

The task of understanding traffic signals on the road for autonomous vehicles includes two parts [76]: traffic sign detection and recognition, and traffic light detection and recognition. Basically, traffic signs and traffic lights can be treated as targets, so the methods of target detection and recognition are universal, but some specific knowledge and properties will be added to reflect the uniqueness of these targets.

Different regions and countries have different ways of placing and changing traffic lights. Therefore, if rule-based algorithms are formulated, different rules will be defined; if deep learning-based model algorithms are formulated, different model frameworks will also exist.

This section introduces methods for traffic sign detection and recognition, while Sect. 5.9 covers the detection and recognition of traffic lights (which can be further improved in the era of vehicle networking technologies).

The traffic sign recognition task typically involves three stages: detection, classification, and tracking, forming a classical visual object recognition pipeline. First, candidate regions likely to contain traffic signs are detected from the scene using methods such as sliding windows, region proposal networks, or feature-based approaches. Next, the detected regions are classified into specific traffic sign categories using classifiers trained on labeled datasets. Finally, temporal tracking is used to maintain the consistency of sign recognition across video frames, enhancing robustness against occlusion or missed detections.

The objects of recognition in this task include various types of traffic signs, such as regulatory signs (e.g., speed limits), warning signs (e.g., sharp turns), and guide signs (e.g., highway exits). These signs differ in shape, color, and symbol, which serve as key cues for detection and classification algorithms. Common examples include circular speed limit signs, triangular warning signs, and rectangular informational signs.

Known TSR systems often face the following challenges [76]:

- The traffic symbols within and between each type of traffic sign are often similar.
- Traffic signs fade or get dirty, changing color.
- Traffic signs bend, no longer perpendicular to the road surface.
- Poor lighting conditions.
- Low contrast.
- In complex environments, other targets are similar to traffic signs.
- Weather changes.

The general process of TSR typically involves three key stages: image segmentation, feature extraction, and final detection. Traditional methods are generally categorized into color-based and shape-based approaches. Color-based methods leverage the distinct hues commonly found in traffic signs (e.g., red, blue, yellow), while shape-based methods utilize geometric constraints such as circles, triangles, and rectangles. However, with the development of machine learning, particularly deep learning, end-to-end solutions have gained popularity. These approaches bypass hand-crafted features by directly learning hierarchical representations from raw images, offering improved accuracy and robustness under complex conditions.

An example of such a deep learning-based TSR system is the one developed jointly by Tsinghua University and Tencent [77]. This system relies on large-scale annotated datasets for training. One key form of annotation involves pixel-level masks of traffic signs, which provide dense supervision to guide the neural network in learning the spatial structure and precise contours of signs. These annotations enable the network to perform fine-grained segmentation, which is essential for high-accuracy detection and classification in real-world driving environments.

A comprehensive dataset of traffic signs used in China serves as the foundational input for training and evaluating recognition models. This dataset includes a wide variety of sign types, such as prohibitory, warning, mandatory, and guide signs, encompassing the full spectrum of traffic signage encountered in Chinese road networks. The diversity and quantity of the signs are critical to ensuring the robustness and generalization capability of the trained models.

The CNN model employed in the joint research by Tsinghua University and Tencent is a modified version of the OverFeat architecture [81], originally developed for general object detection. This modified model has been optimized for the specific task of traffic sign detection and recognition, with enhancements made to improve spatial localization and classification performance. The model incorporates a multi-scale feature extraction strategy and adopts a fully convolutional structure that allows for end-to-end training. Detailed information on the model's size and parameters is provided in the corresponding tabular summary (Table 5.1).

The last layer of the CNN model is divided into two branch layers: one is the pixel layer, and the other is the bounding box layer. Each pixel layer output represents the probability of each image area of 4×4 pixels including the target, and the result of each bounding box layer represents the distance from the area to the four sides of the target box. In addition, a new label layer is added to output an N-dimensional classification vector. In this way, the model can complete detection and classification tasks simultaneously.

5.9 Traffic Light Detection and Recognition

Although there are some common parts between traffic lights and traffic signs, there are still some different situations. Here we discuss the issues in Ref. [82].

The main challenges of traffic light detection and recognition are as follows:

- Color drift and halo interference under atmospheric conditions.
- Obstruction by other objects.
- Incomplete shape (light failure).
- False alarms (other light sources, brake lights, reflections, pedestrian crossing lights, road sign lights).
- Synchronization between the camera shutter speed and the duty cycle of the traffic light LED.

In California, traffic lights are categorized based on color and arrow indicators. Typically, the topmost indicators are red lights, followed by yellow lights in the next row, then green lights. Below the green lights, various green arrows are displayed, which indicate permitted directional movements. Unlike static traffic signs, traffic lights operate in a temporal sequence that reflects dynamic traffic control logic. The state transitions of traffic lights follow a defined order—such as from red to green, then to yellow—often incorporating directional arrows to guide turning traffic. Understanding this sequence is crucial for accurate detection and behavior prediction in autonomous driving systems.

If the location of a traffic light is known in advance, such as from a high-definition (HD) map, the complexity of traffic light detection can be significantly reduced. This strategy is employed by Stanford University's autonomous driving team. By leveraging the known spatial relationship between the camera coordinate system and the vehicle coordinate system, the expected position of the traffic light can be projected onto the image plane. Consequently, the search space for detecting traffic lights can be narrowed, allowing for more efficient and accurate detection by focusing only on the relevant region within the image.

Table 5.1 OverFeat model parameters

Layer	Data	conv1	conv2	conv3	conv4	conv5	conv6	conv7	conv8-Border	conv8-Pixel	conv8-Label
Output size (chan × h × w)	3, 480, 640	96, 118, 158	256, 59, 79	384, 29, 39	384, 29, 39	384, 29, 39	4096, 15, 20	4096, 15, 20	256, 15, 20	128, 15, 20	1000, 15, 20
Output size	3, 480, 640	3, 480, 640	3, 480, 640	3, 480, 640	3, 480, 640	3, 480, 640	3, 480, 640	3, 480, 640	3, 480, 640	3, 480, 640	3, 480, 640
Kernel size/Stride/padding	3, 1, 1	3, 1, 1	3, 1, 1	3, 1, 1	3, 1, 1	3, 1, 1	3, 1, 1	3, 1, 1	3, 1, 1	3, 1, 1	3, 1, 1
Pooling size/stride	2, 2	2, 2	2, 2	2, 2	2, 2	2, 2	2, 2	2, 2	2, 2	2, 2	2, 2
Additional part	Input image	Convolution	Convolution	Convolution	Convolution	Convolution	Fully connected	Fully connected	Output border	Output pixel	Output label

Below we introduce Bosch's method for traffic light detection and recognition [79], which integrates deep learning, stereo vision, and vehicle odometry. The approach follows a "detection–classification–tracking" pipeline. Notably, the classification module is applied twice: once after detection and again during tracking. This dual classification strategy is designed to mitigate tracker drift by verifying the identity of detected traffic lights across frames, ensuring robustness and accuracy in dynamic driving environments.

For this, researchers have established a traffic light database, the BOSCH Small Traffic Lights Dataset. The classifier is based on the CNN model, and its corresponding model structure parameters are shown in Table 5.2. It includes five types, namely, 'background', 'red', 'green', 'yellow', and 'off' where 'off' is for the situation where the traffic light is turned off.

The detector is based on YOLO (you only look once, single-step detector) v1, and the tracking algorithm is based on stereo vision and vehicle odometry estimated vehicle motion. Because stereo vision is used, the computed disparity map can calculate the 3D position of the traffic light target box, and with the motion information from the vehicle odometer, the position of the traffic light at the next moment can be predicted.

The model trained with deep learning (using a spatial transformer network) can eliminate tracking drift. The structural parameters of the Spatial Transformer Network (STN) are shown in Table 5.3.

A CNN-based tracking method is designed to predict the offset of the tracking target's candidate position along with an associated error estimate. Specifically, the network outputs the horizontal and vertical offsets (u, v) indicating the displacement of the target, as well as an error estimate e that reflects the model's confidence in the prediction. This allows the tracker to not only localize the target but also quantify uncertainty, which is crucial for robust tracking performance, especially in complex or ambiguous visual environments.

Table 5.2 Model structure parameters of the traffic light classifier

Channel	Resolution
Input	3×64×64
Convolution	Filter: 32, Kernel: (7,7), Padding: 0, ReLU
Max pooling	Kernel: (2,2)
Convolution	Filter: 64, Kernel: (3,3), Padding: 0, ReLU
Max pooling	Kernel: (2,2)
Convolution	Filter: 128, Kernel: (3,3), Padding: 0, ReLU
Fully connected	Kernel: 256, ReLU
Dropout	p: 0.5
Fully connected	Unit: 128, ReLU
Dropout	p: 0.5
Fully connected	Unit: 5, softmax

Table 5.3 Structural parameters of the spatial transformer network

Channel	Resolution
Input	6×64×64
Convolution	Filter: 16, Kernel: (3,3), Padding: 0, ReLU
Max pooling	Kernel: (2,2)
Convolution	Filter: 32, Kernel: (3,3), Padding: 0, ReLU
Max pooling	Kernel: (2,2)
Convolution	Filter: 64, Kernel: (3,3), Padding: 0, ReLU
Max pooling	Kernel: (2,2)
Convolution	Filter: 32, Kernel: (3,3), Padding: 0, ReLU
Fully connected	Unit: 128, ReLU
Fully connected	Unit: 64, Linear
Fully connected	Unit: 3, Linear

5.10 Drivable Area Segmentation

Segmentation is a challenging task in computer vision, primarily due to the difficulty in acquiring accurate semantic information. However, recent advances in deep learning have significantly advanced the field, particularly in the areas of semantic segmentation, instance segmentation, and the increasingly popular panoptic segmentation. Semantic segmentation involves predicting a semantic label for each individual pixel in the image, enabling dense classification. Instance segmentation goes a step further by distinguishing between different instances of the same object category, identifying the pixel regions belonging to each individual object. Panoptic segmentation unifies the two approaches, assigning both semantic categories and instance identifiers to every pixel in an image, thus achieving a comprehensive, global understanding of the scene.

Several deep learning-based methods have been proposed to address semantic segmentation, as summarized in the literature [83]. These approaches typically involve encoder–decoder architectures, feature pyramid networks, and various attention mechanisms to capture both low-level spatial details and high-level semantic context, thereby improving pixel-wise classification accuracy across complex visual scenes.

It can be said that the work of deep learning in semantic segmentation began with FCN (fully convolution network), that is, true pixel-to-pixel image understanding. Instance segmentation is actually a combination of object detection and semantic segmentation, and its methods are basically divided into two directions: based on object detection and based on semantic segmentation. The former is marked by Mask-R-CNN, while the latter is typically represented by the sequential grouping network (SGN).

Reference [84] was the first to formally propose the concept of panoptic segmentation. This approach unifies semantic segmentation and instance segmentation into a single task, aiming to achieve a comprehensive understanding of the scene. While semantic segmentation assigns a category label to each pixel (e.g., road, sky, vegetation), and instance segmentation distinguishes between individual object instances (e.g., car 1, car 2), panoptic segmentation simultaneously performs both—labeling each pixel with a semantic category and an instance identifier when applicable. This unified strategy provides a holistic view of the scene, enabling the model to recognize both the type and the identity of all elements in the image, including both "stuff" (amorphous background regions) and "things" (countable objects).

The paper jointly published by MIT and Google proposed DeeperLab [85], which represents a typical approach to panoptic segmentation. In this method, a FCN is employed to generate dense predictions for both semantic segmentation and instance-level information simultaneously. The semantic branch assigns a class label to each pixel, while the instance branch encodes pixel embeddings that enable grouping of pixels belonging to the same object instance. A fast and efficient post-processing algorithm is then applied to fuse these two branches of output into a unified panoptic segmentation map, in which every pixel is assigned both a semantic label and an instance ID where applicable. This design allows DeeperLab to achieve real-time performance on high-resolution images while maintaining high segmentation accuracy.

In autonomous driving systems, segmentation plays a crucial role in supporting high-definition map localization and drivable area recognition. One effective approach to drivable road segmentation is introduced in Ref. [85]. This method leverages the unique properties of road areas, which typically exhibit simple textures and are centrally located within images. It demonstrates strong automatic segmentation capabilities under weak supervision. Specifically, feature maps are extracted using a pre-trained dilated ResNet. These feature maps are then refined through superpixel-based image segmentation to better align with image boundaries. Road prior information—such as the typical location and structure of road regions—is used to guide a clustering algorithm, which segments the road area without requiring pixel-level manual annotations. The resulting masks are used to train a semantic segmentation model (SegNet), enabling effective road area prediction in a fully automated manner.

Reference [86] discusses a method of road detection by combining lidar and cameras through deep learning. This method first projects the 3D point cloud onto the image plane, and then upsamples to obtain a dense depth image. This method trains multiple FCNs to complete road detection in different ways, either by a single sensor or by three fusion methods, namely, pre-fusion, post-fusion, and cross-fusion. Pre/post-fusion generally integrates information at some depth layer of the network, while the cross-fusion FCN network learns directly from the data how to integrate information, its method is to train a cross-joint between the lidar branch and the camera branch.

The core network proposed in Ref. [86] is a FCN consisting of 21 layers, where the ELU is used as the activation function instead of the commonly adopted ReLU. To effectively combine multi-sensor data, three fusion mechanisms are designed: (1) pre-fusion, where sensor data are directly concatenated and processed through a unified network from the beginning; (2) post-fusion, in which the output results at the penultimate layer (L_{20}) are concatenated along the depth dimension, and a subsequent convolutional layer performs high-level fusion; and (3) cross-fusion, where the two data processing branches (e.g., camera and LiDAR) are connected at multiple depths through trainable scalar cross-joints. For the j-th layer, two trainable coefficients a_j and b_j are used to modulate the interaction between the inputs from each modality. Specifically, for camera input $\boldsymbol{I}_j^{\text{Cam}}$ and LiDAR input $\boldsymbol{I}_j^{\text{Lid}}$, the fused feature at layer j is computed as follows:

$$\boldsymbol{F}_j = a_j \cdot \boldsymbol{I}_j^{\text{Cam}} + b_j \cdot \boldsymbol{I}_j^{\text{Lid}}$$

where a_j and b_j are learned during training. This formulation enables adaptive, layer-wise sensor fusion that balances modality-specific and shared representations.

5.11 Binocular Vision Perception

Monocular ranging and 3D estimation rely on the target bounding box (BB) detected. If obstacles cannot be detected, the system cannot estimate their distance and 3D posture/orientation. When there is no deep learning, Mobileye (ME), which ranks first in the ADAS market, mainly estimates distance based on BB, camera calibration obtained installation posture and height, and the assumption of a flat road surface.

With deep learning, you can train the NN model based on the true value of the 3D distance to get the 3D size and posture estimation.

The advantage of binocular vision is that it can calculate disparity and depth. Even if obstacles are not detected (because there is additional depth information, the detector will be better than monocular), it will alarm. The problem is that it is not easy for a binocular vision system to estimate disparity, and stereo matching is a typical problem in computer vision. The baseline width is calibrated for the ranging of distant targets, and the baseline is short for the ranging results of near targets, so there is a compromise in the design of the system.

The existing binocular vision system in the ADAS market is the ADAS kit EyeSight from the car company SUBARU. Baidu's Apollo L4 shuttle bus has mass-produced 100 units and installed a binocular system. The EU's autonomous parking project V-Charge also adopted a forward-looking binocular vision system, and the autonomous driving research and development system Berta Benz did the same, and it was post-fused with the radar system. The obstacle detection algorithm Stixel for binocular matching is very famous.

Speaking of the difficulties of the binocular system, in addition to stereo matching, there is also calibration. The system after calibration will have a "drift," so online calibration is a must. The same is true for monocular, because tire deformation and vehicle bumps will cause changes in the external parameters of the camera, and online calibration must be performed to correct some parameters, such as pitch angle and yaw angle.

Binocular online calibration is more complicated, because binocular matching is simplified to 1D search as much as possible, so the optical axis direction of the two lenses needs to be parallel and perpendicular to the baseline through binocular calibration (stereo rectification). Compared with the obtained benefits, if the increased complexity and cost are not cost-effective, merchants will give up.

Recently, binocular vision has gained increasing attention in the industry, particularly following the acquisition of the University of Parma's VisLab by the Silicon Valley chip company Ambarella in 2014. This acquisition enabled Ambarella to develop binocular vision-based ADAS and autonomous driving chips. After showcasing its technology at the 2018 CES (Consumer Electronics Show in Las Vegas, USA), Ambarella began entering the automotive OEM and Tier-1 supplier markets. Since then, it has continued to enhance the performance of its binocular systems. For example, the intelligent driving system developed by DJI's automotive division for SAIC-GM-Wuling employs a forward-facing binocular camera system for driving assistance, supplemented by four surround-view fisheye cameras for parking applications. In contrast, monocular systems based solely on bounding box target detection still carry certain safety risks.

Ambarella's stereo vision system typically consists of multiple (up to six) pairs of stereo cameras installed on the vehicle's roof, with each pair potentially having a different baseline width. This variation allows for coverage of different effective detection distances. The system is also capable of online calibration, enabling real-time adjustment of binocular parameters to maintain performance under varying conditions.

5.11.1 Stereo Matching

Stereo matching refers to the process of estimating disparity or depth from a pair of stereo images. Consider a 3D point $\boldsymbol{X}$ observed by two horizontally aligned cameras with focal length f and baseline distance B (the distance between the optical centers of the two cameras). When $\boldsymbol{X}$ is projected onto the image planes of the left and right cameras, it results in two corresponding 2D image points. The disparity Δx is defined as the horizontal coordinate difference between these two projections. The relationship between disparity and depth z is given by

$$\Delta x = x - x' = \frac{B \cdot f}{z} \tag{5.55}$$

This inverse relationship indicates that the closer the object is to the camera pair, the larger the disparity, and vice versa. This fundamental principle underlies depth estimation in stereo vision systems.

It can be seen that the disparity can be used to calculate the depth value. But the most difficult thing here is how to determine that the images seen by the left and right cameras are the same target, that is, the matching problem.

There are two types of matching methods: global and local. Reference [87] gives four steps of stereo matching:

(1) Calculation of matching cost.
(2) Cost aggregation.
(3) Disparity calculation/optimization.
(4) Disparity correction.

The methods used in the industry basically belong to the local type, because the global type methods have a large amount of calculation and are time-consuming. Deep learning can replace the global method, but it is not very mature yet. The most famous local method is SGM (semi-global point). The methods used in many products are improvements based on this, and many vision chips adopt this algorithm.

SGM approximates a global optimization problem into a combination of multiple local optimization problems. The following formula represents the optimization objective function for 2D matching, and SGM implements it as the sum of multiple 1D optimization paths:

$$E(\boldsymbol{D}) = \sum_{\boldsymbol{p}} C(\boldsymbol{p}, \boldsymbol{D}_{\boldsymbol{p}} + \sum_{q \in N_p} P_1 T[|\boldsymbol{D}_{\boldsymbol{p}} - \boldsymbol{D}_{\boldsymbol{q}}| = 1] + \sum_{q \in N_p} P_2 T[|\boldsymbol{D}_{\boldsymbol{p}} - \boldsymbol{D}_{\boldsymbol{q}}| > 1] \tag{5.56}$$

In this context, $\boldsymbol{D}$ represents the disparity image. The first term is the sum of the matching costs for all disparity pixels in $\boldsymbol{D}$; the second term adds a constant penalty P_1 for all pixels $\boldsymbol{q}$ in the neighborhood $\boldsymbol{N_p}$ of a pixel $\boldsymbol{p}$, where the disparity changes slightly (i.e., by 1 pixel); for larger disparity changes, the third term introduces a greater constant penalty P_2. The following formula represents the path optimization function along a single direction $\boldsymbol{r}$:

$$L'_{\boldsymbol{r}}(\boldsymbol{p}, d) + \min(L'_{\boldsymbol{r}}(\boldsymbol{p} - \boldsymbol{r}, d), \qquad L'_{\boldsymbol{r}}(\boldsymbol{p} - \boldsymbol{r}, d - 1) + P_1 \tag{5.57}$$

$$L'_{\boldsymbol{r}}(\boldsymbol{p} - \boldsymbol{r}, d + 1) + P_1, \min_i L'_{\boldsymbol{r}}(\boldsymbol{p} - \boldsymbol{r}, i) + P_2) \tag{5.58}$$

Here, C denotes the cost, and the remaining part adds the lowest cost of the previous pixel $\boldsymbol{p} - \boldsymbol{r}$ along the path, including an appropriate penalty for discontinuities. This implements behavior along any 1D path. The cost does not enforce visibility or ordering constraints, as these two concepts cannot be achieved on paths that are not epipolar lines. Therefore, this method is more similar to scanline optimization compared to dynamic programming (DP). L' is permanently increased along the path, which can lead to very large values. However, the minimum path cost of the previous pixel can be subtracted from the entire term, thus becoming:

$$L_{\boldsymbol{r}}(\boldsymbol{p}, d) = C(\boldsymbol{p}, d) + \min(L_{\boldsymbol{r}}(\boldsymbol{p} - \boldsymbol{r}, d), \qquad L_{\boldsymbol{r}}(\boldsymbol{p} - \boldsymbol{r}, d - 1) + P_1 \tag{5.59}$$

$$L_{\boldsymbol{r}}(\boldsymbol{p} - \boldsymbol{r}, d + 1) + P_1, \qquad \min_i L_{\boldsymbol{r}}(\boldsymbol{p} - \boldsymbol{r}, i) + P_2) - \min_k L_{\boldsymbol{r}}(\boldsymbol{p} - \boldsymbol{r}, k) \tag{5.60}$$

This adjustment does not alter the actual path through the disparity space. The cost L_r is summed over paths in all directions $\boldsymbol{r}$. The number of paths must be at least 8, and it should be 16 to provide good 2D image coverage.

$$S(\boldsymbol{p}, d) = \sum_{r} L_r(\boldsymbol{p}, d) \tag{5.61}$$

The **Census Transform (CS)** is a non-parametric method used to encode local spatial structure. It converts each 8-bit or 24-bit pixel value into a binary sequence by comparing the intensity of a center pixel with those of its surrounding neighbors within a predefined window, typically of size 5×5. Each neighboring pixel is assigned a 1 or 0 depending on whether it is greater or less than the center pixel, forming a binary pattern. This transformation is robust to illumination changes and commonly used in stereo vision.

A similar binary descriptor is the **Local Binary Pattern (LBP)**, which also encodes local texture information using binary comparisons. Stereo matching algorithms often employ such binary descriptors to simplify the matching process: rather than computing complex similarity measures, they reduce it to a search for minimal Hamming distance between binary strings. This makes the matching process both faster and more robust to intensity variations.

The Census Transform (CS) has the advantage of being highly resistant to interference from radiometric distortions, vignetting, and noise.

PatchMatch (PM) is an efficient algorithm designed to accelerate image template matching, particularly in stereo matching tasks. It operates by randomly initializing the disparity and plane parameters for each pixel, then iteratively refining these estimates through propagation and sampling strategies. The PM algorithm consists of the following five steps:

(1) **Spatial Propagation**: Each pixel compares its disparity and plane parameters with those of its spatial neighbors (typically the left and upper pixels). If adopting a neighbor's estimate reduces the matching cost, the pixel updates its own estimate accordingly.

(2) **View Propagation**: Pixels are transformed to corresponding positions in other camera views (if available), and their disparity and plane estimates are tested. If these alternate estimates offer a lower matching cost, they are adopted.

(3) **Temporal Propagation**: Estimates from temporally adjacent frames (previous and next) are considered. If using these temporal estimates improves the cost, the current estimate is updated.

(4) **Plane Refinement**: Random plane parameter samples are generated and tested. If any of these lead to lower matching costs, the current plane and disparity estimates are updated.

(5) **Post-processing**: Outlier removal is performed via left-right consistency checks, followed by a weighted median filter to smooth the disparity map and preserve edges.

This iterative framework enables PM to quickly converge to high-quality disparity maps with low computational overhead, making it well suited for real-time stereo vision applications.

5.11.2 Binocular Online Calibration

In Ref. [88], an online camera calibration method was proposed that leverages road markings—specifically, pedestrian crossings (commonly known as zebra crossings). This method exploits the geometric regularity of the parallel line pattern formed by the crossings. The algorithm involves detecting the pedestrian crossing within the image and extracting its corner points. Based on these corner correspondences, the method computes a homography that maps the observed pattern onto a planar road surface. Using this homography, the algorithm derives the intrinsic and extrinsic camera calibration parameters. This approach enables online calibration without the need for special calibration objects or manual intervention, making it well suited for autonomous driving systems operating in real-world urban environments.

Another calibration method is based on visual odometry (VO) and simultaneous localization and mapping (SLAM), which is more complex but has the advantage of supporting map-based localization concurrently with calibration. This approach uses SLAM to perform online calibration and is therefore not suitable for high-frequency execution due to its computational cost. The method begins by using stereo vision SLAM to construct a globally continuous map. An initial transformation estimate of the stereo camera is then provided, and the individual maps from multiple stereo cameras are aggregated into a unified global map. Finally, the relative poses between multiple cameras are computed based on the constructed map and their estimated transformations. This approach is particularly useful in multi-camera systems where consistent spatial alignment is critical for scene understanding and autonomous navigation.

Similar to monocular calibration methods, a commonly used approach for binocular systems assumes that the lane lines remain approximately parallel and the road surface is flat. This forms the basis of vanishing point-based calibration techniques. Under this assumption, online calibration can be performed efficiently by estimating the yaw and roll angles of the cameras using the observed vanishing point. The method assumes a flat road model with clear, continuous longitudinal lane lines, and no other edges in the scene that are parallel to these lines. Additionally, it requires that the vehicle is moving slowly and that the stereo camera setup has minimal yaw or roll deviations with respect to the road plane—particularly for the left camera in a left-right configuration. The principle of the method lies in comparing the currently observed vanishing point with an initialized reference vanishing point obtained from earlier calibration. Any drift in the stereo extrinsic parameters can then be inferred from the deviation of the current vanishing point, thus enabling real-time correction of the camera's yaw and roll angles during operation.

5.11.3 Binocular Vision Perception System

Below are some typical binocular autonomous driving systems.

The obstacle detection algorithm Stixels, used in systems like Berta Benz, relies on several key assumptions: objects in the environment are represented as vertical columns (Stixels), each object stands upright on the road due to gravitational stability, and the depth of the upper part of an object is greater than that of the lower part. Based on these assumptions, scene segmentation can be performed by first computing the disparity map using semi-global matching (SGM). The resulting disparity information allows the algorithm to represent objects as narrow vertical rectangles that are perpendicular to the road surface. These stixels serve to simultaneously separate foreground targets from the background and provide a compact and efficient mid-level representation of obstacles in the scene, which can be used for both detection and tracking tasks.

The calculation of Stixels can be enhanced by integrating deep learning-based disparity fusion. This approach typically includes several components: disparity estimation using the Semi-Global Matching (SGM) algorithm, an "Unexpected Obstacle Network" (UON) based on a FCN model for detecting anomalous objects, and a stereo vision-based method called "Fast Direct Planar Hypothesis Testing" (FDPHT) for detecting planar obstacles. These modules work together through a probability fusion framework, which combines the confidence outputs from each detection method to generate a more robust and accurate representation of obstacles. This fusion not only improves the reliability of detection but also refines the mid-level Stixel representation by incorporating both geometric and semantic cues.

VisLab's early binocular obstacle detection algorithm, known as Generic Obstacle and Lane Detection (GOLD), integrates stereo disparity analysis with inverse perspective mapping (IPM) to perform lane line and obstacle detection. The algorithm first uses stereo vision to compute a disparity map between the left and right images. Based on this map, lane markings are detected by transforming the image into a top-down view using IPM, which simplifies the geometric structure of the road and highlights the parallel nature of lane lines. Simultaneously, road obstacles are identified by analyzing disparity discontinuities and depth variations in the stereo image pair. This combination of depth-based obstacle localization and IPM-enhanced lane detection makes the GOLD algorithm suitable for structured road environments in early autonomous driving systems.

In the VisLab Intercontinental Autonomous Challenge (VIAC), the test vehicle was equipped with a binocular camera system for stereo vision and also featured a lidar sensor to assist in road classification tasks. The binocular camera provided depth information and was primarily used for lane and obstacle detection, while the lidar offered complementary 3D point cloud data that improved the robustness and accuracy of road-surface classification, especially in challenging or unstructured environments. The integration of both vision and lidar sensing allowed the VIAC vehicle to handle diverse terrains and complex driving scenarios encountered throughout the long-distance autonomous journey.

The binocular obstacle detection algorithm developed by VisLab follows a structured processing pipeline. It begins with disparity estimation using the Semi-Global Matching (SGM) algorithm, supplemented by a correlation-based approach employing the Sum of Absolute Differences (SAD) metric. To enhance the robustness and accuracy of the depth information, two filters are applied to the disparity space image (DSI) during the post-processing stage. These filters help suppress noise and outliers in the disparity map, improving the overall quality of obstacle detection.

The obstacle detection algorithm incorporates a method developed by the Jet Propulsion Laboratory, which clusters obstacles based on both spatial layout characteristics and the physical characteristics of vehicles. The physical constraints include the vehicle's maximum height, the obstacle's minimum height, and the maximum drivable width of the road. These constraints together define a truncated conical space originating from the camera's viewpoint. Any point that falls within this truncated cone during the clustering process is classified as a potential obstacle.

To improve the computational efficiency of the disparity estimation step, the algorithm divides the DSI into discrete layers. Each layer corresponds to a specific range of disparities, allowing the matching process to be constrained within appropriate depth intervals, thereby reducing search complexity and accelerating the overall computation.

Another method detects free road space by first transforming the original image into an illuminant-invariant representation, which reduces the impact of varying lighting conditions. Using this enhanced image, the algorithm identifies the drivable area by constructing a convex hull around the detected free-space region to generate an ROI. Within this ROI, obstacles on the road are defined based on analysis of the horizontal disparity map obtained from stereo vision. For dynamic scenes involving multiple moving objects, a particle filter is employed to perform multi-target tracking, ensuring robust and consistent obstacle localization over time.

Overall, the methods of binocular obstacle detection are mostly based on disparity maps and road surface disparity. Perhaps with the rapid development of deep learning and the enhancement of computing platforms, binocular autonomous driving systems will also become popular.

5.11.4 Deep Learning Binocular Disparity Estimation

Below is an introduction to the application of deep learning in binocular disparity estimation.

References [89, 90] summarize the methods for estimating depth maps from images (monocular, binocular, and multi-view), including multiple modules of disparity/depth estimation, namely, feature extraction, matching cost calculation, cost volume regularization, and refinement.

For these modules, the application of deep learning can be combined to form a network of stereo vision matching, which can be divided into early methods and end-to-end methods.

Early methods replaced manual features and similarity calculations with deep learning architectures, with their feature extraction modules implemented as multi-branch networks with shared weights. Each branch calculates features based on its input, then matches them using the following methods:

- Fixed correlation layer (implemented as a convolutional layer).
- A fully connected neural network takes the connected features from stereo image pairs as input and produces matching scores.
- Convolutional network composed of convolutional layers and ReLU.

In early methods, convolutional layers and/or fully connected layers enable the network to learn appropriate similarity measures from the data, which is more accurate than using correlation layers, but significantly slower.

Volume, connected features, or cost volume connected with reference image features and/or semantic features (such as segmentation masks or edge maps) as a semantic prior, which is then regularized and outputs a depth/disparity map or depth/disparity distribution.

Segmentation masks and edge maps can be used to compute depth networks for joint training and end-to-end training of disparity depth estimation networks. Adding semantic features to the cost volume can improve the reconstruction of details, especially near object boundaries.

The regularization module is typically implemented as a convolution deconvolution (hourglass) depth network with skip connections between the contracting and expanding paths, or as a simple convolutional network. It can use either 2D or 3D convolution, with the latter requiring fewer parameters. In both cases, the disparity range is predetermined and cannot be adjusted without retraining.

Depth calculation of the regularized cost volume can use a soft argmin operator, which is differentiable, allows subpixel accuracy, but is limited to the unimodal distribution of network output, or subpixel MAP approximation, which can handle multimodal distributions.

End-to-end methods can be divided into the following two approaches:

(1) Using a single encoder–decoder, stack the left and right images into a 6D volume, and regress the disparity map. These methods do not require an explicit feature matching module and are very fast at runtime. However, they require a large amount of training data and it is difficult to obtain results.

(2) Decompose the problem into multiple steps to simulate the traditional stereo vision matching pipeline, each stage is composed of differentiable blocks, allowing end-to-end training.

The performance issues of end-to-end methods are that the disparity map may have low resolution (along width, height, and disparity dimensions), be incomplete, noisy, lack detail, and be overly smooth, especially at object boundaries. Therefore, some variant methods improve resolution, improve runtime processing time, and disparity completion and denoising, etc.

Reference [91] introduces GC-Net, a regression network for disparity estimation that fully leverages geometric features and contextual constraints to enable end-to-end disparity map prediction. The model integrates 3D convolutional layers to capture spatial context and constructs a cost volume that encodes geometric features by combining information from stereo image pairs. Addressing limitations in previous methods—namely, ① the use of discrete operations that prevent subpixel-level disparity estimation, and ② the lack of differentiability which hinders training via backpropagation—the proposed approach introduces a soft minimization mechanism known as *soft argmin*. Specifically, the method first negates the predicted costs c_d for each disparity level d to convert them into unnormalized probabilities. Then, a softmin operation $\sigma(\cdot)$ is applied across the disparity dimension to normalize these values into a probability distribution. Finally, the expected disparity is computed as the weighted sum of all disparity levels, where each d is weighted by its corresponding normalized probability. This formulation is fully differentiable, allowing effective training of the disparity estimation network via gradient descent.

The structure of GC-Net is shown in Table 5.4, where each 2D or 3D convolutional layer represents a network block including convolution, batch normalization (BN), and nonlinear activation ReLU (unless otherwise specified, such as no ReLU or BN); H represents height, W represents width, C represents the number of channels, D represents the maximum disparity, and F represents the feature size.

The network proposed in Ref. [92] is an unsupervised learning model for disparity estimation. Starting from a randomly initialized state, the network is trained using a left-right consistency check as the supervisory signal. During training, the model iteratively selects appropriate stereo matching pairs based on consistency cues to update its parameters, gradually converging to a reliable disparity estimate. The architecture consists of two main branches: the cost volume branch, which constructs a cost volume to represent potential stereo correspondences, and the image feature branch, which performs cost volume filtering using learned image features. To obtain continuous and subpixel-accurate disparity values, the model applies a soft minimization operation over the filtered cost volume, similar to the soft argmin strategy described in other works. This differentiable operation enables end-to-end optimization of the entire network without requiring ground-truth disparity maps.

The iterative unsupervised training framework described in Ref. [92] is composed of four key components: disparity prediction, confidence estimation, training data selection, and network training. Initially, the network predicts disparity maps from stereo image pairs. Then, a confidence estimation module evaluates the reliability of these predictions. Based on a left-right bidirectional consistency check, the system identifies trustworthy disparity values, which are used to construct a reliable training subset. This subset is subsequently used to refine the network in the next training iteration. Through repeated execution of this cycle, the model progressively improves its disparity estimation performance in an unsupervised manner, ultimately achieving stable and accurate results without requiring ground-truth disparity annotations.

Reference [93] proposed the Pyramid Stereo Matching Network (PSM-Net), which is composed of two major components: a Spatial Pyramid Pooling (SPP) module and a 3D convolutional neural network (3D CNN). The SPP module aggregates contextual information from multiple spatial scales and positions, enabling the network to effectively capture global context features across the entire image.

Table 5.4 Parameters of the GC-Net model

Parameters	Layer description	Output tensor dimensions
	Input image	$H \times W \times C$
Unary features	5×5 convolutional layer, 32 features, stride 2	$1/2H \times 1/2W \times F$
	3×3 convolutional layer, 32 features	$1/2H \times 1/4W \times F$
	3×3 convolutional layer, 32 features	$1/2H \times 1/4W \times F$
	Add features of layer 1 and layer 3 (residual connection)	$1/2H \times 1/2W \times F$
	(Repeat layers 2, 3 and residual connection) ×7	$1/2H \times 1/2W \times F$
	3×3 convolution layer, 32 features, (no ReLU or BN)	$1/2H \times 1/2W \times F$
Cost volume	Cost volume	$1/2D \times 1/2H \times 1/2W \times 2F$
Learning regularization	3D convolution layer, 3×3×3, 32 features	$1/2D \times 1/2H \times 1/2W \times F$
	3D convolution layer, 3×3×3, 32 features	$1/2D \times 1/2H \times 1/2W \times F$
	3D convolution layer, 3×3×3, 32 features	$1/2D \times 1/2H \times 1/2W \times F$
	3D convolution layer, 3×3×3, 64 features, stride 2	$1/4D \times 1/4H \times 1/4W \times 2F$
	3D convolution layer, 3×3×3, 64 features	$1/4D \times 1/4H \times 1/4W \times 2F$
	3D convolution layer, 3×3×3, 64 features	$1/4D \times 1/4H \times 1/4W \times 2F$
	3D convolution layer, 3×3×3, 64 features, stride 2	$1/8D \times 1/8H \times 1/8W \times 2F$
	3D convolution layer, 3×3×3, 64 features	$1/8D \times 1/8H \times 1/8W \times 2F$
	3D convolution layer, 3×3×3, 64 features	$1/8D \times 1/8H \times 1/8W \times 2F$
	3D convolution layer, 3×3×3, 64 features, stride 2	$1/16D \times 1/16H \times 1/16W \times 2F$
	3D convolution layer, 3×3×3, 64 features	$1/16D \times 1/16H \times 1/16W \times 2F$
	3D convolution layer, 3×3×3, 64 features	$1/16D \times 1/16H \times 1/16W \times 2F$
	3D convolution layer, 3×3×3, 128 features, stride 2	$1/32D \times 1/32H \times 1/32W \times 4F$
	3D convolution layer, 3×3×3, 128 features	$1/32D \times 1/32H \times 1/32W \times 4F$
	3D convolution layer, 3×3×3, 128 features	$1/32D \times 1/32H \times 1/32W \times 4F$
	3×3×3, 3D transposed convolution, 64 features, stride 2	$1/16D \times 1/16H \times 1/16W \times 2F$
	Add features 33 and 29 (residual connection)	$1/16D \times 1/16H \times 1/16W \times 2F$
	3×3×3, 3D transposed convolution, 64 features, stride 2	$1/8D \times 1/8H \times 1/8W \times 2F$
	Add features 34 and 26 (residual connection)	$1/8D \times 1/8H \times 1/8W \times 2F$
	3×3×3, 3D transposed convolution, 64 features, stride 2	$1/4D \times 1/4H \times 1/4W \times 2F$
	Add features 35 and 23 (residual connection)	$1/4D \times 1/4H \times 1/4W \times 2F$
	3×3×3, 3D transposed convolution, 32 features, stride 2	$1/2D \times 1/2H \times 1/2W \times F$
	Add features 36 and 20 (residual connection)	$1/2D \times 1/2H \times 1/2W \times F$
	3×3×3, 3D transposed convolution layer, 1 feature (no ReLU or BN)	$D \times H \times W \times 1$
Soft minimization	Soft minimization	$H \times W$

This helps the network to better understand the scene structure and improve matching in textureless or ambiguous regions. Following the SPP, the 3D CNN takes the constructed cost volume and performs regularization through a series of stacked hourglass-shaped sub-networks. These sub-networks are designed to refine disparity estimates by repeatedly downsampling and upsampling the cost volume, thereby capturing both coarse and fine-level disparity cues and leading to more accurate and robust depth estimation results.

Reference [94] proposed a stereo matching decomposition model to address the problem of computational cost increasing significantly with higher image resolutions. Instead of performing dense disparity estimation across the entire high-resolution image, the model runs dense matching only at a very low resolution and performs sparse matching at progressively higher resolutions. This hierarchical approach allows the model to efficiently reconstruct disparity details without incurring the full computational burden. The model incorporates an occlusion-aware mask that guides the iterative fusion of disparity maps from different resolution levels. Specifically, sparse and dense disparity estimates from adjacent levels are integrated to recover a full-resolution dense disparity map. This hierarchical fusion process involves three main steps: disparity upsampling, fusion of sparse and dense disparity estimates, and refinement through a dedicated network. Additionally, warping operations are used to align left and right feature maps, while square difference computations between feature maps at adjacent scales assist in the estimation and fusion process. This multi-scale strategy significantly reduces memory and time requirements while maintaining high-quality disparity estimation.

The calculation of **dense matching** is as follows. A full cost body is established for disparity regression at the lowest level, and since the size of the search space can be ignored, this part of the calculation occupies very few resources. Before performing disparity regression with softmax, cost regularization is used to correct the cost body. Cost regularization consists of 8 3D convolutions, all of which are followed by a BN layer.

A mask is used to indicate the position of lost details, which is calculated by a **detail detection network** calculating the square difference between the current layer features and the upsampling of the lower layer features. This network consists of three convolution operations and one sigmoid function, and its learning is guided by an unsupervised loss, i.e., maximizing the difference between adjacent layer features and enforcing the sparsity of details.

Since the lost details are content-aware, their shape and size are dynamic but not fixed, so it is not suitable to represent them with a cost body in **sparse matching**. Instead, the disparity map is directly calculated through cross correlation.

In sparse matching, occlusions can cause serious matching blur, and the detail area contains many edges that are easy to appear in occlusions. Therefore, a so-called **soft fusion** learning mask is used to fuse the sparse disparity map and the upsampled dense disparity map. Specifically, a **regression network** is used to generate a soft mask. The regression network includes three 2D convolution operations and a sigmoid activation function. The input of the regression network is the left features, upsampled dense disparity map, sparse disparity map, mask of the current detail area, and robustness mask of sparse matching cascaded together.

As for the **refinement network**, the features of the right image are distorted by disparity, and then the distorted right feature map, left feature map, and current disparity map are connected to the refinement network. The refinement network contains seven convolution operations, all of which are followed by a ReLU activation function and a BN layer, but only a BN after the last convolution, without ReLU.

ACVNet introduces a novel approach to constructing the cost volume in stereo matching. It enhances matching accuracy by generating attention weights from relevant contextual clues, which serve to suppress redundant information and emphasize discriminative features that are important for accurate correspondence. To produce reliable attention weights, ACVNet proposes a method called multi-level adaptive image block matching (MAPM). This technique improves the distinguishability of matching costs under various disparity conditions, particularly in challenging regions such as textureless areas.

The resulting cost representation, termed the Attention Correlation Volume (ACV), can be seamlessly integrated into a wide range of stereo matching networks. This integration enables such networks to achieve higher accuracy while relying on a more lightweight cost aggregation module. When ACV is used as the cost volume representation, the corresponding stereo matching framework is referred to as ACVNet.

The construction of ACV proceeds in three main steps: (1) building the initial correlation volume, (2) generating attention weights, and (3) applying attention filtering. The attention weights are used to filter the initial correlation volume, effectively suppressing irrelevant or redundant information while enhancing the features that are relevant to disparity estimation. This yields the final attention correlation volume used for cost aggregation. The network architecture employs a three-layer ResNet for feature extraction and multiple stacked hourglass networks for cost aggregation, striking a balance between computational efficiency and disparity estimation accuracy.

The purpose of the attention weights in ACVNet is to filter the initial correlation volume by emphasizing informative regions and suppressing irrelevant or ambiguous content. To achieve this, geometric information is extracted from the correlation between stereo image pairs to guide the generation of attention weights. Traditional correlation volume estimation methods typically compute pixel-to-pixel similarity, which is often unreliable in textureless regions due to insufficient matching cues. To address this issue, ACVNet introduces a more robust correlation volume construction method called multi-level adaptive image block matching (MAPM).

MAPM operates by leveraging multi-scale features extracted from different layers of the feature extraction module. Specifically, three feature maps are obtained, each corresponding to a different level and having channel dimensions of 64, 128, and 128, respectively. For each pixel in a given layer, MAPM computes a matching cost using an atrous (dilated) image patch centered on that pixel. These patches are weighted adaptively using learnable parameters. The dilation factor is controlled in accordance with the scale of the feature map, ensuring that the image block covers a broader receptive field without increasing the number of pixels involved in the similarity computation. The similarity between two corresponding pixels across stereo views is then determined by a weighted sum of the correlations computed within the respective image patches. This approach enhances robustness in challenging areas, such as regions lacking texture, by aggregating more contextual information while preserving spatial precision.

To obtain accurate attention weights for filtering the initial correlation volume across different disparity levels, the learning process is guided using ground-truth disparity supervision. Specifically, the network is trained by minimizing the smooth L_1 loss between the predicted disparity and the ground-truth disparity. This loss function encourages the network to generate attention weights that more effectively highlight informative matching regions while suppressing noise and ambiguity.

Building upon the Attention Correlation Volume (ACV), a real-time variant of the network, termed ACVNet-Fast, is further proposed. ACVNet-Fast retains the same feature extraction backbone as ACVNet but is optimized for efficiency. The key difference lies in the design of the ACV construction module and the cost aggregation strategy, which are streamlined to reduce computational complexity while preserving accuracy. These modifications enable ACVNet-Fast to achieve real-time performance, making it suitable for deployment in latency-sensitive stereo vision applications.

Specifically, ACVNet-Fast constructs MAPM based on 1/8 resolution feature maps, then uses two 3D convolutions and a 3D hourglass network for regularization to obtain 1/8 resolution attention weights. Around the predicted disparity, h ($h = 6$) hypotheses are sampled to narrow the disparity search space, and this predicted disparity is obtained by upsampling the attention weights at 1/2 resolution. Sparse correlation bodies and sample attention weights are constructed based on these hypotheses to obtain sparse attention weights, and then sparse attention correlation bodies are constructed.

For cost aggregation, only two 3D convolutions and a 3D hourglass network are used to regularize the sparse attention correlation body. Because the included matching information is very effective, only a very light aggregation network is needed. In this way, a good balance between accuracy and efficiency is achieved.

5.12 Human Pose Estimation

In pedestrian detection work, there is basically little involvement of body posture, but tracking and behavior prediction of pedestrians require modeling and estimation of human posture. The following discusses methods of human pose estimation based on recent advancements in deep learning in this field.

The human pose skeleton is a graphical representation that captures a person's body orientation and posture. It consists of a set of keypoints, also referred to as parts or joints, which correspond to anatomical landmarks such as the head, shoulders, elbows, hips, knees, and ankles. These keypoints are connected in a predefined topology to form limbs, which represent valid pairwise connections between body parts. Together, the keypoints and limbs form a structured skeleton that encodes spatial relationships within the human body. This representation can be extended to multi-person scenarios by detecting and linking multiple skeletons within the same scene. Such skeletal models are widely used in human pose estimation tasks, enabling high-level understanding of human activities and interactions.

Reference [95] provides a comprehensive review of recent 2D and 3D human pose estimation (HPE) solutions based on deep learning.

Human Pose Estimation (HPE) can be divided into two categories: 2D HPE and 3D HPE. Based on the number of subjects, 2D HPE methods are classified into single-person and multi-person settings. For single-person estimation, deep learning-based approaches generally fall into two types: ① Regression methods, which directly map the input image to body joint coordinates using a deep learning regressor. ② Body-part detection methods, which involve two steps—the first generates heatmaps for body part localization (keypoints, i.e., joints), and the second assembles these detected keypoints into a full-body pose or skeleton.

For multi-person HPE, deep learning-based approaches can also be divided into two categories:

① Top-down methods, which first detect individuals in the image and then apply single-person HPE to predict each person's keypoints and construct their pose.

② Bottom-up methods, which first detect body keypoints without prior knowledge of the number of people and subsequently group these keypoints into individual poses.

Top-down methods rely on a pre-trained human detector to identify bounding boxes in the input image, each corresponding to an individual. A single-person pose estimator is then applied to each bounding box to construct the final multi-person poses. In contrast, bottom-up methods first localize all body joints in the image and then group them into individuals. A major drawback of the top-down pipeline is that its computational cost increases with the number of people in the scene. Bottom-up approaches, by eliminating the need for per-person pose estimation, are typically more computationally efficient and enable faster inference.

For 2D in terms of single-person HPE methods, regression methods and body part detection methods each have their own advantages and limitations. Regression methods can learn the nonlinear mapping from input images to keypoint coordinates through an end-to-end framework, providing a fast learning paradigm and subpixel-level prediction accuracy. However, due to the highly nonlinear problem, it usually only provides suboptimal solutions. The accuracy of keypoint prediction for body part detection methods, especially the widely used heatmap-based framework, depends on the resolution of the heatmap. When using high-resolution heatmaps, computational cost and memory usage significantly increase.

For 2D multi-person HPE methods, on one hand, the top-down pipeline produces better results because it first uses detection methods to detect each person from the image, and then uses single-person-based methods to predict the position of keypoints. On the other hand, bottom-up methods are usually faster than top-down methods because they directly detect all keypoints and group them into individual poses using keypoint association strategies (such as close relationship links, association embedding, and per-pixel keypoint regression).

The challenges faced by 2D HPE include severe occlusion scenarios, computational efficiency, and lack of data for rare poses.

OpenPose is one of the most widely used bottom-up approaches for multi-person human pose estimation, in part due to its high accuracy and the availability of a well-maintained open-source implementation. Like other bottom-up methods, OpenPose first detects all body parts (keypoints) present in the entire image, regardless of the number of individuals. It then groups these detected keypoints into individual human poses based on learned pairwise relationships. This two-step process—keypoint detection followed by part association—enables OpenPose to handle crowded scenes and multiple overlapping persons effectively. The model architecture typically includes a multi-stage convolutional network that jointly predicts confidence maps for keypoint locations and Part Affinity Fields (PAFs), which represent the degree of association between keypoints belonging to the same person.

The OpenPose network first uses the first few layers (VGG-19) to extract features from the image, and then feeds these features into two parallel convolutional layer branches. The first branch predicts a set of 18 confidence maps, each representing a specific part of the human pose skeleton. The second branch predicts a set of 38 PAFs, which represent the degree of association between parts.

The system iteratively refines the predictions of both keypoint detection and part association branches. In the process of constructing the final human pose skeletons, part confidence maps are used to identify candidate keypoints for each body part. A bipartite graph is then constructed to represent possible connections between part pairs. To ensure accurate person-level pose assembly, PAFs are employed to evaluate the strength of each connection. Connections with low PAF responses are considered weak and are pruned from the graph. This pruning step helps eliminate incorrect associations and results in coherent and anatomically plausible skeletons. Ultimately, the refined connections allow the system to estimate the pose of each individual in the image and assign the detected keypoints accordingly.

Facebook proposed DensePose-R-CNN, an extension of the Mask-R-CNN framework, which enables dense estimation of human body part coordinates within the detected human regions in an image. This work builds upon the earlier DenseReg system, which focused primarily on dense 3D surface reconstruction for facial imagery. DensePose-R-CNN generalizes this idea to the full human body, enabling dense mapping from 2D image pixels to the surface of a 3D human body model.

To support this effort, the authors introduced a new human pose dataset that provides dense correspondences between real-world images and the surface coordinates of a canonical 3D human model. These images exhibit natural challenges such as cluttered backgrounds, occlusions, and varying scales. Each image is first segmented into semantic regions corresponding to body parts. Then, for each sampled point within a part region, the corresponding 3D surface location on the human model is manually annotated, establishing a pixel-to-surface mapping.

To handle incomplete annotations and enrich the training signal, an inpainting network is employed to infer missing surface coordinate values, allowing for the generation of more complete and coherent training data. This approach significantly improves the effectiveness of learning dense correspondences from sparse supervision.

The architecture of DensePose-R-CNN is built upon the Mask-R-CNN framework and includes a sequence of modules designed to perform dense pose estimation. It begins with a region proposal network (RPN) that identifies candidate human regions in the input image. These proposals are then processed through a feature pooling mechanism—typically using ROIAlign—to extract fixed-size feature maps for each region of interest. These pooled features are fed into a FCN, which performs dense predictions at the pixel level within each proposed region. The network outputs both discrete part labels and continuous UV coordinates that map each pixel to a corresponding location on the surface of a 3D human body model. This design enables DensePose-R-CNN to establish fine-grained correspondences between image pixels and 3D surface geometry with high accuracy and spatial resolution.

DensePose-R-CNN adopts a cascaded architecture to refine predictions across multiple tasks. After extracting region proposals using a region proposal network (RPN), the RoIAlign module is applied to obtain aligned feature maps for each region of interest. These feature maps are then passed to the DensePose prediction branch as well as to auxiliary branches responsible for related tasks such as instance segmentation (masks) and human keypoint detection.

The model operates in multiple stages: in the first stage, each task-specific branch independently produces initial predictions. These preliminary outputs are then fed into a second stage, where further refinement is performed for each task. This cascaded refinement strategy allows the network to progressively improve accuracy by leveraging intermediate outputs, and to share contextual information across different prediction tasks, ultimately enhancing the quality of dense pose estimation.

3D HPE methods are classified according to the type of input source, divided into monocular RGB images/videos or other sensors (such as IMU). Most of these methods use monocular RGB images and videos, which are further divided into single-view and multi-view methods. Single-view methods are distinguished by single-person and multi-person. Multi-view methods are mainly used for multi-person pose estimation.

Reconstructing 3D human poses from a single view of monocular images and videos is affected by self-occlusion and occlusion by other objects, depth blur, and insufficient training data. This is a severe ill-posed problem, as different 3D human poses can be projected to similar 2D pose outlines. Moreover, for 2D joint-based methods, minor localization errors of 2D body joints can lead to large pose distortions in 3D space. The multi-person situation is more complex compared to the single-person situation.

Single-person 3D HPE methods can be divided into model-free and model-based methods according to whether they use a human body model.

Model-free methods do not use a human body model to reconstruct the 3D human representation and can be further divided into two categories: ① direct estimation methods and ② methods that lift 2D poses to 3D.

The model-based method uses a human body model (kinematic model and body model). The kinematic model is a joint representation that connects the skeleton and joints under motion constraints, which has received increasing attention in 3D HPE in recent years. Based on the prior knowledge of the kinematic model, such as the connection information of the skeleton joints, the rotation characteristics of the joints, and the fixed ratio of the skeleton length, the posture can be reasonably estimated. The volumetric model can restore high-quality human body mesh, providing additional shape information of the human body.

3D HPE multi-person methods, for single-view RGB images or video sources, are similar to 2D HPE multi-person methods, and can be divided into top-down methods and bottom-up methods.

The top-down method first performs human body detection to detect each person, and then for each detected person, the absolute root (center joint of the person) coordinates and 3D root-relative posture are estimated through the 3D posture network, aligning all postures with the world coordinates.

Unlike the top-down method, the bottom-up method first generates all body joint positions and depth maps, and then associates body parts with each person based on root depth and part relative depth. A key challenge for the bottom-up method is how to cluster the body joints belonging to each person.

The top-down method typically relies on human detection followed by single-person pose estimation, and it often achieves strong performance. However, as the number of people increases—particularly in crowded scenes—the computational complexity and inference time can become prohibitively high. Moreover, because the method begins by detecting bounding boxes for each person, it may overlook global contextual information in the scene. Depth estimation within cropped human regions may also be inconsistent with the actual depth ordering (e.g., front–back occlusion), causing predicted body parts such as hands, arms, legs, and feet to overlap incorrectly and produce physically implausible results.

On the contrary, the bottom-up method has linear computation and time complexity. However, if the goal is to recover the 3D human body mesh, then the bottom-up method is not simple to reconstruct the human body mesh. For the top-down method, after each person is detected, the human body mesh of each person can be easily recovered by combining the model-based 3D HPE single-person method. For the bottom-up method, an additional model regression module is needed to reconstruct the human body mesh based on the final 3D posture.

Some 2D HPE methods, such as OpenPose, have been widely used as 2D posture detectors in 3D methods. In addition to 3D posture, 3D human body mesh can also be recovered from images or videos. The challenges faced by 3D HPE include: model generalization, robustness to occlusion, and computational efficiency (same as 2D HPE).

Reference [96] proposed the Multi-Hypothesis Transformer (MHFormer), a framework designed to learn spatio-temporal representations of multiple plausible human pose hypotheses. This approach addresses the inherent ambiguity in 3D human pose estimation, particularly when inferring 3D structure from single-frame 2D observations. To effectively capture the dependencies among different hypotheses and model the relationships between their feature representations, the framework is structured into three sequential stages.

First, the model generates multiple initial hypothesis representations from the input 2D pose. Second, a self-hypothesis communication mechanism is introduced to facilitate interaction among the hypotheses. This stage involves merging the individual hypotheses into a single aggregated representation that encodes shared information, which is then disentangled back into distinct hypothesis features. Third, a cross-hypothesis communication stage is employed to further integrate the hypothesis features and synthesize a final, more accurate 3D pose estimate.

This three-step architecture allows MHFormer to reason collectively over multiple plausible interpretations of pose, thereby enhancing robustness and reducing prediction ambiguity. While the model supports temporal inputs, the framework can also operate in a single-frame setting, where only a 2D pose is given as input for estimating the corresponding 3D posture.

This framework more effectively models the dependencies of multiple hypotheses, while also establishing stronger relationships between the features of the hypotheses.

Specifically, in the **first step**, a multi-hypothesis generation (MHG) module is introduced to model the inherent structural information of human joints and generate multi-level features in the spatial domain. These features contain semantic information of different depths from shallow to deep, so they can be regarded as the initial representations of multiple hypotheses.

In the **second step**, a **self-hypothesis refinement (SHR)** module is proposed to refine each hypothesis feature. Here, two modules are proposed to model the temporal consistency and enhance the rough representations in the temporal domain. The first module is multi-hypothesis **self-attention (MH-SA)**, which independently models the dependencies of single hypotheses, builds self-hypothesis communication, and can pass messages within each hypothesis to enhance features. The second module is a **hypothesis-mixing** multi-layer perceptron (MLP), which exchanges information across hypotheses. Multiple hypotheses are merged into a convergent representation and divided into multiple divergent hypotheses.

In the **final step**, the cross-hypothesis interaction (CHI) module models the interaction between multi-hypothesis features. A key design module of CHI is multi-hypothesis cross-attention (MH-CA), which captures the correlation of multiple hypotheses with each other, establishes cross-hypothesis communication, and can pass information between hypotheses to better model interaction. Subsequently, another module, the hypothesis-mixing MLP, aggregates multiple hypotheses to synthesize the final prediction.

The MHFormer framework is composed of several key modules that together enable multi-hypothesis 3D human pose estimation. The overall design is centered on the principle of generating, refining, and integrating multiple pose hypotheses to handle ambiguity in 2D to 3D mapping.

The first component, the Multi-Hypothesis Generation (MHG) module, processes each input frame to extract structural information from the 2D human joint coordinates and generates multiple distinct hypothesis representations. For temporal inputs, N denotes the number of frames, and a transpose operation T may be applied to align temporal and spatial dimensions appropriately.

Following this, the Self-Hypothesis Refinement (SHR) module operates independently on each hypothesis to improve the quality of its feature representation. This stage allows each hypothesis to be internally strengthened before being exposed to cross-hypothesis influences.

Finally, the Cross-Hypothesis Interaction (CHI) module enables mutual information exchange across all hypotheses. This interaction allows the network to identify consistent patterns and resolve conflicts among different interpretations, culminating in a more accurate and coherent 3D pose estimation. These modular stages ensure both diversity and consistency among hypotheses, contributing to the robustness of the overall system.

MH-SA lacks cross-hypothesis connections, which limits its interaction modeling. To capture the correlation between multiple hypotheses for cross-hypothesis communication, multiple Multi-Head Cross-Attention (MCA) elements composed in parallel are proposed to form MH-CA.

In Ref. [97], the authors proposed a method for jointly inferring the pose and depth of multiple people from a single image. Unlike earlier approaches that reasoned exclusively in the image plane, this method introduces a virtual bird's-eye-view (BEV) representation to explicitly incorporate depth reasoning. The BEV representation analyzes body centers in both the image and depth domains and fuses these cues to estimate 3D body positions. In contrast to previous methods, the BEV framework is formulated as an end-to-end differentiable, single-sample approach. In addition, Ref. [97] presents a 3D human body model space, which allows the BEV method to infer body shapes across different stages of human development, ranging from infancy to adulthood.

SMPL and SMIL models are used to parameterize the 3D body mesh of adults and infants into low-dimensional parameters. The AGORA model linearly blends the SMIL and SMPL template shapes with a weight $\alpha \in [0, 1]$, extending SMPL to support children's bodies, referred to as "age offset." Once the blended template solves the scale and proportion differences between adults and children, AGORA can use the adult shape space regardless of age. SMPL+A also defines a piecewise function in the manner of SMPL, mapping 3D pose, shape, and age offset to a 3D mesh. When the age offset is too large, the SMIL model is used.

The BEV method introduces a multi-head architecture for monocular 3D human pose and mesh estimation. Given a single RGB image as input, the network produces five output images, each serving a specific role in the estimation pipeline. The first four outputs are designed for coarse-to-fine localization and consist of body center heatmaps and localization offset maps in both the front-view (FV) and BEV representations.

To obtain 3D localization, the FV and BEV images are unfolded along the depth and height dimensions, respectively, and then combined to construct 3D center and

3D offset maps. During coarse localization, the rough 3D position of the human subject is extracted from the 3D center image. For fine localization, the system samples the offset vector at the identified 3D center position from the 3D offset map. By adding this offset to the coarse 3D center estimate, a refined 3D translation vector is obtained.

For mesh parameter regression, the estimated 3D translation (x_i, y_i, d_i) and a mesh feature image are used as input. The depth component d_i is transformed into a depth encoding, while the feature vector is sampled from the mesh feature image at the spatial location (x_i, y_i). The depth encoding and sampled feature are then fused and passed to a regression module that predicts the parameters of the human mesh.

Finally, the SMPL+A model is used to convert the predicted parameters into a full 3D mesh representation of the human body. This end-to-end architecture allows the BEV method to accurately recover both body pose and shape in 3D space from a single RGB image.

To effectively establish a 3D representation, the front view and bird's-eye view must work together to estimate the image position and depth of the corresponding target. Parallel independent estimation of the two view images will inevitably lead to misalignment, resulting in failure of detection based on the 3D heat map. To connect these two views, the bird's-eye view is estimated based on the front view (i.e., the center image and offset image). Specifically, to estimate the bird's-eye view, the associated front-view and trunk feature images are taken as input. The front-view 2D body center heat map is used as a form of robust attention to the human body in the image, helping the model to focus on exploring depth in BEV estimation. Then, the front view and BEV view are stripped and combined to generate 3D images. To integrate 2D features from the two views and enhance 3D consistency, 3D convolution is performed on the synthesized 3D image for refinement.

5.13 Driver Monitoring System

Reference [98] describes the state of the driver from five main dimensions: drowsiness, mental burden, distraction, emotional, and affected.

The Driver Monitoring System (DMS) is generally for L2~L3 autonomous driving systems, and it is meaningless for L4, unless the system is still in the testing stage that requires a safety officer.

The purpose of monitoring is to detect driver distraction, fatigue, or drowsiness, or even unexpected situations where driving is impossible, such as deceiving the assisted driving system with mineral water instead of hands on the steering wheel, or arguing and fighting with passengers, etc. In addition, if it is in the development stage of autonomous driving, monitoring the driver can provide first-hand data on driving behavior, and even be used in the simulation system.

Non-intrusive methods are the preferred methods for monitoring, and vision-based systems are more attractive. The main visual cues include facial features, hand features, or body features. Many detection systems only use a single visual cue, and the robustness of these systems is poor, such as being easily disturbed by occlusion or changes in illumination. Therefore, combining multiple visual cues is the key and also challenging.

A driver facial monitoring system is a real-time system that studies the physical and mental condition of the driver based on facial image processing. It can detect the driver's state from eyelid closure, blinking, gaze direction, yawning, and head movement. The system is basically divided into two categories:

(1) Detect the driver's physical and mental condition only from the eye area.

(2) Not only can it detect from the eye area, but it can also detect from other areas such as the face and head.

Specifically, the driver facial monitoring system is designed to detect and analyze facial features—such as eyes, mouth, and head orientation—in order to monitor the driver's state in real time. The system performs a sequence of tasks: it begins with facial detection, followed by the localization and tracking of specific facial landmarks. Based on these tracked changes, it extracts physiological and behavioral cues that indicate symptoms of fatigue, drowsiness, or distraction. By analyzing these features continuously, the system can infer the driver's level of alertness and trigger appropriate warnings or interventions if necessary.

Despite its importance in enhancing driving safety, the development of an effective driver facial monitoring system faces several challenges. These include robustness under varying lighting conditions, accurate detection when the face is partially occluded (e.g., by sunglasses or hands), real-time processing constraints, individual variability in facial features and behaviors, and the ability to generalize across diverse environmental and driving scenarios. Addressing these challenges is essential for ensuring the reliability and practical deployment of such systems in real-world vehicles.

- The first challenge is how to accurately define fatigue and how to measure fatigue. Fatigue has a certain relationship with body temperature, skin resistance, eye movement, breathing rate, heart rate, and brain activity. The first and most important sign of fatigue will appear in the eyes.
- The second challenge is to measure the driver's attention to the road. The driver's attention can be estimated from the driver's head and gaze direction.

Face detection methods can refer to general object detection methods, and deep learning has already demonstrated strong capabilities in this field, greatly exceeding traditional methods. Face detection is an old problem, and the more challenging situations for face detection are as follows:

- In-plane rotation.
- Out-of-plane rotation.
- The presence of cosmetics, beards, and glasses.
- Expressions (happy, crying, etc.).
- Lighting conditions.
- Face occlusion.
- Real-time processing requirements.

The eye area is always first used for driver symptom extraction because the most important psychological activities are related to eye activities. Eye detection is divided into two categories, namely, methods based on infrared spectral imaging and methods based on vision.

In addition to the eyes, other facial components such as the mouth, nose, and salient points on the face can also be detected.

Facial tracking is the main means of analyzing driver psychological activities. This tracking task is similar to general single-target tracking and faces the following challenges:

- Some information is lost when mapping from 3D space to 2D space.
- Having complex shapes or motion patterns, such as complex facial expressions.
- Partial occlusion.
- Changes in environmental lighting.
- Real-time tracking requirements. Symptoms related to fatigue, distraction, and drowsiness include the following:
- Symptoms related to the eye area: closed eyes, distance between eyelids, fast blinking, gaze direction, and jumping movements.
- Symptoms related to the mouth area: open/closed.
- Symptoms related to the head: nodding, head posture, and head fixed unchanged.
- Symptoms related to the face: mainly expressions.

A driver monitoring system based on deep neural networks (DNNs) typically operates in two main stages. The first stage involves face detection and calibration, where the system identifies the driver's face within the video stream and accurately locates key facial landmarks such as the eyes, nose, and mouth. This step ensures that the facial region is properly aligned and normalized for subsequent analysis.

The second stage focuses on drowsiness detection. In this step, the calibrated facial data is processed to extract temporal and spatial features that may indicate signs of fatigue, such as prolonged eye closure, slow blinking, or yawning. By leveraging the representational power of DNNs, the system can learn complex patterns associated with driver drowsiness and make reliable inferences in real time. This two-step architecture enhances the system's ability to detect early warning signs of reduced alertness.

Two types of detection network structures are commonly employed for driver monitoring based on facial regions. The first type utilizes a composite approach by integrating features from the entire face, both eyes (binocular), and the mouth region. This structure aims to capture a comprehensive set of visual cues related to drowsiness and distraction by analyzing multiple facial areas simultaneously. The second type adopts a more lightweight configuration by focusing on a subset of facial regions, specifically the left eye and the mouth. This design reduces computational complexity while still retaining critical information for monitoring eye closure and yawning behaviors, which are key indicators of fatigue. Both network structures are tailored to balance accuracy and efficiency depending on deployment constraints and hardware capabilities.

A facial expression recognition system based on deep learning typically follows a multi-stage pipeline. First, the system detects the face and localizes key facial components such as the eyes, eyebrows, nose, and mouth from the input image. Once these regions are identified, time-domain features are extracted to capture subtle changes in facial muscle movements over a sequence of frames. These dynamic features provide essential cues for distinguishing between different expressions. Finally, a pre-trained deep learning classifier is employed to determine the facial expression, such as happiness, sadness, anger, or surprise. The system is often trained and evaluated on benchmark datasets, such as the CK+ dataset, which contains annotated facial expression sequences captured under controlled conditions.

The overall deep learning model used for facial expression recognition combines a CNN with an LSTM network. In this architecture, the CNN is responsible for extracting spatial features from individual facial frames, capturing local textures and structural patterns. These frame-level features are then passed to the LSTM, which models temporal dependencies across sequences of frames, enabling the system to recognize dynamic changes in facial expressions over time. This hybrid CNN-LSTM structure allows the model to effectively integrate both spatial and temporal information for robust expression classification.

Reference [99] presents a system for recognizing driver distraction symptoms based on body posture analysis. The system is designed to identify a variety of common distraction behaviors, including drinking, adjusting the radio, maintaining proper driving posture, grooming (such as fiddling with hair or applying cosmetics), turning to face backward, conversing with passengers, making phone calls with either the left or right hand, and sending text messages using either hand.

The system's algorithmic framework comprises several key components: face detection, hand detection, and skin region segmentation. For each of these modalities, individual output images are generated and passed through deep learning classifiers. Specifically, five instances each of AlexNet and InceptionV3 networks are independently trained on the outputs corresponding to face, hand, and skin images. The final classification result is obtained by performing a weighted fusion of the predictions from all ten models, leveraging ensemble learning to improve robustness and accuracy in detecting driver distraction symptoms.

Reference [100] proposes a gaze estimation model designed to classify a driver's gaze direction into one of six predefined regions of interest (ROIs) within the vehicle. The model takes as input cropped images of the driver's face and a region surrounding the eyes, combining them into a dual-channel tensor for representation learning. To enhance robustness against variations in individual appearance and gaze behavior, the model incorporates a personalized hourglass network. This network is conditioned on auxiliary inputs that capture each driver's baseline gaze characteristics, enabling personalized feature encoding.

The core architecture includes an encoder–decoder structure based on the hourglass network. The encoder receives the dual-channel input, composed of the cropped face image $\boldsymbol{I}_{\mathrm{c}}$ and eye region image $\boldsymbol{I}_{\mathrm{e}}$, concatenated as $(\boldsymbol{I}_{\mathrm{c}} \oplus \boldsymbol{I}_{\mathrm{e}})$. This input is passed through a series of dilated convolutional layers, downsampling residual blocks, and dense layers to produce a feature embedding $E(\boldsymbol{I}_{\mathrm{c}} \oplus \boldsymbol{I}_{\mathrm{e}})$. The resulting feature vector is then processed by a prediction head P, consisting of two fully connected layers followed by a softmax activation, to output the estimated gaze vector $\boldsymbol{p}_g$.

To further improve learning, a reconstruction module is added in the form of a decoder D, which receives the encoded feature and attempts to reconstruct the original input images. The decoder is implemented using n upsampling blocks with pixel shuffle layers, and skip connections are established between corresponding encoder and decoder layers to preserve spatial detail. The reconstruction loss serves as an auxiliary supervision signal, acting as a regularizer to guide the encoder toward more context-aware feature representations.

This joint training approach—combining classification and reconstruction objectives—helps the model learn more generalizable and robust features. By encouraging the encoder to preserve contextual and appearance cues, the model achieves improved performance on gaze classification tasks, particularly in scenarios involving intra- and inter-subject variability.

A personalized extension of the hourglass framework is employed to improve gaze classification robustness by incorporating individual-specific baseline gaze behavior. This architecture retains the same encoder (E) and decoder (D) modules as in the standard model, but introduces an additional baseline pathway for each driver in the training dataset. Prior to training, baseline face crops $\boldsymbol{B}_c$ and eye region images $\boldsymbol{B}_e$ are computed by averaging multiple samples in which the driver is looking forward. Specifically, for m forward-looking instances of a given driver, the baseline images are calculated as follows:

$$\boldsymbol{B}_e = \frac{1}{m}\sum_{i=1}^{m} \boldsymbol{I}_e, \quad \boldsymbol{B}_c = \frac{1}{m}\sum_{i=1}^{m} \boldsymbol{I}_c$$

During training, the current frame's dual-channel input $(\boldsymbol{I}_{\mathrm{c}} \oplus \boldsymbol{I}_{\mathrm{e}})$ is passed through the encoder to obtain a feature representation $E(\boldsymbol{I}_{\mathrm{c}} \oplus \boldsymbol{I}_{\mathrm{e}})$. Simultaneously, the baseline images $(\boldsymbol{B}_{\mathrm{c}} \oplus \boldsymbol{B}_{\mathrm{e}})$ are encoded to produce $E(\boldsymbol{B}_{\mathrm{c}} \oplus \boldsymbol{B}_{\mathrm{e}})$. The difference between these two representations,

$$E(\boldsymbol{I}_{\mathrm{c}} \oplus \boldsymbol{I}_{\mathrm{e}}) - E(\boldsymbol{B}_{\mathrm{c}} \oplus \boldsymbol{B}_{\mathrm{e}}),$$

quantifies the deviation from the driver's baseline forward-looking gaze behavior.

This residual vector is concatenated with the current frame's representation and passed to the prediction head P, which consists of fully connected layers followed by softmax activation to output the predicted gaze vector $\boldsymbol{p}_g$. Throughout training, the encoder and decoder weights are shared across both the main input stream and the baseline stream. This design enables the model to personalize gaze estimation, improving its ability to adapt to appearance and behavioral differences across individuals.

When the camera type (RGB or NIR), position (steering wheel or rear-view mirror), and interior decoration change, the driver's gaze will be significantly different. To alleviate the problem of domain inconsistency, a multi-domain training mechanism is proposed for the dual-channel hourglass model. This mechanism uses a rich set of annotated training images from one domain to learn domain-invariant features from training samples in the second domain.

Reference [101] presents an integrated in-vehicle monitoring system that simultaneously performs 3D pose estimation, seatbelt segmentation, and seatbelt status classification. The system is designed to enhance occupant safety monitoring by detecting whether the driver and front-seat passenger are wearing seatbelts correctly, using input from a single time-of-flight (ToF) camera.

The proposed 3D pose estimation module directly predicts the absolute 3D coordinates of key semantic points on the bodies of the driver and front passenger in a top-down manner. Keypoint heatmaps are generated using a conventional convolutional neural network (CNN) architecture, enabling reliable localization of joints in 3D space.

For seatbelt segmentation, the system employs a feature pyramid-based architecture. Specifically, it uses a parallel deconvolutional structure that processes the feature maps generated by the backbone network. These processed outputs are combined with high-resolution features to produce precise segmentation masks of the seatbelt regions.

The seatbelt status classification module takes as input the fused outputs of the 3D pose estimation and seatbelt segmentation components. This module consists of a 1×1 convolution layer followed by a fully connected layer. The final classification is performed using a softmax function, which outputs confidence scores indicating whether the seatbelt is worn correctly or abnormally.

By integrating spatial pose information with pixel-level seatbelt segmentation, the system achieves robust classification of seatbelt usage status under varying occupant postures and environmental conditions.

Reference [102] proposes a time-efficient driver monitoring system (DMS) model that intentionally disregards temporal dependencies between video frames. Instead, the model focuses solely on the discriminative power of individual image frames and the contribution of different detection modalities to driving behavior classification.

The architecture of the proposed single-modal DMS model begins with a pre-trained 2D convolutional neural network (CNN) serving as the base encoder to extract spatial features from each input image. To enhance the feature representations, a Multi-Scale Channel Attention Module (MS-CAM) is applied, which introduces channel-wise attention across multiple scales to emphasize important feature channels.

Building on the standard squeeze-and-excitation (SE) block, the architecture adds an additional branch that omits global average pooling, allowing local spatial information to be preserved. The outputs from the original SE path and the added local-preserving branch are fused through element-wise addition. This combined feature map is then passed through a sigmoid activation function to generate channel-wise attention weights.

Finally, the attention-weighted feature representations are fed into a fully connected layer to perform the final classification of driver behavior. This architecture prioritizes per-frame efficiency and is particularly suitable for scenarios with limited computational resources or strict latency requirements.

In the multimodal variant of the driver monitoring system, input data is collected from four synchronized sources: a top-view infrared camera, a top-view depth camera, a front-view infrared camera, and a front-view depth camera. Due to the heterogeneous nature of these modalities, four separate encoder networks are employed—one for each input stream—to extract modality-specific spatial features.

To enable unified fusion of these multimodal features in a single stage, the architecture extends the Multi-Scale Channel Attention Module (MS-CAM) by introducing multiple attention heads. Specifically, the number of attention heads in MS-CAM is expanded to match the number of modalities, producing four separate attention weight matrices. A softmax function is applied across the heads to normalize and assign weights for each modality, allowing the system to adaptively emphasize the most informative channels.

The modified attention structure builds upon the original MS-CAM design, as well as enhancements from Attention Feature Fusion (AFF) and Iterative Attention Feature Fusion (iAFF). In this setup, the traditional MS-CAM modules used within AFF and iAFF are replaced with the new multi-head variant. This allows for a one-shot fusion of all four feature maps, rather than sequential or pairwise fusion steps, thereby improving computational efficiency and maintaining inter-modal consistency. The resulting fused representation is then used for downstream tasks such as driver behavior classification.

5.14 BEV's Visual Perception System

BEV stands for bird's-eye view, also known as the god's-eye view, which is a perspective or coordinate system (3D) used to describe the perception of the world. In autonomous driving systems, planning and control are generally carried out in BEV space. A BEV-based visual perception system represents an end-to-end framework for understanding the environment by transforming visual information from camera perspective projection from image space to BEV space using neural networks.

Tesla AI DAY showcased a BEV-based end-to-end visual perception system, which constructs a perception-map-location end-to-end system from multiple camera and IMU inputs, and uses the Transformer network structure to achieve spatial fusion. This has sparked another innovation in visual perception in the industrial and academic world of autonomous driving.

References [103, 104] provide a comprehensive overview of BEV methods in the context of visual perception. A central component of BEV perception is the transformation from the perspective view (PV), typically captured by cameras, to the top-down BEV representation. The transformation mechanism fundamentally determines the structure and performance of BEV-based systems. Based on the approach used for perspective transformation, existing BEV methods can be broadly classified into two major categories: geometry-based methods and network-based methods.

Geometry-based methods rely on explicit geometric transformations and can be further subdivided into homography-based and depth-based approaches. Homography-based methods utilize planar projective transformations under strong assumptions (e.g., flat ground), while depth-based methods leverage estimated depth information to map perspective features to 3D space. The depth-based approaches can be further categorized into three subtypes: point-based methods, which project depth points individually; voxel-based methods, which discretize the 3D space into voxels; and methods that combine depth estimation with inverse perspective mapping (IPM) for enhanced accuracy.

Network-based methods, by contrast, learn the PV-to-BEV transformation directly through data-driven architectures. These can be divided into MLP-based methods, which use fully connected layers to learn feature mappings, and Transformer-based methods, which employ attention mechanisms for spatial reasoning. Transformer-based approaches are further refined into three types based on their query strategy: sparse query methods, which use a small number of learnable queries; dense query methods, which cover the entire BEV grid; and mixed query methods, which combine both strategies to balance accuracy and efficiency.

This taxonomy provides a structured lens through which to understand and compare BEV methods, guiding the selection of appropriate architectures for specific applications in autonomous driving and scene understanding.

5.14.1 Homography-Based BEV

Points in 3D space can be transformed to image space through perspective projection, and the inverse problem of projecting image pixels to 3D space is ill-posed. Inverse perspective mapping (IPM) solves the mathematically impossible mapping problem based on the additional constraint that the inverse mapping point is on the horizontal plane. The homography matrix can be physically derived from the camera's internal and external parameters.

Some BEV methods integrate inverse perspective mapping (IPM) directly into the network architecture during training, rather than applying it as a separate pre-processing or post-processing step. A representative example is the View Parsing Network (VPN), which is designed for cross-view semantic segmentation. VPN is trained to convert perspective-view observations into pixel-level semantic maps in the bird's-eye view (BEV), providing precise spatial localization of all objects in the scene from a top-down perspective.

A key challenge in this transformation is the lack of ground-truth BEV annotations for real-world data, as top-down labels are difficult to acquire in practice. To address this issue, VPN is trained within a 3D graphics simulation environment where ground-truth BEV labels are readily available. The model learns to perform view transformation and semantic segmentation simultaneously under controlled, synthetic conditions.

To bridge the gap between simulated and real-world data, a domain adaptation strategy is employed. This allows the VPN model, initially trained on synthetic data, to generalize effectively to real-world scenes. The domain adaptation process enables the transfer of learned cross-view representations from simulation to real environments, ensuring accurate and robust semantic mapping in practical applications.

In VPN, the view observation results are first input into the encoder to extract the view feature map, and all view feature maps from different angles and modalities will be transformed. Then they are aggregated into a top-down view feature map in the View Transformer module (VTM). Finally, the aggregated view feature map is decoded into a top-down view semantic map.

VTM primarily learns the correlation of all spatial positions between the view feature map and the top-down view feature map. VTM does not change the shape of the input view feature map, so it can be inserted into any existing encoder–decoder-type network architecture to achieve classic semantic segmentation. VTM consists of two parts: the View Relationship Module (VRM) and the View Fusion Module (VFM).

Due to the substantial geometric distortion and large perspective gap between the front view and the BEV, traditional IPM alone is insufficient for generating accurate, distortion-free BEV representations. This limitation is especially evident in the presence of occlusions or missing information in the front view that cannot be directly recovered through geometric transformations.

To address this issue, the Mono Layout network is proposed as a deep learning-based solution for predicting scene layouts in BEV space from monocular images. Rather than relying solely on geometric cues, Mono Layout employs adversarial feature learning to infer and complete the occluded or "unreal" parts of the scene—regions that are not visible from the perspective view but are essential for understanding the full layout in BEV.

The architecture of Mono Layout consists of three main components: a context encoder, which extracts rich semantic features from the input image; an asymmetric layout decoder, designed to infer amodal (i.e., occlusion-invariant) representations of the scene; and two adversarial discriminators that guide the network to generate plausible BEV layouts through adversarial training. The output is a dense, top-down semantic map of the scene, including occluded areas, allowing the model to reconstruct a coherent and complete layout of the environment in BEV space.

The context encoder extracts multi-scale feature representations from the input monocular image for capturing static and dynamic scene components for subsequent processing. The non-modal static scene decoder decodes the shared context to generate the asymmetric layout of the static scene. This model consists of a series of deconvolution and upsampling layers, mapping the shared context to the static scene BEV. The structure of the non-modal dynamic scene decoder is similar to the road decoder, which can predict vehicle occupancy in BEV. Finally, the two discriminators regularize the predicted static/dynamic layout, whose distribution is similar to the real distribution of reasonable road geometry and the real ground vehicle occupancy rate.

5.14.2 Depth-Based BEV

Another method of viewpoint transformation is based on depth prediction. The point-based method directly uses depth estimation to convert pixels into point clouds, scattering in continuous 3D space. This method is more direct and easier to integrate mature experiences of monocular depth estimation and 3D detection based on lidar. However, this method faces the problem of model complexity and low performance due to the difficulty of generalizable end-to-end training.

Another viewpoint transformation based on depth prediction is the voxel-based method. Compared with point clouds distributed in continuous 3D space, voxels construct a uniform structure for feature transformation by discretizing 3D space, providing a more effective representation for 3D scene understanding, and subsequent BEV-based modules can be directly attached. Despite sacrificing local spatial accuracy, it is still more effective in covering large-scale scene structure information and is compatible with the end-to-end learning paradigm of view transformation.

Early work used the outer product of 2D feature maps and the corresponding predicted depth distribution to achieve this goal. It assumes that the distribution is uniform, i.e., all features along the ray are the same, such as the monocular 3D object detection method—OFT introduced in Sect. 5.4.

In contrast to methods that attempt direct geometric or learned perspective-to-BEV transformations, an alternative paradigm explicitly predicts depth distributions and constructs 3D features based on these estimates. This class of methods is highly regarded for its computational efficiency and modeling flexibility. A representative example is the Lift-Splat-Shoot (LSS) approach.

LSS is an end-to-end framework that directly derives BEV scene representations from camera image inputs. One of the primary challenges in monocular vision-based BEV perception is the inherent ambiguity of depth—each image plane pixel corresponds to an entire ray in 3D space, and the true depth along this ray is unknown. To address this, LSS decomposes the transformation process into three stages: lift, splat, and shoot.

In the "lift" stage, the model predicts a categorical probability distribution over discrete depth bins for each pixel, as well as a corresponding context feature vector. The outer product of the depth distribution and the context vector yields a depth-aware volumetric representation along each perspective ray. This formulation better captures the uncertainty and spread of possible depth values, leading to a more accurate and physically plausible representation of the scene in 3D space. The subsequent stages project these features into the BEV coordinate system and perform downstream tasks such as semantic segmentation or occupancy prediction.

The system has n images Xk input, each image has an external matrix $\boldsymbol{E}_k$ and an internal matrix $\boldsymbol{I}_k$, the external matrix and the internal matrix together define the mapping of each camera from the reference coordinates (x, y, z) to the local pixel coordinates (h, w, d), $d \in D$. During training or testing, no depth sensor is required. Let p be an image pixel with image coordinates (h, w). Associate $|D|$ points $\{(h, w, D)\}$ with each pixel, where D is a set of discrete depths, defined by $\{d_0 + \Delta, \cdots, d_0 + |D|\Delta\}$. The context vector of each point in the point cloud is parameterized, matching the concepts of attention and discrete depth inference. At pixel p, the network predicts the context c and depth distribution α of each pixel. The feature c_d associated with point p_d is defined as the context vector of pixel p.

The LSS framework enhances robustness to calibration errors by fusing predictions from multiple cameras into a unified BEV representation of the scene. The method takes as input n images from different camera viewpoints, along with their corresponding intrinsic and extrinsic calibration parameters.

Each image is first processed independently in the "lift" stage, where it is transformed into a 3D feature frustum by predicting depth distributions and associating them with learned image features. This produces a pseudo-point cloud within each camera's frustum, representing possible locations of scene elements in 3D space.

In the subsequent "splat" stage, all camera-specific frustums are projected into a shared BEV grid, enabling a unified spatial representation of the scene. This step aggregates features across views, facilitating spatial alignment and redundancy reduction across overlapping regions.

Finally, a convolutional neural network operating in the BEV domain processes the fused representation. This BEV CNN supports various downstream tasks such as BEV semantic segmentation, object detection, or trajectory planning. The modular and geometry-aware design of LSS allows for accurate, real-time scene understanding in multi-camera autonomous systems.

The Category Depth Distribution Network (CaDDN) addresses monocular 3D object detection by leveraging category-aware depth distributions to more accurately project contextual image features into 3D space. Instead of relying on a single estimated depth value per pixel, CaDDN predicts a depth probability distribution conditioned on the semantic category of each pixel. This enables richer and more reliable mapping of 2D features into appropriate depth ranges along the viewing frustum.

The architecture of CaDDN is composed of three main modules for generating 3D feature representations and a final detection module. First, given an input image I and an estimated depth distribution D, the network constructs frustum-aligned features G, which capture the spatial uncertainty of each pixel along its depth axis. These frustum features are then aggregated and converted into a 3D voxel grid representation V, which encodes features in discretized 3D space.

Next, the voxel features V are projected (or "folded") into a BEV feature map B using an efficient projection operation. The BEV representation provides a compact and structured view of the scene suitable for fast processing. Finally, a single-stage 3D object detector operates on the BEV features to generate the final object predictions, including location, orientation, and category. By combining depth-aware feature lifting with efficient BEV detection, CaDDN achieves competitive accuracy while maintaining real-time performance.

The frustum feature network serves as a bridge between 2D image features and their corresponding 3D representations by leveraging depth estimation. For each image pixel (u, v), the associated feature vector $F(u, v)$ is distributed across multiple depth bins using the predicted depth probability distribution $D(u, v)$. This results in a set of frustum features $G(u, v)$, where each feature is weighted according to the likelihood of the corresponding pixel appearing at a particular depth. This process effectively constructs a volumetric frustum representation that encodes both spatial and semantic information in 3D space.

To convert these frustum features into a regular 3D voxel grid, a frustum-to-voxel conversion module is employed. This module utilizes the known camera intrinsics and extrinsics to project each voxel center into the corresponding location in the frustum space. To populate the voxel grid V, trilinear interpolation is applied to sample the values from the frustum feature volume G at the projected coordinates. This operation is implemented in a differentiable manner, allowing end-to-end gradient flow during training. The result is a dense 3D voxel representation V that faithfully preserves the geometric and semantic structure of the scene, enabling downstream tasks such as BEV-based object detection.

FIERY is a future prediction model that estimates the motion and instance segmentation of dynamic entities in the BEV space using input from multiple monocular RGB cameras. It predicts probabilistic future trajectories of moving objects in a non-parametric form, providing a unified framework that integrates the perception, fusion, and motion forecasting components typically found in modular autonomous driving stacks.

Rather than explicitly separating 3D reconstruction, tracking, and trajectory prediction, FIERY processes raw images from multiple calibrated monocular cameras and directly outputs future BEV segmentation maps over a prediction horizon. These maps capture both the spatial extent and future motion of dynamic agents, such as vehicles and pedestrians, by modeling the probability distributions of occupancy and movement over time.

The model operates in a pipeline consisting of several key stages. First, features are extracted from each input image using a shared image backbone. These features are then lifted into 3D space using depth-aware projections and fused across views to construct a BEV representation of the scene. A temporal module is used to encode the dynamics of the scene, often leveraging recurrent or Transformer-based architectures. Finally, a prediction head outputs a sequence of future BEV instance segmentation maps, which are decoded into temporally consistent non-parametric trajectories of individual objects.

By unifying perception and forecasting within a single end-to-end learnable framework, FIERY simplifies the pipeline while maintaining accuracy and real-time applicability, making it particularly effective for monocular-based autonomous driving systems.

(1) Past moments $\{1, 2, \cdots, t\}$ predict the depth probability distribution of pixels and elevate the camera input $(O_1, O_2, \cdots, O_t)$ to 3D based on known camera intrinsic and extrinsic parameters.

(2) Features are projected onto BEV$(x_1, x_2, \cdots, x_t)$. The spatial transformer module S transforms the BEV features into the current reference frame (time t) based on past self-motion $(a_1, a_2, \cdots, a_{t-1})$.

(3) The 3D convolutional temporal model learns the spatio-temporal state $\boldsymbol{s}_t$.

(4) Parameterize two probability distributions: current and future distributions. The current distribution is conditioned on the current state $\boldsymbol{s}_t$, and the future distribution is conditioned on the current state $\boldsymbol{s}_t$ and future labels $(y_{t+1}, y_{t+2}, \cdots, y_{t+H})$.

(5) A latent code η_t is sampled from the future distribution in training and the current distribution in inference. The current state $\boldsymbol{s}_t$ and latent code η_t are inputs to the future prediction model, recursively predicting future states $(\hat{\boldsymbol{s}}_{t+1}, \hat{\boldsymbol{s}}_{t+2}, \cdots, \hat{\boldsymbol{s}}_{t+H})$.

(6) States are decoded into BEV future instance segmentation and future motion $(\hat{y}_{t+1}, \cdots, \hat{y}_{t+H})$.

Depth probability acts as a form of self-attention mechanism, modulating features based on predicted depth planes. Based on known camera intrinsic and extrinsic parameters (relative to the vehicle), the image and depth from each camera form tensors $(\boldsymbol{u}_t^1, \boldsymbol{u}_t^2, \cdots, \boldsymbol{u}_t^n)$, which are elevated to 3D space under a common reference coordinate system (the inertial center of the car at time t).

The output features produced by the backbone and temporal modules are passed to a bird's-eye view decoder D, which generates multiple prediction heads tailored to various scene understanding tasks. Each of these heads is responsible for producing a specific type of output relevant to object-level perception and motion forecasting in the BEV space.

The set of outputs includes the following components:

- Instance centrality: a heatmap indicating the likelihood of each pixel being the center of an instance, aiding in instance localization.
- Semantic segmentation: per-pixel classification in the BEV domain, assigning semantic labels such as vehicle, pedestrian, or background.
- Instance offset: a vector field pointing from each pixel to its corresponding instance center, used for grouping pixels into distinct object instances.
- Instance future flow: a dense motion prediction field that estimates the future displacement of each pixel, encoding the expected motion trajectory over time.
- Instance segmentation: a refined mask output that combines centrality and offset information to delineate object boundaries in BEV.

These outputs work in concert to support downstream tasks such as trajectory forecasting, occupancy prediction, and interactive scene understanding. By jointly learning all components in a unified architecture, the model achieves temporally consistent, spatially accurate, and semantically rich predictions directly from monocular camera inputs.

The instance segmentation results produced by the model can be interpreted through the following key steps:

1. **Instance Center Detection**: Instance centers are identified using non-maximum suppression (NMS) applied to the instance centrality heatmap. This process selects local maxima as the most probable object centers, effectively suppressing nearby redundant responses.
2. **Pixel-to-Instance Assignment**: Each pixel in the BEV map is associated with its nearest detected instance center based on the predicted offset vectors. These offset vectors indicate the relative direction and distance from each pixel to its corresponding object center, allowing pixels to be grouped into distinct object instances.
3. **Temporal Consistency via Future Flow**: To ensure consistent instance tracking across time steps, the model uses the predicted future flow from time t to $t + 1$. This flow field allows the instance centers at time t to be warped forward, and then compared to the actual predicted centers at $t + 1$. The matching of warped and predicted centers enables stable instance identification over time, facilitating the generation of temporally coherent trajectories.

This multi-stage process ensures that instance segmentation not only captures spatial grouping at a single time frame but also maintains consistency across future predictions for reliable tracking and forecasting.

5.14.3 BEV Based on MLP

In contrast to geometry-based methods that rely on explicit camera calibration and projection operations, another class of approaches models the transformation from perspective view (PV) to bird's-eye view (BEV) in a data-driven manner. These methods learn to approximate the view transformation using neural networks, which act as flexible mapping functions that implicitly encode geometric relationships within the camera system.

Multi-layer perceptrons (MLPs), in particular, have been adopted as generic function approximators capable of learning mappings between inputs and outputs that differ in modality, dimensionality, or spatial configuration. In the context of PV-to-BEV transformation, MLPs can be trained to implicitly learn camera calibration and spatial warping functions, allowing for flexible and efficient view conversion without requiring hand-crafted projection models.

HDMapNet is a representative example of this paradigm. It is designed for online high-definition map learning, where the goal is to construct detailed, vectorized BEV maps directly from sensor data. HDMapNet encodes features from multiple modalities, including camera images and LiDAR point clouds, and aggregates them in the BEV space to predict semantic map elements such as lane boundaries, crosswalks, and road dividers. Unlike traditional offline mapping pipelines that rely heavily on precise sensor calibration and sequential processing, HDMapNet performs map learning end-to-end in real time, enabling rapid adaptation to dynamic environments.

This shift from rule-based projection to learnable view transformation highlights the growing role of neural architectures in unifying perception and spatial reasoning in autonomous driving systems.

The traditional pipeline for annotating and maintaining HD maps presents several significant challenges. First, Mobile Mapping Systems (MMS), which collect raw data for HD maps using multiple sensors (e.g., LiDAR, cameras, GNSS), require precise sensor calibration and continuous maintenance. This sensitivity to sensor drift and calibration errors makes MMS-based solutions difficult to scale for widespread deployment. Second, small errors in SLAM algorithms used for point cloud stitching can accumulate, resulting in substantial geometric deviations in the final composite point cloud.

Moreover, even after point cloud fusion, extensive manual or semi-automatic annotation is needed to extract the road's geometric layout, semantic elements (e.g., lanes, road boundaries, traffic signs), and the topological structure of the drivable area. This annotation process is labor-intensive and resource-demanding. Finally, achieving robust localization under diverse environmental conditions, especially in poor weather or low-light scenarios, remains an open problem, further limiting the practicality and scalability of HD map generation.

To address these limitations, HDMapNet is proposed as a data-driven framework for online HD map learning. It supports inputs from one or more images and LiDAR point clouds and is capable of predicting semantic segmentation, instance embeddings, and directional vectors directly in the BEV space. These outputs are subsequently converted into a vectorized representation of the HD map, significantly reducing the need for manual annotation and improving adaptability to real-world driving environments.

The HDMapNet model is composed of four key neural network components: the perspective image encoder and the neural view transformer from the image branch, a LiDAR encoder based on a pillar-based structure, and a map element decoder. These modules work together to extract, transform, and fuse spatial features for HD map generation in the BEV domain.

Depending on the input modalities, HDMapNet can operate in three variants: HDMapNet(Surr) when using only surround-view RGB images, HDMapNet(LiDAR) when using only LiDAR point clouds, and HDMapNet(Fusion) when both modalities are provided. Each variant processes sensor-specific features and projects them into a unified BEV representation for downstream mapping tasks.

The neural view transformation module plays a central role in the image-based processing pipeline. It begins by extracting deep image features from perspective-view images using a convolutional image encoder. These features are then transformed into the camera coordinate system using an MLP, effectively disentangling geometric distortions caused by the perspective projection. Each camera-specific feature map covers a particular region of the physical environment based on the camera's field of view and orientation.

Subsequently, these transformed feature maps from all cameras are fused and reprojected into a common vehicle-centric coordinate system using known camera calibration parameters. The resulting BEV feature map aligns spatial information across views, enabling consistent semantic reasoning. This unified representation is then passed to the map element decoder, which outputs semantic segmentation, directional vectors, and instance embeddings that are further vectorized into a high-definition map structure.

The point cloud encoder is a variant of the PointPillar network with dynamic voxelization, which divides the 3D space into multiple pillar structures and learns feature maps from the pillar features of the pillar point cloud.

The map is a complex graph network, including instance and direction information of lane dividers and lane boundaries. Lane lines need to be vectorized, not pixel-level representation, for autonomous vehicles to follow.

Therefore, the BEV decoder not only outputs semantic segmentation but also predicts instance embedding and lane direction. The post-processing process clusters instances from the embedding and vectorizes them. The BEV decoder is a fully convolutional network (FCN) with three branches, namely, the semantic segmentation branch, the instance embedding branch, and the direction prediction branch.

The Fishing Net is a multimodal perception framework composed of a set of neural networks that process heterogeneous sensor inputs and convert them into a unified BEV semantic representation. Its design enables the integration of data from various sensing modalities—such as cameras, LiDAR, and radar—into a common top-down semantic grid, facilitating scene understanding and downstream decision-making in autonomous systems.

In the visual processing branch, the Fishing Net employs a modified orthographic feature transformation (OFT) module to map features from the PV to the BEV space. This OFT variant builds upon and corrects limitations of the earlier VPN approach, offering improved spatial consistency and semantic alignment across views. Instead of relying solely on depth or geometry-based projection, the OFT module learns to transform image plane features into a top-down layout in a differentiable and end-to-end trainable manner.

By enabling direct conversion from PV features to BEV features, the Fishing Net achieves efficient and accurate top-down semantic perception, supporting complex tasks such as 3D semantic mapping, motion prediction, and multi-agent coordination.

The Fishing Net framework also includes specialized branches for processing LiDAR and millimeter-wave radar data, each designed to project sensor-specific features into a unified BEV representation. The architecture of these branches leverages the inherent spatial alignment of 3D point cloud data, which allows for more direct and geometrically consistent mapping to BEV space.

Compared to the image-based branch described previously, where features from perspective-view images must undergo complex view transformations, the BEV conversion for LiDAR and radar is relatively more straightforward. This is because LiDAR and radar sensors naturally capture data in 3D space or near-horizontal planes, minimizing the need for elaborate geometric rectification. As a result, the transformation pipeline for these modalities primarily focuses on voxelization, pillar encoding, or direct projection into BEV grids, followed by convolutional processing.

This contrast highlights the added challenge in visual BEV transformation, where the network must infer depth and correct for perspective distortion before achieving spatial alignment. In Fishing Net, these differences are carefully accounted for, with each modality-specific pipeline optimized to produce semantically rich and spatially aligned BEV features for sensor fusion and downstream tasks.

LiDAR and millimeter-wave radar rely on a structure similar to UNet. The encoder consists of a set of five blocks, a pair of convolutional layers, and average pooling layers. The decoder consists of five blocks, three convolutional layers, and batch normalization (BN). The network also includes UNet style skip connections.

The visual network architecture consists of a traditional encoder–decoder scheme based on a fully convolutional ResNet backbone. A three-layer ResNet block, with four blocks each in the encoder and decoder. An OFT layer is used between the encoder and decoder.

All modalities share the same output reference frame and representation, so an aggregation function is applied to the softmax values along the modality dimension. The two aggregation functions evaluated are average pooling and priority pooling. The average is the normal arithmetic average, used to reduce output variance. The priority is defined as follows: pedestrian: 3, vehicle: 2, background: 1. If the imaging devices are inconsistent in pixel classification, choose the imaging device with the highest priority for the predicted category.

Given a set of sensor modalities that provide perception data from the current time step t_0 and p previous time steps $t-p, \ldots, t-1$, the objective is to infer a sequence of top-down semantic grid representations for both the current frame t_0 and future time steps $t_1, \ldots, t_f$. These semantic grids capture the spatial distribution and class-specific segmentation of dynamic and static entities in the scene.

The semantic grid categorizes each cell into one of three target classes: ① **Vulnerable Road Users (VRU)**, which includes pedestrians, cyclists, and motorcyclists. ② **Vehicles**, such as passenger cars, trucks, and buses. ③ **Background**, including static infrastructure and unoccupied areas.

This predictive task requires the model to perform temporal reasoning over multimodal sensor inputs, spatially align observations into a unified BEV, and generate semantically meaningful forecasts of agent motion and scene evolution. The goal is to support downstream applications such as planning and risk assessment in autonomous driving systems.

In typical visualization results, large rectangular bounding boxes correspond to predicted vehicle positions, while smaller boxes indicate VRUs such as pedestrians and cyclists. These visualizations highlight the model's ability to reason about dynamic agents and predict their future states in a spatially coherent manner.

5.14.4 Transformer-Based BEV

In addition to MLP, another solution in data-driven methods that do not explicitly use camera model is the Transformer (with cross-attention) network, which has caused performance breakthroughs in natural language processing (NLP) and computer vision. It is currently used to achieve the best performance in mapping PV-BEV.

Tesla also adopted this method, first using position encoding to design a set of BEV queries, and then performing view transformation through the cross-attention mechanism between BEV queries and image features. Since then, many methods have been proposed to use Transformer or, more specifically, to model view transformation with cross-attention.

These methods are divided into three categories according to the granularity of the learnable slots (referred to as queries) in the Transformer decoder: methods based on sparse queries, methods based on dense queries, and methods based on hybrid queries. The following mainly introduces several top-performing methods, which come from both sparse and dense query approaches.

1. Based on Sparse Queries

For methods based on sparse queries, query embedding allows the network to directly generate sparse perception results without explicitly performing dense transformations of image features. This design choice is natural for target-centric perception tasks (such as 3D object detection), but it is not easy to extend it to dense perception tasks (such as segmentation).

ETH's work on STSU (Structured Bird's-Eye View Traffic Scene Understanding) proposes a Transformer-based framework for structured scene perception in BEV space, following the sparse query paradigm. Inspired by DETR (DEtection TRansformer), STSU extracts a structured and interpretable directed graph that captures the local road network topology from a single perspective image.

The model employs two distinct sets of sparse queries: one dedicated to detecting road centerlines and the other to dynamic objects such as vehicles. By jointly processing these queries within a Transformer architecture, the network is able to model the interactions between traffic participants and the underlying road structure. This joint reasoning allows for more accurate and context-aware predictions of both road geometry and object trajectories.

Specifically, the Transformer encoder–decoder processes road centerline queries to output detection confidence scores, control points defining the centerline geometry, and additional structural features related to the road layout. Simultaneously, the target queries are used to infer class probabilities and oriented 3D bounding box parameters for dynamic objects.

An extension of this framework, TopologyPL, further enhances performance by explicitly incorporating topological constraints. It maintains minimal cycles within the predicted road network graph, ensuring coherent and navigable connectivity structures in the output. Together, STSU and TopologyPL represent a significant advancement in structured scene understanding from monocular inputs, enabling interpretable and planning-aware representations in the BEV domain.

STSU processes the output of Transformer in two branches:

The first is the **lane branch**, which has four parts: detection head (centerline existence probability), control head (Bezier control points), association head (associated feature vector), and association classifier (associated centerline probability). During training, the centerline control points and detection probabilities are output first, and the Hungarian matching algorithm is used between the estimated centerline and GT centerline. The association step is performed on the estimated match. In inference, the detection probability of the centerline is thresholded, and the association step is performed on the active centerline.

The second is the **target branch**, which processes the proposal vector from the Transformer, including two modules and an optional post-processing network. The first module is the detection head: it processes the Transformer output to produce a class probability distribution, including the "no detection" class; the second module is the 5-params head: an MLP+sigmoid layer, which produces normalized parameters for the oriented target box; the post-processing network is the Refinement Net: the refinement network converts instance estimates into semantic segmentation. The structure of the refinement network is similar to the BEV decoder, where the lower-resolution input is magnified to provide a fine-grained segmentation map.

Since Transformer has no concept of position, position embedding (PE) is used. The system uses two different position embeddings, the first encodes image domain spatial information, using a sine function in normalized cumulative position; the second position embedding encodes the corresponding BEV position for a given pixel.

Here, two **position embeddings** (image and BEV) are designed so that they are half the size of the input feature map channels. Therefore, the image position embedding is added to half of the channels, and the BEV position embedding is added to the other half. This dual position embedding is called split positional encoding.

Another typical example of sparse query BEV is DETR3D. DETR's contribution is to introduce a defined set-to-set loss in 2D object detection, and DETR3D also uses this loss for BEV-based 3D object detection. This method replaces the cross-attention mechanism with a geometry-based feature sampling process. It first predicts 3D reference points from learnable sparse queries, then projects the reference points onto the image plane using a calibration matrix, and finally samples the corresponding multi-view, multi-scale image features for end-to-end 3D bounding box prediction.

DETR (DEtection TRansformer) is a fully end-to-end object detection framework that eliminates the need for traditional components such as anchor generation, region proposal networks, and non-maximum suppression (NMS). Instead, DETR formulates object detection as a direct set prediction problem, predicting all targets simultaneously using a fixed number of queries and optimizing with a set-to-set loss.

The model is trained using a bipartite matching loss, which uniquely assigns each predicted target to a corresponding ground-truth target. This matching ensures that each query is either associated with a ground-truth object or explicitly learns to predict a "no object" class, thereby enabling DETR to handle a fixed-size output regardless of the number of actual objects in the image.

Architecturally, DETR consists of a standard CNN backbone that extracts a 2D feature representation from the input image. This feature map is flattened into a sequence and augmented with learned positional encodings before being passed to the Transformer encoder. The Transformer decoder receives a fixed number of learned object queries—position embeddings that guide the model to attend to specific regions—and attends to the encoded features to produce object-specific embeddings.

Each output embedding from the decoder is fed into a shared FFN that independently predicts the object class and the normalized bounding box coordinates. If no object is assigned to a given query, the model learns to predict a "no target" label for that query. This architecture provides a streamlined, interpretable, and fully differentiable pipeline for object detection.

DETR3D extends the DETR framework to 3D object detection from multi-view images by combining sparse query-based reasoning with geometric projection. The model takes as input a set of synchronized multi-view images, which are processed through a ResNet backbone enhanced with deformable convolutions, followed by a feature pyramid network (FPN) to extract multi-scale 2D features.

At the core of DETR3D is a set of learnable sparse target queries, each corresponding to a hypothesized 3D object. These queries are decoded into 3D reference points in the camera coordinate system. To associate image features with the 3D queries, the reference points are projected into each camera's image plane using known intrinsic and extrinsic calibration parameters. This projection enables the model to sample and aggregate 2D features from multiple views and align them with the corresponding 3D query.

These image-aligned features are used to refine the query representations through cross-attention mechanisms in the Transformer decoder, ultimately yielding predictions for object class and 3D bounding box parameters. DETR3D is trained end-to-end using a set-to-set loss based on bipartite matching, similar to the original DETR. This formulation enables DETR3D to detect objects in 3D space directly from image inputs without relying on dense 3D voxelization or explicit depth estimation.

DETR3D adopts an iterative working mode, generating bounding box estimates from 2D feature maps through L network layers calculated based on set-to-set. Each layer includes the following steps:

(1) Predict a set of bounding box centers associated with the target query.

(2) Project these centers into all feature maps using the camera transformation matrix.

(3) Sample the features through bilinear interpolation and merge them into the target query.

(4) Use multi-head attention mechanism to coordinate target interaction and refine the target query.

To address the complex and computationally expensive feature sampling process in DETR3D, PETR (Position Embedding Transformer for Multi-view 3D Object Detection) introduces a more efficient design by incorporating 3D positional information directly into 2D multi-view image features. Instead of projecting 3D reference points into image space during runtime for feature extraction, PETR pre-encodes 3D position embeddings—derived from known camera intrinsics and extrinsics—into the 2D feature maps.

This design enables the sparse object queries to interact directly with position-aware 2D features through cross-attention in the Transformer decoder, eliminating the need for explicit sampling operations. As a result, PETR streamlines the spatial alignment process between 2D observations and 3D object hypotheses, leading to faster inference and improved representation consistency.

By embedding geometric priors into the feature space, PETR simplifies the interaction between image and query features while maintaining the accuracy and flexibility of transformer-based 3D object detection. This contrasts with DETR3D, which requires dynamically projecting queries and sampling features from multi-view images at each iteration.

The PETR architecture begins by inputting multi-view images into a shared backbone network, such as ResNet, to extract 2D feature maps from each camera view. These features are then used in conjunction with geometric information to produce a 3D spatially aware representation of the scene.

In the 3D coordinate generation module, the shared frustum space of all cameras is discretized into a 3D grid. Each grid point is transformed from its original camera-centric coordinate system into a unified world coordinate system using the corresponding intrinsic and extrinsic camera parameters. This transformation aligns spatial positions across all views into a consistent global reference frame.

The resulting 3D coordinates are then fused with the multi-view 2D image features in the 3D position encoder. This module encodes both appearance and geometric information to produce 3D position-aware features that embed context from both the visual and spatial domains.

Target queries, generated by a dedicated query generator, interact with these 3D position-aware features within the Transformer decoder. Through iterative cross-attention and self-attention mechanisms, the queries are progressively refined. The updated query embeddings are finally passed through prediction heads to estimate the 3D bounding boxes and class labels of the detected objects.

By directly encoding spatial priors and avoiding explicit feature sampling, PETR achieves efficient and accurate multi-view 3D object detection in an end-to-end trainable manner.

The architecture of the 3D position encoder in PETR is designed to integrate geometric information with visual features in a spatially coherent manner. The process begins with the multi-view 2D image features, which are first passed through a 1×1 convolutional layer to reduce their channel dimensionality while preserving spatial resolution. This step ensures computational efficiency and facilitates subsequent feature fusion.

Simultaneously, the 3D coordinate generator produces a dense grid of 3D points representing discretized positions in the unified camera frustum space. These 3D coordinates are then passed through an MLP, which transforms them into high-dimensional 3D position embeddings. The resulting embeddings encode spatial context and geometric priors that are crucial for aligning features across views.

Each 3D position embedding is shaped to match the dimensions of its corresponding 2D image feature map. The embeddings are then added element-wise to the 2D image features of the same view, effectively injecting spatial awareness into the visual features. This fusion produces the 3D position-aware feature maps, which are rich in both semantic and geometric information.

Finally, the 3D position-aware features are flattened into a sequence and used as the input to the Transformer decoder. These enriched features allow the decoder to attend to geometrically meaningful locations, enabling more accurate and spatially aligned 3D object detection.

For temporal modeling, the main issue is how to align the target positions in different frames in 3D space. PETR implicitly encodes 3D positions into 2D image features, making explicit feature transformation impossible. Since PETR has already demonstrated the effectiveness of 3D PE in 3D perception, is 3D PE still applicable for temporal alignment?

In PETR, grid points of the camera frustum space (shared across different views) are transformed into 3D coordinates through camera parameters, and then the 3D coordinates are input into a simple multi-layer perceptron (MLP) to generate 3D PE. In practice, it is found that simply aligning the 3D coordinates of the previous frame with the current frame allows PETR to work well under temporal conditions.

The overall architecture of PETRv2 builds upon the original PETR framework, extending its capabilities by incorporating temporal modeling and BEV segmentation. The model processes multi-view image inputs using a 2D backbone network, such as ResNet-50, to extract spatial image features. In parallel, 3D coordinates are generated from the discretized camera frustum space using known camera parameters, as in PETR.

To enable temporal reasoning, PETRv2 accounts for ego-motion by transforming the 3D coordinates from the previous frame at time step $t-1$ into the coordinate system of the current frame t through pose transformation. This alignment ensures spatial consistency across time steps. The 2D image features and corresponding 3D coordinates from adjacent frames are then concatenated and fed into the FPE, which encodes both visual and geometric information. The FPE outputs key and value tensors that are used in the cross-attention mechanism of the Transformer decoder.

PETRv2 employs two types of queries in the Transformer decoder: detection queries and segmentation queries. Detection queries are initialized from learnable 3D anchors, while segmentation queries are derived from a fixed grid of BEV points. These queries interact with the multi-view image features through attention, allowing the model to reason about both object-level and scene-level semantics.

The updated queries are then passed to two separate output heads. The detection head predicts 3D object classes and bounding boxes, while the segmentation head generates BEV semantic segmentation maps. By unifying detection and segmentation within a temporally aware transformer-based framework, PETRv2 achieves robust and comprehensive 3D scene understanding from monocular and multi-view images.

2. Based on Dense Queries

For methods based on dense queries, each query is pre-allocated a position in 3D space or BEV space. The number of queries is determined by the rasterized spatial resolution, usually larger than methods based on sparse queries. Dense BEV representation can be achieved by interacting dense queries with image features of multiple downstream tasks (such as 3D detection, segmentation, and motion prediction).

Due to the high memory complexity of the attention operation in each Transformer decoder layer among a large number of queries and key elements, the image resolution and BEV resolution are usually limited to reduce memory overhead, which may hinder the scalability of the model in many cases. Recently, this problem has been solved in methods based on dense queries, namely, by adopting deformable attention. This attention mechanism combines the sparse spatial sampling of deformable convolution and the relationship modeling capability of the attention mechanism. Focusing only on sparse locations can significantly reduce the memory overhead of the ordinary attention mechanism.

BEVFormer incorporates a deformable attention mechanism to efficiently aggregate spatio-temporal information from multi-camera inputs. It introduces a dual-structured Transformer framework, composed of a spatial Transformer and a temporal Transformer, to generate a unified BEV feature representation.

In the spatial domain, BEVFormer uses learnable BEV queries to attend to multi-view image features. These queries are refined through deformable cross-attention, which dynamically samples a sparse set of key points from the image feature maps, guided by 3D geometry. This approach reduces computational cost while maintaining high spatial precision.

In the temporal domain, BEVFormer introduces temporal queries that enable the model to aggregate features across sequential frames. This temporal attention mechanism allows the network to maintain motion consistency and leverage historical observations to enhance scene understanding.

By jointly processing queries in both spatial and temporal domains, BEVFormer fuses multi-view and multi-frame information into a consistent BEV representation. This spatio-temporal fusion significantly enhances the performance of downstream perception tasks such as 3D object detection and semantic mapping, making BEVFormer particularly effective for real-world autonomous driving applications.

The overall architecture of BEVFormer consists of a multi-layer Transformer encoder with six stacked encoding layers. Each layer follows the conventional Transformer structure, but incorporates three custom-designed components tailored for BEV generation: BEV query, spatial cross-attention, and temporal self-attention. Notably, BEVFormer does not employ the deformable attention mechanism within these layers, instead relying on structured attention to integrate spatial and temporal information.

The BEV query is a grid-shaped set of learnable parameters representing spatial locations in the BEV space. These queries serve as anchors for aggregating relevant features from both current multi-camera inputs and historical BEV representations. In each encoding layer, spatial cross-attention is used to associate each BEV query with image features extracted from different camera views. This allows the model to capture multi-view visual context aligned with the BEV grid.

Temporal self-attention is applied across BEV features from consecutive frames, enabling the model to retain temporal consistency and model dynamic changes in the scene. By alternating between spatial and temporal attention, the network iteratively refines the BEV feature map, fusing spatial information from the present frame with temporal cues from the past.

Through this layered attention mechanism, BEVFormer effectively generates temporally and spatially coherent BEV representations, supporting high-performance 3D object detection and other downstream autonomous driving tasks.

During the inference process, at timestamp t, multi-camera images are fed into the backbone network (such as ResNet-101), and features $\boldsymbol{F}_t = \{\boldsymbol{F}_t^i\}$ of different camera views are obtained, where $\boldsymbol{F}_t^i$ is the feature of the i-th view. At the same time, the BEV feature $\boldsymbol{B}_{t-1}$ at timestamp $t - 1$ is retained. In each encoder layer, the temporal information is first queried from the previous BEV feature $\boldsymbol{B}_{t-1}$ using the BEV query Q through temporal self-attention. Then, the spatial information is queried from the multi-camera features $\boldsymbol{F}_t$ using the BEV query Q through spatial cross-attention. After the feed-forward network, the encoder layer outputs refined BEV features as the input for the next encoder layer. After six stacked encoder layers, a unified BEV feature $\boldsymbol{B}_t$ is generated at the current timestamp t. With BEV feature Bt as input, the 3D detection head and map segmentation head can, respectively, predict perception results such as 3D bounding boxes and semantic maps.

3. Introduction of Geometric Clues

In the Transformer method of viewpoint transformation, the convergence speed of training the network is relatively slow, and a large amount of data is needed. Now more methods are trying to use 3D geometric constraints to achieve rapid convergence or data efficiency.

Deformable attention methods (such as BEVFormer) usually rely on the camera projection matrix to calculate 2D reference points for feature sampling, which helps the network focus on appropriate areas on the image and abandon global attention operations.

In addition, the Position Embedding Transformation PETRv2 model can generate 3D position embeddings using the camera calibration matrix to enrich image features with geometric priors, and help the Transformer learn the mapping from PV-BEV through implicit geometric reasoning.

Although Transformer-based methods do not necessarily need pixel-by-pixel depth for view transformation, depth information is still important for Transformer's geometric reasoning.

5.14.5 *Extension of BEV Framework*

BEVDepth is a depth-based BEV perception framework that enhances voxel-based 3D object detection by explicitly incorporating depth supervision and projection-aware design. Unlike purely geometry-free or implicit depth learning approaches, BEVDepth encodes camera intrinsic and extrinsic parameters to guide depth estimation with explicit supervision signals. To address the inaccuracies introduced by noisy depth ground truth due to projection errors, BEVDepth further introduces a depth correction sub-network. This module refines depth predictions and compensates for calibration or sampling inconsistencies, thereby improving spatial alignment and overall detection performance.

To overcome the computational bottleneck associated with projecting image view features into BEV space, BEVDepth employs an optimized view transformation technique known as efficient voxel pooling. This operation accelerates the mapping of depth-aware image features to 3D voxels and enables the construction of a high-resolution BEV representation in real time.

The architecture of BEVDepth comprises four key components:

1. **Image Encoder**: A backbone network (e.g., ResNet) is used to extract 2D feature maps from N synchronized input images $\mathbf{I}$ captured from multiple camera views.
2. **DepthNet**: A dedicated depth estimation network predicts dense depth maps for each input image, enabling depth-aware feature lifting.
3. **View Transformer**: This module projects 2D features into a 3D voxel grid using the estimated depth and camera parameters, followed by efficient voxel pooling to aggregate features into a unified BEV representation.
4. **3D Detection Head**: The BEV feature map is passed to a detection head—such as the CenterPoint head—which predicts object categories, 3D bounding box offsets, sizes, orientations, and other relevant attributes.

By combining accurate depth modeling with efficient feature transformation and strong detection heads, BEVDepth achieves state-of-the-art performance in camera-based 3D object detection, bridging the gap between image-based and LiDAR-based methods.

In the regular BEVDepth, the only supervision for the depth module comes from the detection loss. However, due to the difficulty of monocular depth estimation, a single detection loss is far from enough to supervise the depth module. It is suggested to supervise the intermediate depth prediction value with the true depth derived from point cloud data. Given the rotation $\boldsymbol{R}$ and translation $\boldsymbol{t}$ from the vehicle coordinate system to the i-th view camera coordinate system, and the internal parameters of the i-th camera, the true depth of the image at a certain moment t for camera i can be calculated.

With the help of depth supervision, DepthNet should be able to generate reliable predicted depth. Due to the violent motion of the vehicle, the calibrated camera parameters $\boldsymbol{R}$ and $\boldsymbol{t}$ may sometimes be inaccurate, leading to spatial deviation between image features and depth truth. When the receptive field of DepthNet is limited, the deviation problem becomes more serious.

Therefore, a deep correction module is proposed to alleviate this problem. In addition, the deep correction module increases the receptive field of DepthNet to solve the localization error problem, rather than adjusting the depth truth based on self-vehicle interference. Specifically, multiple residual blocks are stacked in DepthNet, followed by deformable blocks. The architecture of DepthNet, where the context features are directly generated from the image features, and the designed SE (squeeze-and-excitation) class layer deforms and aggregates with the image features to better estimate the depth.

Thanks to the improvement of the receptive field, the offset depth truth can handle these image features at the correct position.

Voxel pooling is key to the BEV detector, with the aim of aggregating 3D multi-view features into a complete BEV feature. Generally, it divides the self-vehicle space into several evenly distributed grids, and then summarizes the frustum features falling into the same grid to form the corresponding BEV feature. For this, LSS uses a cumulative sum trick (cumsum trick), that is, all frustum features are sorted according to the corresponding BEV grid ID, all features are accumulated and summed, and then subtracted from the cumulative sum of the grid boundaries. This implementation brings a lot of additional calculations because it needs to sort a large number of BEV coordinates. In addition, the sequential prefix sum used by this trick is inefficient.

Both of these shortcomings will affect the overall running speed of the detector. Of course, you can choose to increase parallelism to speed up the prefix sum, while BEV Depth provides a simpler solution, which makes full use of the high parallelism of GPU devices, called **efficient voxel pooling**, as shown in Fig. 4 of [55]. The main idea is to assign a CUDA thread to each frustum feature, which can add this feature to the corresponding BEV grid. By replacing the original voxel pooling with improved efficient voxelization, BEVDepth can be accelerated by three times.

The compact and effective BEV representation derived from the PV-BEV transformation method is friendly to many downstream tasks, such as object detection, map segmentation, prediction, and motion planning. Sharing the backbone network can greatly save computational costs and improve efficiency. Therefore, some work tries to perform multiple tasks simultaneously using a unified framework.

BEVerse designs an iterative process to effectively generate future states and jointly infer 3D detection, semantic map reconstruction, and motion prediction tasks. Specifically, BEVerse first performs shared feature extraction and lifting, generating 4D BEV representation from multi-timestamp and multi-view images. After self-motion compensation, the BEV features are further extracted using a spatio-temporal encoder. Finally, multiple task decoders are added for joint inference and prediction. In the decoder, a **grid sampler** is proposed to generate BEV features that support different ranges and granularities for different tasks. In addition, an **iterative flow** method is designed to achieve memory-efficient future prediction.

The traditional paradigm is to stack these subtasks in sequence, where the output of one subtask is fed as input to the next subtask. Due to repeated feature extraction and propagation, the traditional paradigm itself brings additional computational burden. The proposed BEVerse framework for joint perception and prediction, through shared feature extraction and parallel multi-task inference, achieves a better balance between performance and efficiency.

BEVerse obtains M surrounding camera images from N timestamps and takes the corresponding self-motion and camera parameters as input. Through multi-task inference, the output includes the current frame's 3D bounding box and semantic map, as well as the future instance segmentation and motion trajectory of the subsequent T frames. BEVerse consists of four sub-modules, namely, the image view encoder, view transformer, spatio-temporal BEV encoder, and multi-task decoder.

The **image view encoder** uses the Swin Transformer as the backbone network, and the **view transformer** uses the method of LSS, such as **FIERY spatio-temporal BEV encoder**, which consists of a set of temporal blocks.

The **multi-task decoder** consists of a parallel and independent set of decoders, each of which includes a grid sampler, task encoder, and task head.

Since different tasks may require specific ranges and granularities, the spatial range and resolution of the input BEV features cannot be directly used for decoding. For example, learning the semantic map requires fine-grained features because the traffic lines in 3D space are narrow. Therefore, the **grid sampler** crops the specific task area and converts it to the ideal resolution through bilinear interpolation. To improve efficiency, the BEV grid is generally set to be large and coarse in experiments.

After feature sampling, a lightweight **task encoder** is used to encode specific task features on the corresponding BEV grid. Here, the ResNet basic block is used to build the backbone, and it is combined with the multi-scale features similar to the image view encoder. The output features are upsampled to the input resolution and sent to the **task head**.

Since the BEV feature representation is built from multiple camera videos, the dimension gap between monocular and lidar methods has disappeared. Therefore, the detection head designed for lidar can be directly used without modification. Therefore, the first step of the two-step method **CenterPoint** is used. first stage) As the **3D target detection head**.

Two ordinary convolutional layers with BatchNorm and ReLU are used to construct the semantic map reconstruction head, and the output channel is the number of classes in the semantic map Cmap.

With the rapid development of BEV perception algorithms, a more interpretable image and point cloud fusion method is to transfer image features to BEV and fuse features from two modalities based on the physical correspondence on BEV.

BEVFusion fully retains the dense semantic information and spatial geometric information of the image during the fusion stage, and proposes an effective BEV pooling operation to accelerate inference. RRF defines a 3D body for image features through projection and bilinear sampling, connects the rasterized radar BEV image, and reduces the vertical dimension, finally obtaining the BEV fusion feature map.

Given different perception inputs, first apply modality-specific encoders to extract their features; secondly, transform multimodal features into a unified BEV representation, while retaining geometric and semantic information; next, the existing view transformation efficiency bottleneck can be accelerated by precomputation and interval reduction; then, applying the convolution-based BEV encoder to the unified BEV features can alleviate the local bias between different features; finally, add some task-specific heads to support different 3D scene understanding tasks.

Regrettably, the efficiency and speed of BEV pooling are astonishingly low. This is because the number of camera feature point clouds is very large, constituting a typical workload. To eliminate this efficiency bottleneck, BEVFusion proposes to optimize the BEV pooling process through precomputation and interval reduction.

The conversion from the camera to BEV is a key step in sensor fusion in the unified BEV space. The effective BEV pooling method accelerates grid association by precomputation to reduce intervals and speed up. The execution speed of the view transformation module has increased by 40 times.

The first step of BEV pooling is to associate each point of the camera feature point cloud with the BEV grid. Unlike lidar point clouds, the coordinates of the camera feature point cloud are fixed (as long as the camera's internal and external parameters remain unchanged, usually after proper calibration). Based on this, precompute the 3D coordinates of each point and the BEV grid index. Sort all points according to the grid index and record the point ranking. During the inference process, all feature points need to be resorted according to the precomputed sorting. This caching mechanism can reduce the latency of grid association from 17ms to 4ms.

After grid association, all points in the same BEV grid will be continuous in the tensor representation. The next step of BEV pooling is to aggregate the features within each BEV grid through some symmetric functions (such as average, maximum, and sum). The existing implementation method is to first calculate the prefix sum of all points, and then subtract the boundary value where the index changes. However, the prefix sum operation requires tree reduction on the GPU and generates many unused partial sums (because only boundary values are needed), both of which are inefficient. To accelerate feature aggregation, BEVFusion implements a dedicated GPU kernel, which directly parallelizes on the BEV grid, assigns a GPU thread to each grid, calculates its interval sum, and writes back the result. This kernel eliminates dependencies between outputs (so no multi-level tree reduction is needed) and avoids writing partial sums to DRAM, thereby reducing the latency of feature aggregation from 500ms to 2ms.

5.14.6 Problems with BEV

1. Perception Resolution

The BEV resolution determines the representation granularity of the BEV features, where each feature can be precisely traced back to the grid in world coordinates. High resolution is needed to better represent small-scale targets, such as traffic lights and pedestrians. In recent years, with the rapid development of graphics card computing capabilities, PV image resolution and BEV grid size have significantly increased. These BEV-based methods still have a serious problem of high computational burden caused by high input resolution, which needs further exploration.

2. Sampling Range and Frequency of Image Features

The setting of the sampling range determines how much of the viewing frustum behind the image plane will be sampled into the BEV space. By default, this range is equal to the effective range of lidar annotation. When efficiency has a higher priority, the upper z-axis part of the viewing frustum may be affected, because, in most cases, it only contains unimportant information, such as the sky. The sampling frequency determines the utility of image features. A higher frequency ensures that the model accurately samples the image features corresponding to each BEV position at a higher computational cost.

3. Post-processing

Although BEV detection eliminates the burden of multi-camera target-level fusion, further post-processing can still be beneficial. According to the nature of BEV transformation, repetitive features may be sampled at different BEV positions along the line of sight to the camera center. This can lead to repeated false detections of a foreground target, each with a different depth, but all can be projected back to the same foreground target in the image space. To alleviate this problem, it is beneficial to use 2D detection results to remove duplicates from 3D detection results, where 2D bounding boxes and 3D bounding boxes achieve bipartite matching.

4. Multi-task Training (MTL)

Although CNN benefits from joint optimization with multiple related tasks, joint training of 3D object detection and BEV segmentation often does not bring improvement. The detection performance is often affected, the improvement of segmentation performance is not consistent among different categories, and more efforts are needed to explore the dependencies between different perception tasks to achieve joint improvement.

5.15 Occupancy Network of Visual Perception

Bird's-eye-view (BEV) perception is the mainstream perception mode. Its advantages are with absolute scale and no occlusion, which can be used to describe the environment. BEV perception provides a unified representation space for multi-source information fusion and not a few downstream applications. However, BEV perception has some shortcomings:

- The depth of the distant view close to the horizon is inconsistent.
- It is still a 2D, not a 3D representation.
- It cannot distinguish between obstacles and vacancy in the height direction.
- It cannot identify unknown object categories.
- The shape of obstacles can only be framed in a rectangle, and deformed or shaped objects are difficult to locate.
- There is no prediction of suspended objects at height.
- It is difficult to distinguish between static and dynamic obstacles.
- Occlusion visualization is inconsistent in the time domain (ghosting).

In a sense, the occupancy network is an extension of the BEV network. Vision-based 3D occupancy perception can estimate the spatial occupancy state and semantic category of the 3D voxel grid around the autonomous vehicle based on the image input, and can provide fine-grained representation and robust detection for undefined long-tail obstacles in the 3D space.

Occupancy representation originated in the field of robotics, where 3D space is divided into voxel units to binarily predict whether a voxel is occupied by an object, thereby achieving effective collision avoidance. Mescheder [105] proposed an occupancy network, which implicitly represents a 3D surface through a continuous decision boundary of a deep neural network classifier and uses an occupancy grid map to determine whether the grid is occupied. Peng [106] further introduced a convolutional occupancy network, a new shape representation that combines the expressive power of convolutional neural networks with the advantages of implicit representation. At the CVPR 2022 Autonomous Driving Workshop, Tesla introduced occupancy networks into autonomous driving, triggering a research boom in vision-based 3D occupancy perception [107]. Tesla engineers were inspired by the idea of position coordinate modeling in NeRF (Neural Radiance Field), also the 3D Gaussian Splatting method that emerged after NeRF. This type of neural rendering method for 3D reconstruction emphasizes the geometric quality and visual appearance of the scene. In contrast, voxel-level occupancy perception has lower requirements on resolution and visual appearance but focuses on the occupancy distribution and semantic understanding of the scene.

Two recent surveys on occupancy networks [108, 109] provide a detailed summary of the work in this area. The survey [108] focuses on visual perception, while the survey [109] focuses on multimodal fusion. This chapter session focuses on visual perception and, according to the category way of [108], mainly introduces three approaches, namely: feature enhancement, deployment-friendly, and label-efficient. Feature enhancement methods alleviate the difference between three-dimensional spatial output and two-dimensional spatial input by optimizing the feature representation ability of the network. Deployment-friendly methods aim to significantly reduce resource consumption while ensuring performance by designing a simple and efficient network architecture. Label-efficient methods expect to achieve satisfactory performance even when ground-truth annotations are insufficient or completely missing.

5.15.1 *Definition of Occupancy Network*

Given the images from multi-view cameras $\mathbf{I} = \mathbf{I}_{1,}\mathbf{I}_{2,\ldots,}\mathbf{I}_{n}$, where n is the number of camera views, model $\mathcal{F}_\theta$ needs joint estimation of the occupancy state of each voxel in the scene in range $[H_l, W_l, Z_l, H_r, W_r, Z_r]$. That is, classify the voxel into two states: empty or occupied $O_o \in \{0, 1\}^{H\times W\times Z}$, where H, W, Z denote the number of voxels along the x-, y-, z-axes, respectively, and each voxel has the shape of $\left[\frac{W_l-W_r}{W}, \frac{H_l-H_r}{H}, \frac{Z_l-Z_r}{Z}\right]$. If a voxel is predicted to be occupied, its semantic class also needs to be predicted among predefined classes $\mathbf{O}_c \in 1, 2, \ldots, C^N$, where C denotes the number of predefined categories and N the number of occupied voxels. The two outputs are combined as $\mathcal{F}_\theta(\mathbf{I}) = \mathbf{O}_o \in 0, 1, 2, \ldots, C^{H\times W\times Z}$. For each input frame, there generally provides a visibility mask $\mathbf{M} \in 0, 1^{H\times W\times Z}$, and the invisible voxels are omitted in the evaluation process. The training of a vision-based 3D occupancy perception network can be described as: to minimize a total loss $\mathcal{L}$, which is generally a classification loss, across a set of multiple-view cameras images and its corresponding annotated occupancy label $\mathbf{y} \in 0, 1, 2, \ldots, C^{H\times W\times Z}$ from the dataset $\mathcal{D}$:

$$\min_\theta \mathcal{L}\left(\mathcal{F}_\theta \odot \mathbf{M},\ \mathbf{y} \odot \mathbf{M}\right), \quad \text{where } (I, y, M) \in \mathcal{D}$$

5.15.1.1 Ground-Truth Generation of Occupancy Network

Although many 3D perception datasets (such as nuScenes and Waymo) provide LiDAR point segmentation labels, these labels are sparse and difficult for use to supervise dense 3D occupancy network tasks.

The ground-truth labels in the 3D occupancy prediction task indicate whether each voxel in the 3D space is occupied, as well as the semantic labels of the occupied voxels. Due to the huge number of voxels in the 3D space, it is very difficult to manually annotate each voxel. A common practice is to voxelize the ground truth of the existing 3D point cloud segmentation task and then generate the ground truth of the 3D occupancy prediction by voting based on the semantic labels of the voxel midpoints. However, the ground truth generated in this way is relatively sparse, for example, there are still a large number of occupied voxels in places such as roads that are not labeled as occupied. Using such sparse ground truth to supervise the model will lead to a compromise in model performance. Therefore, it is necessary to study the automatic or semi-automatic generation of high-quality dense 3D occupancy annotations.

As shown in Fig. 5.1, generating dense 3D occupancy annotations usually involves the following four steps [108]:

- Obtain continuous raw LiDAR frames and segment the LiDAR points into static background and dynamic foreground.

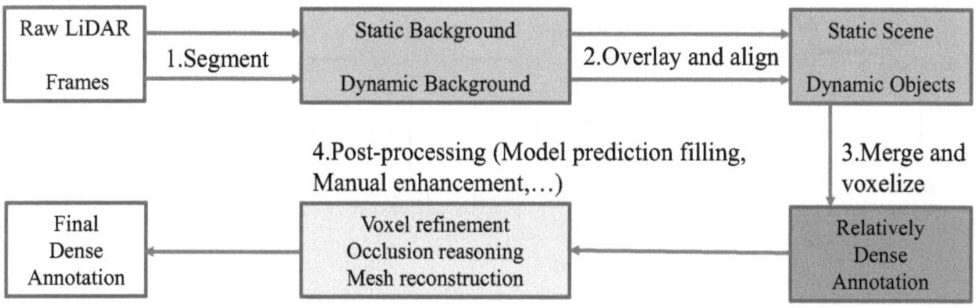

Fig. 5.1 The 3D occupancy annotation pipeline

- Overlay the continuous LiDAR frames on the static background and perform motion compensation based on the position information to align the multi-frame point clouds, which obtain a denser point cloud. Overlaying the continuous LiDAR frames on the dynamic foreground and aligning the point cloud of the dynamic foreground based on the target frame and target ID can make it denser. It should be noted that although the point cloud is relatively denser, there are still some gaps after voxelization, which require further processing.
- Merge the point clouds of the foreground and background, then voxelize them and use a voting mechanism to determine the semantics of the voxels to obtain a relatively dense voxel annotation.
- Refine the voxels obtained in the previous step through post-processing to obtain a denser and more detailed annotation as the ground truth.

5.15.1.2 The Feature Enhancement Method of Occupancy Networks

The vision-based 3D occupancy perception task involves estimating the occupancy state and semantic information of the 3D voxel space from the 2D image space, which poses a key challenge to obtain perfect 3D features from 2D visual input. From the perspective of feature enhancement, there are different methods to learn, namely, from the bird's-eye view (BEV), tri-perspective (TPV), and 3D voxel representation.

An effective method for learning occupancy is based on the bird's-eye view (BEV), which provides features that are insensitive to occlusion and contain certain depth geometry information (such as FB-Occ [110] and OccTransformer [111]). By learning a powerful BEV representation, robust 3D occupied scene reconstruction can be achieved. First, a 2D backbone network is used to extract image features from the visual input, and then the BEV features are obtained through viewpoint transformation. Finally, 3D occupancy perception is completed based on the BEV feature representation.

The FB-Occ [110] framework is shown in Fig. 5.2: FB-BEV adopts a unified design and utilizes both forward–backward projection methods to improve the perceptual effect of each method while overcoming their limitations; forward projection

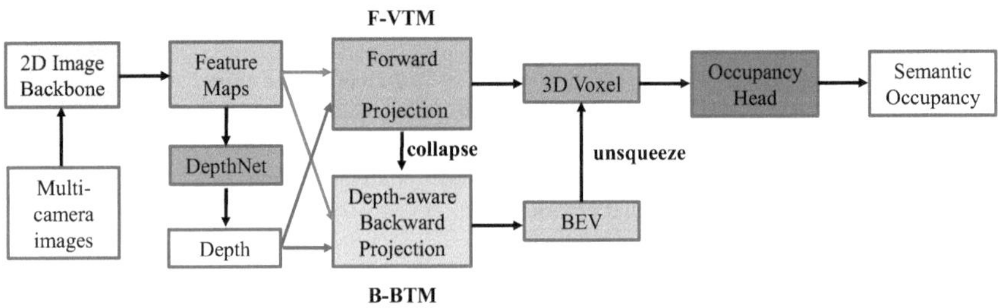

Fig. 5.2 FB-Occ framework

generates an initial 3D voxel representation, similar to Lift-Splat-Shift which adopts the uncertainty of depth estimation; then, the 3D voxel representation is compressed into a flat BEV feature map; the BEV feature map is regarded as a query of the BEV space and focuses on the image encoder features to obtain dense geometric information; the fused features of the 3D voxel representation and the optimized BEV representation are fed into the downstream task head; unlike BEVFormer which uses randomly initialized parameters as the BEV query, the backward projection method here compresses the obtained 3D voxel representation into a BEV representation, thereby considering stronger semantic priors; in addition, the backward projection method utilizes the depth distribution in the projection stage to achieve more accurate modeling of the projection relationship.

While BEV-based representations have some advantages over images because they provide a top-down projection of 3D space, they lack the ability to describe the fine-grained 3D structure of a scene using only a single plane. Tri-perspective view (TPV)-based methods use three orthogonal projection planes to model the 3D environment, further enhancing the representational power of visual features in occupancy perception (e.g., TPVFormer [112] and PointOcc [113]). First, image features are extracted from the visual input using a 2D backbone network. Subsequently, these image features are lifted to a tri-perspective space, and finally 3D occupancy perception is completed based on feature representations from three projected viewpoints.

TPVFormer [112] is shown in Fig. 5.3: The TPV representation can model the full scale of the 3D space without suppressing any axis and avoiding cubic complexity. It takes camera images as input, and the adopted TPVFormer model is trained only with sparse LiDAR semantic annotations but can effectively predict the semantic occupancy of all voxels during inference. In the TPVFormer framework, an image backbone network extracts multi-scale features from multi-camera images; then, a cross-attention mechanism is performed to adaptively promote 2D features to TPV space, and a cross-view mixed attention mechanism is used to realize the interaction between TPV planes. To predict the semantic occupancy of a 3D spatial point, a lightweight prediction head is applied after summing the projected features of the three TPV planes.

In addition to converting 3D space into projected perspective (e.g., BEV or TPV), there are also some methods that operate directly on 3D voxel representations (e.g., MonoScene [114] and VoxFormer [115]). A key advantage of these methods is that they are able to learn directly from the original 3D space, thereby minimizing information loss. By leveraging raw 3D voxel data, these methods can effectively capture and utilize complete spatial information, resulting in a more precise and comprehensive understanding of occupancy. First, a 2D backbone network is used to extract

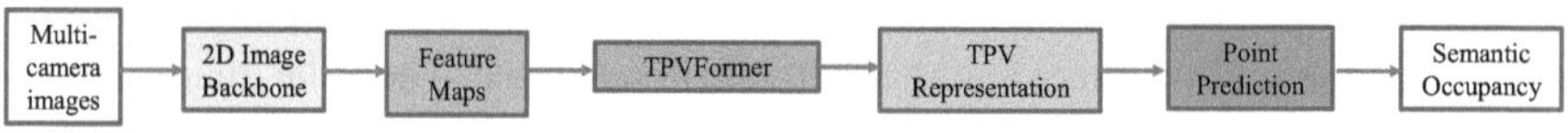

Fig. 5.3 TPVFormer

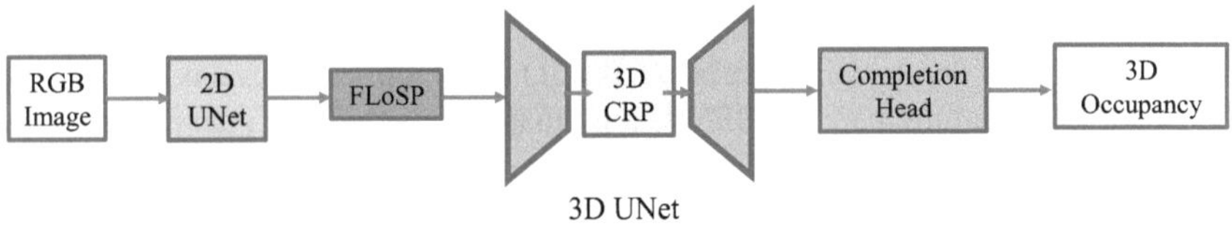

Fig. 5.4 MonoScene

image features. Then, a special convolution-based mechanism is used to bridge the 2D and 3D representations, or a query-based method is used to directly obtain the 3D representation. Finally, based on the learned 3D representation, a 3D occupancy head is used to complete the final perception.

MonoScene [114] is shown in Fig. 5.4: it infers 3D SSC (semantic scene completion) from a single RGB image, uses 2D UNet, 3D UNet, and feature lines of sight projection (FLoSP) to bridge 2D features to seemingly reasonable 3D positions, improve information flow, and achieve 2D-3D decoupling. FLoSP utilizes feature perspective projection to project 2D features onto the 3D space and calculates the coordinates of each point in the 3D feature space on the 2D feature based on the imaging principle and camera parameters for sampling the features in the 3D feature space. A 3D contextual relationship prior (3D CRP) captures long-range semantic context and enhances spatial-semantic awareness. This provides the network with a global receptive field and increases spatio-semantic awareness due to the relations discovery mechanism. Based on the standard cross-entropy, the scene-class affinity loss improves global semantics and geometric structure, while the frustum proportion loss enforces category distribution in the local frustum, providing supervision beyond occlusion.

5.15.1.3 Deployment-Friendly Methods of Occupancy Network

It is mainly divided into two categories: viewpoint decomposition and coarse-to-fine.

By separating the viewpoint information from the 3D scene features or projecting it into a unified representation space, the computational complexity can be effectively reduced, making the model more robust and more generalizable. The core idea of viewpoint decomposition (such as FlashOcc [116] and FastOcc [117]) is to decouple the representation of the 3D scene from the viewpoint information, thereby reducing the number of variables that need to be considered in the feature learning process and reducing the computational complexity. The decoupling of viewpoint information can make the model have better generalization ability and adapt to different viewpoint transformations without relearning the entire model.

The FastOcc [117] process is shown in Fig. 5.5: first, the backbone network is used to extract multi-camera features from the image input. Then, the image features are transformed into 3D space according to the Lift-Splat-Shoot strategy. The voxel features are folded into BEV form and decoded in a 2D representation. Subsequently,

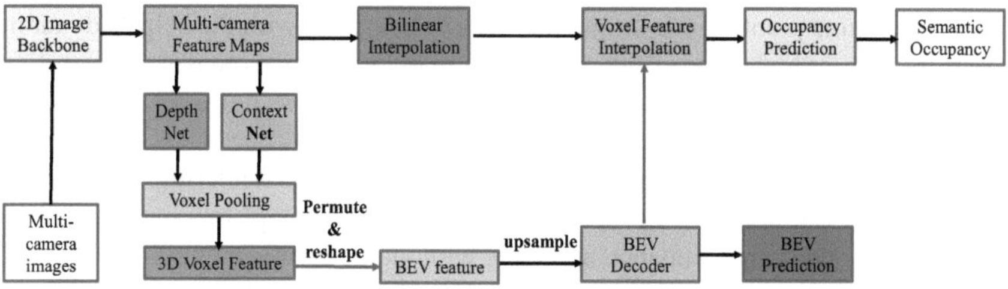

Fig. 5.5 Fast-Occ pipeline

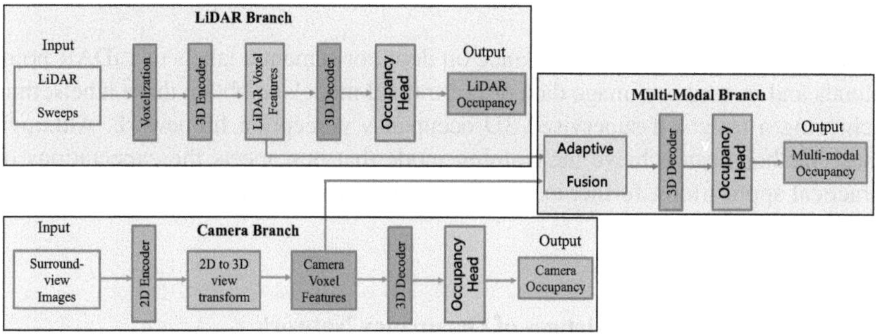

Fig. 5.6 The overall architecture of OpenOccupancy

the BEV features are upsampled, repeated, and supplemented with voxel features interpolated from image features. BEV semantic segmentation is supervised as an auxiliary loss function.

Some methods have begun to explore the coarse-to-fine feature learning paradigm (such as OpenOccupancy [118] and SurroundOcc [119]). Specifically, the network first learns a coarse representation from the image and then refines and recovers a fine-grained representation of the entire scene. These two steps help to predict the occupancy of the scene more accurately and effectively.

The overall architecture of OpenOccupancy [118] is shown in Fig. 5.6: the LiDAR branch uses a 3D encoder to extract voxelized lidar features, the camera branch uses a 2D encoder to learn surround-view features, and then transforms them to generate 3D camera voxel features; in the multimodal branch, the adaptive fusion module dynamically integrates features from two modalities. All three branches use a 3D decoder and an occupancy head to produce semantic occupancy.

5.15.1.4 The Label-Efficient Methods of Occupancy Networks

There are two basic steps in the existing methods for creating accurate occupancy labels. The first step is to collect LiDAR point clouds corresponding to multi-view images and perform semantic segmentation and annotation. The second step is to fuse multi-frame point clouds through complex algorithms with the help of dynamic target tracking information. Both steps are computationally expensive, which limits the ability of occupancy networks to utilize the massive multi-view images in autonomous driving scenarios. In recent years, neural radiance fields (NeRF) and Gaussian Splatting (GS) have been widely used for 2D image rendering. Some methods render the estimated 3D occupancy positions into 2D maps in a manner similar to NeRF/GS, and train occupancy networks without the involvement of fine-grained annotations (such as UniOcc [120], GaussianOcc [121]) or LiDAR point clouds (such as OccNeRF [122], OG-Gaussian [123]), thereby significantly reducing the cost of data annotation. For a detailed theoretical introduction to NeRF and GS, please refer to Chap. 15 of this book.

These methods bypass the reliance on depth or semantic labels of LiDAR point clouds and instead use image data or pre-trained models to obtain these labels, thus achieving a truly self-supervised 3D occupancy perception framework. Although these methods can achieve the training mode that best meets the expectations of practical applications, further exploration is still needed to achieve satisfactory performance.

5.15.1.5 The Main Challenges of Occupancy Network

To summarize, the vision-based 3D occupancy perception framework is shown in Fig. 5.7 [109]: it first extracts the front-view feature map from the multi-camera images, and then performs 2D to 3D conversion (spatio-temporal alignment) to generate 3D volume features, or TPV/BEV features to obtain the missing depth dimension in the front view; these are integrated through spatial and temporal information fusion, and finally feed into an occupancy head to infer the 3D occupancy result of the environment.

Vision-based 3D occupancy perception still faces limitations in feature representation, practical application, and annotation costs. There are three key technical challenges for this task:

- It is difficult to obtain perfect 3D features from 2D visual input. The goal of vision-based 3D occupancy perception is to achieve a detailed understanding of

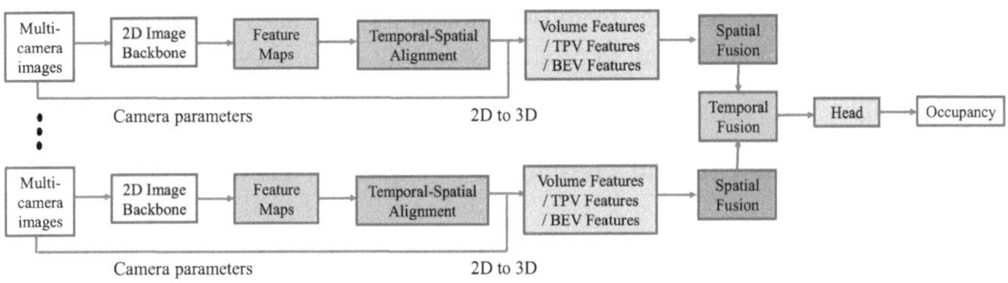

Fig. 5.7 Vision-based 3D occupancy perception network methods

the 3D scene from image input alone, but the lack of inherent depth and geometric information in images poses a major challenge to directly learning 3D feature representations from them.

- The computational load of 3D space is heavy. 3D occupancy perception usually requires the use of 3D voxel features to represent the environment space, which inevitably involves operations such as 3D convolution for feature extraction, greatly increasing computational and memory overhead and hindering practical deployment.
- Expensive fine-grained annotation. Three-dimensional occupancy perception involves estimating the occupancy state and semantic category of high-resolution voxels, but achieving this goal usually requires fine-grained semantic annotation of each voxel, which is time-consuming and expensive, becoming a bottleneck for this task.

5.16 Summary

This chapter introduces the key part of autonomous driving—the perception module, which involves the detection, tracking, and segmentation of perception targets, and analyzes the calibration, target ranging, depth estimation, and fusion tasks of the camera as the main sensor, and discusses some problems of visual sensors, such as lane line, traffic signal light, and traffic sign detection and recognition. In addition, the work of the driver monitoring system is supplemented. Finally, the latest BEV perception network and occupancy perception network are introduced. Chapter 6 will introduce the HD map technology of autonomous driving.

References

1. R ISHIKAWA et al., LiDAR and Camera Calibration using Motion Estimated by Sensor Fusion Odometry, arXiv preprint arXiv:1804.05178, 2018.
2. WANG X, XU L, SUN H, et al. "On road Vehicle Detection and Tracking Using MMW Radar and Monovision Fusion", IEEE Trans. on Intelligent Transportation Systems, 2016, 17(7):1–10.
3. SHEEHAN M., HARRISON A., and NEWMAN P., Automatic self-calibration of a full field-of-view 3D n-laser scanner, Experimental Robotics, KHATIB, O., KUMAR, V., SUKHATME, G. (eds) Experimental Robotics, Springer Tracts in Advanced Robotics, Springer, Berlin, Heidelberg, 2014, 79: 165–178.
4. D BENDER, M SCHIKORA, J STURM et al., INS-Camera Calibration without Ground Control Points, IEEE ISIF Workshop on Sensor Data Fusion: Trends, Solutions, Applications, Bonn, Germany, October, 2014:1–6.
5. G PANDEY, Automatic Targetless Extrinsic Calibration of a 3D Lidar and Camera by Maximizing Mutual Information, Proceedings of AAAI Conference on Artificial Intelligence, Toronto, Canada, July 2012: 2053–2059.

6. TAYLOR Z, NIETO J, A Mutual Information Approach to Automatic Calibration of Camera and Lidar in Natural Environments, Australasian Conference on Robotics and Automation, Wellington, New Zealand, 2012: 1–8.
7. B REDDY et al., Real-time Driver Drowsiness Detection for Embedded System Using Model Compression of Deep Neural Networks, IEEE Conference on Computer Vision and Pattern Recognition Workshops, Honolulu, HI, USA, June 2017: 121–128.
8. N SCHNEIDER et al, RegNet: Multimodal Sensor Registration Using Deep Neural Networks, arXiv preprint arXiv:1707.03167, 2017.
9. G YAN, L ZHUOCHUN, C WANG, C SHI, P WEI, X CAI,T MA, Z LIU, Z ZHONG, Y LIU, M ZHAO, Z MA and Y LI, OpenCalib: A Multi-sensor Calibration Toolbox for Autonomous Driving, arXiv preprint arXiv:2205.14087, 2022
10. J. LIN and F. ZHANG, Loam livox: A fast, robust, high-precision LiDAR odometry and mapping package for LiDARs of small FoV, arXiv preprint arXiv:1909.06700, Sep. 2019.
11. T WANG et al., Integrating Millimeter Wave Radar with a Monocular Vision Sensor for On-Road Obstacle Detection Applications, MDPI Sensors,2011, 11(9): 8992–9008.
12. G P STEIN et al., Vision-based ACC with a Single Camera: Bounds on Range and Range Rate Accuracy, IEEE Intelligent Vehicles Symposium, Columbus, OH, USA, June 2003: 120–125.
13. K Y PARK, S-Y HWANG, Range Estimation with a Monocular Camera for Vision-Based Forward Collision Warning System, The Scientific World Journal,2014, 2014(1): 1–9.
14. D F LLORCA, A H MARTÍNEZ, and I G DAZA,Vision-based Vehicle Speed Estimation: A Survey, arXiv preprint arXiv:2101.06158, 2021
15. J ZHU and Y FANG,Learning Object-Specific Distance From a Monocular Image, Int. Conf. on Computer Vision, Seoul Korea, Dec. 2019: 3839–3848.
16. A SAXENA, M SUN, A Y. NG, Make3D: Learning 3-D Scene Structure from a Single Still Image, IEEE Trans. on Pattern Analysis and Machine Intelligence, 2008, 31(5): 824–840.
17. D HOIEM, A. EFROS, M HEBERT, Automatic Photo Pop-up, ACM Transactions on Graphics, July 2005, 24(3): 577–584.
18. J KONRAD et al., 2d-to-3d image conversion by learning depth from examples, IEEE Conference on Computer Vision and Pattern Recognition Workshops, Providence, RI, USA, July 2012: 16–22.
19. K KARSCH et al., Depth Transfer: Depth Extraction from Video Using Non-parametric Sampling, IEEE Trans. on Pattern Analysis and Machine Intelligence, 2014, 36(11): 2144–2158.
20. Y LI,T CHEN,M KABKAB,R YU,L JING,Y YOU and H ZHAO, R4D: Utilizing Reference Objects For Long-range Distance Estimation, Int. Conf. Learning Representation,Virtual, 2022.
21. B LI et al., Depth and surface normal estimation from monocular images using regression on deep features and hierarchical CRFs, IEEE Conference on Computer Vision and Pattern Recognition, Boston, USA, June 2015: 1119–1127.
22. R GARG, KUMAR V, CARNEIRO G et al., Unsupervised CNN for Single View Depth Estimation: Geometry to the Rescue, European Conference on Computer Vision, Amsterdam, The Netherlands, Sep. 2016: 740–756.
23. C GODARD et al., Unsupervised Monocular Depth Estimation with Left-Right Consistency, IEEE Conference on Computer Vision and Pattern Recognition, Honolulu, HI, USA, June 2017: 6602–6611.
24. D XU et al., Multi-Scale Continuous CRFs as Sequential Deep Networks for Monocular Depth Estimation, IEEE Conference on Computer Vision and Pattern Recognition, Honolulu, HI, USA, June 2017: 161–169.
25. C WANG et al., Learning Depth from Monocular Videos using Direct Methods,IEEE Conference on Computer Vision and Pattern Recognition, Salt Lake City, UT, June 2018: 2022–2030.
26. X QI et al., GeoNet: Geometric Neural Network for Joint Depth and Surface Normal Estimation, IEEE Conference on Computer Vision and Pattern Recognition, Salt Lake City, UT, June 2018: 283–291.

27. S LEE, S IM,S LIN and I KWEON,Learning monocular depth in dynamic scenes via instance-aware projection consistency. arXiv preprint arXiv:2102.02629, 2021
28. K SAUNDERS, G VOGIATZIS and L J. MANSO,Dyna-DM: Dynamic Object-aware Self-supervised Monocular Depth Maps, arXiv preprint arXiv:2206.03799, 2022
29. X WANG, Z ZHU, G HUANG, X CHI, Y YE, Z CHEN and X WANG, Crafting Monocular Cues and Velocity Guidance for Self-Supervised Multi-Frame Depth Learning, arXiv preprint arXiv:2208.09170, 2022
30. H CHO et al., A Multi-Sensor Fusion System for Moving Object Detection and Tracking in Urban Driving Environments, IEEE Int. Conference on Robotics and Automation, Hong Kong, China, May-June 2014: 1836–1843.
31. A MOUSAVIAN et al., 3D Bounding Box Estimation Using Deep Learning and Geometry,IEEE Conference on Computer Vision and Pattern Recognition,Honolulu, HI, USA, June 2017: 7074–7082.
32. Z YIN, J SHI, GeoNet–Unsupervised Learning of Dense Depth, Optical Flow and Camera Pose, IEEE Conference on Computer Vision and Pattern Recognition, Salt Lake City, UT, June 2018: 1983–1992.
33. H-N HU et al., Joint Monocular 3D Vehicle Detection and Tracking, arXiv preprint arXiv:1811.10742, 2018
34. T RODDICK, A KENDALL, R CIPOLLA, Orthographic Feature Transform for Monocular 3D Object Detection, arXiv preprint arXiv:1811.08188, 2018
35. B XU, Z CHEN,Multi-Level Fusion based 3D Object Detection from Monocular Images, IEEE Conference on Computer Vision and Pattern Recognition, Salt Lake City, UT, June 2018: 2345–2353.
36. Z QIN,J WANG,Y LU,MonoGRNet: A Geometric Reasoning Network for Monocular 3D Object Localization, arXiv preprint arXiv:1811.10247, 2018
37. H BADINO, D HUBER, T KANADE, Integrating LIDAR into Stereo for Fast and Improved Disparity Computation, Int Conf. on 3D Imaging, Modeling, Processing, Visualization and Transmission, Washington DC, USA, May 2011:405–412.
38. Z LIU, Z WU, and R T LOTH. 3D Object Detection from Images for Autonomous Driving: A Survey, arXiv preprint arXiv:2202.02980, 2022
39. Y WANG, W-L CHAO, D GARG et al., Pseudo-LiDAR from Visual Depth Estimation: Bridging the Gap in 3D Object Detection for Autonomous Driving, IEEE Conference on Computer Vision and Pattern Recognition, Long Beach, CA, USA, June 2019: 8445–8453.
40. T HE, S SOATTO,Mono3D++: Monocular 3D Vehicle Detection with Two-Scale 3D Hypotheses and Task Priors, Proceedings of AAAI Conference on Artificial Intelligence, Honolulu, HI, USA, Feb. 2019: 8409–8416.
41. A YILMAZ, O JAVED, M SHAH, Object tracking: A survey,ACM Computing survey, 2006, 38(4): 13–23.
42. A SMEULDERS et al., Visual Tracking: An Experimental Survey, IEEE Trans. on Pattern Analysis and Machine Intelligence, 2014, 36(7): 1442–1468.
43. N WANG et al., Understanding and Diagnosing Visual Tracking Systems, IEEE Int. Conf. on Computer Vision, Santiago, Chile, Dec. 2015: 3101–3109.
44. W LUO et al., Multiple Object Tracking: A Literature Review, arXiv preprint arXiv:1409.7618, 2017.
45. P PATEL, C-H P HUANG, J TESCH et.al. AGORA: Avatars in geography optimized for regression analysis. IEEE Conference on Computer Vision and Pattern Recognition, Virtual, June 2021: 13468–13478.
46. P DAI,R WENG,W CHOI,C ZHANG,Z HE,and W DING, Learning a Proposal Classifier for Multiple Object Tracking, arXiv preprint arXiv:2103.07889, 2021
47. S GUO, J WANG, X WANG,and D TAO,Online Multiple Object Tracking with Cross-Task Synergy, arXiv preprint arXiv:2104.00380, 2021
48. J PINTO, G HESS, W LJUNGBERGH, Y XIA, H WYMEERSCH, and L SVENSSON,Can Deep Learning be Applied to Model-Based Multi-Object Tracking? arXiv preprint arXiv:2202.07909, 2022

49. Y LIU, Y YIXUAN and M LIU,Ground-aware Monocular 3D Object Detection for Autonomous Driving, arXiv preprint arXiv:2102.00690, 2, 2021
50. S KREBS et al., Survey on Leveraging Deep Neural Networks for Object Tracking, IEEE International Conference on Intelligent Transportation Systems, Yokohama, Japan, October, 2017:411–418.
51. J KU, A HARAKEH, S WASLANDER, In Defense of Classical Image Processing: Fast Depth Completion on the CPU, arXiv preprint arXiv:1802.00036, 2018
52. D FERSTL et al., Image Guided Depth Upsampling using Anisotropic Total Generalized Variation, IEEE Int. Conf. on Computer Vision, Sydney, Australia, Dec. 2013: 993–1000.
53. N SCHNEIDER et al., Semantically Guided Depth Upsampling, arXiv preprint arXiv: 1608.00753, 2016
54. H XUE et al., Depth Image Inpainting: Improving Low Rank Matrix Completion with Low Gradient Regularization, IEEE Trans. on Image Processing,2017, 26(9): 4311–4320.
55. Y. LI, Z. GE, G. YU, J. YANG, Z. WANG, Y. SHI, J. SUN, and Z. LI, BEVdepth: Acquisition of reliable depth for multi-view 3d object detection, arXiv preprint arXiv:2206.10092, 2022
56. F MA,S KARAMAN,Sparse-to-Dense: Depth Prediction from Sparse Depth Samples and a Single Image,IEEE Int. Conference on Robotics and Automation,Brisbane, QLD, Australia, May 2017: 1-8.
57. F MA et al., Self-Supervised Sparse-to-Dense: Depth Completion from LiDAR and Mono Camera, arXiv preprint arXiv:1807.00275, 7, 2018
58. Y ZHANG, T FUNKHOUSER,Deep Depth Completion of a RGB-D Image,IEEE Conference on Computer Vision and Pattern Recognition, Salt Lake City, UT, June 2018: 175–185.
59. M JARITZ et al., Sparse and Dense Data with CNNs: Depth Completion and Semantic Segmentation, arXiv preprint arXiv:1808.00769 ,8 ,2018
60. J HU, C BAO, M OZAY, C FAN, Q GAO, H LIU and T L LAM,Deep Depth Completion from Extremely Sparse Data: A Survey, arXiv preprint arXiv:2205.05335, 2022
61. Y. LIN, T. CHENG, Q. ZHONG, W. ZHOU and H. YANG,Dynamic Spatial Propagation Network for Depth Completion. Proceedings of AAAI Conference on Artificial Intelligence, Virtual, Feb. 2022: 1638–1646.
62. J QIU et al., DeepLiDAR: Deep Surface Normal Guided Depth Prediction from LiDAR and Color Image, arXiv preprint arXiv:1812.00418 ,12, 2018
63. AYCARD O, BAIG Q, BOTA S et al., Intersection Safety using Lidar and Stereo sensors, IEEE Intelligent Vehicles Symposium, Baden-Baden, Germany, June 2011: 863–869.
64. R O CHAVEZ-GARCIA, O AYCARD, Multiple Sensor Fusion and Classification for Moving Object Detection and Tracking, IEEE Trans. on Intelligent Transportation Systems,2014, 17(2): 1–10.
65. Z WANG et al., Fusing Bird's Eye View LIDAR Point Cloud and Front View Camera Image for Deep Object Detection, arXiv preprint arXiv:1711.06703, 2018
66. ABOUELNAGA Y, ERAQI H, MOUSTAFA M, Real-time Distracted Driver Posture Classification, arXiv preprint arXiv:1706.09498, 2018.
67. J ZHANG, S SINGH, Visual-lidar Odometry and Mapping: Low-drift, Robust, and Fast, IEEE Int. Conference on Robotics and Automation, Seattle, WA, USA, May 2015: 2174-2181.
68. K SHIN et al., RoarNet: A Robust 3D Object Detection based on Region Approximation Refinement, arXiv preprint arXiv:1811.03818, 2018
69. C QI et al., Frustum PointNets for 3D Object Detection from RGB-D Data, arXiv preprint arXiv:1711.08488, 2018.
70. X CHEN et al., Multi-View 3D Object Detection Network for Autonomous Driving, arXiv preprint arXiv:1611.07759, 2017.
71. K HUANG,B SHI,X LI,X LI,S HUANG,and Y LI, Multi-modal Sensor Fusion for Auto Driving Perception: A Survey, arXiv preprint arXiv:2202.02703, 2022
72. D FROSSARD, R URTASUN, End-to-end Learning of Multi-sensor 3D Tracking by Detection, arXiv preprint arXiv:1806.11534, 2018.
73. H MA et al., Multiple Lane Detection Algorithm Based on Optimized Dense Disparity Map Estimation, arXiv preprint arXiv:1808.09128 , 2018.

74. C. READING, A. HARAKEH, J. CHAE, and S. L. WASLANDER, Categorical depth distribution network for monocular 3d object detection,IEEE Conference on Computer Vision and Pattern Recognition, Virtual, June 2021: 8551–8560.
75. J KIM, C PARK, End-to-End Ego Lane Estimation based on Sequential Transfer Learning for Self-Driving Cars, IEEE Conference on Computer Vision and Pattern Recognition, Honolulu, HI, June 2017: 30–38
76. S LEE et al., VPGNet: Vanishing Point Guided Network for Lane and Road Marking Detection and Recognition, Int. Conf. on Computer Vision, Venice, Italy, Dec. 2017:1947–1955.
77. D NEVEN et al., Towards End-to-End Lane Detection: an Instance Segmentation Approach, arXiv preprint arXiv:1802.05591, 2018
78. X PAN et al., Spatial as Deep: Spatial CNN for Traffic Scene Understanding, arXiv preprint arXiv:1712.06080, 2017
79. Z WANG, W REN Q QIU, LaneNet: Real-Time Lane Detection Networks for Autonomous Driving, arXiv preprint arXiv:1807.01726, 2018
80. Y ZHANG, Z LU, X ZHANG, J-H XUE, and Q LIAO, Deep Learning in Lane Marking Detection: A Survey,IEEE Trans. on Intelligent Transportation Systems,2022, 23(7): 5976-5992.
81. J WANG,Y MA,S HUANG et al.,A Keypoint-based Global Association Network for Lane Detection, IEEE Conference on Computer Vision and Pattern Recognition, New Orleans Louisiana, USA, June 2022: 1392-1401.
82. Z ZHU et al., Traffic-Sign Detection and Classification in the Wild,IEEE Conference on Computer Vision and Pattern Recognition, Las Vegas, NV, USA, 2016: 2110-2118.
83. W LIU et al., Center and Scale Prediction: A Box-free Approach for Object Detection, arXiv preprint arXiv:1904.02948, 2019
84. S. LIU, J. JIA, S. FIDLER, R. URTASUN. SGN: Sequential grouping networks for instance segmentation. Int. Conf. on Computer Vision, Venice, Italy, Dec. 2017: 3496-3504.
85. T-J YANG et al., DeeperLab: Single-Shot Image Parser, arXiv preprint arXiv:1902.05093, 2019
86. S TSUTSUI et al., Minimizing Supervision for Free-space Segmentation, arXiv preprint arXiv:1711.05998, 2018
87. L CALTAGIRONE et al.,LIDAR-Camera Fusion for Road Detection Using Fully Convolutional Neural Networks, arXiv preprint arXiv:1809.07941 , 2018
88. M BLEYER et al.,PatchMatch Stereo - Stereo Matching with Slanted Support Windows,British Machine Vision Conference, U. of Dundee, UK, Sept. 2011:14.1–14.11.
89. F YAN, M NIE, X CAI,et al.,ONCE-3DLanes: Building Monocular 3D Lane Detection,IEEE Conference on Computer Vision and Pattern Recognition, New Orleans Louisiana, USA, June 2022: 17143–17152.
90. H LAGA, A Survey on Deep Learning Architectures for Image-based Depth Reconstruction, arXiv preprint arXiv:1906.06113, 2019
91. WANG B, FREMONT C, FLOREZ S,Multiple Obstacle Detection and Tracking using Stereo Vision: Application and Analysis, Int. Conf. on Control Automation Robotics & Vision, Singapore, Dec. 2014: 1074–1079.
92. A KENDALL et al., End-to-End Learning of Geometry and Context for Deep Stereo Regression, IEEE Int. Conf. on Computer Vision, Venice, Italy, Dec. 2017: 66–75.
93. C ZHOU et al., Unsupervised Learning of Stereo Matching, IEEE Int. Conf. on Computer Vision, Venice, Italy, Dec. 2017: 1576–1584.
94. H LAGA, L V JOSPIN, F BOUSSAID, and M BENNAMOUN,A Survey on Deep Learning Techniques for Stereo-based Depth Estimation, arXiv preprint arXiv:2006.02535, 2020
95. G XU, J CHENG, P GUO, and X YANG,Attention Concatenation Volume for Accurate and Efficient Stereo Matching,IEEE Conference on Computer Vision and Pattern Recognition, New Orleans Louisiana, USA, June 2022: 12981-12990.
96. C ZHENG, W WU, T YANG, S ZHU, C CHEN, R LIU, J SHEN, N KEHTARNAVAZ and M SHAH, Deep Learning-Based Human Pose Estimation: A Survey, arXiv preprint arXiv:2012.13392, 2021

97. W LI,H LIU,H TANG,P WANG,and L VAN GOOL,MHFormer: Multi-Hypothesis Transformer for 3D Human Pose Estimation, arXiv preprint arXiv:2111.12707, 2021
98. P DAI, Y FENG, R WENG, and C ZHANG,Joint Spatial-Temporal and Appearance Modeling with Transformer for Multiple Object Tracking, arXiv preprint arXiv:2205.15495, 2022
99. T WANG et al., Integrating Millimeter Wave Radar with a Monocular Vision Sensor for On-Road Obstacle Detection, MDPI Sensors, 2011, 11(9): 8992–9008.
100. A HALIN, J G. VERLY and M V DROOGENBROECK, Survey and synthesis of state of the art in driver monitoring, arXiv preprint arXiv:2110.00472, 2021
101. S BANERJEE, A JOSHI, J TURCOT, B REIMER, and T MISHRA, Driver Glance Classification In-the-wild: Towards Generalization Across Domains and Subjects, arXiv preprint arXiv:2012.02906, 2021
102. G KIM, H KIM, K KIM, S-S CHO, Y-H PARK, and S-J KANG, Integrated In-vehicle Monitoring System Using 3D Human Pose Estimation and Seat Belt Segmentation, arXiv preprint arXiv:2204.07946, 2022
103. G IYER et al., CalibNet: Self-Supervised Extrinsic Calibration using 3D Spatial Transformer Networks, arXiv preprint arXiv:1803.08181, 2018
104. Y MA, T WANG, X BAI, H YANG, Y HOU, Y WANG, Y QIAO, R YANG, D MANOCHA, and X ZHU, Vision-Centric BEV Perception: A Survey, arXiv preprint arXiv:2208.02797, 2022
105. L Mescheder et al., "Occupancy Networks: Learning 3D Reconstruction in Function Space", arXiv:1812.03828, 2018
106. L. Peng, J. Xu, H. Cheng, et al., "Learning Occupancy for Monocular 3d Object Detection", arXiv:2305.15694, 2023.
107. Ashuk Elluswamy, "Occupancy Networks" (Tesla), Workshop on Autonomous Driving,in conjunction with CVPR, https://cvpr2022.wad.vision, 2022
108. Y Zhang et al., "Vision-based 3D occupancy prediction in autonomous driving: a review and outlook", arXiv:2405.02595, 2024
109. H Xu et al., "A Survey on Occupancy Perception for Autonomous Driving: The Information Fusion Perspective", arXiv :2405.05173, 2024
110. Z Li, Z Yu, D Austin, et al. "FB-Occ: 3d occupancy prediction based on forward-backward view transformation". arXiv:2307.01492, 2023
111. J Liu, S Zhang, C Kong, et al., "OccTransformer: Improving BEVFormer for 3D camera-only occupancy prediction". arXiv:2402.18140, 2024
112. Y Li, Z Yu, C Choy, et al. "VoxFormer: Sparse voxel transformer for camera-based 3d semantic scene completion". IEEE/CVF CVPR, 2023
113. Y Huang, W Zheng, Y Zhang Y, et al. "Tri-Perspective view for vision-based 3d semantic occupancy prediction" (TPVFormer). IEEE/CVF CVPR, 2023
114. A Q Cao, D R Charette. "MonoScene: Monocular 3d semantic scene completion". I IEEE/CVF CVPR. 2022
115. S Zuo, W Zheng, Y Huang Y, et al., "PointOcc: Cylindrical tri- perspective view for point-based 3d semantic occupancy prediction". arXiv:2308.16896, 2023
116. Z Yu, C Shu, J Deng, et al. "FlashOcc: Fast and memory-efficient occupancy prediction via channel-to-height plugin". arXiv:2311.12058, 2023
117. J Hou, X Li, W Guan, and et al. "FastOcc: Accelerating 3d occupancy prediction by fusing the 2d bird's-eye view and perspective view". arXiv preprint arXiv:2403.02710, 2024
118. X Wang, Z Zhu, W Xu, and et al. "OpenOccupancy: A large scale benchmark for surrounding semantic occupancy perception". arXiv:2303.03991, 2023
119. Y Wei, L Zhao, W Zheng, et al., "SurroundOcc: Multi- camera 3d occupancy prediction for autonomous driving". ICCV. 2023
120. M Pan, L Liu, J Liu, and et al., "UniOcc: Unifying vision-centric 3d occupancy prediction with geometric and semantic rendering". arXiv:2306.09117, 2023
121. W Gan, F Liu, H Xu, and et al., "GaussianOcc: Fully Self-supervised and Efficient 3D Occupancy Estimation with Gaussian Splatting", arXiv:2408.11477, 2024

122. C Zhang, J Yan, Y Wei, and et al. "OccNeRF: Self-supervised multi-camera occupancy prediction with neural radiance fields". arXiv:2312.09243, 2023
123. Y Shen, X Zhang, Y Duan, et al., "OG-Gaussian: Occupancy Based Street Gaussians for Autonomous Driving", arXiv: 2502.14235, 2025
124. J LEVINSON, S THRUN, Automatic Online Calibration of Cameras and Lasers, Conference on Robotics: Science and Systems, Stanford, California, June, 2013
125. B LI et al., Pitch Angle Estimation Using a Vehicle Mounted Monocular Camera for Vehicle Target Range Measurement, The Journal of The Institute of Image Information and Television Engineers, 2015, 69(4): 169–176.
126. M BAIG et al., Im2Depth: Scalable Exemplar Based Depth Transfer, IEEE Winter Conference on Applications of Computer Vision, Steamboat Springs, CO, USA, March 2014: 145–152.
127. D EIGEN, R FERGUS, Predicting Depth, Surface Normals and Semantic Labels with a Common Multi-Scale Convolutional Architecture, IEEE Int. Conf. on Computer Vision, Santiago, Chile, Dec. 2015: 2550–2558.
128. WANG P, SHEN X, LIN Z et al., Towards Unified Depth and Semantic Prediction From a Single Image, IEEE Conference on Computer Vision and Pattern Recognition, Boston, USA, June 2015: 2800–2809.
129. F LIU et al., Learning Depth from Single Monocular Images Using Deep Convolutional Neural Fields, IEEE Trans. on Pattern Analysis and Machine Intelligence, Oct. 2016 38(10): 2024–2039.
130. WANG P, SHEN X, RUSSELL B et al., SURGE: Surface Regularized Geometry Estimation from a Single Image, Proceedings of International Conference on Neural Information Processing Systems, December 2016: 172–180.
131. B LI et al., Single image depth estimation by dilated deep residual convolutional neural network and soft-weight-sum inference, arXiv preprint arXiv:1705.00534, 2017
132. Y KUZNIETSOV et al., Semi-Supervised Deep Learning for Monocular Depth Map Prediction, IEEE Conference on Computer Vision and Pattern Recognition, Honolulu, HI, USA, June 2017: 6647–6655.
133. T ZHOU et al., Unsupervised Learning of Depth and Ego-Motion from Video,IEEE Conference on Computer Vision and Pattern Recognition, Honolulu, HI, USA, June 2017: 6612–6619.
134. FU H, GONG M, WANG C et al., Deep Ordinal Regression Network for Monocular Depth Estimation, arXiv preprint arXiv:1806.02446 , June 2018
135. X LI et al., A Survey of Appearance Models in Visual Object Tracking, arXiv preprint arXiv:1303.4803, 2013.
136. Y WU, J LIM, M-H YANG, Online Object Tracking: A Benchmark, IEEE Conference on Computer Vision and Pattern Recognition, Portland, OR, USA, June 2013: 2411–2418.
137. B-N VO et al., Multi-target Tracking, Wiley Encyclopedia of Electrical and Electronics Engineering, edited by WEBSTER J. G., Sept. 2015
138. J UHRIG, SCHNEIDER N, SCHNEIDER L et al., Sparsity Invariant CNNs, Int. Conf. on 3D Vision (3DV),Qingdao, China, August, 2017:11–20
139. R FAN, N DAHNOUN, "Real-Time Stereo Vision-Based Lane Detection System", arXiv preprint arXiv:1807.02752, 2018.
140. N GARNETT et al., 3D-LaneNet: E2E 3D multiple lane detection, arXiv preprint arXiv:1811.10203, 2018
141. A MØGELMOSE, M TRIVEDI, T MOESLUND, Vision-Based Traffic Sign Detection and Analysis for Intelligent Driver Assistance Systems: Perspectives and Survey, IEEE Trans. on Intelligent Transportation Systems, 4, 2012, 13(4): 1484–1497.
142. M B. JENSEN et al., Vision for Looking at Traffic Lights: Issues, Survey, and Perspectives, IEEE Trans. on Intelligent Transportation Systems, July, 2016, 17(7): 1–16.
143. LEVINSON J, ASKELAND J, DOLSON J et al., Traffic Light Mapping, Localization, and State Detection for Autonomous Vehicles, IEEE Int. Conference on Robotics and Automation, Shanghai, China, May 2011: 5784–5791.

144. BEHRENDT K, NOVAK L, BOSTROS R, A Deep Learning Approach to Traffic Lights: Detection, Tracking, and Classification,IEEE Int. Conference on Robotics and Automation,May-June,Singapore, 2017: 1370–1377.
145. SERMANET P., EIGEN D, ZHANG X, et al., Overfeat: Integrated recognition, localization and detection using convolutional networks. arXiv preprint arXiv:1312.6229, 2014
146. R. GIRSHICK et al., Region based convolutional networks for accurate object detection and segmentation. IEEE Trans. on Pattern Analysis and Machine Intelligence, 2015, 38(1): 142–158.
147. R. GIRSHICK. Fast R-CNN. IEEE Int. Conf. on Computer Vision, Santiago Chile, Dec. 2015: 1440–1448.
148. REN S., HE K, GIRSHICK R, SUN J., Faster R-CNN: Towards real-time object detection with region proposal networks. IEEE Trans. on Pattern Analysis and Machine Intelligence, June 2017, 39(6): 1130–1149.
149. W. LIU et al., SSD: Single shot multi-box detector. arXiv preprint arXiv:1512.02325, 2015.
150. C.-Y. FU et al., DSSD: Deconvolutional single shot detector. arXiv preprint arXiv:1701.06659, 2016
151. J. REDMON et al., You only look once: Unified, real-time object detection. IEEE Conference on Computer Vision and Pattern Recognition, Las Vegas, NV, USA, June 2016: 779–788.
152. T.-Y. LIN et al., Feature pyramid networks for object detection. IEEE Conference on Computer Vision and Pattern Recognition, Honolulu, HI, USA, June 2017: 2117-2125.
153. J. REDMON, A. FARHADI, YOLO9000: Better, faster, stronger. IEEE Conference on Computer Vision and Pattern Recognition, Honolulu, HI, USA, June 2017: 7263–7271.
154. T-Y LIN et al., Focal Loss for Dense Object Detection, arXiv preprint arXiv:1708.02002, 2017.
155. X. Wang et al., A-Fast-RCNN: Hard Positive Generation via Adversary for Object Detection, arXiv:1704.03414, 2017.
156. J REDMON, A FARHADI, YOLOv3: An Incremental Improvement, arXiv preprint arXiv:1804.02767, 2018
157. H LAW, J DENG, CornerNet: Detecting Objects as Paired Keypoints, arXiv preprint arXiv:1808.01244, 2018
158. X ZHOU et al., Bottom-up Object Detection by Grouping Extreme and Center Points, arXiv preprint arXiv:1901.08043, 2019
159. C ZHU, Y HE, M SAVVIDES. Feature Selective Anchor-Free Module for Single-Shot Object Detection. arXiv preprint arXiv:1903.00621, 2019.
160. Z TIAN et al., FCOS: Fully Convolutional One-Stage Object Detection. arXiv preprint arXiv:1904.01355, 2019.
161. K TAO et al., FoveaBox: Beyond Anchor-based Object Detector. arXiv preprint arXiv: 1904.03797, 2019.
162. J WANG et al., Region proposal by guided anchoring. arXiv preprint arXiv:1901.03278, 2019
163. X ZHOU, D WANG, P KRÄHENBÜHL, Objects as Points, arXiv preprint arXiv:1904.07850, 2019
164. K DUAN et al., CenterNet: Object Detection with Keypoint Triplets, arXiv preprint arXiv:1904.08189, 2019
165. H LAW et al., CornerNet-Lite: Efficient Keypoint Based Object Detection, arXiv preprint arXiv:1904.08900, 2019
166. A GARCIA et al., A Review on Deep Learning Techniques Applied to Semantic Segmentation, arXiv preprint arXiv:1704.06857, 2017
167. J. LONG, E. SHELHAMER, T. DARRELL. Fully convolutional networks for semantic segmentation. IEEE Conf. Computer Vision and Pattern Recognition, Boston, USA, June 2015: 640–651.
168. K. HE, G. GKIOXARI, P. DOLLA R, R. GIRSHICK. Mask R-CNN. IEEE Int. Conf. on Computer Vision, Venice, Italy, Dec. 2017: 386–397.
169. A KIRILLOV, K. HE, R. GIRSHICK, C. ROTHER, P. DOLLA R. Panoptic segmentation. arXiv preprint arXiv:1801.00868, 2018.

170. D SCHARSTEIN,R SZELISKI, A Taxonomy and Evaluation of Dense Two-Frame Stereo Correspondence Algorithms,Int. J. Computer Vision, 2002, 47(3): 7–42.
171. H. HIRSCHMULLER, Stereo processing by semiglobal matching and mutual information, IEEE Trans. on Pattern Analysis and Machine Intelligence, Feb. 2008, 30(2): 328–341.
172. R. ZABIH, J. WOODFILL, Non-parametric Local Transforms for Computing Visual Correspondence, European Conf. on Computer Vision, Stockholm, Sweden, May 1994: 151–158.
173. XU D, ZENG Q, ZHAO H et al., Online Stereovision Calibration Using On-road Markings, IEEE International Conference on Intelligent Transportation Systems,Qingdao, China, October 2014: 245–252.
174. L HENG et al., Self-Calibration and Visual SLAM with a Multi-Camera System on a Micro Aerial Vehicle,Autonomous Robotics,2015, 39(3): 259–277
175. S NEDEVSCHI et al., Online Extrinsic Parameters Calibration for Stereovision Systems Used in Far-Range Detection Vehicle Applications, IEEE Trans. on Intelligent Transportation Systems,2007, 84(4): 651-660.
176. J ZIEGLER et al., Making Bertha Drive—An Autonomous Journey on a Historic Route,IEEE Intelligent Transportation Systems Magazine, 2014, 6(2): 8-20.
177. H BADINO, U FRANKE, D PFEIFFER. The Stixel world—a compact medium level representation of the 3d-world. DAGM (German Annual Symposium of Pattern Recognition), Jena, Germany, September 2009: 51–60.
178. S RAMOS et al.,Detecting Unexpected Obstacles for Self-Driving Cars: Fusing Deep Learning and Geometric Modeling, arXiv preprint arXiv:1612.06573, 2016
179. M. Bertozzi and A. Broggi, GOLD: A parallel real-time stereo vision system for generic obstacle and lane detection, IEEE Transactions on Image Processing, vol. 7, no. 1, pp. 62–81, 1998.
180. A BROGGI et al.,Stereo obstacle detection in challenging environments: the VIAC experience, IEEE Int. Conf. on Intelligent Robots and Systems, San Francisco, USA, Sept. 2011: 1599–1604.
181. R MANDUCHI et al.,Obstacle Detection and Terrain Classification for Autonomous Off-Road Navigation,Autonomous Robots,2005, 18(1): 81–102.
182. J R CHANG, Y S CHEN, Pyramid Stereo Matching Network, arXiv preprint arXiv: 1803.08669, 2018.
183. B RAJ, An Overview of Human Pose Estimation with Deep Learning.
184. Z CAO,G HIDALGO, T SIMON,S WEI,Y SHEIKH, OpenPose: real time multi-person 2D pose estimation using Part Affinity Fields, arXiv:1812.08008, 2018
185. R GUELER et al., DensePose: Dense Human Pose Estimation In The Wild, arXiv preprint arXiv:1802.00434, 2018
186. R GUELER et al., DenseReg: Fully convolutional dense shape regression in-the-wild, arXiv preprint arXiv:1612.01202, 2017
187. S ZAFEIRIOU, C ZHANG, Z ZHANG,A Survey on Face Detection in the wild: past, present and future, Computer Vision and Image Understanding,April 2015, 138(C): 1–24.
188. B KO,A Brief Review of Facial Emotion Recognition Based on Visual Information, MDPI Sensors, 2018, 18(2): 401–420.
189. H LI, C SIMA, J DAI, et.al, Delving into the Devils of BEV: A Review, Evaluation and Recipe, arXiv preprint arXiv:2209.05324, 2022
190. B PAN, J SUN, A ANDONIAN, A OLIVA, and B ZHOU. Cross-view semantic segmentation for sensing surroundings. arXiv preprint arXiv:1906.03560, 2019
191. K. MANI, S. DAGA, S. GARG, S. S. NARASIMHAN, M. KRISHNA, and K. M. JATAVALLABHULA, Monolayout: Amodal scene layout from a single image, IEEE Winter Conf. on Applications of Computer Vision, Snowmass CO, USA, 2020: 1689–1697.
192. N. HENDY, C. SLOAN, F. TIAN, P. DUAN, N. CHARCHUT, Y. XIE, C. WANG, and J. PHILBIN, Fishing Net: Future inference of semantic heatmaps in grids, arXiv preprint arXiv:2006.09917, 2020.
193. Q. LI, Y. WANG, Y. WANG, and H. ZHAO, HDMapNet: An online HD map construction and evaluation framework, arXiv preprint arXiv:2107.06307, 2021.

194. J PHILION and S. FIDLER, Lift, splat, shoot: Encoding images from arbitrary camera rigs by implicitly unprojecting to 3D, European Conference on Computer Vision, Glasgow, UK, Nov. 2020: 194-210.
195. A HU, Z. MUREZ, N. MOHAN, S. DUDAS, J. HAWKE, V. BADRINARAYANAN, R. CIPOLLA, and A. KENDALL, FIERY: Future instance prediction in BEV from surround monocular cameras, IEEE Int. Conf. on Computer Vision, Virtual, Dec. 2021: 15273–15282.
196. Y. B. CAN, A. LINIGER, D. P. PAUDEL, and L. V. GOOL, Structured Bird-Eye-View traffic scene understanding from onboard images, IEEE Int. Conf. on Computer Vision, Virtual, Dec. 2021: 15661–15670.
197. N. CARION, F. MASSA, G. SYNNAEVE, N. USUNIER, A. KIRILLOV, and S. ZAGORUYKO. End-to-end object detection with transformers. European Conference on Computer Vision, Glasgow, UK, Nov. 2020: 213–229.
198. WANG, V. C. GUIZILINI, T. ZHANG, Y. WANG, H. ZHAO, and J. SOLOMON, DETR3D: 3d object detection from multi-view images via 3d-to-2d queries, arXiv preprint arXiv:2110.06922, Conference on Robot Learning, 2022: 180–191.
199. Y. LIU, T. WANG, X. ZHANG, and J. SUN, PETR: Position embedding transformation for multi-view 3d object detection, arXiv preprint arXiv:2203.05625, 2022
200. Y. LIU, J. YAN, F. JIA, S. LI, Q. GAO, T. WANG, X. ZHANG, and J. SUN, PETRv2: A unified framework for 3d perception from multi-camera images, arXiv preprint arXiv:2206.01256, 2022.
201. J. HUANG, G. HUANG, Z. ZHU, and D. DU, BEVDet: High-performance multi-camera 3d object detection in BEV, arXiv preprint arXiv:2112.11790, 2021.
202. Z. LIU, H. TANG, A. AMINI, X. YANG, H. MAO, D. RUS, and S. HAN, BEVfusion: Multi-task multi-sensor fusion with unified BEV representation, arXiv preprint arXiv:2205.13542, 2022.
203. Z. LI, W. WANG, H. LI, E. XIE, C. SIMA, T. LU, Y. QIAO, and J. DAI, BEVformer: Learning BEV representation from multi-camera images via spatiotemporal transformers, arXiv preprint arXiv:2203.17270, 2022.
204. Y. ZHANG, Z. ZHU, W. ZHENG, J. HUANG, G. HUANG, J. ZHOU, and J. LU, BEV-erse: Unified perception and prediction in BEV for vision-centric autonomous driving, arXiv preprint arXiv:2205.09743, 2022.
205. E. XIE, Z. YU, D. ZHOU, J. PHILION, A. ANANDKUMAR, S. FIDLER, P. LUO, and J. M. ALVAREZ, M2BEV: Multi-camera joint 3d detection and segmentation with unified BEV representation, arXiv preprint arXiv:2204.05088, 2022.
206. D. NAZIR, A. PAGANI, M. LIWICKI, D. STRICKER and M. AFZAL, SemAttNet: Towards Attention-based Semantic Aware Guided Depth Completion. arXiv preprint arXiv:2204.13635, 2022.
207. Z. LIU and F. ZHANG, Balm: Bundle adjustment for lidar mapping, IEEE Robotics and Automation Letters, 2021, 6(2): 3184–3191.
208. C ZHAO, Q SUN, C ZHANG, Y TANG and F QIAN, Monocular Depth Estimation Based On Deep Learning: An Overview, arXiv preprint arXiv:2003.06620, 2020
209. W ZIMMER, E ERCELIK , X ZHOU , X J D ORTIZ, and A KNOLL,A Survey of Robust LiDAR-based 3D Object Detection Methods for Autonomous Driving, arXiv preprint arXiv:2204.00106, 2022
210. A H. LANG, S. VORA, H. CAESAR, L. ZHOU, J. YANG, and O. BEIJBOM, PointPillars: Fast Encoders for Object Detection From Point Clouds, IEEE Conference on Computer Vision and Pattern Recognition,Long Beach, CA, USA, June 2019: 12697–12705.
211. C R. QI,H SU,K MO,and L J. GUIBAS,PointNet: Deep Learning on Point Sets for 3D Classification and Segmentation, arXiv preprint arXiv:1612.00593, 2016
212. T. YIN, X. ZHOU, and P. KRAHENBUHL, Center-based 3d object detection and tracking, IEEE Conference on Computer Vision and Pattern Recognition, Virtual, June 2021: 11784–11793.
213. Y LI,A W YU,T MENG,B CAINE,J NGIAM,D PENG, J SHEN,B WU,Y LU,D ZHOU,Q LE,A YUILLE,and M TAN,DeepFusion: Lidar-Camera Deep Fusion for Multi-Modal 3D Object Detection, arXiv preprint arXiv:2203.08195, 2022

214. Y LI,X QI,Y CHEN,L WANG,Z LI,J SUN,and J JIA, Voxel Field Fusion for 3D Object Detection, IEEE Conference on Computer Vision and Pattern Recognition, New Orleans Louisiana, USA, June 2022: 1120–1129.
215. C YAO, Y JIA, H DI, P LI, and Y WU, A Decomposition Model for Stereo Matching, arXiv preprint arXiv:2104.07516, 2021
216. Y SUN,W LIU,Q BAO,Y FU,T MEI,and M J. BLACK,BEV: Putting People in their Place: Monocular Regression of 3D People in Depth, arXiv preprint arXiv:2112.08274, 2021
217. M LOPER, N MAHMOOD, J ROMERO, G PONSMOLL, and M J. BLACK. SMPL: A skinned multi- person linear model. ACM Trans. on Graphics, 2015, 34(6): 1–16.
218. N HESSE, S PUJADES, J ROMERO, et al., Learning an infant body model from RGB-D data for accurate full body motion analysis. In. Conf. on Medical Image Computing and Computer-Assisted Intervention, Granada, Spain, Sept. 2018: 792–800.
219. Y MA, V SANCHEZ, S NIKAN, D UPADHYAY, B ATOTE and T GUHA, Real-Time Driver Monitoring Systems through Modality and View Analysis, arXiv preprint arXiv:2210.09441, 2022
220. Z LIU et al., Swin Transformer: Hierarchical Vision Transformer using Shifted Windows, arXiv preprint arXiv:2103.14030, 2021.
221. W Zheng, W Chen, Y Huang, et al., "OccWorld: Learning a 3D Occupancy World Model for Autonomous Driving", arXiv:2311.16038, 2023

Chapter 6
High-Definition Map of Autonomous Driving

The concept of HD maps was first introduced by Mercedes-Benz in 2010 and then contributed to the Bertha Drive project in 2013. In the Bertha Drive project, the Mercedes-Benz S500 completed the Bertha-Benz Memorial Route in a fully autonomous mode, using a highly accurate and information-rich 3D route map, which was later named by a map company called HERE that participated in the project as an HD real-time map. HD maps are highly valued in the field of autonomous driving because they can provide prior knowledge, reduce the difficulty of perception, and make planning targeted.

HD maps are not new, and some map and autonomous driving companies have started to produce and use HD maps. However, the production of HD maps and the information and accuracy they contain still have no unified consensus.

This chapter mainly introduces the content related to HD maps, where Sect. 6.1 introduces the basic types, levels, formats, and production methods of HD maps, Sect. 6.2 focuses on various generation methods of semantic maps, Sect. 6.3 emphasizes the production methods of lane-level HD maps, and Sect. 6.4 introduces SLAM methods based on deep learning.

6.1 HD Maps

Unlike previous navigation maps, autonomous driving requires redefined HD maps to locate vehicles. To describe this driving environment, these HD maps need to achieve a centimeter-level accuracy that has never been achieved before, while the accuracy of general navigation maps is at the 10-meter level.

HD maps store all the key static features of the road and surrounding environment that are essential for autonomous driving, such as roads, buildings, traffic lights, and traffic signs. They also include semantic elements that may not be directly

Y. Huang and Z. Yang, *Autonomous Driving Handbook*,
https://doi.org/10.1007/978-981-95-3725-9_6

observable by onboard sensors due to occlusion. In recent years, HD maps have become a defining component of autonomous driving technology, noted for their high accuracy as well as their rich geometric and semantic content. Closely tied to vehicle localization, they interact continuously with sensors including lidar, radar, and cameras, thereby serving as an integral part of the perception module in autonomous driving systems.

The basic principles of HD maps are as follows.

- Pre-computation (mapping as pre-computation): It is completed offline.
- Safety (mapping to improve safety): Supports rules for safe driving, such as traffic rules.
- Sensor (map as a unique sensor): Extension of perception, super vision of prediction.
- Global Sharing (map as global shared state): Data can be exchanged in the cloud to achieve real-time sharing.

Figure 6.1 is the layered structure of HD maps given by the American ride-hailing company Lyft [22], and the content and role of each layer of HD maps are introduced one by one below.

(1) Geometric Map Layer: Mainly 3D information The original sensor data collected during mapping–such as lidar, cameras, GPS, and IMU–are processed by a SLAM algorithm to reconstruct the 3D structure of the observed area. Simultaneously, the SLAM output is aligned with the 3D point cloud to recover the vehicle's motion trajectory. The resulting point clouds are then further processed to extract various elements for storage in the geometric map. Among these, two key components

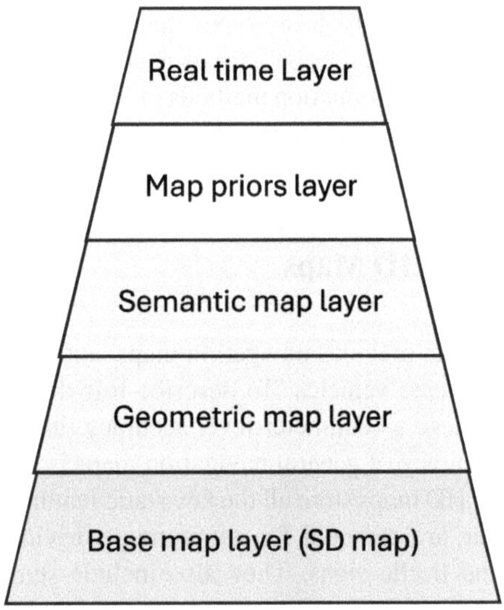

Fig. 6.1 Layered structure of HD maps from Lyft

are the voxelized geometric map and the ground map. The ground map is particularly important, as it provides the foundation for aligning additional maps in later stages, such as the semantic map.

(2) Semantic Map Layer: Adding semantic objects to the geometric map.

Semantic targets include a wide range of 2D and 3D traffic elements, such as lane boundaries, intersections, crosswalks, parking spaces, stop signs, and traffic lights. These targets are enriched with metadata, including speed limits and lane-turning restrictions. Each semantic target has a defined position and bounding box, which can typically be extracted using object detection algorithms to generate semantic hypotheses. However, these automated outputs alone are insufficient for producing high-precision maps. To ensure accuracy and reliability, manual post-processing with visualization and annotation tools is required to validate quality and correct errors.

However, high-definition maps that only include the static and physical parts provided by the geometric map and semantic map layers are not enough, sometimes the dynamic and behavioral parts of the environment are also important.

(3) Map Priors Layer: Includes information about dynamic elements, even driving behavior.

The information in the map priors layer involves the semantic map layer and the geometric map layer. For example, the change order and time interval of traffic lights, and the probability of parking on the lanes in the parking lot.

Unlike the geometric map layer and the semantic map layer, the map priors layer mainly plays a suggestive role.

(4) Real-time Knowledge Layer: At the top of the high-definition map, readable and writable, updatable.

The real-time knowledge layer includes real-time traffic information, such as speed, congestion, and newly discovered construction zones. The main function of the real-time knowledge layer is to support autonomous vehicles in real-time collection and exchange of global information.

The layers of the above high-definition map are made independently. They need to be calibrated together before being provided to unmanned vehicles.

There are currently two paths for high-definition maps, namely the high-cost mapping method mainly based on lidar scanning and the low-cost mapping method mainly based on cameras. Large companies generally adopt high-cost methods, such as HERE, TomTom, Google, Apple, Baidu, Gaode, Siwei Map News and ZENRIN, etc., and some start-ups also follow this route, such as DeepMap, Civil Maps and Carmera, etc.

Reference [1] provides a summary of the methods for making high-definition maps.

Data acquisition/collection is the first step in generating high-definition maps. Data collection is done with a mobile mapping system (MMS). The MMS is a mobile vehicle equipped with map sensors, including GNSS (Global Navigation Satellite System), inertial IMU, lidar, cameras, and radar, used to collect geospatial data. Commercial high-definition map providers use crowdsourcing technology to collect data to build and maintain high-definition maps.

Once the initial sensor data is collected, it is usually fused and sorted to generate an initial map, mainly used for precise localization. The initial map is mainly generated using 3D laser sensors. However, it can be fused with other sensors, such as IMU, GPS, odometer, and visual odometer, to make more accurate state estimates in the high-definition map. INS and GPS sensors provide direction and position information, they update the map position within the range of centimeter accuracy.

These point cloud maps have high accuracy and can help ground vehicles achieve centimeter-level precise localization in 3D space. After calibrating the map with point clouds, a vector map will be created from the PCL map. Point cloud calibration aligns multiple overlapping point clouds to generate a detailed and accurate map.

The vector map contains information related to lanes, sidewalks, intersections, roads, intersections, traffic signs, and traffic lights. This key feature is later used for detecting traffic signs and traffic lights, route planning, global planning, and local path planning. Undoubtedly, map generation is a component of high-definition map production and can be defined as the basic geometric map layer of a high-definition map.

Map generation technology can be divided into online maps and offline maps. Offline map data is all collected at a central location. These data use satellite information or data stored by lidar/cameras. After collecting the data, the map will be generated offline. In addition, lightweight modules in online maps automatically generate maps. In addition to the type of map generation, map drawing technology can also be classified by sensor or sensor fusion method.

In addition to using the NMS (Non-Maximum Suppression) system to collect data to make high-definition maps, there are other methods that can be used, such as 2D aerial imagery and satellite images.

The following takes the Japanese map company ZENRIN as an example to introduce the basic process of lidar mapping [23]. Figure 6.2 shows the flowchart for making high-definition maps, inputting vehicle position and posture, as well as point data and images, and completing point cloud calibration, point cloud-image calibration, and map feature extraction in a GPU-accelerated environment, and finally obtaining high-definition maps through annotation. Among them, the image-based deep learning model can support detection classification tasks in a GPU environment.

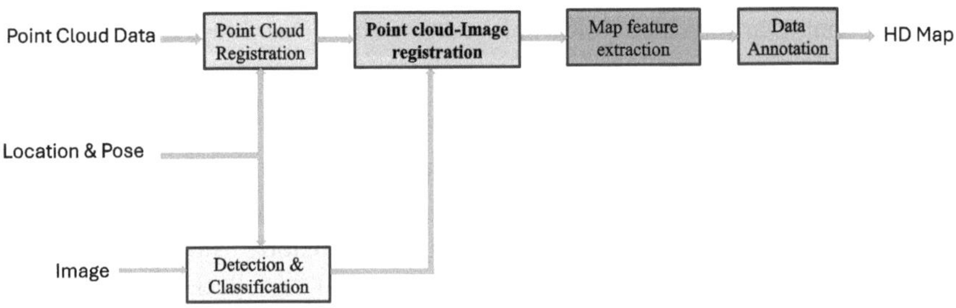

Fig. 6.2 The process of high-definition mapping by the Japanese company ZENRIN

Semantic targets include traffic lights, lane lines, traffic signs, and road boundaries. After detecting traffic lights and estimating 3D positions based on the camera, they are matched with the positions of traffic lights on the high-definition map.

Low-cost methods are also adopted by many companies, such as Intel Mobileye, Mapbox, Tesla, Lvl5, Mapper, Ushr, Atlatec, Geo++, MOMENTA, Kuandeng Technology, DeepMotion, and Wayz.ai. Generally, low-cost methods use crowdsourcing to obtain data, and the update cost is low.

The low-cost visual map-making method is based on SLAM technology. References [2, 3] summarize dynamic environment SLAM, dividing these methods into the following three major categories.

(1) Localization and reconstruction, divided into two parts, namely motion segmentation, localization, and 3D reconstruction.

(2) Dynamic object segmentation and tracking, consisting of two components: dynamic object segmentation and 3D dynamic object tracking.

(3) Joint estimation of motion segmentation and reconstruction, primarily based on factorization methods, encompassing both non-rigid and multibody objects.

Taking the road experience management (REM) framework defined by Intel Mobileye as an example, also known as roadbook, the annotated landmarks are extracted visually, including lane lines, road boundaries, traffic signs, road markings, etc.

6.1.1 Map Formats

Two popular map formats are OpenDRIVE [24, 25] and NDS [26].

OpenDRIVE is an open format for logically describing road networks, defined using XML syntax and primarily applied in simulation environments. An OpenDRIVE file encodes the geometric characteristics of roads and lanes, as well as associated elements such as road signs and traffic signals. The format is structured in a node-based hierarchy, with support for user-defined extensions to these nodes, thereby enabling flexible data exchange across different applications.

The road elements of OpenDRIVE include three categories: reference lines, lanes, and features (attachments), The road network in OpenDRIVE is modeled along a reference line, which serves as the core element of each road. Roads, lanes, and elevation profiles are defined relative to this reference line. Road features such as traffic signals can be positioned either with respect to the reference line or in the global coordinate system. In visualization, the reference line (typically shown as the central blue line) represents the center of the road, while the lanes (blue and light blue) are attached to it.

OpenDRIVE can be regarded as a construction tool for modeling different roads. Multiple road segments can be combined and interconnected to form a complete road network. OpenDRIVE specifies all static elements of the road network, enabling simulated vehicles to navigate within it. However, to render a realistic environment, additional description formats are required. Static roadside elements such as trees and buildings are described using dedicated formats, while road surface characteristics are captured in the OpenCRG file format. The dynamic aspects of driving simulation, including vehicle maneuvers and scenarios, are defined using OpenSCENARIO. Together, these three standards complement one another, providing comprehensive coverage of both static and dynamic elements in closed-loop vehicle simulation applications.

OpenDRIVE Manager (OdrManager) is responsible for real-time management of how to read OpenDRIVE data.

OpenDRIVE uses the road as its fundamental data management unit, with lanes, road features, and markings defined as attached attributes under each road entity. Geometrically, these attached elements are referenced to the road's reference line and therefore cannot exist independently of it.

NDS (Navigation Data Standard) is a standardized format for the automotive navigation database, using SQLite database, supporting step-by-step updates, and providing a variety of tools for users to use.

- NDS SQLite: Reference engine.
- Relational DataScript (RDS) tool: Describes the NDS format language, supports Java/C++/Python language to generate API tools and HTML documents.
- Certification Bench: Verifies the compatibility of the NDS database based on detailed tests and certificates.
- Investigation Modules: Test the execution framework of the NDS database on hardware.
- Database Inspector: Desktop visualization tool for the NDS database.

NDS already has the description function of high-definition maps. The structure of NDS related to autonomous driving includes the following modules: road geometry, road topology, high-definition lane model, localization landmarks, and obstacles.

6.2 Semantic Map

The concept of a semantic map is somewhat ambiguous. In this context, semantics refers to the presence of real-world entities that can be detected and classified, such as landmark buildings, traffic signs, and traffic lights. Currently, these semantic elements are typically obtained using deep learning–based detection classifiers and instance-level or panoptic segmentation methods.

Reference [4] analyzes the method of semantic SLAM. Traditional scene understanding methods ignore the semantic part and only focus on the geometric reconstruction of the environment.

Semantic understanding goes beyond processing isolated 3D points; it involves assigning higher-level meaning to the collected data, which forms the basis of semantic SLAM. Real-world environments contain numerous structures and objects that carry semantic information and can serve as landmarks in the SLAM process. Incorporating semantics not only aids scene reconstruction by inferring missing information but also enriches the reconstructed environment with supplementary contextual details.

The semantic segmentation technology based on deep learning has constructive significance for promoting the research and development of semantic SLAM. Combining model-based SLAM methods with semantic segmentation based on spatio-temporal CNN can provide richer localization feature representation for the SLAM model.

Information obtained from the camera can also be integrated with GPS, INS, and wheel speedometers as a self-motion estimation system. This model essentially uses deep learning to capture temporal motion dynamics. The direct method of using visual odometry (VO) is to use the grayscale information of the input image. However, these methods cannot guarantee optimality compared to feature-based methods. Deep neural networks can further learn rigid body motion, using raw point cloud data as input to predict rigid body transformations.

For visual SLAM, Ref. [5] provides a summary highlighting three main issues: the extraction and association of semantic information, the application of semantic information, and the advantages of semantic visual SLAM. A probabilistic target association strategy is considered one of the most effective approaches to enhance the perception capability of SLAM systems; however, its robustness and generalizability remain limited. Moreover, due to the complexity of real-world environments, the accuracy of semantic information extraction and association is highly susceptible to degradation.

Semantic information can help algorithms distinguish between static and dynamic targets, improving localization accuracy and robustness in dynamic environments. Using semantic information to segment moving targets and filter out feature points associated with moving targets is a common method that can improve system localization in dynamic environments.

In order to better solve the map problem, it is becoming increasingly important to establish accurate and reliable 3D semantic maps. Early semantic maps often used a priori target CAD model databases to construct 3D semantic maps, which can recover real scenes and save a lot of space for storing dense point clouds. However, CAD models are limited to targets in predefined databases.

Currently, semantic information can be used in all stages of the traditional SLAM algorithm framework, including initialization, backend optimization, relocalization, and loop closure detection. However, the contradiction between limited computing resources and growing computing demands greatly hinders the development of semantic visual SLAM. For example, in the extraction of semantic information, the system needs to obtain semantic information in real time, and needs to filter and associate semantic information in a timely manner. It should be noted that semantic visual SLAM is still in the development stage, and many hidden problems need to be solved. For example, incorrect target association makes the system more vulnerable in target-level SLAM.

The following discusses typical semantic map making methods through several references. Reference [6], under the premise of input stereo images, does depth map estimation and object segmentation, combined with sparse scene flow (a kind of 3D optical flow), to reconstruct the scene.

The system diagram of the mapping includes preprocessing, instance processing, and reconstruction. The preprocessing part obtains dense depth map estimation, target segmentation, and visual odometry results; the instance processing part gives the scene flow vector of the mask, 3D target tracking, mask color and depth, and static background color and depth; the reconstruction part gives instance reconstruction, regularization reconstruction based on voxel garbage collection algorithm, and static background reconstruction.

Here are some implementation details.

(1) Preprocessing: Calculate dense depth maps and sparse scene flows based on the input stereo sequence, and perform instance (target) semantic segmentation based on the left image. Here, the depth map estimation uses the deep learning trained model DispNet, and the semantic segmentation uses multi-task network cascades (MNC).

(2) Calculate visual odometry from the scene flow, using the open-source library libviso2.

(3) Segment the input color, depth, and scene flow to output the background and each moving target.

(4) Estimate the 3D motion of each newly detected target through scene flow and semantic segmentation, and compare it with the camera's own motion (ego-motion) to get the status of each target, i.e., static, dynamic, or uncertain.

(5) For each rigid body of interest, initialize/update its reconstruction result. Here, the SLAM method borrows the open-source tool InfiniTAM (a depth map fusion and tracking SLAM platform).

(6) Update the reconstruction of the static map.

(7) Complete the collection of voxel garbage, removing those unreasonable voxels caused by depth map distortion.

The method in Ref. [7] is a SLAM that combines geometric, semantic, and data-related parts. It adds semantic targets to the pure geometric model, describes them as structured models of intermediate layer component features, and uses probabilistic data-related methods, rather than directly establishing a connection between semantic features and targets.

The measurement data of inertia, geometry, and semantics (top left corner) will reconstruct the sensor's trajectory, as well as detect the target (car) and estimate the position of car components (i.e., doors and wheels).

This method adopts the keyframe approach, each keyframe extracts semantic landmarks and geometric landmarks based on ORB (a fast feature point extraction and description algorithm).

In order to obtain the pose of the target related to the semantic landmarks, a part-based target model (similar to the famous Deformable Parts Model, DPM) is used, introducing structural constraints between the landmark features of the target, such as the root position and the relative translation of the target parts.

The problem of SLAM is to solve the optimization problem of sensor trajectory and semantic landmark position, and the factor graph is a probabilistic graphical model, which can convert this optimization problem into a graph of random variables (nodes) with constraints (factors).

The method of solving data-related and SLAM together is the expectation-maximization (E-M) algorithm. The E-step calculates the related distribution given the sensor trajectory and semantic landmark position, and the M-step updates the sensor trajectory and semantic landmark position given the data-related distribution.

The method in Ref. [8] is to add the results of deep learning to monocular semi-dense SLAM to form a semantic map.

An exemplary algorithm flowchart includes three processes: keyframe selection and refinement, 2D semantic segmentation, and 3D reconstruction and semantic segmentation. The DeepLab-v2 model is used for semantic segmentation, the previous LSD-SLAM (large-scale direct monocular SLAM) method is used for SLAM, and the Conditional Random Field (CRF) model is responsible for 3D fusion.

6.3 High-Definition Map Based on Lane Lines

The first feature of a high-definition map is that it has map accuracy at the lane line level. The following discusses (mainly the camera vision system) how to build a lane line map, which is also some methods of special semantic maps.

Reference [9] introduces a method to directly build a map based on aerial images. The production of landmark maps from aerial images requires a combination of image classification and landmark description. Reference [9] proposes a rotation-invariant feature vector as an image description for estimating landmarks and road boundaries.

The image descriptor used for dense aerial image classification needs to meet the following requirements.

(1) Suitable for color images. Color information is an important basis for landmark detection (such as vegetation). Although color is highly dependent on lighting changes, aerial images can usually be used for almost invariant large areas.

(2) No contrast normalization. For dense classification of high-contrast pixels and low-contrast pixels, local contrast is a unique image feature. In this case, contrast normalization can be used to explain the changes in lighting.

(3) No scale invariance. In orthorectified aerial images, ideally, all landmarks in the image have similar sizes. If scale invariance is not required, on the contrary, the target size can be used as a feature.

(4) Rotation invariance. Landmarks may appear at any rotation. Although classifiers can be trained with a large number of different rotations of landmarks, rotation-invariant feature descriptors can significantly reduce the manual training workload.

(5) Regular, from the algorithm of estimating lane boundaries from landmarks, experimental results show that it significantly reduces the workload of manual correction in the later stage.

Based on Haar wavelets, the author proposes a fast rotation-invariant image descriptor, To simplify, take a subset of 20×20 sample points as an example, first, calculate the vertical and horizontal Haar wavelet responses $d_{y,i}$ and $d_{x,i}$ to pixel i; then, convert the wavelet response of pixel i into the absolute value di and direction φ_i relative to the center point; Next, the sampling points are divided into 8 concentric sub-regions of different sizes; Finally, the absolute values of the sampling points are accumulated in an 8×8 histogram according to the 8 different distances r and 8 different directions φ from the center, outputting a 64-dimensional feature vector.

To incorporate color information, this descriptor is calculated in parallel for the hue (H), saturation (S), and brightness (I) channels. Experiments have shown that this method produces better results than color descriptors in the RGB channel.

Based on this descriptor, each pixel is identified as part of a landmark or non-landmark, and the recognizer uses a simple SVM.

In Ref. [10], the requirements for a high precision lane level road map are as follows.

- Each road is divided into a sequence of road segments.
- The number of lanes in each road segment remains the same.
- The lanes of each road segment are implicitly connected in the same direction.
- The lanes of different roads can intersect without being connected.
- The lanes of each road segment can be defined as analytic curves.
- The expected accuracy of the lane curve diagram is at the decimeter level.

Here, a linearly parameterized cubic Hermite spline (CHS) method is used to describe the lane lines. The segmentation method based on CHS is: First, an initial decomposition is provided based on the intersection points of the roads, assuming $(n - 1)$ vertices and n basis functions between two nodes; Then estimate the position of the vertices and the tangential direction.

Reference [11] defines an enhanced map (Emap), which aims to describe roads more completely and accurately than standard maps to meet the requirements of advanced driver assistance systems (ADAS). The Emap represents each lane and describes the topological connections between lanes. According to this method, enhanced maps have been constructed in France, Germany, and Sweden.

Vehicle localization is generally divided into the following three levels.

- Macroscale: 10-meter accuracy, matching GPS and road database.
- Microscale: Sub-meter accuracy, map without absolute coordinates.
- Mesoscale: Lane-level accuracy, map with absolute coordinates.

The construction of Emap belongs to mobile mapping, a method of directly collecting road network geometry in the mapping area. Most map providers use dedicated vehicles equipped with calibrated sensors to collect and process information, with the aim of locating and mapping all surrounding roads, lane markings, and road signs.

Compared to the most common method based on polylines, Emap defines road elements through clothoids, which are more suitable for actual road shapes, reducing the amount of information to be stored in the database.

Emap not only includes a set of road segments describing the geometric shape of the road, but also includes the topological information of these road segments. Road segments are connected to their adjacent road segments, and Emap assumes that vehicles can transition from any point on one road segment to another, without artificially restricting to the nodes of the road segment, providing greater flexibility for the localization system.

Emap provides topological information of line segments including:

- Features of left, right, and front adjacent road segments.
- The relative lateral position of the lane segments on each road.
- Determine the connectivity between two road segments.
- Establish complex connectivity graphs in a short time.

In Ref. [12], Lanelets record the geometric and topological characteristics of the autonomous driving environment. It is a core, referring to those interconnected driving area road segments, mainly used for the behavior layer (note: the German high-definition map company Atlatec uses Lanelets to edit map results).

Lanelets are polylines with left and right boundaries, approximating the geometry of lanes with a certain accuracy, and determining the driving direction.

Path planning can be executed based on connected Lanelets.

Reference [13] uses a 2D LiDAR (single line), which is cheaper than the multi-line radar used by general high-definition maps. Road sign recognition is a prerequisite for establishing a lane map. An example of lane line recognition here is where the LiDAR faces the ground and performs a 180° flat scan, with a maximum distance of 80 m and a possible range error of 5 cm. The algorithm requires first performing road surface estimation (polyline representation), then extracting lane lines by setting a threshold through road surface reflection values, and finally using differential GPS and IMU for data calibration, transforming the vehicle position to the UTM absolute coordinate system.

An example of lane extraction: First, road centerline segmentation (RCS) uses a simplified polyline segment representation; then, based on the Douglas-Peucker algorithm, the centerline segment is divided into multiple parts in different directions and a set of shape points is obtained; finally, lanes are obtained through a Radon transform similar to the Hough transform.

The map defined in Ref. [14] contains three elements: roads, lanes, and lane lines. However, it is a high-definition map based on LiDAR, made with ArcGIS. This map provides road-level and lane-level planning capabilities, helping intelligent vehicles pass through intersections.

Note: ESRI (Environmental Systems Research Institute) is an international provider of GIS (Geographic Information System) software, and ArcGIS is a software tool for commercial location services.

This map is used to describe the topological structure of lane networks and road networks, and also provides a geometric description of lane lines, where each lane line is composed of a set of lane segment lines, i.e., a set of digital descriptions.

Reference [15] establishes a three-layer lane-level map for vehicle path planning, namely the road-level layer, intermediate layer, and lane-level layer. The road layer retains most of the traditional mathematical expressions of the current existing map model, including roads and intersections, aiming to utilize mature routing algorithms based on road layer maps; the intermediate layer acts as a bridge between the upper and lower layers, storing some relationships between sets for routing planning and other applications; the lane layer is designed to express lane-level details, not only including the collection of lanes on the road and more detailed intersections, but also geometric elements such as high-precision points of lane centerlines and lane lines.

Reference [16] discusses a system that generates a high-precision lane map,which includes three parts: data acquisition, data processing, and road modeling. The system uses 3D LiDAR and GPS combined with an Inertial Navigation System (INS) to collect data. Data processing includes road surface extraction, accumulation of 3D LiDAR data, lane line segmentation, skeletonization, and sorting. Road modeling uses segmented polynomial curves.

Reference [17] proposes a method for automatically constructing a high-definition map based on lane semantic segmentation. It uses a single-lens system to detect lane lines through FCN semantic segmentation, then extracts lane features for loop closure detection.

The system uses lane lines and wheel encoders to establish graph node maps. There are two types of nodes: landmark nodes and vehicle posture nodes. Finally, the map is generated using the iSAM (incremental smoothing and mapping) method based on graph optimization.

The pixels closest to the vehicle are set as the seed points of the lane. Points connected to the seed points are selected as points from the same lane. However, in real road conditions, many lanes are deleted or interrupted, so it is difficult to determine whether the lanes are connected using 4-connectivity or 8-connectivity methods. Since lanes are connected laterally, a method of making lateral connectivity judgments along image columns is used here. This way, even if the lane is interrupted or erased, the connectivity of the lane can be calculated. The feature points of the lane are extracted at a certain distance from the seed point, set at 40 pixels, approximately 0.4m, and the final feature points are extracted from the end of the lane.

In Ref. [18], a map management system is proposed.The system locates multiple vehicles simultaneously through a shared remote map accessed by mobile communication. The "appearance-based landmark selection" module can retrieve landmarks from the map that meet appearance conditions and effectively utilize existing bandwidth.

Once a vehicle completes a drive, the collected map data is uploaded to the backend and merged into the map. Subsequently, the backend's "map summarization" module ensures that the scale of the map does not exceed the maximum number of landmarks, also ensuring the maximum storage capacity condition, which is more beneficial for long-term map maintenance.

Map updating is an important task in map management. First, the newly collected data needs to be located on the available map. If the location accuracy is too low, new landmarks are established from the dataset and added to the map. Then there is a summarization step, limiting the number of landmarks in the map to not exceed a fixed number. In another case, all landmarks observed in the location are re-counted, but no new landmarks are added.

6.4 Deep Learning-Based SLAM Methods

The following analyzes the semantic SLAM of deep learning through some references.

The idea of DeepSFM (see Ref. [19]) comes from traditional bundle adjustment (BA), introducing two cost volumes through a deep learning model, corresponding to depth map estimation and pose estimation.

Traditional computer vision uses cost volumes to enhance photometric and geometric consistency between multiple frames. Generally, deep learning will estimate depth and pose separately. DeepSFM, however, enhances photometric and geometric consistency through two cost volumes (depth and pose).

In the architecture diagram of the DeepSFM model. A 2D CNN is used to extract photometric features to construct cost volumes. Initial source depth maps and camera poses are used to introduce photometric and geometric consistency. A series of 3D CNN layers are applied to the two cost volumes, D-CV and P-CV. Finally, a context network and depth regression operation are used to generate the predicted depth map of the target image.

Feature extraction uses a 7-layer CNN (3×3 convolution) and SPP (spatial pyramid pooling) layer, with the four pooling kernel sizes being 4×4, 8×8, 16×16, and 32×32. Finally, a feature map of the same resolution is obtained through 2D CNN and upsampling, which is used to construct the cost volume.

D-CV is based on the traditional plane sweep approach, with inputs including three elements: target image features, warped source image features, and homogeneous depth consistency map. Using hypothesis sampling, L virtual planes (perpendicular to the forward axis) samples are uniformly generated in the inverse depth space. This allows the feature map and depth map from the source viewpoint to be back-projected into the 3D space of the target viewpoint, thereby generating the cost volume.

For P-CV, its cost volume is based on camera pose. Its input elements are the same as D-CV, but generating the cost volume requires uniformly generating a set of candidate camera poses around the initial camera pose.

Similarly, based on these poses, the feature map and depth map from the source viewpoint are warped to the target viewpoint to form the cost volume.

Camera Pose Sampling (HCPS) Ref. [20] proposes a dense mapping framework, CodeMapping, that uses compact scene representation for sparse visual SLAM systems. Today's sparse visual SLAM systems can accurately and reliably estimate camera trajectories and landmark positions. Although these sparse maps are useful for localization, they cannot be used for other tasks such as obstacle avoidance or scene understanding.

CodeMapping takes the camera pose, keyframes, and sparse points generated by the SLAM system as input, and predicts a dense depth image for each keyframe. It uses a Variational Autoencoder (VAE), conditioned on grayscale, sparse depth, and reprojection error images from sparse SLAM, to predict uncertainty-aware dense depth maps.

The VAE can perform multi-view optimization to improve dense depth images, thereby enhancing the consistency of overlapping frames. The mapping module runs in a loosely coupled manner in parallel with the SLAM system on a single thread. This design can be integrated with any scale of sparse SLAM system without delaying the SLAM process. The dense mapping module can be used not only for local mapping but also for globally consistent dense 3D reconstruction through TSDF (Truncated Signed Distance Function) fusion, with ORB-SLAM3 as an example in the experiment.

An overview of the CodeMapping system: The dense mapping module subscribes to the sparse point set and camera pose obtained by BA (Bundle Adjustment), and runs sparse-dense prediction and multi-view optimization in the dense mapping thread.

The system starts a SLAM thread (tracking, localization, and global map thread) and a continuously parallel running dense mapping thread until the input data stream stops. This work uses ORB-SLAM3 as the sparse SLAM system, supporting multi-modal sensor input. The SLAM thread is as follows:

- Tracking. It uses sensor information to estimate the camera's pose relative to the existing map. ORB-SLAM3 runs pose-only BA to minimize the reprojection error between the current image and the map. This process is completed in real-time for each input frame, determining whether the current frame is a keyframe.
- Mapping. It runs BA for pose and geometric refinement. ORB-SLAM3 runs local and global BA for this; local BA optimizes with only a few keyframes, while global BA uses the entire pose graph. After each local BA, the SLAM thread transfers data from a window with 4 keyframes to the dense mapping thread; the 4 keyframes include the latest keyframe and the previous 3 co-visible keyframes selected based on the ORB-SLAM3 co-visibility criterion. Keyframe data consists of their respective camera poses, sparse maps, and reprojection error images; each sparse map is obtained by projecting 3D landmarks from the keyframe, and the reprojection error image is obtained by averaging the distance of the corresponding matching ORB feature positions obtained by projecting each 3D landmark to all observable frames. For newly registered keypoints with no matches to other frames, their reprojection error is initialized to a large value (such as 10).

The dense mapping thread is as follows:

- Sparse-Dense Prediction. The dense mapping thread receives keyframe data from the SLAM thread, checks whether this keyframe data has been processed, and if not, uses the sparse map and reprojection error input image to run the depth completion VAE, predicting the initial dense depth map. Here, the depth completion VAE generates a low-dimensional latent code for the predicted dense depth map, which is used for subsequent multi-view optimization.
- Multi-view Optimization. After the initial prediction, the depth map is refined through multi-view optimization of the sliding window camera pose provided by the SLAM system, which uses a factor graph. The graph considers different types of factors between overlapping frames (photometric factors, reprojection factors, and sparse geometric factors). Multi-view optimization only optimizes each frame code, not the camera pose. This is because the pose from sparse SLAM is assumed to be accurate enough, and pixel-by-pixel dense alignment is unlikely to further improve accuracy.

Compared to inliers, outliers often have higher average reprojection error values. Therefore, the input of the U-Net includes a sparse reprojection error map, providing the network with per-point uncertainty. The network at the bottom is a VAE that depends on decomposed features. This network generates latent codes, dense depth predictions, and uncertainty maps. The loss function of this network consists of depth reconstruction loss and KL divergence loss.

DROID-SLAM (see Ref. [21]) is a deep learning-based SLAM system that iteratively updates camera pose and pixel depth through a dense BA layer. Although trained on monocular videos, it has the potential to improve performance if stereo vision or RGB-D videos are used.

The schematic diagram of DROID-SLAM includes recurrent iterative updates, and has the following two improvements based on the optical flow estimation model:

(1) Multi-frame optimization is used instead of two frames.

(2) The update is based on a differential dense BA (DBA) layer.

DROID-SLAM uses a frame graph to represent co-visibility between frames, which is dynamically constructed during the training inference process. After each pose or depth update, visibility can be recalculated to update the frame graph. If the camera returns to a previously mapped area, long-distance connections can be added to the frame graph to perform loop closure optimization. Feature extraction mainly comes from the optical flow estimation model, which also includes two networks: the feature network and the context network. The feature network inputs the before and after frame image pairs to construct a set of correlation volumes, while the context network inputs the current frame image, and the extracted context features are injected into the optical flow network during the update operation.

The core component of the SLAM system is the learning update operation. The update operation is a 3×3 ConvGRU with hidden state, which operates on the edges of the frame graph, predicts the flow correction mapped to the depth, and updates the pose through the DBA layer.

The correlated features provide information about the visual similarity near each pixel location in the dense flow field, enabling the network to learn to align visually similar image areas. However, correspondences are sometimes ambiguous. The flow field provides a complementary source of information, allowing the network to exploit smoothness in the motion field for robustness.

The correlated features and flow features are mapped through two convolutional layers before being injected into the GRU. In addition, the context features extracted by the context network are injected into the GRU through an element-wise addition operation.

The ConvGRU is a local operation with a small receptive field. A global context is extracted by averaging the hidden states along the spatial dimension, and this feature vector is used as an additional input to the GRU. Global context is important in SLAM because incorrect correspondences (such as those caused by large moving objects) can degrade system accuracy. It is important for the network to detect and correct these errors. The DBA layer maps the set of flow corrections to a set of poses and updates the depth element-wise.

The entire system consists of two asynchronously running threads. The front-end thread receives new frames, extracts features, selects keyframes, and performs local BA, while the back-end thread performs global BA on the entire history of keyframes simultaneously.

6.5 Summary

This chapter introduces the basic principles of high-definition maps, including format, hierarchy, and production methods, with a particular emphasis on the discussion of the semantic map layer, especially low-cost production methods. Lane lines are the most important semantic objects in road semantic maps, and camera-based autonomous driving systems place particular emphasis on the extraction and matching of lane lines, such as Mobileye's Rem. Finally, it introduces the SLAM method based on deep learning.

On this basis, Chap. 7 will focus on the introduction of localization technology based on high-definition maps, which is also the primary function of high-definition maps.

References

1. Z BAO, S HOSSAIN, H LANG, and X LIN, High-Definition Map Generation Technologies For Autonomous Driving: A Review, arXiv preprint arXiv:2206.05400,2022
2. M SAPUTRA, A MARKHAM, N TRIGONI, Visual SLAM and Structure from Motion in Dynamic Environments: A Survey, ACM Computing Survey, 2019. 51(2): 1–36.
3. C CADENA et al., Past, Present, and Future of Simultaneous Localization and Mapping: Toward the Robust-Perception Age, arXiv preprint arXiv:1606.05830,2016

4. T LAI, A review on visual-SLAM: Advancements from geometric modelling to learning-based semantic scene understanding, arXiv preprint arXiv:2209.05222,2022
5. K CHEN, J ZHANG, J LIU, Q TONG, R LIU, S CHEN, Semantic Visual Simultaneous Localization and Mapping: A Survey, arXiv preprint arXiv:2209.06428,2022
6. I BARSAN et al., Robust Dense Mapping for Large-Scale Dynamic Environments, IEEE Int. Conference on Robotics and Automation, Brisbane, QLD, Australia, May 2018: 7510–7517.
7. N ATANASOV et al., A Unifying View of Geometry, Semantics, and Data Association in SLAM, Int. Joint Conf. Artificial Intelligence, Stockholm, Sweden, July 2018: 5204–5208.
8. X LI, R BELAROUSSI, Semi-Dense 3D Semantic Mapping from Monocular SLAM, arXiv preprint arXiv:1611.04144,2016
9. O PINK, C STILLER, Automated Map Generation from Aerial Images for Precise Vehicle Localization, IEEE Conf. on Intelligent Transportation Systems, Funchal, Portugal, Sept. 2010: 1517–1522.
10. A CHEN, A RAMANANDAN, and J A. FARRELL, High Precision lane-level road map building for vehicle navigation, IEEE Position, Location and Navigation Symposium,Indian Wells, CA, USA, May 2010: 1035–1042.
11. D BETAILLE, R TOLEDO-MOREO, Creating Enhanced Maps for Lane-Level Vehicle Navigation, IEEE Trans. on Intelligent Transportation Systems, 2011, 11(4): 786–798.
12. P BENDER, J ZIEGLER, and C STILLER, Lanelets: Efficient Map Representation for Autonomous Driving, IEEE Intelligent Vehicles Symposium, Dearborn, MI, June 2014: 420–425.
13. D KIM, T CHUNG, and K YI, Lane Map building and Localization for Automated Driving Using 2D Laser RangrFinder, IEEE Intelligent Vehicles Symposium, Seoul Korea, June 2015: 680–685.
14. L LIU et al., A Smart Map Representation for Autonomous Vehicle Navigation, Int. Conf. Fuzzy Systems and Knowledge Discovery, Zhangjiajie, China, Jan. 2015: 2308–2313.
15. C LIU et al., Design of a Multi-layer Lane-Level Map for Vehicle Route Planning, MATEC Web of Conferences, 6th International Conference on Transportation and Traffic Engineering, Jan. 2017: 1–5.
16. G-P GWON et al., Generation of a Precise and Efficient Lane-Level Road Map for Intelligent Vehicle Systems, IEEE Trans. on Vehicle Technology, 2017. 66(6): 4517–4533.
17. W JANG et al., Road Lane Semantic Segmentation for High Definition Map, IEEE Intelligent Vehicles Symposium, Changshu, China, June 2018: 1001–1006.
18. M BU RKI et al., Map Management for Efficient Long-Term Visual Localization in Outdoor Environments, IEEE Intelligent Vehicles Symposium, Changshu, China, June 2018: 682–688.
19. X WEI, Y ZHANG, Z LI, Y FU, and X XUE, DeepSFM: Structure From Motion Via Deep Bundle Adjustment, European Conference on Computer Vision, Glasgow, UK, Aug. 2020: 230–247.
20. H MATSUKI, R SCONA, J CZARNOWSKI and A J. DAVISON, CodeMapping: Real-Time Dense Mapping for Sparse SLAM using Compact Scene Representations, arXiv preprint arXiv:2107.08994,2021
21. Z TEED, J DENG,DROID-SLAM: Deep Visual SLAM for Monocular, Stereo, and RGB-D Cameras, arXiv preprint arXiv:2108.10869,August 2021.
22. K CHELLAPILLA, Rethinking Maps for Self-Driving, October, https://medium.com/@LyftLevel5/https-medium-com-lyftlevel5-rethinking-maps-for-self-driving-a147c24758d6, 2018.
23. S IWATA, Highly Efficient HD Map Creation: Accelerating Mapping Process with GPUs, ZENRIN CO. Ltd. 2017.
24. Wikipedia OpenDrive. https://en.wikipedia.org/wiki/OpenDRIVE_(specification)
25. OpenDrive. http://www.opendrive.org/project.html
26. Wikipedia NDS. https://en.wikipedia.org/wiki/Navigation_Data_Standard

Chapter 7
Localization Module for Autonomous Driving

For autonomous vehicles on the road, one of their primary tasks is to position the vehicle on the road. To do this, the vehicle needs to consider information from multiple sensors and fuse this information with data from road maps. The road localization problem can be broken down into three parts [24]: The first part is to determine the road on which the vehicle is currently traveling. In fact, the Global Navigation Satellite System is not accurate enough to infer this information on its own, so a filtering step is needed. The second part is to estimate the vehicle's position in the lane. The third part is to evaluate the lane in which the vehicle is currently traveling. The discussion here is mainly limited to the last two parts, as the first part already has mature solutions in localization based on navigation maps, such as map matching (matching the vehicle's position with the road network on the map), etc.

There are two technologies suitable for this task [23]. The first technology depends on the precise location of landmarks stored in the high-definition map (such as lane markings), and the system must match the detected landmarks with the landmarks stored in the map. The second technology relies entirely on onboard sensors to detect visual landmarks, such as all lanes on the road. However, due to the obstruction of other vehicles, sometimes not all lane markings on the road can be detected. Therefore, the extracted relevant road-level features need to be fed into a high-level fusion framework for evaluating the number of lanes and the lane in which the vehicle is traveling. This chapter only discusses the first technology.

Map-based localization is actually divided into two stages [25]. In the place recognition stage, the output of the visual sensor is compared with a set of geographic markers of interest in the map area to determine the initial position of the vehicle in the map. In the map metric localization stage, the output of the visual sensor is continuously aligned with the current map area passed by, so that the vehicle is tracked as it moves on the map.

Y. Huang and Z. Yang, *Autonomous Driving Handbook*,
https://doi.org/10.1007/978-981-95-3725-9_7

Therefore, localization requires data input from the perception module and matching with map information. Similar to the perception module, the localization module also adopts the idea of sensor fusion, that is, the combination of GPS/DGPS (such as RTK), IMU, camera, lidar, and high-definition map.

GPS localization is a classic method, but the propagation of electromagnetic waves in the environment will cause time errors due to weather, atmosphere, high-rise buildings, and hills, and cause estimated distance errors. To compensate for some of the errors, Differential GPS, or DGPS, such as RTK technology, was proposed. Qianxun Network, backed by China North Industries Group Corporation and Alibaba Group, provides services in this area nationwide.

There are also two other localization algorithms, namely, the odometer method and the inertial navigation method. The odometer method is a widely used method that estimates its position by recording the vehicle's movement process through a photoelectric encoder installed on the wheel. Inertial navigation uses accelerometers and gyroscopes for localization. For odometers, wheel slip will bring errors, and the integral cumulative error of inertial navigation will also cause trouble. The localization accuracy of odometers and inertial navigation is generally poor, and it is best to use them in conjunction with GPS or lidar/camera to achieve dead reckoning.

The method of using lidar point cloud data for localization requires the perception input of lidar point cloud (some methods allow reflection maps) and the matching of high-definition map point cloud. If the ICP algorithm is used directly, the calculation amount is too large. Among the familiar algorithms, commonly used ones include Normal Distributions Transform (NDT), and Stanford University's histogram filter and particle filter, etc. The algorithm requires the vehicle system to have a lidar sensor that can collect real-time point cloud data on site. There are also fast matching localization methods that compress point cloud data, such as TomTom's RoadDNA, Civil Maps' fingerprint map, but there will be compromises in localization difficulty and accuracy.

Camera localization is a method used by visual odometry or visual SLAM. There have been many studies on the problem of "image-based re-localization" in the past, which is used for SLAM loop closure detection and tracking failure recovery, etc. SLAM methods include sparse, dense, and semi-dense methods, but in the scenario of autonomous driving, the sparse method is the most suitable. However, it is not enough to rely solely on visual feature points for localization, and the assistance of semantic targets, i.e., semantic maps, is also needed. In autonomous driving, the most practical semantic map is the map based on lane lines.

This chapter is a continuation of Chap. 6, first introducing low-cost lane-line map-based localization technology, then discussing laser radar-based localization algorithms, followed by an analysis of multi-sensor fusion localization methods, and finally introducing deep learning-based localization methods.

7.1 Lane Map-Based Localization

This section focuses on several localization methods based on lane line maps.

Reference [1] proposes a method for detecting defined symbolic road markings (SRM) and stop lines, which is part of a low-cost vehicle localization system. The output of this method collaborates with the Global Navigation Satellite System (GNSS), Inertial Measurement Unit (IMU), and Extended Digital Map (EDM) for vehicle localization. The detection results of SRM and stop lines match with the Extended Digital Map (EDM), compensating for the localization error of GNSS and the cumulative error of IMU.

Since vehicle positioning relies not only on the detection of SRMs and stop lines but also on GNSS and IMU measurements, the corresponding detectors do not need to capture all possible targets; instead, minimizing false alarms is of greater importance. Furthermore, the SRM and stop-line detectors should be designed to consume minimal computational resources, as the remaining resources after detection are allocated to vehicle localization tasks.

SRM contains 9 types, namely straight, left turn, right turn, straight/left turn, straight/right turn, no straight, no left turn, no right turn, and diamond, plus stop lines, a total of 10 types of road signs can be used for localization.

Reference [2] proposes a method for automatically recognizing road markings and uses the features detected therein to calculate the position of the vehicle. If the road markings have been measured before, they can provide absolute global localization; otherwise, they only provide relative position information.

The road sign detection system is based on the features of the road sign templates learned from training images. The selected template feature is the FAST (Accelerated Segment Test) corner point. During operation, the following process is used to match these templates: first, promising feature matches are selected, and then structural matching is performed. The template is learned from training images containing true-value bounding boxes of road signs. First, the training images are corrected to compensate for lens and perspective distortion; then, FAST corner points are detected in the area of interest, and the bounding box and corner points and their HOG (Histogram of Gradients) descriptors are stored as templates for that specific road sign. Each type of road sign class (stop line, bicycle lane, turn arrow, etc.) may contain multiple templates corresponding to various views, roads, and lighting conditions. During operation, inverse perspective correction, MSER detection, FAST corner point detection, and HOG descriptor inference are performed on each test image. Road signs in the test image are detected and recognized based on corner point features. First, the HOG descriptors based on corner points find the assumed corner point match pairs, and then the results are refined through a 2D geometric structure matching algorithm that matches corner points within road signs. Geometric matching considers the possibility of multiple road signs in the image and the possibility that some features of the template cannot be detected due to lighting and perspective changes.

The road signs here include arrows, zebra crossings, and speed limit signs.

Two methods of pose estimation are analyzed: the PnP algorithm and the IPM method. The PnP algorithm requires knowledge of the 3D position and features of the road sign. For this, the Visual Odometry (VO) method is used, based on stereo vision triangulation and local BA to track points and calculate their 3D positions. Each pixel in the IPM image corresponds to a physical distance on the ground, so given the pixel positions of the corner points within the road sign in the IPM image, the relative pose of the camera to these corner point features can be estimated. However, the results of the PnP method are worse than those of the IPM image.

Therefore, the focus is on the IPM method.

Reference [3] calibrates the edges of road signs and a lightweight 3D map based on Chamfer matching (based on Chamfer distance), where road signs are represented as a set of sparse points.

This localization algorithm that only matches the geometric structure of road signs enhances robustness in terms of ambient light and appearance changes. The algorithm considers vehicle odometry and epipolar geometry constraints and proposes a nonlinear optimization problem to estimate the 6 DoF camera pose.

In the process of localization, the detected road markings and the road markings in the 3D map are projected onto the image plane for matching. The camera pose is estimated using epipolar geometry and visual odometry as constraints to solve the optimization function.

Reference [4] uses a commonly used Optical Character Recognition (OCR) system, TESSERACT, to classify symbols. If odometry or location information is available, spatial filtering and mapping can be performed. The obtained information can be used to improve localization and provide further information for planning or generating planning maps.

The sensor uses two system configurations, namely monocular and binocular. Monocular uses vanishing point geometry constraints, and binocular directly completes localization through 3D reconstruction. IPM generates a top view, the watershed algorithm realizes image segmentation, estimates the candidates of traffic signs, and performs OCR. In addition, "reprojection to the road surface" is optional, and OCR needs to train the recognizer model offline.

If binocular is used, 3D information can be directly reconstructed through stereo vision triangulation.

In Ref. [5], a stereo camera system and a high-precision map containing road edges and road signs are used to achieve centimeter-level localization. The map is created in advance using an extended sensor setup. GNSS localization is only used for initialization and is not needed during the localization process. The map data can be projected onto the image plane when the vehicle pose is known.

Reference [5] is based on the data collected by the lidar and camera sensors for the road boundary and lane lines to complete the map localization.

The sensor setup used in the map making process includes binocular and GPS-IMU systems. The vehicle's position is described by a point position and a direction in a fixed coordinate system, and the localization method is to use lane line and road boundary data as observation vectors under the Kalman filter framework.

The map contains line segments representing road signs or road edges, and measurements are obtained in the form of points. The goal of map matching is to achieve the best match between point measurements and line segments, which means lateral and longitudinal correlation.

Therefore, each map line segment is sampled to map points. For each of these points in the map, the nearest measurement point is searched to extract the error, which achieves lateral and longitudinal error estimation.

Reference [6] proposes a system framework for vehicle global localization and 3D point cloud reconstruction that integrates stereo visual odometry, satellite imagery, and road maps within a particle filter architecture. The framework is designed for vehicle localization scenarios in which moving objects are present and operates without relying on GPS. Its key feature is the use of road maps and stereo reconstruction to generate accurate overhead views, which are then matched against satellite images to eliminate drift and achieve precise global localization.

Binocular vision achieves visual odometry and dense depth maps, and gets the reconstructed point cloud; after the point cloud generates the top view, it matches with the threshold-processed satellite map, then enters the particle filter (PF) with the distance from the road map, obtains the pose, and combines with the reconstructed point cloud to generate the point cloud map, finally generating map coverage.

The system uses the Google Static Maps API to obtain satellite images and street maps through HTTP requests. Create a custom map style called road map, remove all non-road features and text from the street map, select the foreground color as white, the road color as black, the resulting binary map image is suitable for image processing algorithms to recover road positions.

Before the Particle Filter (PF) performs camera and satellite image matching, the satellite image area corresponding to the road is thresholded by its intensity value, resulting in a binary image that only contains road visual features. In addition, a route map is obtained using distance transformation, and a smooth declining value of nearby road pixels is obtained. PF stores samples of the probability distribution of possible vehicle states on the map and resamples the particles based on particle weights through importance resampling. Each particle has a hypothesis about the vehicle's position, posture, and speed. The particle state is (x, y, θ, v), where x and y represent the position of the vehicle relative to a fixed origin, in meters; parameter θ represents the absolute heading of the vehicle relative to the map (in degrees); parameter v corresponds to the linear/tangential speed of the vehicle in the heading direction, in m/s.

Reference [7] proposes a method for coarse map localization using low-cost sensors (GPS and camera-based lane recognition) and maps with lane-level accuracy, while updating the perceived road network and map, called LaneSLAM.

Overall, this factor graph has five different types of factors when modeling observation errors.

(1) Factor ψ_1: Assuming normally distributed noise, it connects the actual posture with the observed posture.

(2) Factor ψ_2: Assuming the observation error is normally distributed, it connects the actual postures at two consecutive moments through the odometer.

(3) Factor ψ_3: It connects a lane line on the map with the corresponding actual lane line, and its deviation satisfies a normal distribution.

(4) Factorψ_4: The observed lane line and the corresponding actual lane line and its actual posture are connected together, and the observation error still satisfies a normal distribution.

(5) Factor ψ_5: A pair of actual lane lines and their binary indicator variables are connected together.

Reference [7] uses the loopy belief propagation (LBP) algorithm to solve a SLAM problem of a factor graph, where message passing is the actual iterative algorithm.

Reference [8] proposes a method based on landmark SLAM, called Road-SLAM, which fully utilizes landmarks obtained from camera images for global localization. To achieve loop closure detection with landmark matching, it defines the features composed of landmarks and surrounding lanes as sub-maps, and uses the random forest method to improve matching accuracy. The equipment used by the system includes stereo cameras, inertial navigation, wheel speed encoders, and RTK differential GPS.

The main observation of Road-SLAM is landmarks. It needs to perform image segmentation and classification.

In the flowchart of the Road-SLAM algorithm first, the camera image and inertial navigation, odometer are input together, the image with landmarks is converted to the road surface (i.e., IPM, inverse perspective mapping) and generates a point cloud; then, to segment the point cloud belonging to the landmark, adaptive binarization is performed on the IPM image; next, the segmented landmarks enter the classification module (using random forest), and are divided into 6 categories; then, based on submap matching (using the generalized ICP algorithm), loop closure is detected; finally, optimization is performed in the pose graph (using the known iSAM algorithm).

In the point cloud segmentation algorithm, the process begins with 3D point filtering, which includes voxel filtering and radius outlier removal. The first segmentation step then removes elements such as centerlines, stop lines, and dashed lane lines while retaining other road markings. After lane line detection, a second segmentation is performed to remove the lane lines themselves, leaving only features such as numbers or crosswalks. This process ultimately produces the set of classification candidates.

The six classification categories are road markings, numbers, arrows, lane lines, crosswalks, and unknown.

7.2 Lidar-Based Localization

7.2.1 Normal Distribution Transform Localization

The most classic method of normal distribution transformation localization should be the Maximum Distribution Transformation (NDT) [19–21]. First, the point cloud is divided into standard cells, and the mean and variance of each cell are calculated; then, the environment map is represented as a sum of normal distributions.

Each distribution can be seen as an approximation of a local surface, that is, its position, orientation, and smoothness. The orientation and smoothness are the eigenvectors and eigenvalues of the variance matrix. Accordingly, a distribution can be a point, a straight line, an ellipse, or a circle in 2D space. In 3D space, if the eigenvalues of this point cloud distribution are equal, or one is much smaller than the other, it means that the local area is a plane. Similarly, the NDT characteristics of other geometric surfaces, such as spheres and ellipsoids, can be obtained.

In localization, the initialization of LiDAR point cloud matching is generally carried out through GPS, IMU or odometer (such as speedometer). Given two point cloud data (usually one from the map and the other from the vehicle to be located), the matching estimation steps based on NDT are as follows.

(1) Establish the NDT $(q_i, \sum_i)$ of the first point cloud.

(2) Initialize the 3D transformation parameters p (such as 0 or IMU data).

(3) Each point x_i of the second point cloud data is mapped to the first point cloud coordinate system $x_i^{'}$ according to the transformation parameters.

(4) Determine the corresponding normal distribution of each mapped point.

(5) The fraction of this transformation parameter depends on the sum of the distribution estimates of each mapped point.

(6) Perform the iterative step of the optimization algorithm of this score, such as the Newton method, to get the updated parameter estimate.

(7) Jump back to step (3) until the convergence condition is met.

The score in step (5) is actually a measure of correlation, and the calculation formula is as follows:

$$score(p) = \sum_i exp(\frac{-(x_i{'} - q_i)^T \sum_i^{-1} (x_i{'} - q_i)}{2}) \tag{7.1}$$

where, p is the transformation parameter to be estimated; $x_i^{'}$ is the point xi in the second point cloud mapped to the first point cloud coordinate system according to the transformation parameters; qi is the mean of the first point cloud; $\sum_i$ is the variance matrix of the first point cloud.

7.2.2 Particle Filter Localization

Particle filtering is a Monte Carlo simulation [18] method for solving state space problems. Its idea is to use a set of particles to represent probabilities, and the state particles randomly drawn from the posterior probability represent its distribution, also known as sequential importance sampling. Simply put, particle filtering is a method to approximate the probability density by propagating a set of random samples, replacing the integral operation with the sample mean, and obtaining the minimum variance distribution.

In localization tasks, the 3D posture is its state vector. It can be assumed that the variables to be estimated are the horizontal position and the deviation angle (x, y, yaw), while the other two angles, roll and pitch, are considered to be accurate enough. The transformation of particle filtering is also based on GPS/IMU for prediction, and the observation value will affect the importance weight of the particles and serve as the probability of resampling. To avoid catastrophic localization errors, some particles will be selected based on the current GPS/IMU posture estimate. GPS/IMU can also be used for observation value calculation, and the prediction value of GPS/IMU can also be used for localization calculation.

Particle filtering can operate without GPS or IMU, using the map as the sole environmental reference. In Ref. [9], the map employed for localization is primarily a road feature map, meaning that non-road elements are discarded during the localization process.

7.2.3 Histogram Filter Localization

Another localization method is to represent the likelihood distribution of offset (x, y) on the plane as a $2D$ histogram filter. Similarly, this filter includes the following two parts [22].

(1) Motion prediction, to reduce the credibility of estimates based on motion.

(2) Observation update, to increase the credibility of estimates based on sensor data.

By integrating the IMU update, a smooth coordinate system can be "immune" to changes in GPS posture. The motion model of the filter is simple, and it is not necessary to model the uncertainty of vehicle motion, but only to model the uncertainty of the smooth coordinate system and the global coordinate system. Observation update means that new lidar data can correct the vehicle's position estimate.

Instead of treating each individual lidar reflection as an observation, a rolling grid is constructed from accumulated sensor data. This representation allows direct comparison between sensor data and map data, while avoiding overemphasis on regions with dense reflections, such as trees or areas with numerous dynamic objects.

Finally, by combining motion prediction and observation updates, that is, multiplying the likelihood distribution and the prior distribution, the posterior distribution and posterior estimation of pose estimation are formed.

7.3 Localization Based on Sensor Fusion

As previously introduced, LiDAR and GPS/IMU can be fused together. Below, other methods of sensor fusion localization are introduced. Reference [10] is a continuation of the work in Ref. [1], where the main focus is on the use of road markings and parking lines for localization, while the work in Ref. [10] combines GPS, inertial navigation, wheel speedometers, and maps for vehicle localization.

The localization algorithm uses particle filtering to fuse information from various sensors. Each particle represents a state vector, including 2D vehicle position and its forward angle (heading angle), i.e.,

$$\mathbf{x}_t^n = \left[x_t^n y_t^n \theta_t^n\right]^T, 1 \leq n \leq N \tag{7.2}$$

where n, N are the index and number of particles, respectively.

Particle initialization comes from the GPS pose, and the state prediction equation (motion) is defined as

$$\hat{\mathbf{x}}_{t+1}^n = \begin{bmatrix} \hat{x}_{t+1}^n \\ \hat{y}_{t+1}^n \\ \hat{\theta}_{t+1}^n \end{bmatrix} = \begin{bmatrix} x_t^n - \frac{\tilde{v}}{\tilde{w}\sin\theta} + \frac{\tilde{v}}{\tilde{w}\sin(\theta+\tilde{w}\Delta t)} \\ y_t^n - \frac{\tilde{v}}{\tilde{w}\cos\theta} + \frac{\tilde{v}}{\tilde{w}\cos(\theta+\tilde{w}\Delta t)} \\ \theta_t^n + \tilde{w}\Delta t + \tilde{\gamma}\Delta t \end{bmatrix} \tag{7.3}$$

$$\tilde{v} = v + sample(\sigma_v^2), \tilde{\omega} = \omega + sample(\sigma_\omega^2), \tilde{\gamma} = \gamma + sample(\sigma_\gamma^2) \tag{7.4}$$

where $\hat{\gamma}$ is the random noise of the heading angle; sample(σ^2) generates a random disturbance with a mean of 0 and a variance of σ^2; v and ω are the vehicle speed and angular speed, respectively, coming from the wheel speedometer and inertial navigation.

In particle filtering, the particle weight is important, mainly depending on the observation value, here referring to GPS, lane lines, road sign markings (SRM), and maps based on SRM. Here the weight ω_t^n is defined as

$$\omega_t^n = \omega_{t-1}^n \cdot \omega_{t,GPS}^n \cdot \omega_{t,LINE}^n \cdot \omega_{t,SRM}^n \tag{7.5}$$

The first multiplier on the right side of Eq. 7.5 is the weight at the previous moment, and the latter three depend on GPS, lane lines, and road sign detection, respectively.

The weight calculation of GPS is as follows:

$$w^n_{t,GPS} = \frac{1}{2\pi\sqrt{det(\sum_{GPS})}} \times exp - \frac{1}{2}(\hat{p}^n_t - c)^T \sum^{-1}_{GPS}(\hat{p}^n_t - p_{GPS}) \tag{7.6}$$

where$\sum_{GPS}$ is a 2×2 matrix, p_{GPS} is the position given by GPS, and the error between it and the particle prediction position forms a Gaussian distribution.

The weight calculation of the lane line is as follows:

$$\omega^n_{t,LANE} = \frac{1}{\sqrt{2\pi\sigma^2_{LANE}}} \times exp\left\{-\frac{1}{2}(\hat{l}^n_t - l_{LANE})^2/2\sigma^2_{LANE}\right\} \tag{7.7}$$

where l_{LINE} is the lateral position detected by the lane line, and the error between it and the particle prediction lateral position forms a Gaussian distribution.

The detection of SRM and the matching of the map get its map position. If multiple SRMs are detected in one image at the same time, their matching positions in the map can be averaged as the final estimate pSRM, and the error between it and the particle prediction position satisfies a Gaussian distribution. Thus, the weight calculation of SRM is as follows:

$$\omega^n_{t,SRM} = \frac{1}{2\pi\sqrt{det(\sum_{SRM})}} \times exp\left\{-\frac{1}{2}(\hat{p}^n_t - p_{SRM})^T \sum^{-1}_{GPS}(\hat{p}^n_t - p_{SRM})\right\} \tag{7.8}$$

The above constitutes the particle filtering localization platform.

Reference [11] mentions the localization system used by Baidu's Apollo open-source system, which includes a 64-line LiDAR (velodyne HDL-64E), high dynamic range (HDR) camera, long-range radar, GPS (NovAtel ProPak6), and an inertial navigation system composed of a gyroscope and accelerometer (NovAtel IMU-IGM-A1).

It estimates the optimal position, velocity, and attitude (PVA) by combining sensor inputs (as introduced above) and the map established by the LIDAR. The PVA estimated by the GPS and LIDAR is used as the observation for the Kalman Filter (KF), and the KF provides the predicted PVA prior knowledge. The strap-down inertial navigation system (SINS) integrates IMU data (from accelerometers and gyroscopes) as the prediction model for the KF, and the accelerometer and gyroscope biases and PVA errors estimated by the KF are used as correction signals input to the SINS.

The LIDAR-based localization algorithm is shown as below. The estimation of the forward angle uses the visual method of Lucas-Kanade feature tracking, and the horizontal localization uses the histogram filter mentioned earlier.

Algorithm: LIDAR-based localization
Input: Prior map **m**, online point cloud z, coarse transformation
$T_0 = (x_0, y_0, \phi_0, \theta_0, h_0)$
and search space X, Y

Output: Best calibration $(\hat{x}, \hat{y}, \hat{a}, h)$ and covariance matrix C_{xy}
1. $\hat{h} \leftarrow$ Heading degree estimation
2. $\hat{a}_0 \longleftarrow m(x_0, y_0)$* to get the height in the map
3. Transform point cloud z, through $(x_0, y_0, \hat{a}_0, \theta_0, h_0)$
4. for $x_i y_i \in x_0 + X, y_0 + Y$ do
5. $P_r \leftarrow SSD_r(x_i, y_i, z, m)$
6. $P_a \leftarrow SSD_a(x_i, y_i, z, m)$
7.$P(z|x_i, y_i; m) \leftarrow (P_r)^{\gamma} \cdot (P_a)^{1-\gamma}$
8. $P(x_i, y_i) \leftarrow P(z|x_i, y_i, m) \cdot (\bar{P}(x_i, y_i))^{1/k}$
9. end for
10. $(\hat{x}, \hat{y}) \leftarrow P(x_i, y_i)$
11. $C_{xy} \leftarrow P(x_i, y_i)$
12. $\hat{a} \longleftarrow P(x_i, y_i)$ *to get the height in the map
13. return $(\hat{x}, \hat{y}, \hat{a}, \hat{h}, C_{xy})$

The GPS localization system uses RTK, i.e., the differential GPS method. The sensor fusion localization method combines GPS, LIDAR, and IMU localization under the framework of the Kalman filter. Since the IMU error is cumulative, the state variable of this filter is the error signal, i.e., the error-state Kalman filter. Due to transmission and computation delays, the problem of observation delay needs to be solved. Here, a dual filter is used, along with a fixed-size buffer for storing filter states, filter observations, and IMU data in chronological order. The first filter calculates real-time PVA and its variance through time updates and IMU data integration; the second filter handles delayed observations.

Reference [12] refers to cameras, LIDAR, and IMU as local sensors, and GPS, magnetometers, and barometers as global sensors. Local sensors are accurate in localization but will drift; global sensors are messy in localization but do not drift.

The idea it proposes is to combine local sensors and global sensors for localization. The local estimator is mainly visual odometry (VO) and visual inertial odometry (VIO); the global estimator is based on pose graph optimization. The combination of the two is to align the local estimate with the global coordinate system and eliminate drift.

The method in Ref. [13] comes from previous visual inertial odometry (VIO), which aims to estimate the camera's pose in a sliding window through IMU data and image features.

The pipeline operation includes the initialization module, measurement preprocessing module, local visual-inertial odometry module with relocalization, and global pose graph optimization module. The initialization module includes visual SFM and visual-inertial alignment; the measurement preprocessing module performs feature detection tracking and inertial pre-integration; the local visual-inertial odometry module with relocalization includes VIO, keyframe detection, loop closure detection, feature search, etc.; the details of the global pose graph optimization module are seen below.

Each posture (including position and orientation) constitutes a node in the posture graph, and the density of the nodes depends on the sensor with the lowest working frequency. The edges between two adjacent nodes represent local constraints from local estimation (VO/VIO), which limit the relative posture transformation between nodes, while other edges are global constraints of global sensors.

Posture graph optimization is a maximum likelihood estimation (MLE) problem, which includes the joint probability distribution of vehicle posture over a period of time. The algorithm for solving it is the famous bundle adjustment (BA).

7.4 Localization Method Based on Deep Learning

Reference [14] focuses on visual localization methods, summarizing from two directions: structure-based and regression-based. Structure-based localization estimates camera posture by matching 2D pixel features with 3D point scene coordinates. Its methods are divided into two types: matching-based localization and scene coordinate regression-based localization. The first method is based on descriptor matching, and the second method is based on a trainable localization pipeline.

Regression-based methods can be divided into absolute posture regression (APR) and relative posture regression (RPR). For APR-based localization, a 2D map built from an end-to-end neural network directly regresses the posture; for RPR-based localization, the relative posture is regressed using the retrieval method of a geo-fenced dataset or other methods.

In the localization pipeline based on structural features, Structure from Motion (SFM) or SLAM can build a 3D point cloud model. Compared with regression-based methods, the pipeline based on structural features relies more on the prior information of the 3D scene model. After establishing the correspondence between the 3D point cloud of the scene model and the query image, the camera posture is recovered through geometric constraints, the camera posture is calculated using PnP solution, and outliers are removed using the RANSAC method.

According to the network input, APR can be divided into three types: monocular images, a pair of images assisted, and image sequences/videos.

Unlike the method of monocular images, APR that uses image pairs for auxiliary learning learns absolute posture by estimating the relative posture of auxiliary constraints. This may involve globally consistent posture prediction, which can improve localization performance, that is, reduce localization errors and improve localization robustness.

The APR method based on image sequences/videos has a similar pipeline, and it estimates translation and orientation through CNN feature extraction and localization regression.

The following is an analysis of neural network-based localization methods based on some references. Reference [15] proposes an online meta-learning algorithm, where the Visual Odometry (VO) network can continuously adapt to new environments in a self-supervised manner.

Then it is updated according to the loss and infers the frame at the next time. The network learns to find a set of weights that perform well for both image sequences

Di and Di+1. During online learning, spatiotemporal information is merged by convLSTM, and fast adaptation is achieved through feature alignment.

The VO network θ_i takes N consecutive frames in the image sequence Di in the sliding window, and estimates the posture $\hat{T}_t^{t-1}$, depth D^t, D^{t-1}, and mask M^t of the image sequence D_i in a self-supervised manner. DepthNet predicts depth, PoseNet takes two consecutive frames and consecutive depth images to regress posture, and MaskNet predicts a mask based on the distortion residuals brought by the image synthesis of depth and posture. At each iteration i, the network parameters are updated according to the loss, and the image sequence Di+1 is inferred next time. The network learns to find a set of weights θ_i^* that perform well for both images D_i and D_{i+1}. During online learning, spatiotemporal information is aggregated through convLSTM, and fast self-adaptation is achieved by aligning the feature distributions t,$i + 1$ at different times.

The Model-Agnostic Meta-Learning (MAML) is used, which aims to minimize the evaluation (adaptive) error on the validation set, rather than minimizing the training error on the training set. In the inner loop gradient update, the performance of VO in the image sequence D_i is evaluated by self-supervised loss and updating parameters θ_i. In the outer loop gradient update, the performance of the updated model θ_{i+1} on the subsequent frame sequence D_{i+1} is evaluated.

This continuous adaptation process is simulated during the training and online testing stages. During the training process, by minimizing the total loss of all sequences in the training dataset, the network learns the base weight θ, thereby achieving fast online adaptation. This meta-learning scheme solves the stochastic gradient problem in online learning.

The network architecture of DepthNet, PoseNet, and MaskNet, embeds recursive units into the encoders of DepthNet and PoseNet, allowing convolutional networks to use not only spatial information but also temporal information for depth and pose estimation. The length of ConvLSTM is the number of frames in the image sequence, and ConvLSTM acts as the memory of the network. When a new frame is processed, the network can remember and learn from past experiences, thus updating parameters quickly to adapt to unseen environments. This method can enhance the correlation between different time steps, learning the temporal dynamics of the moving camera from video input.

Feature alignment enhances the correlation of non-stationary feature distribution in changing environments by adjusting the feature distribution at different times, extending the domain adaptive method to the online learning environment. The learning algorithm performs well when the feature distribution of the test data is the same as the training data. When changing to a new environment, even though the extracted features are different, the feature distribution of the two domains should be the same.

Even when running VO in the open world, the view will change, but the depth image sequence will be observed continuously over time, and the feature distribution should be similar. This feature normalization and alignment method, as a regularization means, simplifies the learning process and keeps the weights of learning consistent in non-stationary environments.

Reference [16] proposes a DA4AD visual localization framework for autonomous driving based on attention-aware features, aiming to achieve centimeter-level localization accuracy. Using the attention mechanism, an end-to-end deep neural network searches for significant, unique, and stable features.

The network architecture and system workflow of the visual localization framework based on end-to-end deep attention-aware features can be divided into three different stages: training stage, map generation stage, and online localization stage. To achieve the best performance, different algorithms or strategies are chosen for the same module when the system is at different stages.

The entire architecture can be decomposed into three main modules: ① Local Feature Embedding (LFE); ② Attention Key Point Selection (AKS); ③ Weighted Feature Matching (WFM).

The training stage includes all three modules: LFE, AKS, and WFM. First, given the predicted pose, select the map image closest to it in Euclidean distance. Next, the LFE module extracts dense features from online and map images and correspondingly extracts attention heatmaps from map images. The AKS module selects good features from the map image as keypoints based on the attention scores of the heatmap. Then, their corresponding 3D coordinates are obtained from the projection of the LiDAR point cloud. Finally, these 3D keypoints and feature descriptors are inputted, and the WFM module searches for the optimal pose offset in the 3D cost body, comparing the optimal pose offset with the truth value to get the training loss.

After training, there is a designated map generation step, calling a subset of the network. Considering the LiDAR scan and true pose, the global 3D coordinates of the LiDAR points can be easily obtained. Then, the keypoints and their descriptors and 3D coordinates are saved in the map database.

In the online localization stage, the feature maps of different resolutions in the online image are estimated again by inference through the LFE sub-network. Given the predicted camera pose, keypoints and their feature descriptors and global 3D coordinates are collected from the nearest map image. Given the candidate poses sampled in the cost body constructed in the WFM module, these keypoints are projected onto the online image. Three feature matching networks of different resolutions are cascaded to achieve coarse-to-fine inference and output the estimated vehicle pose.

The LFE module is designed to extract high-quality local feature descriptors along with task-related weights (attention scores), represented as heatmaps over the image. Ideally, the extracted descriptors should remain robust under appearance variations caused by changes in lighting or seasonal conditions. The attention scores are expected to emphasize reliable objects while suppressing interference and noise in the scene.

The AKS module uses a "describe-select" method to select a set of keypoints suitable for long-term matching and saves them in the map database.

Given the dense features extracted from the online image and the 3D keypoints associated with their 2D feature descriptors in the map image, the WFM module estimates the best pose by sampling a set of candidate poses around the prior pose and evaluating the matching cost of each candidate.

The Ref. [17] proposes a VO online adaptive framework based on scene-agnostic geometric computation and Bayesian inference. FlowNet predicts the dense optical flow $F_{k,t}$, $F_{t,k}$ and photometric uncertainty Γ_t between the keyframe I_k and the current frame It, while DepthNet provides the prior of the current scene geometry by estimating the depth D_k^0 and uncertainty $\sum_t$ of the keyframe. Here, the DepthNet uses the Monodepth2 model, and the FlowNet uses the RAFT model.

The relative pose Ttk is solved from the selected flow correspondences by the fundamental matrix or PnP. In the online adaptive process, the sparse depth Dsk is first recovered by a differentiable triangulation module, and this sparse result can be depth-filtered to obtain the densified depth Ds'k through a patch-based representation; then, the prior depth D0k is continuously improved in subsequent depth estimation within a Bayesian update framework.

The current frame I_t and its warped frame $\hat{I}_t$ according to the predicted T_k^t obtain the weighted photometric loss L_{pho}, which is minimized by a differentiable Gauss-Newton layer; finally, the optimized depth D_k' and $F_{k,t}'$ are used as pseudo ground truth poses for online supervised learning of DepthNet and FlowNet, where the loss function includes depth term L_d and optical flow term L_{flow}.

7.5 Summary

This chapter discusses the localization technology of high-definition maps, from the most basic lane line maps to high-cost lidar point cloud maps, analyzes the localization methods of multi-sensor fusion, including GPS, IMU, lidar, and cameras, and finally introduces the localization methods based on deep learning. Chapter 8 will introduce another challenging topic in autonomous driving, namely planning and decision-making.

References

1. J K SUHR, H G JUNG, Fast Symbolic Road Marking and Stop-line Detection for Vehicle Localization, IEEE Intelligent Vehicles Symposium, Seoul, Korea, June 2015: 186–191.
2. T WU, A RANGANATHAN, Vehicle Localization using Road Markings, IEEE Intelligent Vehicles Symposium, Gold Coast, QLD, Australia, June 2013: 1185–1190.
3. Y YU, H ZHAO, H ZHA, Monocular Localization in Urban Environments using Road Markings, IEEE Intelligent Vehicles Symposium, Los Angeles, USA, June 2014: 468–474.
4. M SCHREIBER, F POGGENHANS, C STILLER, Detecting Symbols on Road Surface for Mapping and Localization using OCR, IEEE Conf. on Intelligent Transportation Systems, Qingdao, China, Oct. 2014: 597–602.
5. M SCHREIBER, C KNOEPPEL, U FRANKE, LaneLoc: Lane Marking based Localization using Highly Accurate Maps, IEEE Intelligent Vehicles Symposium, Gold Coast, QLD, Australia, June 2013: 449–454.
6. T SENLET,A ELGAMMAL,A Framework for Global Vehicle Localization Using Stereo Images and Satellite and Road Maps, Int. Conf. on Computer Vision Workshop, Barcelona, Spain, Nov. 2011: 2034–2041.
7. T HEIDENREICH, J SPEHR,C STILLER,LaneSLAM - Simultaneous Pose and Lane Estimation Using Maps with Lane-Level Accuracy, IEEE Conf. on Intelligent Transportation Systems, Gran Canaria, Spain, Sept. 2015: 2512–2517.
8. J JEONG, Y,CHO,A KIM,Road-SLAM : Road Marking based SLAM with Lane-level Accuracy, IEEE Intelligent Vehicles Symposium, Los Angeles, CA, USA, Sep. 2017: 1736–1743.
9. J LEVINSON S THRUN, Map-Based Precision Vehicle Localization in Urban Environments, Conf. on Robots, Science and Systems, Georgia Inst. of Tech., June 2007.
10. G WAN et al., Robust and Precise Vehicle Localization based on Multi-sensor Fusion in Diverse City Scenes, arXiv preprint arXiv:1711.05805,2017.
11. J K SUHR et al., Sensor Fusion-Based Low-Cost Vehicle Localization System for Complex Urban Environments, IEEE Trans. on Intelligent Transportation Systems, 2017, 18(5): 1–9.
12. T QIN et al., A General Optimization-based Framework for Global Pose Estimation with Multiple Sensors,arXiv preprint arXiv:1901.03642, 2019.
13. T. QIN, P. LI, S. SHEN, Vins-mono: A robust and versatile monocular visual-inertial state estimator, IEEE Trans. on Robotics, 2018. 34(4): 1004–1020.
14. M XU, Y WANG, B XU, J ZHANG, J REN, S POSLAD,P XU,A Critical Analysis of Image-based Camera Pose Estimation Techniques, arXiv preprint arXiv:2201.05816,2022.
15. S LI,X WANG,Y CAO,F XUE,Z YAN,H ZHA,Self-Supervised Deep Visual Odometry with Online Adaptation, arXiv preprint arXiv:2005.06136,2020.
16. Y ZHOU, G WAN, S HOU, L YU, G WANG, X RUI, S SONG, DA4AD: End-to-End Deep Attention-based Visual Localization for Autonomous Driving, arXiv preprint arXiv:2003.03026, 2020.
17. S LI,X WU,Y CAO,H ZHA,Generalizing to the Open World: Deep Visual Odometry with Online Adaptation, arXiv preprint arXiv:2103.15279, 2021.
18. S THRUN, W BURGARD, D FOX, Probabilistic Robotics, MIT Press, 2006.
19. P. BIBER et al., The normal distributions transform: A new approach to laser scan matching, IEEE/RSJ Conf. On Intelligent Robots And Systems,Las Vegas, NV, USA, Oct. 2003: 2743–2748.
20. M. MAGNUSSON et al., Scan registration for autonomous mining vehicles using 3D-NDT, Journal of Field Robotics, 2007, 24(10): 803–827.
21. C ULAS,H TEMELTAS,3D Multi-Layered Normal Distribution Transform for Fast and Long Range Scan Matching, J. Intelligent and Robotic Systems,2013. 71(1): 85–108.
22. J LEVINSON S THRUN, Robust Vehicle Localization in Urban Environments Using Probabilistic Maps, IEEE Int. Conference on Robotics and Automation, Anchorage, AK, USA, May 2010: 4372–4378.

23. GINGRAS D, An overview of positioning and data fusion techniques applied to land vehicle navigation systems, Automotive Informatics and Communicative Systems: Principles in Vehicular Networks and Data Exchange, edited by GUO H, IGI Global, Hershey, PA, Jan 2009: 219–246.
24. J LACONTE, A KASMI, R AUFRERE, M VAIDIS AND R CHAPUIS,A Survey of Localization Methods for Autonomous Vehicles in Highway Scenarios,MDPI Sensors,December,2021. 22(1): 247–276.
25. M ELHOUSNI, X HUANG, A Survey on Visual Map Localization Using LiDARs and Cameras,arXiv preprint arXiv:2208.03376,2022.

Chapter 8
Planning Module for Autonomous Driving

The planning module in autonomous driving is used to determine the vehicle's actions. The planning module outputs planning results based on the input perception results, and the planning results will be input into the control module to determine the actual operation of the vehicle. As mentioned earlier, "perception" is equivalent to the "eyes" of autonomous driving, then "planning" is equivalent to the "brain," which is a very important module.

Section 8.1 first introduces the basic planning theory, divided into three levels, namely task planning, behavior planning, and motion planning.

Section 8.2 discusses driving behavior modeling and prediction, mainly based on data-driven machine learning methods; Sect. 8.3 extends to the modeling and prediction of pedestrian behavior; finally, Sect. 8.4 discusses how to learn the driving skills of experienced drivers, which is an in-depth discussion of Sect. 8.2.

8.1 Basic Planning Theory

The planning of an autonomous vehicle can be divided into three levels: task planning, behavior planning, and motion planning.

1. Task Planning

Task planning (also known as route planning) is the highest level, defining the start point (pickup) and end point (drop-off), as well as the road from the start to the end point. The most classic algorithms are mainly Dijkstra and A*.

The Dijkstra algorithm is suitable for calculating the shortest route problem where the road weight is non-negative, and can give the shortest route from one node to all other nodes in the graph. Since the input is a large sparse matrix, this algorithm has the disadvantages of long time consumption and large space occupation. The essence of the Dijkstra algorithm is breadth-first search, which is a divergent method,

Y. Huang and Z. Yang, *Autonomous Driving Handbook*,
https://doi.org/10.1007/978-981-95-3725-9_8

so both space complexity and time complexity are relatively high, and its algorithm complexity is $O(n^2)$, where n is the number of nodes. The heuristic search algorithm A* is depth-first search. By introducing a cost function, it speeds up the search speed and improves the accuracy of local search. It is a popular shortest path algorithm. The storage space occupied by the A* algorithm is less than that of the Dijkstra algorithm. Its time complexity is O(bd), where b is the average out-degree of the nodes, and d is the search depth of the shortest path from the start point to the end point.

Dijkstra and A* are only suitable for route search in small environments. Large environment road networks require more advanced route search algorithms.

2. Behavior Planning

Behavior planning (also known as decision-making) makes some specific decisions to correctly interact with other intelligent entities, follows rule constraints, and generates local goals, such as lane changing, overtaking, and passing through intersections. This is mostly a decision-making process of a finite state machine (FSM) and a behavior tree, but it is difficult to handle when undefined states appear.

The finite state machine model describes different driving states and the transition relationships between states by constructing a finite directed connection graph (DCG), and generates driving actions based on the transition of driving states. However, this type of model ignores the dynamics and uncertainty of the environment. Moreover, when there are many driving scene features, the division and management of states become difficult to handle, making it unsuitable for complex urban road environments.

The Behavior Tree (BT) model also selects different driving actions based on the attributes of the current driving state, but it turns the driving state and actions into a tree structure, and uses a "query" mechanism for strategy search. This type of decision model needs to define decision routes offline for each driving scene, and is not suitable for large state spaces and action spaces. It also cannot consider the uncertainty in the environment. Behavior-aware planning requires an interactive and collaborative decision-making process, but the uncertainty will increase exponentially. There are three methods that can be thought of to solve this problem.

(1) Better description of the dynamic environment, including dynamic obstacles.

(2) Modeling of interactive and collaborative conditions.

(3) Use of joint distribution modeling.

One behavioral decision-making method is to use a large amount of driving data to model driving behavior through machine learning, which can predict the trajectory of other vehicles and also plan the behavior of the vehicle itself, which will be discussed in Sect. 8.2.

3. Motion Planning

Motion planning, also known as local planning, generates appropriate paths or/and a series of actions to achieve local goals. Motion planning can take the form of paths or trajectories. In the framework of route planning, the solution path is represented as a function $\sigma(\alpha)$: $[0, 1] \rightarrow \chi$, where χ is the vehicle's state configuration space. In the framework of trajectory planning, the control execution time is explicitly considered. This consideration allows direct modeling of vehicle dynamics and dynamic obstacles. The solution trajectory is represented as a time-parameterized function $\pi(t)$: $[0, T] \rightarrow \chi$, where T is the planning time range. Unlike route planning, trajectory planning specifies how the vehicle state configuration evolves over time.

The problem of path planning is to find a path $\sigma(\alpha) : [0, 1] \rightarrow \chi$ in the configuration space χ, which begins at the initial state configuration and terminates in the target region, while satisfying specified global and local constraints. The quality of a solution path is commonly described using the terms feasible and optimal. Feasible path planning seeks any solution that satisfies the problem constraints, without regard to solution quality. In contrast, optimal path planning aims to identify a path that minimizes (or maximizes) a defined quality criterion while still adhering to the given constraints.

The optimal path planning problem can be formally stated as follows. Let χ be the state configuration space, $\sum(\chi)$ represents the set of all continuous functions $[0, 1] \rightarrow \chi$, the vehicle's initial state configuration is $x_{init} \in \chi$, and the path must end in the target area $X_{goal} \subseteq \chi$. The set of all allowed state configurations of the vehicle is called the free state configuration space, and is represented as χ_{free}. Usually, free state configurations are those that will not lead to collisions with obstacles, but the set of free state configurations can also represent other complete constraints on the path. The differential constraints on the path are represented by $D(x, x', x'', \ldots)$, which can be used to constrain the smoothness of the path to some extent, such as the path curvature and/or the rate of change of curvature limits. For example, in the case of $X \subseteq R^2$, the differential constraints can enforce the maximum curvature K of the path using the Frenet–Serret formula:

$$D(x, x', x'', \cdots) \Leftrightarrow \frac{\|x' \cdot x''\|}{\|x'\|^3} \tag{8.1}$$

In addition, let $J(\sigma) : \sum(\chi) \rightarrow R$ be the cost function, then the optimal solution of the path planning problem can be summarized as follows, given (χ_{free}, x_{init}, X_{goal}, D, J), find $\sigma*$:

$$\arg \min_{\sigma \in \sum(\chi)} J(\sigma) \tag{8.2}$$

subject to

$$\sigma(0) = x_{init} \ and \ \sigma(1) = X_{goal} \tag{8.3}$$

$$\sigma(\alpha) \in \chi_{free} \forall \alpha \in [0, 1] \tag{8.4}$$

$$D(\sigma(\alpha), \sigma'(\alpha), \sigma''(\alpha), \cdots) \forall \alpha \in [0, 1] \tag{8.5}$$

The numerical methods of path planning can be roughly divided into the following three categories.

(1) Variational methods. The path is represented as a function parameterized by a finite-dimensional vector, and the vector parameters are optimized through nonlinear continuous optimization techniques to seek the optimal path. These methods can quickly converge to a local optimum, however, unless an appropriate initial guess is provided, these methods usually lack the ability to find a global optimum.

(2) Graph search methods. The state configuration space is discretized into a graph, where vertices represent a finite set of state configurations, and edges represent transitions between vertices. The desired path is found by searching for the minimum-cost path in this graph. Graph search methods are not easily trapped in local minima; however, they are limited to optimizing on a finite set of paths.

(3) Incremental search method. This method samples the state configuration space and gradually builds a reachability graph, usually a tree structure, maintaining a set of discrete reachable state configurations and their possible transitions. Once the graph is large enough to have at least one node in the target area, the desired path is obtained by tracing the edges pointing to that node from the starting state configuration. The sampling-based method gradually increases the size of the graph until a satisfactory solution path is found in the graph.

Motion planning problems in dynamic environments or with dynamic constraints are formulated in the trajectory planning framework, where the solution to the problem is a trajectory, i.e., a time-parameterized function $\pi(t) : [0, t] \longrightarrow \chi$ that describes the gradual change of the vehicle's state configuration over time.

Let $\prod(\chi, T)$ represent a set of all continuous functions $[0, T] \rightarrow \chi$, x_{init} is the initial state configuration, and the target area is X_{goal}. The set of all allowable state configurations at time $t \in [0, T]$ is represented as $\chi_{free}(t)$, which encodes complete constraints, such as avoiding collisions with static and possible dynamic obstacles on the path. The differential constraints on the trajectory are represented by $D(x, x', x'', \cdots)$, which are used to impose dynamic constraints on the trajectory. In addition, let $J(\pi) : \sum(\chi) \longrightarrow R$ be the cost function. Under these assumptions, the optimal solution to the trajectory planning problem can be formulated as follows.

Given $(\chi_{free}, x_{init}, X_{goal}, D, J, T)$, find $\pi*$:

$$\arg \min_{\pi \in \prod(\chi, T)} J(\pi) \tag{8.6}$$

subject to

$$\pi(0) = x_{init} \quad \text{and} \quad \pi(1) = X_{goal} \tag{8.7}$$

$$\pi(\alpha) \in \chi_{free} \ for\ all\ t \in [0, T] \tag{8.8}$$

$$D(\pi(t), \pi'(t), \pi''(t), \cdots) \ for\ all\ t \in [0, T] \tag{8.9}$$

Trajectory planning in dynamic environments has been proven to be more difficult than path planning, and numerical methods have become a popular choice for this task. The trajectory planning problem can be numerically solved directly in the time domain using some variational methods, or the trajectory planning problem can be transformed into a path planning problem in the state configuration space with a time dimension.

Motion planning has been proven to be a computationally complex problem, with the famous dimensionality curse. An optional method is discretization, which transforms the continuous space model into a discrete space model. This transformation includes two common approaches: combinatorial planning, which builds a discrete representation that accurately represents the original problem, and sampling-based planning, which uses a collision detection module to perform discrete searches on samples from the state configuration space. The former includes partially observable Markov decision processes, and the latter includes rapidly exploring random trees and lattice planners.

The Rapidly exploring Random Tree (RRT) algorithm begins by initializing a tree with the starting configuration as its root. In each iteration, the algorithm randomly samples a configuration from the configuration space and identifies the nearest existing node in the tree to this sample. A control input is then computed to steer the system from the nearest node toward the sampled configuration. If this motion results in a collision or produces a configuration that lies too close to an existing node, the sample is discarded, and a new one is drawn. Otherwise, a new configuration is generated within a small time interval and added to the tree as a new node, with an edge connecting it to the nearest node.

This process of sampling, extending, and updating continues iteratively, causing the tree to expand rapidly throughout the configuration space. The algorithm terminates when the tree reaches the target region, at which point the constructed tree encodes a feasible path from the initial state to the goal.

RRT is probabilistically complete, ensures motion flexibility, is easy to implement in real time, can handle general dynamic models, and is used in many autonomous driving scenarios. It can quickly expand free space, but the downside is that the generated routes are raw, and collision detection needs to be done every time it expands. If there are many obstacles, it will bring high computational complexity.

The state lattice, also known as the lattice planner, constructs a discrete search space that can achieve relevant state continuity, deterministically obtain the goal state, and satisfy the vehicle's differentiable constraints. The state lattice can be seen as an extension of the grid, which discretizes the continuous space by repeating rectangles or squares. Similarly, connecting the vehicle's acceptable original paths under position, curvature, or time axis states, when repeated in a rule-based manner, constructs the state lattice. Thus, the planning problem is simplified to a boundary value problem connecting the original state and the final state.

The state lattice overcomes the efficiency limitations of grid technology without increasing computational power. At the same time, it also reduces the computational cost of precomputation, unless this grid will follow the shape of the lane or road and bend. The state lattice is usually very suitable for non-complete, highly constrained environments, such as driving road environments.

The grid planner is fully solution-ready, meaning that every solution variation can automatically adjust the control space and is continuously explorable. The grid planner can also ensure optimality and smoothness, as discontinuities related to back-pointers will not occur. However, the problem of direction angle discretization can cause oscillations between heading angles, and exhaustive sampling can lead to unnecessary computational complexity.

The global fixed state grid requires flat terrain and fixed fluidity, which limits vehicle movement, as a small difference in a grid set trajectory can lead to a significant increase in search time and computation. The plans generated by the grid planner are mostly close to the actual movement of the vehicle, but because it is not possible to quickly consider alternative goal states, it cannot effectively execute evasion maneuvers.

In addition, in planning work, there are methods that integrate perception and planning, forming end-to-end planning. However, for autonomous driving, this approach is relatively risky.

The more difficult is planning in dynamic environments, where there is at least one moving obstacle, and there is system uncertainty. Collision avoidance and obstacle avoidance are the main tasks. The classic solution is the Partially Observable Markov Decision Process (POMDP) model, which plans in time-space dimensions considering the position, speed, and trajectory of obstacles.

POMDP is a generalization of the Markov Decision Process (MDP). POMDP is actually a modeling of an intelligent entity decision process: assuming that the system dynamic equation conforms to the Markov decision process, but the intelligent entity cannot directly observe the underlying state; instead, based on observations and observation probability sets and the Markov decision process, the system must maintain a probability distribution over all possible state sets.

In planning, POMDP views the task as a state transition process, defines a reward mechanism for possible events in the process, such as receiving rewards when reaching the destination and getting punished when a collision occurs, and finds the strategy to achieve the maximum reward through the corresponding state update process.

A simplified method of trajectory planning is the decoupling strategy in optimization, which decomposes the problem into several low-dimensional optimization sub-problems, reducing the difficulty of solving the planning problem.

8.1.1 Frenet Coordinate System

Real traffic roads are curved, and to simplify the parameter expression of the optimization problem, the Frenet coordinate system can replace the Cartesian coordinate system.

In the Frenet coordinate system, the centerline of the road is used as the reference line, and a coordinate system is established using the tangent vector t and normal vector n of the reference line, with the self-car as the origin, and the two coordinate axes are perpendicular to each other, that is, the s direction (i.e., the reference line direction, called longitudinal) and the d direction (or L direction, i.e., the reference line normal, called lateral). The Frenet coordinate system is also called the SL coordinate system.

The Frenet coordinate system describes the position of the self-car relative to the road. The longitudinal coordinate indicates the driving distance on the road, and the lateral coordinate indicates the distance deviating from the centerline. Under the Frenet coordinate system, the trajectory of the vehicle moving forward and staying within the lane is calculated as a straight line, which greatly simplifies the difficulty of trajectory planning.

The Frenet coordinate system is just a way of description, and the expression of the road curve has nothing to do with the coordinate system. Any curve expression can be used under the Frenet coordinate system, such as polynomials, B-splines, Bezier curves, and Cycloids. Here, taking the polynomial as an example, the conversion method between the Frenet coordinate system and the Cartesian coordinate system is derived as follows.

Frenet → Cartesian: That is, calculate the (x, y) coordinates known as (s, d). Parameterize the road center reference line with a cubic spline curve, that is, use the arc length s as the independent variable, and the (x_r, y_r) coordinates of the reference line as the function of s, that is, $x_r = x_r(s)$, $y_r = y_r(s)$. According to s, calculate the (x_r, y_r) coordinates of the reference line, then derive the spline curve to calculate the unit tangent of the reference line at (x_r, y_r), so rotate counterclockwise by 90° to get the unit normal vector n, finally $(x, y) = (x_r, y_r) + dn$.

Cartesian → Frenet: That is, calculate the (s, d) coordinates known as (x, y). It is equivalent to finding a point on the curve that minimizes the distance from a fixed point outside the curve. The curve (reference line) is obtained by interpolating discrete reference points with a cubic spline curve. It is assumed that the curve between each two reference points is either a straight line or convex, and there will be no two twists. Suppose the coordinates of the fixed point outside the curve are (p_x, p_y), and the equation of the cubic polynomial curve is $y = ax3 + bx2 + cx + e$, the square of the distance from the fixed point to the curve $g(x) = (x - p_x)^2 + (a_x^3 + b_x^2 + c_x + e - p_y)^2$, let the derivative of $g(x)$ be 0, find up to five points, and find the one with the smallest distance.

8.1.2 EM Planner

Reference [38] provides a motion planning method based on optimization theory, the EM planner. This planner covers multi-lane and single-lane autonomous driving scenarios in a hierarchical manner: ① The top level of the system is a multi-lane strategy, comparing parallel computing lane-level trajectories, and handling lane-changing situations. ② Within the lane-level trajectory generator, the path and speed are optimized iteratively based on the Frenet coordinate system. ③ For path and speed optimization, a combination of dynamic programming (DP) and spline-based quadratic programming (QP) is proposed, constructing a scalable and easy-to-adjust framework, while dealing with constraints such as traffic rules, obstacle decisions, and trajectory smoothness.

At the top of the EM planner, the data center module collects and synchronizes information; after data collection, the reference line generator will generate some candidate lane-level reference lines, as well as information about traffic regulations and targets; this process is based on high-definition maps and navigation information from the routing module; during the lane-level motion planning process, the Frenet coordinate system is first constructed based on the specified reference line and traffic rules, and then the relationship between the ego vehicle and its surrounding environment is evaluated; in addition, the reconstructed information is passed to the lane-level optimizer module to perform path optimization and speed optimization; during the path optimization process, the surrounding environment information is projected onto the Frenet frame (E-step), and a smooth path is generated based on this (M-step); similarly, during the speed optimization period, once the path optimizer has generated a smooth path, obstacles will be projected on the fixed point-time (station-time) graph (E-step), and then a smooth speed curve is generated (M-step); combining the path and speed curves, a smooth trajectory for the specified lane is obtained; finally, all lane-level optimal trajectories are sent to the reference line trajectory decision maker, based on the current car state, rules, and trajectory cost calculation, the trajectory decision will determine the best trajectory for the ego vehicle's maneuver.

In the EM iterative pipeline for lane-level planning, operations include two E-steps and two M-steps, and trajectory information is iteratively updated between planning cycles. In the first E-step, obstacles are projected onto the lane Frenet coordinate system, including static obstacle projection and dynamic obstacle projection, where static obstacles are directly projected based on the Cartesian–Frenet coordinate transformation, and the intent of dynamic obstacles is described by moving trajectories.

First, the estimated positions of the dynamic obstacles and the ego vehicle at each time point are evaluated, and then the overlap of the dynamic obstacles and the ego vehicle at each time point is mapped into the Frenet coordinate system. In addition, the appearance of dynamic obstacles during the path optimization process will eventually lead to the vehicle gradually nudging, so for safety considerations, the SL (Frenet coordinate system) projection of dynamic obstacles only considers low-speed traffic and oncoming obstacles. For high-speed traffic, the parallel lane-changing strategy of the EM planner will cover this scenario.

In the second E-step, all obstacles, including high-speed, low-speed, and oncoming obstacles, are evaluated based on the generated path in the fixed point-time (ST) coordinate system. If the obstacle trajectory overlaps with the planned path, the corresponding area in the ST coordinate system will be generated. In the two M-steps, the path and speed are generated through a combination of dynamic programming and quadratic programming. Although obstacles have been projected onto the SL frame and ST frame, the optimal solution for path and speed still lies in the non-convex space. A coarse solution is first obtained using dynamic programming, and this solution can provide obstacle decisions, such as nudging, yielding, and overtaking. The coarse decision determines the convex hull based on the quadratic programming spline optimizer, and then the optimizer finds the solution within the convex hull.

The M-step path optimizer optimizes the path in the Frenet coordinate system, which can be represented as finding an optimal estimation function $l = f(s)$ for the lateral coordinate relative to the fixed point (station) coordinate in the non-convex SL space, for example, nudging from left to right, which may be two local optimal estimations.

Therefore, the path optimizer includes two steps: path decision based on dynamic programming (DP) and path planning based on spline curves. This DP path decision step provides a coarse path with feasible channels and obstacle fine-tuning decisions.

The spline QP path step is a refinement of the DP path decision step. First, in the DP path step, a feasible channel is generated based on the selected path; then, the spline-based QP path step generates a smooth path within this feasible channel.

In the pipeline of the QP path step, the QP spline solver optimizes the objective function with linearized constraints to generate the spline QP path; the objective function of the QP path step is a linear combination of smoothness cost and guideline cost, where the guideline in this step refers to the DP path, which provides an estimate of the nudging distance for obstacles; the constraints in the QP path step include boundary constraints and dynamic feasibility, and in the EM planner, the ego-vehicle dynamics use a simple ego-vehicle model.

The speed optimizer generates a speed curve in the ST graph, which is represented as a fixed point function $S(t)$ relative to time. Similar to the path optimizer, finding the optimal speed curve on the ST graph is a non-convex optimization problem.

Here, DP is combined with spline QP to find a smooth speed distribution on the ST graph.

The DP speed step includes the cost function, ST graph grid, and DP search, with the goal of optimizing the cost function in the ST graph under constraints. The cost includes items such as speed maintenance, smoothness, and obstacles.

The DP search space is also within the vehicle dynamic constraints, and its search algorithm is simple and clear. Some necessary pruning based on vehicle dynamic constraints can speed up this process. Dynamic constraints include acceleration, jitter limits, and monotonicity constraints (reversing can only be performed when parking or in other specified situations).

Since the piecewise linear speed curve cannot meet the dynamic requirements, the spline QP speed step is needed to fill this gap. The spline QP speed step includes three parts: cost function, linearized constraints, and spline curve QP solver. The spline QP optimization is within the linearized constraints, and the linearized constraints include feasible channels, vehicle dynamics, and traffic regulation constraints.

8.2 Driving Behavior Model and Prediction

For the vehicle itself, its driving action analysis cannot be separated from kinematic and dynamic theories, as well as surrounding obstacles, and also the road conditions (slope, curvature) and rules (traffic lights, traffic signs, lane lines, intersections, etc.). All these factors combined can reflect the behavior patterns of traffic participants, and learning these behavior models is the necessary path to mastering veteran driver techniques in autonomous driving.

When the driving behavior prediction results are used by downstream planners, the following features are also crucial to the overall system performance.

(1) Compatibility: The trajectory predictions of different entities in the scene should be mutually compatible in a single joint prediction. Specifically, collisions between predicted trajectories should be rare.

(2) Trackable joint trajectory prediction: The future motion of entities can be multimodal. As the number of modes grows exponentially with the number of entities, the planner will soon be overwhelmed. Alternatively, the motion planner can take a conservative approach, avoiding all predicted trajectories, but often at the expense of planning performance (for example, if all plans seem to conflict, keep the ego-vehicle stationary). Therefore, it is hoped that the multimodal joint prediction of all entities has a limited but fully representative number of modes, so that the planner can carry out contingency planning caused by accidents.

(3) Temporal consistency: For downstream planners, motion planning largely depends on prediction results. To ensure smooth motion planning, if the scene itself does not undergo drastic changes during this period, the prediction should not undergo significant changes in subsequent times. Therefore, data sampling should be avoided, as predictions may change significantly at time steps, resulting in discontinuous motion planning, thereby affecting planning performance and safety.

(4) Conditioning: Conditioning is to fix one or more future trajectories of entities and predict the final distribution of other entities' future trajectories. Conditioning prediction helps understand motion planning (conditioned on the motion planning of the ego vehicle) and entity interaction. Ideally, the conditioned distribution can be generated without changing the model structure.

8.2.1 Driving Behavior Prediction

In order to operate safely and effectively on the road, autonomous vehicles should not only understand the current state of nearby road participants but also actively predict their future behavior (also known as motion or trajectory). Part of this universal problem is predicting the behavior of pedestrians or vulnerable road users (VRUs); another part is predicting the expected behavior of other vehicles on the road. Compared with pedestrians, vehicle behavior has higher inertia, driving rules, and road geometry restrictions. However, new challenges come from the interdependence of vehicle behaviors, traffic rules, the impact of the driving environment, and the multimodality of vehicle behaviors. In addition, the limitations of actually observing the surrounding environment and the computational resources required to perform prediction algorithms also increase the difficulty of the problem.

Whether the motion structure of vehicles (such as cars and trucks) is reasonable is governed by driving rules and environmental conditions. In addition, vehicles cannot instantly change their trajectories due to inertia. Despite this, predicting vehicle behavior is not an easy task. Firstly, there is an interdependence between vehicle behaviors, and the behavior of any vehicle will affect the behavior of other vehicles. Therefore, predicting the behavior of a vehicle requires observing the behavior of surrounding vehicles.

Secondly, the geometry of the road and traffic rules can change vehicle behavior. For example, placing a give-way sign at an intersection can completely change the behavior of a vehicle approaching the intersection. Therefore, models trained in a specific driving environment without considering traffic rules and road geometry will be limited in performance in other driving environments.

Finally, vehicle behavior is multimodal, which means that there may be more than one possible future behavior given the vehicle's motion history. For example, when a vehicle slows down at an intersection without changing its direction of progress, it can be expected to turn right and left. The behavior prediction module of an autonomous vehicle should recognize all possible next steps of motion in order to operate reliably.

In addition to the three challenges mentioned above, there are also practical limitations in the behavior prediction module of autonomous vehicles. For example, due to sensor constraints such as occlusion, limited sensing range, and measurement noise, an autonomous vehicle can only perceive a partial view of its surrounding environment.

Moreover, the onboard facilities of autonomous vehicles have limited computational resources. Therefore, autonomous driving needs to consider certain assumptions when implementing behavior prediction.

Vehicle behavior prediction models can be divided into three types: physics-based, maneuver-based, and interaction-aware models. Methods that use the vehicle's dynamics or kinematics model for prediction are classified as physics-based methods. Models that predict vehicle maneuvers as output are considered maneuver-based methods.

Models that consider vehicle interactions in the input are called interaction-aware models. Physics-based methods are somewhat outdated and will not be discussed further here.

In driving prediction models, the choice of coordinate system to represent input and output can be roughly divided into the following two categories.

(1) Agent-centric models transform input and perform inference in an agent-centric coordinate system. These models are essentially translation and rotation invariant among scene elements, but their scale is quadratic in the number of agents and scene elements.

(2) Scene-centric models use a fixed coordinate system to handle all agents, with the advantage of sharing representations among all agents. Their computation scales

linearly with the number of agents. However, these models must learn the translation and rotation invariance among scene elements, and they usually perform worse than agent-centric models.

Driving interaction behaviors can be divided into two main categories: model-based and data-driven. Model-based methods can be further divided into the following four forms.

(1) Utility-based models. These integrate physical distance-related information into the objective/cost function, expressing driving interaction behavior as an optimization problem that can be solved with existing dynamic and linear programming algorithms. Commonly used models include optimal swarms, game theory models, imitation learning, and Markov decision processes.

(2) Probabilistic generative models. On one hand, interactions can be represented through probabilistic conditional distributions or conditional behavior predictions; on the other hand, interaction behaviors are also seen as (latent) probabilistic generative processes or conditional probability models, which capture the driver's interaction preferences and consider the prediction of surrounding driving behaviors, so the autonomous vehicle will maneuver at the current time step.

(3) Latent/risk fields. Physical distance-related metrics effectively dictate interactions in some learnable and interpretable functions (called latent functions), which can embed domain knowledge of traffic rules and driving scene context. Meanwhile, the derivatives of the potential function relative to the coordinate system (e.g., x and y directions) result in scaling virtual forces that "push" or "pull" the ego vehicle to minimize the cost of local vehicle planning while interacting with surrounding driving vehicles.

(4) Cognitive models. Using relative distance to characterize driving styles, some models reveal the interaction process among multiple agents, such as psychological theories and information accumulation measures; from a behavioral science and psychology perspective, interaction models can also simulate human driving behavior.

Data-driven interaction modeling methods can be divided into the following three categories.

(1) Deep neural networks. Spatio-temporal state feature tensors, spatial occupancy grids, and dynamic insertion domains (interaction behaviors) are three commonly used feature representations in deep learning model interaction modeling. Here, the attention mechanism is a very effective quantification method, representing the relationships among features. There are generally the following entity representations:

①Temporal attention.
②Agent-centric attention.
③Grid-based spatial attention.
④Dynamic Insertion Area (DIA) based attention.
⑤Graph-based attention.

(2) Graph neural networks. Graph models provide an intuitive interface for modeling highly interactive variable sets and data structures. Moreover, graph-based models can usually be interpreted as structured relationships among human subjects. Using graph models, new models can be envisioned for specific environments.

(3) Topological models. Topological models encode spatio-temporal multi-agent interaction behaviors into compact representations from an algebraic and geometric perspective, and recognize deformations under the condition of topological invariance of agent trajectories. Two types of topological models are often used: topological weaving of interaction behavior encoding and topological invariance of agent trajectories.

The main difficulties faced by prediction methods are: In order to reliably estimate the risk of traffic conditions, it is necessary to perform advanced reasoning on a group of interacting and maneuvering intelligent entities, while considering the uncertainties related to data and models. This advanced reasoning computation is large and is not always compatible with real-time risk estimation. For this reason, some recent algorithms do not need to predict all possible future trajectories of all vehicles on the scene to check for collisions. Instead, they focus only on the most relevant trajectories to speed up computation, or use other risk indicators, such as contradictions between maneuvering intentions. The choice of risk assessment method is closely related to the choice of motion model. Therefore, a method that jointly handles vehicle motion modeling and risk assessment will bring significant improvements in this field.

From the perspective of deep learning, the key factors affecting behavior prediction can be divided into three aspects: scene context, expected tasks or goals of motion, and factors of prediction uncertainty.

(1) The road scene is a highly structured environment, and this road structure contains consistent traffic rules. At the same time, road users (pedestrians, vehicles, and cyclists, etc.) should abide by road rules. All these static and dynamic semantics in the road scene constitute the scene context for safe driving. Therefore, the first aspect of influencing factors is the scene context, which usually involves joint modeling of road static semantics and dynamic participants, considering the links and interactions of these components.

(2) Behavior intention prediction is usually influenced by different expected goals and should be inferred with interpretability. Considering this, the importance of road participants is different and changes with the driving scene. For future predictions conditioned on expected goals, “goals” are modeled as the future states expected by intelligent entities (defined as destination coordinates or types of motion). Imitation learning estimates the likelihood of trajectory prediction by satisfying goals determined by human experts, and multiple untrustworthy predictions are omitted through model training. Usually, the goal is similar to the destination, and the motion goal or destination is considered as a prior. For known goals, goal-conditioned predictions are interrelated through inverse optimal control (IOC) or inverse reinforcement learning (IRL).

(3) Due to changes in the surrounding driving scene context, prediction uncertainty is a natural factor in behavior intention prediction. Inherent multimodality, partial observability, short-term scale, data imbalance, domain gap, and insufficiency all lead to uncertainty. In addition, due to the generalization problem of deep learning models, the distribution of predictions may involve bias, so there are two types of uncertainty in future state prediction: the arbitrary/data uncertainty of expected goal latent decision variables (aleatoric/data uncertainty), and the cognitive/model

uncertainty caused by lack of knowledge in evaluating model generalization (epistemic/model uncertainty). The following discusses examples of prediction models in detail with reference to some literature.

Reference [8] is Uber's driving prediction work at the Toronto Research Institute. First is the joint goal detection and prediction system (fast and furious, FAF).The data here is the point cloud of lidar, using multi-frame bird's-eye view projection as the input of the single-step detector model, to detect, track and short-term motion predict the future trajectory of the target; the input representation is a 4D tensor, encoding the occupancy grid of 3D space within multiple time frames; using 3D convolution of space and time to produce fast and accurate predictions. Due to the inherent sparsity of point cloud data in 3D space, this method saves a lot of computation compared to performing 4D convolution in 3D space and time.

Reference [35] proposes a unified representation method–Rules-of-the-Road encodes high-level semantic information in a spatial grid, allowing deep convolutional models to integrate complex scene context. In this way, in the overall temporal model of intelligent behavior, the learning of interactions between entity-entity and entity-environment can be simply forward-calculated at each time step.

This method includes: ① a new input representation of the entity and its surrounding world context; ② a neural network that maps this past and present world representation to future behavior; ③ several future behaviors that can be integrated into the planning system are one of the output representations. Note that this model is an entity-centered model, which means it models a single "target entity" but takes into account all other world environments, including other entities.

The road network information (lanes, sidewalks, traffic lights, and traffic signs, etc.) is mapped to geometric elements and rendered into RGB images in a top-down grid representation, where each element type corresponds to a unique color, forming a tensor with color 3 channels. Any time step forms a tensor for the target entity, with channels of dimensions such as position, speed, and acceleration, and this scalar is encoded at the center position of the entity, corresponding to the road map tensor in space. To model entity interactions, all other entities are aggregated into an encoded tensor in the same way; these tensors are 7 channels. Additional scene context is encoded as an RGB image of size 3 channels, containing all entities in the scene colored by type (one of bicycle, vehicle, pedestrian) oriented bounding boxes, encoding the range and direction of these targets. It also includes the rendering of intersection traffic lights: The mask intersection shows the connection of each road, rendering the states of allowing (green light), yielding (unprotected), or prohibiting (red light).

A good output representation must have the following characteristics: ① The probability distribution in the entity state space at each time step, for a safety-critical system, a single most likely point estimate is not enough; ② Multimodal, because it is important to cover all potential actions that the entity may take (such as ways of passing through intersections); ③ One-time, for efficiency reasons, it is necessary to predict the complete trajectory (i.e., the time series of state distributions) without the need to iteratively apply recursive steps.

In the experiment, one output representation is the discrete k-Gaussian mixture trajectory distribution of future time, and the other is the occupancy grid map, representing the existence probability of the corresponding state in the future at this grid location.

An encoder–decoder architecture is used for modeling, where the encoder maps the 4D input tensor (time×space×channel) to some internal latent representations, and the decoder uses this representation to model the output distribution of the state at a predetermined set of real-time offsets.

Google's Waymo proposed a method to imitate and learn human driving strategies through driving data, ChauffeurNet, which opposes purely simulating all data, but adds disturbances in real data, creating some interesting scenarios, such as collisions, leaving the driving road or unsmooth trajectories, etc., thus adding some losses on the imitation loss to penalize unexpected events and encourage progress (these disturbances provide important information to enhance the robustness of the model). Experiments show that the ChauffeurNet model can handle complex situations.

The driving model consists of several parts, which is the FeatureNet, used to receive input data and create a context feature representation shared with other networks; These features are used by the agent RNN, which iteratively predicts continuous points in the driving trajectory; Each point in the trajectory is characterized by its position, heading, and speed; The agent RNN predicts the future agent box heatmap, which is used to calculate the loss function.

In a more detailed description of the model's operation, the rendered inputs are fed into the large receptive field FeatureNet with skip connections, and the output captures the feature map F of environmental context and intent; Based on these features and the iterative number k, the previous memory M_{k-1}, the past agent position from the agent RNN, and the predicted previous agent box (heatmap) B_{k-1} as input, the agent RNN can get the predicted position pk and predicted agent box B_k; Memory M_k is an additional storage composed of single-channel images, in iteration k, the memory increases by 1 at the position pk predicted by the agent RNN, and then this memory is fed into the next iteration.

The agent RNN outputs the next posture heatmap of the agent, and the rough predicted position pk can be obtained from this heatmap using the argmax operation. Then, the agent RNN uses a prediction network composed of fully connected layers, which refines the predicted posture at the sub-pixel level and estimates the agent's heading and speed.

Note: Reference [11] specifically mentions an example of successfully avoiding a parked car on the side of the road in the experiment.

VectorNet is a hierarchical graph neural network (GNN). First, it uses vectors to represent the spatial locality of each road component, and then models the high-level interactions between all components. The vectorized HD map and agent trajectory avoid the steps of lossy rendering and computing dense ConvNet encoding. To further enhance VectorNet's ability to learn context features, a new auxiliary task is adopted to restore randomly erased map entities and agent trajectories based on context.

The geographical range of road features can be points, polygons, or curves in geographical coordinates. For example, lane boundaries contain multiple control points for constructing spline curves; Crosswalks are polygons defined by several points; Parking signs are represented by a point. All these geographical entities can be approximated as polylines defined by multiple control points and their attributes. Similarly, the dynamics of moving agents can also be approximated as polylines based on their motion trajectories. All these polylines can be represented as a set of vectors.

GNN is used to merge these sets of vectors. Each vector is viewed as a node in the graph, and the node features are set as the start and end positions of each vector, as well as other attributes, such as the polyline's group ID and semantic labels. Context information from the HD map and the trajectories of other moving agents are propagated to the target agent nodes through the GNN, and then the future trajectory of the target agent is decoded using the output node features corresponding to the target agent.

Specifically, to learn competitive representations with GNN, the connectivity of the graph is constrained based on the spatial and semantic proximity of the nodes.

The overview of VectorNet is a hierarchical graph structure, with vectors of the same semantic label and the same polyline connected and embedded in polyline features, while all polylines are fully connected to exchange information. The first layer is the local polyline subgraph, representing individual graphic elements; The second layer is the global interaction graph, reflecting the relationships between elements. Here, a multilayer perceptron is used to implement the local graph, and a self-attention mechanism is used to implement the global graph.

The polyline subgraph network can be seen as a generalization of the 3D PointNet. However, this method is more suitable for encoding structured map annotations and agent trajectories because it has the following three features: ① Embedding sorting information into vectors; ② Constraining the connectivity of the subgraph based on polyline clustering; ③ Encoding attributes as node features. In addition to behavior prediction, an auxiliary graph completion task is added.

The goal is, more specifically, to randomly mask the input node features belonging to the scene context or the intelligent body trajectory, and require the model to reconstruct the masked features. The intuition is to encourage the graph network to better capture the interactive behavior between the intelligent body dynamics and the scene context.

MultiPath is proposed by Google's Waymo, which uses a set of fixed future state sequence anchors, corresponding to different modes of the trajectory distribution. Based on the top-down scene representation, the scene CNN extracts intermediate features and encodes the states of each intelligent body and their interactions. For each intelligent body in the scene, a view of the intermediate feature representation is cropped centered on the intelligent body, and the probability of the fixed predefined anchor trajectory is predicted. For each anchor, the offset of the anchor waypoint and the uncertainty are regressed, and a Gaussian mixture distribution is generated at each time step (note: MultiPath++ has been further improved recently).

Here, the concept of uncertainty is broken down into independent quantifications. Intention uncertainty models the uncertainty of the rough intention or expected goal of the intelligent body. Under the condition of intention, there is still control uncertainty, which describes the uncertainty of the state sequence that the intelligent body will follow to satisfy its intention. Both intention and control uncertainty depend on past observations of the static and dynamic world environment. In MultiPath, each intention is represented as an anchor trajectory, but its prior is obtained through k-means clustering.

In MultiPath, the historical representation of dynamic and static scene context is from a 3D data array rendered from a top-down orthogonal perspective view. The first two dimensions represent the spatial position in the top-down image, and the channels of the depth dimension contain a fixed number of previous time static and time-varying (dynamic) content. Intelligent body observations are rendered as binary images of oriented bounding boxes, with each time step corresponding to a channel. Other dynamic environments, such as traffic light states, and static environments of the road (lane connections and types, stop lines, speed limits, etc.) form additional channels.

TNT (target-drive N trajectory, driving target N trajectory prediction) prediction is also a work of Google's Waymo, which is a method to generate target trajectory state sequences based on historical data (i.e., interactions between multiple intelligent bodies and the environment), and obtain a compact set of trajectory predictions based on likelihood estimation. The difference is that previous work defined the intelligent body's intention as a latent variable and relied on sampling at test time to generate various trajectories.

For an overview of the TNT model, first, the scene context is encoded as model input. Then, follow the three core stages of TNT: ① target prediction, proposing an initial set of M targets; ② motion estimation, estimating the trajectory of each target; ③ trajectory scoring, sorting the trajectory hypotheses and outputting the final set of K predicted trajectories.

Modeling the scene context is the first step in trajectory prediction, capturing the interactive behavior between the intelligent body and the road and between intelligent bodies. TNT can use any suitable context encoder. When high-definition maps are available, the latest hierarchical graph neural network VectorNet is used to encode the context. If only top-down image representation of the scene context is available, such as ChauffeurNet, ConvNet is used as the context encoder. Similarly, like MultiPath, the uncertainty of future states is divided into target or intention uncertainty and control uncertainty.

In vehicle trajectory prediction, points are uniformly sampled from the lane centerline in the high-definition map and used as target candidates, assuming that the vehicle will never leave the lane. For pedestrians, a virtual grid is generated around the intelligent body, and the grid points are used as target candidates. For each target candidate, the TNT target predictor generates a triplet, namely, position deviation and position discrete distribution.

Compared with direct regression, the biggest advantage of modeling future states as a discrete target set is that it will not be affected by modal averaging, which is exactly the main factor hindering multimodal prediction.

LaneGCN is an open-source motion prediction model proposed by Uber ATG's Toronto lab, which uses a structured map representation to interact with traffic participants. In order to capture the complex topology and long-distance correlation of the lane graph, LaneGCN extends with multiple adjacency matrices and graph convolutions dilated along the lane.

In the interaction model, a fusion network is composed of four types of interactions, namely participant-lane, lane-lane, lane-participant, and participant-participant. ActorNet extracts features from the trajectories of traffic participants, while LaneGCN extracts map features from the lane graph generated from the high-definition map. Finally, the fusion network models interactive behavior to predict motion trajectories.

The overall architecture of motion prediction consists of 4 modules: ① ActorNet receives past participant trajectories as input and uses one-dimensional convolution to extract participant node features. ② MapNet builds a lane graph from the high-definition map and uses LaneGCN to accurately determine lane node features. ③ FusionNet is a stack composed of 4 interaction blocks. The participant-lane block fuses real-time traffic information of participant node-lane node; the lane-lane block propagates information on the lane graph and updates lane features; The lane-participant block merges the updated map information from the lane node to the participant node; The participant-participant block performs interactions between participants. Another LaneGCN is used as the lane-lane block, and the airspace attention layer is used as other blocks. The prediction head uses the post-fusion participant features to generate multimodal trajectories.

ActorNet uses Feature Pyramid Network (FPN), layer normalization, and ReLU.

MapNet consists of two parts: the lane graph and LaneGCN. There are four types of lane connections: predecessor, successor, left neighbor, and right neighbor.

The architecture of LaneGCN is a stack composed of four multi-scale LaneConv residual blocks. Each residual block consists of a LaneConv (1,2,4,8,16,32) and a linear layer with a residual connection, followed by layer normalization and ReLU. LaneConv is a multi-scale graph convolution layer with dilation.

The prediction head includes classification and regression. Classification estimates the modality of the trajectory, and regression is the trajectory of the modality and its confidence score.

In Ref. [23], knowledge distillation techniques were developed between two probability-based motion prediction models and applied to narrow the performance gap between ego-centric models and scene-centric models.

For the teacher–student distillation method, on the left, the teacher and ego-centric models are repeatedly and independently applied to each ego in the scene, with all model inputs and outputs represented in each ego-centric coordinate system; on the right, the student and scene-centric models are applied only once to the entire scene, without the need for each ego to repeat calculations. Although the scene-centric model is faster, it is often less accurate because it must also understand and model the per-ego invariance built into the ego-centric method. To take advantage of the computational efficiency of the scene-centric method while benefiting from the accuracy of the ego-centric method, a knowledge extraction method is proposed here, using the predicted trajectories of the ego-centric or teacher model to train the scene-centric or student model.

The output of trajectory prediction is a Gaussian Mixture Model (GMM), which can be interpreted as the "confidence" of the predicted trajectory. In the ego-centric model, the coordinates of road elements (such as lanes, crosswalks) and the states of other egos are described relative to the ego's pose, so this representation is invariant to the ego's global position and direction. In addition, the scene-centric model encodes the states of all egos in a shared fixed coordinate system. The models run in this framework are usually "top-down" or BEV representations, discretizing the world into spatial grid units and applying a CNN backbone to encode the scene. After such processing, the prediction heads of these models perform global-to-local transformations to decode the trajectories in the egos.

The ego-centric coordinate system model (ACM) serves as the teacher. The ego-centric model encodes, processes, and infers the world from the perspective of each ego. This representation requires all scene information to be transformed from the global coordinate system to the ego coordinate system. Therefore, with the ego-centric method, inference time and memory requirements increase with the number of egos.

The ACM architecture input includes the following four types of data: road map, traffic light information, motion history (i.e., ego state history), and ego interaction. For road map information, ACM encodes road elements in the 3D HD map as multi-segment polylines through an multi-layer perceptron (MLP); for traffic light information, ACM uses a separate LSTM as the encoder; for motion history, ACM uses an LSTM to process a series of past observations, with the last iteration of the hidden state used as the history embedding; for ego interaction, an LSTM encodes the motion history of nearby vehicles in the ego-centric framework, and aggregates all nearby vehicle information through max pooling to achieve a single interaction. This is a simple form of modeling full-connection neighboring vehicle interaction. Finally, these four encodings are concatenated to create an embedding for each ego in the ego-centric coordinate system. An MLP-based decoder transforms the final embedding into a GMM.

The scene-centric coordinate system model (SCM) serves as the student. In the SCM architecture, input data is represented in a global coordinate system, shared among all egos. As mentioned above, one advantage of this method is that it can process the scene as a whole, resulting in efficient inference invariant to the number of egos.

The input to SCM is three types of data: The road information is represented as points enhanced by semantic attributes, the intelligent body information is represented as points sampled from each intelligent body toward the frame, and the traffic light information is represented as points enhanced by semantic attributes. SCM uses the PointPillars encoder and 2D convolution backbone to encode all these input points. A subgraph is cropped from a position in the feature map that maps to the current position of the scene intelligent body, and a final per-intelligent body embedding is extracted. Note that even if a per-intelligent body embedding is finally achieved, all upstream processing is done for the entire scene at once. Like ACM, an MLP-based decoder converts this final per-intelligent body embedding into a GMM.

In this distillation method, the student model is trained to match the complete trajectory set output of the teacher model. Ignoring the covariance of the model output GMM, taking the modality of each component, a set of trajectories is obtained. The weights of this set of trajectories are given by the output of the neural network. The distillation loss is divided into two parts: The first part, all the teacher's predicted trajectories are used as multiple pseudo-real trajectories to train the student; Here, it is hoped that the kth teacher trajectory is the maximum likelihood of the kth modal distribution learned by the student; The second part, imposing cross-entropy loss, encourages the student's trajectory modal distribution to match the teacher's modal distribution.

A trajectory prediction model ScePT based on policy planning is proposed in reference [21], which explicitly enforces scene consistency and learns intelligent body interaction strategies that can be used to constrain predictions. ScePT is a discrete CVAE (conditional variational autoencoder) model, which outputs joint trajectory predictions for multiple intelligent bodies in the scene, and infers the motion strategy of each intelligent body and the influence of its surrounding intelligent bodies to ensure high scene consistency in its predictions.

To maintain scene consistency, ScePT is a scene-centered model, i.e., its output prediction is the joint trajectory of multiple nodes in the scene. Given a scene with multiple nodes, a spatio-temporal scene graph is generated, where nodes represent intelligent bodies and edges represent interactions between intelligent bodies.

The scene graph is determined by the adjacency matrix and divided into small cliques with the maximum scale (fixed as a parameter). With the cliques, the state (node) and relative state (edge) historical data of the intelligent bodies can be encoded into feature vectors through LSTM. The encoder models the joint latent distribution, which is a Gibbs distribution composed of node factors and edge factors.

The decoder is designed inspired by the motion planning process, i.e., each intelligent body is viewed as a motion planner, and its planning process is simulated to output trajectory predictions.

The inputs to the policy network are the current state of the clique nodes, the reference trajectory, and the clique latent variables. The reference trajectory is generated by a GRU network with state history encoding, map encoding, and latent variables as inputs, and then the current node state is compared with the reference trajectory to obtain the tracking error and the next waypoint in the local coordinate system. To model the edges, pair the states of its two nodes together and feed the state pair into a pre-encoding network (fully connected), then input it into an LSTM unit. To encode the variable neighbors, compress all the edges of the node into a single observation encoding through an attention network, then connect the observation encoding, latent variables, and tracking error and feed them into a fully connected action network to obtain the node's control action prediction.

For the overall structure of ScePT, the encoder uses LSTM-encoded state and edge history and CNN-encoded local maps, and generates a discrete Gibbs distribution in the clique latent variables; The latent variables, together with the state history and map encoding, are used by the GRU to generate the desired trajectory for each node;

Then the desired trajectory and latent variables are passed to the policy network to obtain the closed-loop trajectory prediction.

ScePT does not require any structural changes to produce constrained predictions, as it directly learns the interaction strategies of intelligent bodies. Simply fix the trajectory of the constrained intelligent body and output the trajectory prediction of other intelligent bodies in the clique to generate constrained predictions. Since the fixed future trajectory will not fall into any latent modality, any factors related to the constrained node can be removed from the Gibbs distribution factor graph.

To demonstrate the performance of ScePT integrated with downstream planners, its predictions are fed back to a downstream MPC-based planner. The MPC planner considers multimodal trajectory predictions and performs contingency planning through branching.

8.2.2 Learning Driving Behavior

The work of the University of California, Berkeley, is to obtain an integrated planning and control framework through imitation learning, combining a two-layer hierarchical structure of machine learning and optimization theory [10]. The first layer of this framework is the policy layer, which learns the long-term optimal MPC driving strategy through a neural network; the second layer is the execution layer, which mainly tracks the reference trajectory generated by the previous policy layer, and is an optimization-based short-term controller that guarantees short-term safety and feasibility. Notably, the first layer uses online imitation learning, where dataset aggregation is borrowed.

Methods such as aggregation and DAgger can rapidly and continuously improve the policy layer. The two-layer hierarchical architecture comprises perception, decision-making, integrated planning, and control modules. At each time step, the perception module senses the surrounding environment and measures or estimates the full state of the autonomous vehicle. Based on this perceived information and the predefined driving tasks, the decision-making module specifies the reference lane and lateral objective, which then guide the subsequent planning and control stages.

The extracted features include the following three parts.

- **Motion variables of the map**: the current position and velocity of the self-driving car within the abstract map representation.
- **Target variables**: the instantaneous distance from the self-driving car to the designated target point.
- **Safety variables**: the current relative position and velocity of the self-driving car with respect to surrounding obstacles or vehicles.

In order to formulate a smooth and efficient driving strategy, the long-term MPC will penalize the self-driving car for driving as fast as possible within the speed limit, while considering all constraints of system kinematics feasibility (equation constraints), dynamics feasibility, and collision-free safety (inequality constraints). The execution layer then tracks the reference trajectory given by the policy layer.

In order to learn the policy neural network, while avoiding completely imitating and training set data-like policies, an improved method combining data aggregation and imitation learning is proposed. A data aggregator method adopted, called sampled DAgger, contains the following steps.

(1) Run the long-term expert MPC on randomly generated simulated scenarios, collect the initial training set D_0, and train the network to generate the initial policy π_0. Initially, D_0 contains about 20,000 scenarios.

(2) Run the two-layer hierarchical structure, learn the policy π_0 in the policy layer, and at the same time, run the long-term expert MPC in parallel at a slow speed, periodically marking features with the best output. The long-term MPC runs once every M time steps, i.e., $T_{MPC}^{interval} = Mdt$, the interval is set long enough to solve the optimization problem of the long-term MPC.

(3) Compare the optimal approach trajectory given by π^* with the policy trajectory given by π_0 (k = 0, 1, 2, ...). If the normalized Euclidean distance between the two trajectories is greater than the predefined safety standard, label the feature ft with the corresponding optimal output and push them into the new training set D'. Once the size of D' reaches the predefined threshold, proceed to Step (4).

(4) Aggregate the previous training set with D', i.e., $D = D \cup D'$, and retrain the network to produce a new policy π_{new} and set $\pi_0 = \pi_{new}$.

(5) Repeat Steps (2) (4) until the learned policy π_0 converges.

Compared with the original DAgger, the sampled DAgger is data efficient because it does not ask the expert MPC for all policy data, but only queries data with severe output deviations. Although it uses a predefined safety standard, it adopts an additional downsampling step, which allows online querying of the time-consuming expert MPC policy, helping to reduce offline re-labeling work.

In reference [10], a generalization treatment of the learned policy was done, which can be used in complex driving scenarios. Generalization is to reduce a continuous driving problem to a series of abstract scenarios, and each abstract scenario can directly apply the learned policy model to solve.

Example 1: Both lane markings II and III can be crossed. In this case, virtual roadside is introduced to extract the abstract dual-lane scenario. When the self-driving car is at position I, and the driving decision indicates it to stay on "Reference Lane I," "Scenario I" is extracted by setting the roadside as "virtual roadside"; similarly, when the self-driving car is at position II and the reference lane is set to "Reference Lane II," "Scenario II" is used for the policy layer; and if the reference lane is set to "Reference Lane III," "Scenario III" is formulated by introducing a virtual roadside.

Example 2: Lane markings cannot be crossed. In this case, not only is the virtual roadside introduced, but also a virtual neighboring car is introduced to prevent the self-driving car from making aggressive overtaking maneuvers. When the self-driving car is at position I and chooses "Reference Lane I," a virtual neighboring car moving at the same speed as the self-driving car is introduced for feature extraction, which converts "Scenario I" into "Scenario IV"; similarly, when the self-driving car is at position II and sets "Reference Lane III," "Scenario III" becomes "Scenario V";

the introduction of the virtual neighboring car changes the features of the policy layer and makes the learned policy "think" that overtaking and staying on the reference lane is unsafe.

The Neural Motion Planner (NMP) is Uber's ATG proposed a model for learning autonomous driving in complex urban scenarios, including handling traffic signals, yielding, and interacting with multiple road users. This holistic model takes raw lidar data and high-definition maps as input, generates interpretable intermediate representations in the form of 3D detection and their future trajectories, defines the cost body of each position that the autonomous vehicle can take within the planning distance, and then samples a set of physically feasible trajectories, choosing the one with the lowest learning cost. Here, the cost body naturally captures multimodality.

The entire model is trained end-to-end according to multi-task objectives. The planning loss encourages planning similar to human driving trajectories to take the lowest cost. Note: this loss is sparse because a true trajectory only occupies a small part of the space. Another perception loss is introduced here, encouraging the intermediate representation to produce accurate 3D detection and motion prediction. This ensures the interpretability of the intermediate representation and makes learning faster.

The task of this motion planning is defined as a deep structured minimization problem:

$$s* = \arg\min_{s} \sum_{t} c^t(s^t) \tag{8.10}$$

where, c^t is the learning cost at time t; and s^t is the BEV position at time t. This minimization is approximated by sampling a set of physically feasible trajectories $s = s^0, s^1, \ldots, s^{T-1}$, and the one with the smallest cost is chosen.

In NMP, the trunk network takes lidar data and maps as input, outputs the borders of other participants for the future (perception), and uses a set of T filters (T corresponds to the cost body for planning). Next, for each trajectory proposal of the sampler, its cost will be indexed from different filters of the cost body and summed together. The trajectory with the lowest cost will be the final plan.

The perception head consists of two convolutional layers, one for classification and the other for regression. The cost body head consists of several convolutional layers and deconvolutional layers. To generate the cost body c with the same resolution as the BEV input, two deconvolutional layers are applied to the trunk output. Each deconvolutional layer is followed by a convolutional layer, and then the final convolutional layer is applied. Each filter generates a cost body c_t for a future time t. This allows the cost body c to be simply indexed to evaluate the cost of any trajectory.

The curve representation method of the trajectory is divided into three steps: Clothoid representation, speed contour estimation, and trajectory representation.

Clothoid curves, also known as Euler or Cornu spirals, are used to represent the 2D paths of autonomous vehicles. First, a set of parameters of the Clothoid are sampled to determine the shape of the trajectory; then the speed description is sampled to determine the driving speed of the autonomous vehicle along this trajectory. Finally, combining these two, a spatio-temporal trajectory is obtained.

The SafeNet work comes from the Lyft Level5 team, which adds a rule-based fallback layer to perform a sanity check on machine learning (ML) decisions, which can reduce the collision rate of the ML planner. In the basic framework of SafeNet, the planner's predicted trajectory is checked for trajectory in the fallback layer.

In the architecture diagram of the SafeNet model, a hierarchical graph network—similar to VectorNet—is employed. The PointNet module locally encodes input information, including vehicle pose and size, the pose, size, and type of surrounding agents, as well as HD map features encompassing both static and dynamic elements together with route planning. A Transformer is then applied for global embedding to capture interactions between agents and map features. To ensure the physical feasibility of the predicted trajectories, a kinematic decoder based on the ego-vehicle model is introduced, which transforms the embeddings into control signals.

Network training is based on imitation learning with added random perturbations, expanding the distribution to reduce the impact of covariance shift. The loss function considers factors of curvature and jitter to improve comfort.

In the fallback layer, several dimensions are considered for trajectory evaluation.

(1) Dynamic feasibility: includes constraints on longitudinal jerk, longitudinal acceleration, curvature, curvature rate, lateral acceleration, and steering jitter (curvature rate X speed).

(2) Legality: traffic rules, such as stop signs, red lights, leaving the road, or yielding.

(3) Collision probability: collision detection. This marks the trajectory as [Feasible, Infeasible], and finally produces a feasible result closest to the ML predicted trajectory.

8.3 Pedestrian Behavior Model and Prediction

Understanding human motion is a key skill for autonomous driving to coexist and interact with humans as traffic participants, involving information representation, perception, and motion analysis. Prediction plays an important role in human motion analysis: Predicting how scenarios involving multiple pedestrians and vehicles dynamically change, and actively integrating this knowledge, involves enhanced active perception, predictive planning, model predictive control (MPC), or human–machine interaction, etc.

The challenge of accurately predicting human motion comes from the complexity of human behavior and the diversity of internal and external stimuli. Motion behavior may be driven by factors such as: self-goal intent, the presence and behavior of surrounding intelligent entities, social relationships between intelligent entities, social rules and norms, or an environment with certain topological structures, geometric structures, bearing capacity, and semantics. Most of these factors are not directly observable and need to be inferred from noisy perceptual clues or modeled from contextual information. Moreover, for effective practice, the execution of motion prediction should be robust and real time.

From the perspective of predictive modeling, there are three types of methods: ① Physics-based methods, which use explicit dynamic models based on Newton's laws of motion to predict the motion state of intelligent entities; ② Pattern-based methods, which learn the original trajectory from observed intelligent entity behavior to predict future motion; ③ Planning-based methods, which reason about possible goals in some form and calculate possible ways to achieve these goals.

Contextual clues are defined as all relevant internal and external stimuli that influence motion behavior, and prediction methods based on them can be grouped into three categories: ① Target-agent clues, which include attributes such as posture, motion state, and semantic properties (e.g., age and gender). ② Static-environment clues, which may be unperceived (open environment), obstacle-perceived (individual obstacles), map-perceived (geometric and topological features), or semantic-perceived (e.g., crosswalks, traffic signals, prohibitions). ③ Dynamic-environment clues, which may be unperceived (open), alone-perceived (accounting for immediate surroundings), or group-perceived (considering the presence and behavior of groups).

Next, we will discuss the system framework for pedestrian behavior analysis.

Next is a end-to-end multi-task learning system proposed by Google [34], which utilizes human behavior information and visual features interacting with the surrounding environment.

The model encodes individuals using semantic features that capture their visual appearance, body motion, and interactions with the surrounding environment. To facilitate training, the network incorporates auxiliary tasks aimed at predicting future activity locations. In one such task, a discretized Manhattan grid is introduced, serving as the prediction target for location forecasting.

A joint prediction model of activity and path is also designed, which has the following two benefits.

(1) Learning activity and path together may benefit future path prediction.

(2) Considering the semantic context in the video improves the understanding of future paths and activities.

The model has the following four key components.

(1) The person–behavior module extracts visual information from the person's behavior sequence.

(2) The person–interaction module is designed to model the interactions between individuals and their surrounding environment.

(3) The trajectory generator summarizes the encoded visual features and predicts future trajectories through an LSTM decoder with focal attention.

(4) Activity prediction uses visual semantics to predict future activity labels.

In addition, the Manhattan grid divides the scene into multi-scale discretized grids for classification and regression, providing activity location prediction.

Given a sequence of frames for predicting human behavior, the person–behavior module and the person–interaction module encode visual semantic information into feature tensors.

To model changes in a person's appearance, a pre-trained object detection model with RoIAlign functionality is used to extract fixed-size CNN features for the bounding box.

In a visual representation of the person–behavior module, first, the features of each person's spatial dimensions are averaged; then, they are input into the LSTM encoder; finally, a Tobs × d feature representation is obtained, where d is the hidden size of the LSTM. To capture body motion, a person keypoint detection model is used to extract person keypoint information. The keypoint coordinates are embedded using a linear transformation and then input into the LSTM encoder.

In the visual representation of the person–interaction module, the person–target feature captures relational cues such as the distance between a person and another individual or vehicle, while the person–scene feature encodes contextual information, for example, whether the person is near a sidewalk or positioned on grass.

The person interaction module uses an LSTM decoder to directly predict the future trajectory of x-y coordinates. The hidden state is initialized with the last state of the pedestrian trajectory LSTM encoder. In the auxiliary task, i.e., the prediction of activity location, at each moment, the x-y coordinates are calculated based on the decoder state and a fully connected layer.

The focus attention mechanism projects multiple features into related spaces, capturing discriminative features.

In order to bridge the gap between trajectory generation and activity marking prediction, the activity location prediction (ALP) module predicts the final location of future activities. Activity location prediction includes two tasks: location classification and location regression.

The Spatio-Temporal Interaction Network (STINet), developed by Google, is an end-to-end two-stage network. In addition to modeling the 3D geometry of pedestrians, it also incorporates temporal information for each individual. In the first stage, the network predicts current and past locations, linking frames of the same pedestrian. In the second stage, it captures comprehensive spatio-temporal dependencies. Furthermore, an interaction graph is employed to model relationships between pedestrians, aggregating information from adjacent individuals to enhance interaction-aware prediction.

In an overview, STINet takes a sequence of point clouds as input, performing pedestrian detection while simultaneously predicting future trajectories. The point clouds are first processed by a pillar feature encoding module to generate pillar features, which are then passed through a ResUNet backbone to extract high-level representations. The Temporal Region Proposal Network (T-RPN) leverages these features to generate temporal proposals that include both past and current bounding boxes for each target. Finally, the Spatio-Temporal Interaction (STI) feature extraction module learns representations for each temporal proposal, enabling accurate pedestrian detection and trajectory prediction.

In the architecture of the STI feature extraction module, it extracts local geometric, local dynamic, and historical path features given a temporal proposal. For local geometric and local dynamic features, the dashed box area is used for feature extraction. Cross-proposal local features are used for related reasoning to generate interaction features.

Social-STGCNN does not aggregate various learned pedestrian states, but models pedestrian interactions graphically, proposing a kernel function to embed pedestrian social interactions into an adjacency matrix.

In Social-STGCNN's work, pedestrian social interactions and their temporal dynamics are represented by a spatio-temporal graph.

Social-STGCNN consists of two parts: one is the Spatio-Temporal GCNN (STGCNN), and the other is the Time Extrapolation CNN (TXP-CNN). The former performs spatio-temporal convolution on pedestrian trajectory graph representation to obtain features, and the latter uses these features to predict pedestrian trajectories.

A set of spatial graphs G_t is established to represent the positions of pedestrians in the scene at each moment. G_t is defined as $G_t = (V_t, E_t)$, where the graph node set $V_t = v_t^i | i \in 1, 2, \ldots, N$, and the graph edge set $E_t = e_t^{ij} | i, j \in 1, 2, \ldots, N$, where:

$$e_t^{ij} = \begin{cases} 1, v_t^i \text{ and } v_t^j \text{ are connected,} \\ \quad 0, \text{otherwise} \end{cases} \tag{8.11}$$

In order to model the degree of influence between nodes in the graph, a kernel function is used to calculate its parameter a_t^{ij}, which is used as an element of the adjacency matrix Et, defined as follows:

$$a_{sum,t}^{ij} = \begin{cases} \frac{1}{\left\| v_t^i - v_t^j \right\|_2}, & \left\| v_t^i - v_t^j \right\|_2 \neq 0 \\ 0, & otherwise \end{cases} \tag{8.12}$$

For the Social-STGCNN model, first, given T frames, a spatio-temporal graph $G = (V, A)$ is constructed; then, a time-space embedding is created by a Spatio-Temporal Graph Convolutional Neural Network (STGCNN) G. After this, the TXP-CNN predicts future trajectories. The TXP-CNN consists of a series of CNNs with residual connections; only the first layer of the TXP-CNN does not have a residual connection when receiving.

The TXP-CNN directly acts on the time dimension of the graph embedding and expands it for prediction. Because the TXP-CNN relies on convolution operations in the feature space, it has fewer parameters compared to recurrent units. One property to note is the TXP-CNN layer, which is not permutation invariant, because changes in the graph embedding before the TXP-CNN will lead to different results. In addition, if the order of pedestrians is arranged from the input of Social-STGCNN, the prediction remains unchanged.

Overall, there are two main differences between Social-STGCNN and STGCNN: first, Social-STGCNN constructs graphs in a completely different way using a kernel function; second, in addition to the spatio-temporal graph convolution layer, the flexibility of the TXP-CNN to manipulate the time dimension is added. The original design intention of STGCNN was for classification. Through the TXP-CNN, the model can use the graph embedding derived from STGCNN to predict future trajectories.

8.4 Driving Behavior Cloning/Learning

The end-to-end planning method mentioned earlier is a method to obtain human-like driving behavior by using machine learning to treat the input of the sensor and the planning output as a "black box." The methods of machine learning mainly include imitation learning and reinforcement learning.

Firstly, imitation learning is a type of method that is based on the learning strategy of the intelligent body based on expert trajectories, usually providing expert decision-making and control information. There are three widely used methods: The first is a passive method, called Behavioral Cloning (BC); The second is based on BC, called Direct Policy Learning (DPL); The last one depends on the task, called Inverse Reinforcement Learning (IRL).

The goal of Behavior Cloning (BC) is to learn a target policy by passively observing complete expert demonstrations, under the assumption that all state–action pairs in the trajectory are independent. Owing to its straightforward structure, imitation learning can reduce uncertainty across subsystems and shorten feedback latency. However, this simplicity also leads to a key drawback: limited interpretability, as the model cannot provide adequate reasoning to explain its decisions. To mitigate this limitation, many researchers have proposed introducing intermediate representation layers, as exemplified by Conditional Imitation Learning (CIL).

A defining feature of BC is that only expert demonstrations can be used to generate training examples. This constraint leads to a critical limitation: the training set is merely a subset of the states encountered during policy execution. As a result, when the dataset is biased or overfitted, the generalization ability of BC is significantly restricted. Furthermore, when the agent encounters unfamiliar states, it struggles to infer the correct behavior.

The main advantage of DPL is to use expert trajectories to guide the intelligent body on how to recover from current mistakes. In this way, DPL alleviates the BC restrictions caused by insufficient data. Here, the method based on DAgger reduces the dependence on the dataset and improves the learning efficiency. However, these methods cannot distinguish between good or bad expert trajectories, nor can they ignore inappropriate behavior as learning samples. DPL is an iterative online learning strategy that alleviates the requirements for the quantity and distribution of datasets, and because it effectively eliminates incorrect strategies, it promotes the continuous improvement of strategies.

Inverse Reinforcement Learning (IRL) aims to avoid the disadvantages of the above methods by inferring the underlying causes between inputs and outputs. Similar to previous methods, IRL initially needs to collect a set of expert trajectories. However, it does not simply learn the state-action mapping, but first infers these expert trajectories, and then optimizes the driving behavior strategy based on complex reward functions. IRL also has a long-tail problem, and how to effectively improve the robustness and interpretability of IRL is an open question. In summary, IL methods require a large amount of annotated data and diversified data distribution. When encountering the same situation, different drivers may make completely different decisions, which will lead to uncertainty during the training process.

Next, let's talk about Reinforcement Learning (RL). The goal of this method is to optimize cumulative rewards through trial and error.

Through continuous interaction with the environment, the intelligent body gradually obtains the best strategy knowledge to achieve the goal. RL methods are divided into four parts: Value-based reinforcement learning, policy-based reinforcement learning, Hierarchical Reinforcement Learning (HRL), and Multi-Agent Reinforcement Learning (MARL).

Value-based reinforcement learning attempts to estimate the value of different actions under a given state, and learns the expected return obtained by taking actions in this state, so that a value can be assigned to each action. The intelligent body learns to associate rewards with state-action in the environment and uses this information to make the best decisions. Among them, Q-learning became famous because of Google's Go software AlphaGo. Value-based methods are limited to providing discrete commands. However, autonomous driving is a continuous process, and continuous methods are better for vehicle control.

Policy-based reinforcement learning has the potential for high-performance limits in the high-dimensional action space of continuous control commands. This method shows better convergence and exploration than value-based methods, such as DDPG and PPO. Training policies from scratch in RL is usually time-consuming and difficult. Combining RL with methods such as Imitation Learning (IL) and Curriculum Learning (CL) may be a feasible solution.

Hierarchical Reinforcement Learning (HRL) decomposes the entire problem into a hierarchy of subtasks, each with its own goals and strategies. Subtasks are organized in a hierarchical manner, and higher level subtasks provide context and guidance for lower level subtasks. This hierarchical organization allows the intelligent body to focus on smaller sub-problems, reducing the complexity of the learning problem and making it easier to handle.

Multi-Agent Reinforcement Learning (MARL), simply put, aims to learn the decision-making strategies of multiple intelligent bodies in the environment. In summary, Reinforcement Learning (RL) methods are adaptable, but they require a large amount of computational resources.

Next, I will introduce some specific end-to-end planning learning work. Recently, end-to-end autonomous driving research has received increasing attention, and there is a trend to replace the previous modular autonomous driving research. There are two reviews in this area [41, 43] that summarize recent work in this direction.

If this perception–prediction-planning is unified under a differentiable network architecture, it can be regarded as an end-to-end system for autonomous driving planning. Compared with the classical paradigm, this end-to-end autonomous driving system provides several advantages [41]:

1. It simply combines perception, prediction, and planning into a single model that can be trained jointly.
2. The entire system, including its intermediate representations, is optimized toward the final planning task.
3. The shared backbone network improves computational efficiency.
4. Data-driven optimization has the potential to improve the system's capabilities by simply expanding training resources.

The last step of prediction is to generate trajectories. This is usually achieved by decoding the captured latent features. Trajectory generation in deep learning mainly discusses two aspects [43]: decoding rules and multimodal modeling.

Decoding rules. They are basically divided into three categories: one-shot, two shots, and N shots. It is more common to directly predict future trajectory points.

1. One-shot methods are further divided into trajectory regression decoding and anchor trajectory classification. Trajectory regression directly uses NN to decode hidden layer features into trajectories, typically MLP. In particular, based on the Transformer structure, trajectory regression can be performed directly. Anchor trajectory classification uses pre-set anchor trajectories as prior knowledge for classification. Determining the anchor trajectory can ensure the feasibility of the prediction and allow stronger kinematic constraints to be imposed. However, this method lacks flexibility because it cannot handle unexpected situations.
2. Two-shot methods include two types: (1) endpoints-and-completion method and (2) initial-trajectory-and-refinement strategy. The endpoint-and-completion method is based on the assumption that the trajectory is mainly defined by its endpoints, which can be regressed or classified as nodes of the road map from a predefined set; usually, classification-based methods also regress the offset of the classified endpoints. The initial-trajectory-and-refinement method combines the idea of anchor trajectory classification with the idea of endpoint-and-completion. It first classifies the initial trajectory instead of the endpoint; then, the refinement step adjusts the trajectory, for example by performing offset regression on each predicted waypoint.
3. N shots methods are either initial-trajectory-and-refinement or autoregression. The variants of the initial-trajectory-and-refinement method are similar to the two shots method; the only difference is the refinement step. The autoregression method uses recursive decoding to gradually advance into the future. It iteratively predicts the next waypoint and adds information about the resulting scene to the latent features.

Multimodal modeling. Because the observed intentions of the autonomous vehicle are unknown, future behavior is inherently uncertain and multimodal. Multimodal modeling approaches can be distinguished into two categories: discrete trajectories or continuous distributions. Discrete trajectories can be sampled from intermediate descriptions or obtained during model design. Continuous distributions, unlike discrete trajectories, are very suitable for direct use for cost functions in planning.

Integral end-to-end planning methods that can be divided into three categories [43]:

1. Monolithic planning refers to methods that consist of a single end-to-end planner that maps multiple state inputs to trajectories. Therefore, the future behavior of surrounding vehicles and their interactions with each other and with the ego-vehicle are not explicitly modeled. Therefore, no explicit predictions are inferred or used. However, the end-to-end planner learns to imitate the behavior of experts during training, based on reasoning about the future. By learning to imitate behavior, the model will also behave in a way that is implicitly based on such reasoning.
2. Interpretable end-to-end planning, which formulates prediction as an auxiliary learning task. Explicit predictions are additional model outputs that are jointly trained with the planning task. Typically, a shared backbone encodes features for both tasks. Separate heads then decode their respective output representations. The additional objective provides a learning signal to the backbone. It acts as a regularizer and can make the learning process more sample efficient, and can improve generalization. Explicit predictions increase interpretability and facilitate introspection.
3. Manual integration of the planner and prediction module, meaning that a separate subsystem is used for each task. The interaction between the two tasks is manually designed based on domain knowledge. The most widely adopted integration is a sequential pipeline, where the prediction module is first inferred, and its output is then passed to the subsequent planning module.

The work in Ref. [12] is to integrate an open simulation platform Torcs with the open-source ROS, focusing on how to develop and evaluate human-like autonomous driving algorithms based on Inverse Reinforcement Learning (IRL). The system includes four sets of hardware modules, namely two platforms (simulator and real vehicle), a learning and planning library, and an additional ROS module for motion prediction and path tracking.

Both platforms are connected to the motion prediction and tracking module through ROS. It is assumed that each platform has the following built-in functions.

(1) Lane line detection or localization and mapping, allowing the vehicle to be positioned relative to the road/track and other vehicles.

(2) Detection and Tracking of Moving Objects (DATMO), which estimates the relative positions and velocities of surrounding vehicles.

(3) Possessing inherent information of the car such as the odometer and engine status (such as revolutions per minute, current gear).

(4) Control, i.e., the interface for steering angle, acceleration, braking, and gear shifting required for autonomous driving.

The core of the IRL learning and planning algorithm is an independent library called BEhavior learning library (BELLY), which is also responsible for motion planning. The BELLY library consists of four modules: ① feature calculation; ② cost function calculation; ③ motion planning; ④ IRL algorithm (learning).

There are also two modules that are unrelated to IRL itself, namely motion prediction and path tracking.

(1) Motion prediction: The motion prediction module receives tracking data (i.e., position, direction, and speed) and outputs H grids, representing the posterior probability of occupying space in the future at $t_1, t_2, \cdots, t_K$. At this time, a Kalman filter is used to complete the prediction, and then it is "rasterized" into these grids. This representation makes it easy to introduce more advanced prediction algorithms.

(2) Path tracking: The output of the planning algorithm is a series of time-stamped waypoints of the vehicle, containing the vehicle's posture and speed information. The path tracking module is responsible for controlling the vehicle along these waypoints, including the following two sub-modules.

① Speed controller. Used to calculate the gear, acceleration, and disconnection instructions required to maintain speed.

② Posture controller. Used to calculate the required steering angle to maintain position and direction.

The purpose of Ref. [13] is to learn a general vehicle motion model, which combines an FCN-LSTM architecture, can learn from large-scale crowdsourced vehicle action data, and uses available scene segmentation auxiliary tasks to improve system performance.

This general model, the input is pixels and the vehicle's historical status and today's status, the prediction is the likelihood function of future motion, defined as a set of vehicle actions or motion granularity (discrete and continuous), similar to the language model of NLP.

In Ref. [14], predicting driver actions in advance can help avoid the consequences of unsafe driving behavior and avoid possible car accidents. Here, driver action prediction is turned into a prediction problem of time-series anomalies. For example, a deep bidirectional RNN model is used to learn the correlation between sensor input and upcoming driver actions, to achieve accurate and high field of view driver action prediction (DAP).

The sensor module processes sensor data, and the feature extraction module independently extracts features that can be used to identify and predict driver actions. The driver action prediction module continuously predicts future driver actions by running a prediction network (PN) model on the extracted features. In addition, the driver action recognition module runs a recognition network (RN) model on the features to identify driver actions. Finally, the training module retrospectively combines the predicted features and recognized action labels to generate training samples, which are then used to train the PN model.

The DBRNN framework focuses on modeling the transition between negative states (non-action states) and positive states (action states) in the feature sequence. The neural network takes the observed time series as input and generates a vector sequence through nonlinear activation. This output vector is a latent variable describing the observed distribution. Unlike traditional feedforward neural networks, RNNs also consider historical information or so-called "memory." The recognized labels are determined by the nonlinear mapping of these latent variables.

A disadvantage of traditional RNNs is that they can only utilize previous context. However, using the window method can obtain the entire time context, and there is no reason not to utilize the subsequent context. The bidirectional recurrent neural network processes data in two directions using two separate hidden layers, and then feeds these two hidden layers forward to the same output layer.

The Ref. [15] points out that MIT has collected a large amount of actual driving data in its research on autonomous vehicle technology (MIT-AVT), including high-definition videos, to enhance the development of internal and external perception systems based on deep learning. To understand the overall interaction between humans and automotive automation technology, video data and vehicle status data, driver characteristics, psychological models, and self-reported experiences are integrated. The recorded data includes IMU/GPS/CAN bus messages, driver's face, cockpit, forward road, and equipment group.

The data pipeline of the MIT-AVT platform includes data unloading, cleaning, synchronization, and knowledge extraction. The left side of the figure is the asynchronous distributed computing platform with relevant restrictions, the middle is the high-level program for knowledge extraction, and the right side is the categories of generated data.

The study in Ref. [16] proposes a machine learning method that generates driving paths by integrating lidar point clouds, GPS-IMU data, and Google Maps navigation information, and uses an FCN model to learn perception and driving paths from real-world driving sequences. The output is close to vehicle control and understandable, helping to bridge the gap between low-level scene decomposition and end-to-end "behavioral reflection" methods.

Reference [25] proposes an autonomous driving system LookOut, which perceives the environment, predicts various futures of the scene, and estimates the trajectory of the autonomous vehicle by optimizing a set of emergency plans. In particular, LookOut learns the diverse joint distribution of future trajectories of multiple agents in traffic scenarios, efficiently sampling a variety of future modalities.

To extract features useful for both detection and motion prediction, a convolutional backbone network is used, taking voxelized lidar scans and rasterized high-definition maps as inputs, both located in the BEV space centered on the autonomous vehicle.

A shallow convolutional head performs multi-class object detection, estimating the presence, BEV pose, and size of vehicles, pedestrians, and bicycles. It then applies rotated RoIAlign to extract cropped features from the surrounding scene context at each participant's location.

An Actor CNN with max pooling reduces each participant's feature map to a single feature vector.

A diversity sampling function is learned, mapping the participant context from each sensor data to a set of compact scene latent samples, whose decoded trajectories achieve good coverage. Then, this sampler will replace the Monte Carlo sampling from the prior distribution in inference.

Similar to NMP, a set of K future trajectory realizations for all participants in the scene is obtained through sampling and decoding. The goal of this diversity sampler is to generate a set of diverse futures while well recovering the true value observations.

Different sets of K future trajectory realizations provide the coverage needed for safe motion planning. However, for accurate risk assessment, it is necessary to estimate the probability distribution of each realization in the future trajectory set.

An implicit prediction decoder, parameterized as a spatially-aware graph neural network (GNN). Through a GNN, the participant features and all K sample future scenes are input, enhancing the model to output future realization scores. Then it is easy to recover the distribution of such scores by renormalization.

During training, a posterior or GNN-parameterized encoder is introduced to approximate the true posterior distribution, thereby facilitating the learning of a stronger decoder. For participant presence recognition, binary cross-entropy loss with hard negative mining is employed, while for regression tasks (i.e., pose and dimensions), the Huber loss is applied.

Instead of looking for a single motion plan for multiple futures, it is better to generate a single universal action, then generate a set of future trajectories, each corresponding to a future realization of the scene. Here, a contingency planner is proposed, which finds a safe action related to all possible realizations in the future realization set, and easily transitions into a set of contingency trajectories. The decision to postpone such decisions avoids overly conservative behavior while maintaining safety before obtaining more information. Safe motion planning is only possible when a set of predicted future scenarios is diverse and covers possible realizations (including low probability events).

The cost function of this planner is a linear combination of various sub-costs, which encodes different aspects of driving, including safety, comfort, traffic rules, and route-related items.

A planner, DriveIRL, which uses inverse reinforcement learning to learn how to drive, was proposed in Ref. [24].

DriveIRL generates a set of ego-vehicle motion trajectories and evaluates whether they are safe; it first carefully constructs these proposed trajectories, then uses a lightweight safety filter to ensure each trajectory avoids collisions; the learning part of the model is entirely focused on expert demonstrations to reasonably score these trajectories; this design is not only for creating good trajectories, but also for guiding the model to subtle differences in behavior that are difficult to specify (such as speed curves, vehicle spacing), and can avoid obvious unsafe behavior.

At the input stage, intermediate representations are used to encode the environment surrounding the autonomous vehicle. Assuming the ego vehicle is positioned on a high-definition map, targets are detected and tracked by the perception system. Other road users–such as cars, bicycles, and pedestrians–are represented by attributes including type, oriented bounding box, and speed. The high-definition map supplies additional semantic information, such as lane centerlines, road boundaries, traffic signal positions, crosswalks, and speed limits. It also provides a planned route, specifying the lanes that the ego vehicle should follow to reach its destination.

The output is the planner generating multiple ego-vehicle trajectories and scoring each trajectory based on its match with expert driving data in the given scene context.

The trajectory generation module synthesizes a set of possible future motions for the ego vehicle based on the scene context. Important considerations for the ego-vehicle trajectory are: ① dynamically feasible; ② meets all requirements for low-level tracking and actuator control (i.e., continuity level, minimum turning radius, minimum acceleration for stopping). Secondary considerations are: the trajectory complies with map constraints (such as staying on the road). Although these considerations do not rule out the use of a learned trajectory generation module, it was found that a manually designed trajectory generator best meets the above considerations.

The trajectory generator uses the current ego state S, path, and map to create a set of different ego-vehicle trajectories, integrating the required acceleration curves along the route ahead of the ego vehicle.

Before scoring the candidate trajectories, an interpretable safety filter is used to ensure basic safety (i.e., no collisions). It includes:

(1) A set of world assumptions for predicting the behavior of non-ego road users.

(2) A set of trajectory correctors applied to the ego-vehicle trajectory.

(3) A set of safety checks that the modified ego-vehicle trajectory needs to pass.

In order for a candidate trajectory to be considered safe, it must pass all safety checks under the given trajectory modification and assumptions about other road users.

The trajectory score is computed by a deep neural network trained through maximum entropy IRL loss. Expert demonstration data is collected from skilled drivers of the driving vehicle. The loss tends to favor trajectories that are closest to expert demonstrations in feature space.

The features of each proposed trajectory can serve as inputs to the neural network. These features can be based on any combination of the proposed trajectory, ego state, other road users, map, route, and history. Features include:

- Time to Collision (TTC). The shortest second count before the ego-vehicle collides with other road users in the (predicted) future. Evaluated at multiple points.
- ACCInfo. Ego speed, distance to the road user ahead, speed of the road user ahead, and relative speed of the road user ahead. Evaluated at multiple points.
- MaxJerk. The maximum jerk (m/s3) along the trajectory.
- MaxLateralAccel. The maximum lateral acceleration (m/s^2) along the trajectory.
- PastCoupling. The concatenation of the future trajectory with the past one second ego posture, maintaining consistency between past, present, and future trajectories.
- SpeedLimit. The degree to which the trajectory adheres to the speed limit. Evaluated at multiple points.

For example, an architecture of the trajectory scoring network is designed to score a trajectory, in one architecture, by processing the extracted features separately before they interact through a masked self-attention mechanism. In this architecture, each input feature, as a time series of relevant vehicle-environment interaction data, is first normalized through a BatchNormalD layer, then fed into an LSTM module. The output of the LSTM becomes the input to the feedforward module, followed by a self-attention mechanism with two heads and 120 embedding dimensions. Zero-masking is used here to encode positions.

The self-attention mechanism allows for the consideration of other features, generating a "corrected" output embedding for each feature. This is passed to a feedforward network, which transforms it into a scalar, and then activates tanh to generate feature scores. The final score of the trajectory is the sum of these feature scores multiplied by the corresponding learnable feature weight parameters.

ST-P3 is a spatio-temporal feature learning scheme proposed in reference [36], which can provide a set of more representative features for perception, prediction, and planning tasks simultaneously. Specifically, it proposes an ego-centric-aligned accumulation technique that preserves geometric information in 3D space before perception BEV conversion; it designs a dual-pathway model that takes into account past motion changes for future prediction; and it introduces a time-domain refinement unit to compensate for planning based on visual element recognition.

The overall framework of ST-P3 is given a set of surrounding camera videos, which are input into the main trunk to generate preliminary front-view features. An auxiliary depth estimation is performed to convert 2D features into 3D space. The ego-centric-aligned accumulation scheme first aligns past features to the current view coordinate system; then aggregates current and past features in 3D space, preserving geometric information before converting to BEV representation. In addition to the commonly used prediction time-domain model, performance is further improved by constructing a second path to interpret past motion changes. This dual-pathway modeling ensures stronger feature representation and infers future semantic results. To achieve the final goal of trajectory planning, the network integrates early feature prior knowledge and designs a refinement module to generate the final trajectory with high-level commands in the absence of high-definition maps.

The ultimate objective is to generate a safe and comfortable trajectory that guides the vehicle to the target point. The motion planner samples a set of candidate trajectories and selects the one that minimizes a learned cost function. Unlike conventional approaches, an additional optimization step is introduced to integrate target-point and traffic-light information through a time-domain model. During planning, prior knowledge is incorporated and refined, with the overall cost map decomposed into two sub-cost components. The final minimum-cost trajectory is further refined using front-view features, which aggregate vision-based information from camera inputs.

Trajectories with excessive lateral acceleration, abrupt maneuvers, or sharp curvature are penalized, while forward-moving trajectories are rewarded to encourage efficient progress toward the destination. However, these cost terms do not inherently incorporate target information typically provided by the route map. To address this, high-level commands—such as forward, left turn, and right turn—are introduced, and trajectories are evaluated only with respect to the corresponding command.

In addition, traffic lights play a critical role in autonomous driving. To account for them, trajectory optimization is performed using a GRU network. The hidden state of the GRU is initialized with front-camera features extracted by the encoder module, and each sampled cost term is sequentially fed as input, enabling traffic-aware trajectory refinement.

Recently, end-to-end autonomous driving research has received increasing attention, and there is a trend to replace the previous modular autonomous driving research. There are two reviews in this area [41, 43] that summarize recent work in this direction.

If this perception-prediction-planning is unified under a differentiable network architecture, it can be regarded as an end-to-end system for autonomous driving planning. Compared with the classical paradigm, this end-to-end autonomous driving system provides several advantages [41]:

1. It simply combines perception, prediction, and planning into a single model that can be trained jointly.
2. The entire system, including its intermediate representations, is optimized toward the final planning task.
3. The shared backbone network improves computational efficiency.
4. Data-driven optimization has the potential to improve the system's capabilities by simply expanding training resources.

The last step of prediction is to generate trajectories. This is usually achieved by decoding the captured latent features. Trajectory generation in deep learning mainly discusses two aspects [43]: decoding rules and multimodal modeling.

Decoding rules. They are basically divided into three categories: one-shot, two shots, and N shots. It is more common to directly predict future trajectory points.

1. One-shot methods are further divided into trajectory regression decoding and anchor trajectory classification. Trajectory regression directly uses NN to decode hidden layer features into trajectories, typically MLP. In particular, based on the Transformer structure, trajectory regression can be performed directly. Anchor trajectory classification uses pre-set anchor trajectories as prior knowledge for classification. Determining the anchor trajectory can ensure the feasibility of the prediction and allow stronger kinematic constraints to be imposed. However, this method lacks flexibility because it cannot handle unexpected situations.
2. Two shots methods include two types: 1) endpoints-and-completion, 2) initial-trajectory-and-refinement strategy. The endpoint-and-completion method is based on the assumption that the trajectory is mainly defined by its endpoints, which can be regressed or classified as nodes of the road map from a predefined set; usually, classification-based methods also regress the offset of the classified endpoints. The initial-trajectory-and-refinement method combines the idea of anchor trajectory classification with the idea of endpoint-and-completion. It first classifies the initial trajectory instead of the endpoint; then, the refinement step adjusts the trajectory, for example by performing offset regression on each predicted waypoint.
3. N shots methods are either initial-trajectory-and-refinement or autoregression. The variants of the initial-trajectory-and-refinement method are similar to the two shots method; the only difference is the refinement step. The autoregression

method uses recursive decoding to gradually advance into the future. It iteratively predicts the next waypoint and adds information about the resulting scene to the latent features.

Multimodal modeling. Because the observed intentions of the autonomous vehicle are unknown, future behavior is inherently uncertain and multimodal. Multimodal modeling approaches can be distinguished into two categories: discrete trajectories or continuous distributions. Discrete trajectories can be sampled from intermediate descriptions or obtained during model design. Continuous distributions, unlike discrete trajectories, are very suitable for direct use for cost functions in planning.

Integral end-to-end planning methods that can be divided into three categories [43]:

1. Monolithic planning refers to methods that consist of a single end-to-end planner that maps multiple state inputs to trajectories. Therefore, the future behavior of surrounding vehicles and their interactions with each other and with the ego-vehicle are not explicitly modeled. Therefore, no explicit predictions are inferred or used. However, the end-to-end planner learns to imitate the behavior of experts during training, based on reasoning about the future. By learning to imitate behavior, the model will also behave in a way that is implicitly based on such reasoning.
2. Interpretable end-to-end planning, which formulates prediction as an auxiliary learning task. Explicit predictions are additional model outputs that are jointly trained with the planning task. Typically, a shared backbone encodes features for both tasks. Separate heads then decode their respective output representations. The additional objective provides a learning signal to the backbone. It acts as a regularizer and can make the learning process more sample efficient, and can improve generalization. Explicit predictions increase interpretability and facilitate introspection.
3. Manual integration of the planner and prediction module, meaning that a separate subsystem is used for each task. The interaction between the two tasks is manually designed based on domain knowledge. The most widely adopted integration is a sequential pipeline, where the prediction module is first inferred, and its output is then passed to the subsequent planning module.

Unified Autonomous Driving (UniAD) [39] is a comprehensive framework that integrates the full stack of driving tasks into a single network. Its design is goal-oriented and hierarchical. The key component is the query-based design that connects all nodes. Compared with bounding box representations, queries allow for more flexible representations and encode rich social dynamics and cross-agent interactions. As shown in Fig. 8.1: UniAD consists of four Transformer decoder-based perception and prediction modules and a planner at the end. The query Q plays the role of connecting the pipeline to model multi-agent and scene-level context;

1. A series of multi-camera images are input into the feature extractor, and then the original perspective view features are converted into unified BEV features using the off-the-shelf BEV encoder from the multi-view BEVFormer;

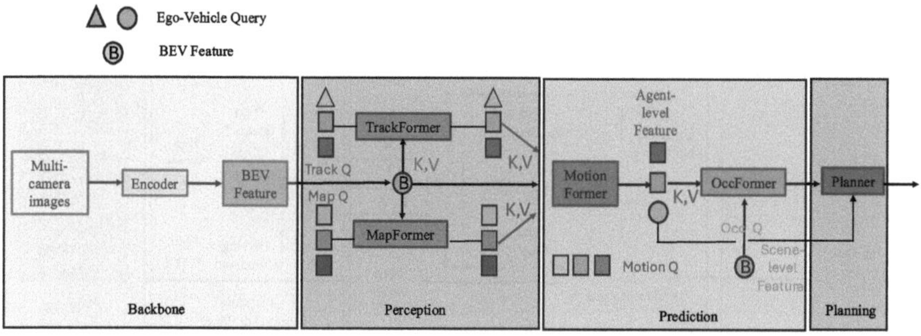

Fig. 8.1 The UniAD framework

2. In TrackFormer, the learnable embedding of the track query queries the agent's information from the BEV features to detect newly generated objects and track previously existing objects;
3. MapFormer updates the map query features to perform panoramic segmentation of road elements such as lanes and dividers; with effective agent features and map features;
4. MotionFormer It can capture the interaction between agents and maps and predict the future trajectory of each agent. Since the behavior of agents in the scene can have a significant impact on other agents, this module generates a scene-centric joint prediction and uses motion anchors extracted from the empirical future trajectory as queries. At the same time, an ego-vehicle query is designed to explicitly capture ego-vehicle information and learn interactions with other agents in this scene-centric paradigm.
5. OccFormer uses BEV features as queries (Q) and agent-level features as keys (K) and values (V) to explore future scene representations and predict the occupied grid map while preserving the identity of the agent.
6. Finally, the planner decodes the rich ego-vehicle query in MotionFormer into a planning result and keeps itself away from the occupied area predicted by OccFormer.

The overall architecture of VAD [40] is shown in Fig. 8.2: the entire pipeline of VAD [40] is divided into four stages; the backbone includes an image feature extractor and a BEV encoder that projects image features into BEV features; vectorized scene learning aims to encode scene information into agent queries and map queries, and represents the scene with motion vectors and map vectors; in the inference phase of planning, VAD uses self-queries to extract map and agent information through query interactions and outputs a planning trajectory (represented as a self-vector). The proposed vectorized planning constraints regularize the planning trajectory in the training phase.

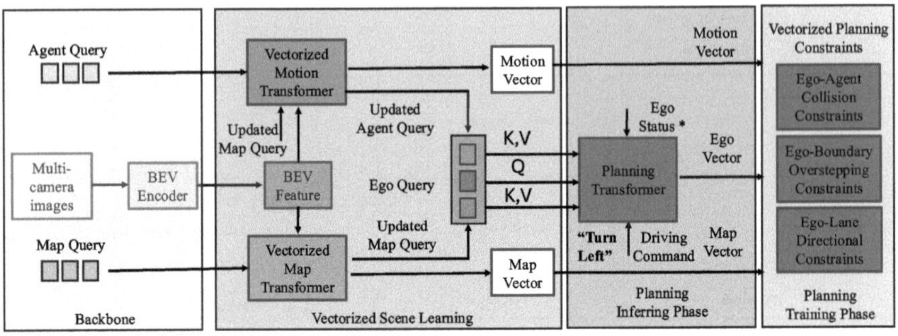

Fig. 8.2 VAD architecture

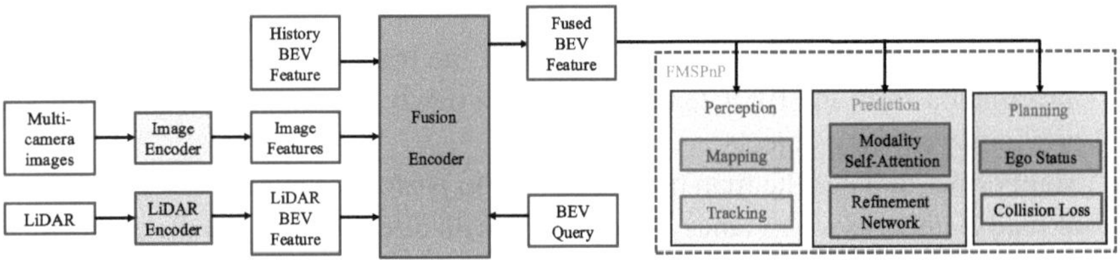

Fig. 8.3 FusionAD architecture

In the network architecture of FusionAD [42], as shown in Fig. 8.3, first, the camera images are mapped to the bird's-eye view (BEV) space using a BEVFormer-based image encoder; then they are combined with the LiDAR features in the BEV space; after temporal fusion, the fused BEV features are used for detection, tracking, and mapping tasks through a query-based approach. Subsequently, the tokens are forwarded to the action and occupancy prediction and planning tasks. The fusion-assisted modality-aware prediction and state-aware planning modules are referred to as FMSPnP.

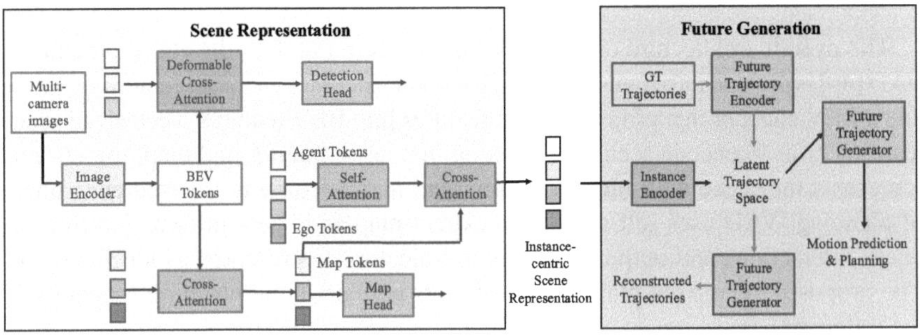

Fig. 8.4 GenAD framework

In the generative end-to-end autonomous driving framework GenAD [44] shown in Fig. 8.4, given an image of the surroundings as input, the image backbone is used to extract multi-scale features, and then the BEV encoder is used to obtain BEV tokens. Afterward, the BEV tokens are converted into map tokens and agent tokens using cross-attention and deformable cross-attention, respectively. With additional ego-vehicle tokens, self-attention is used to achieve ego-agent interaction, and cross-attention is used to further incorporate map information to obtain an instance-centric scene representation. This representation is mapped to a structured latent trajectory space, which is jointly learned using the ground truth (GT) future trajectory. Finally, the future trajectory generator is used to generate future trajectories while completing motion prediction and planning.

As shown in Fig. 8.5, the overall architecture of VADv2 [45] takes multi-view image sequences as input in a streaming manner, converts sensor data into environment tokens embedding, outputs the probability distribution of actions, and samples an action to control the vehicle; large-scale driving demonstrations and scene constraints are used to supervise the predicted distribution.

The overall framework of SparseDrive [46] is shown in Fig. 8.6. It consists of three parts: image encoder, symmetric sparse perceptron, and parallel motion planner. Given a multi-view image, the image encoder (including the backbone network and the bottleneck network) first encodes the image into a multi-view multi-scale feature map I. In the symmetric sparse perceptron module, the feature map is aggregated into two groups of instances to learn a sparse representation of the driving scene. The two groups of instances, representing the surrounding agents and map elements, are input into the parallel motion planner to interact with the initialized self-instance. The parallel motion planner predicts the multimodal trajectories of the surrounding agents and the ego-vehicle, and selects a safe trajectory as the final planning result through a hierarchical planning selection strategy.

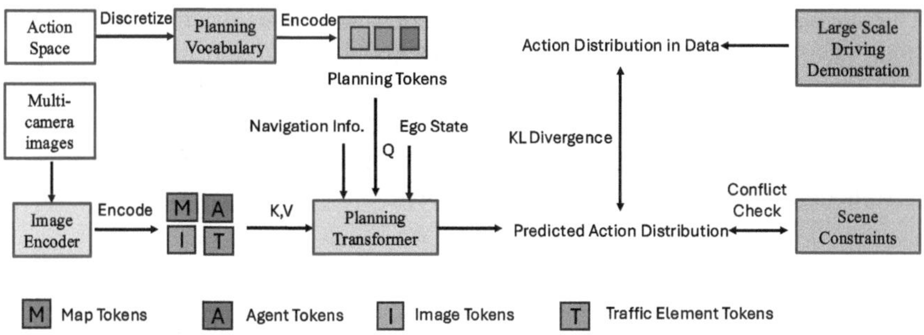

Fig. 8.5 VADv2 architecture

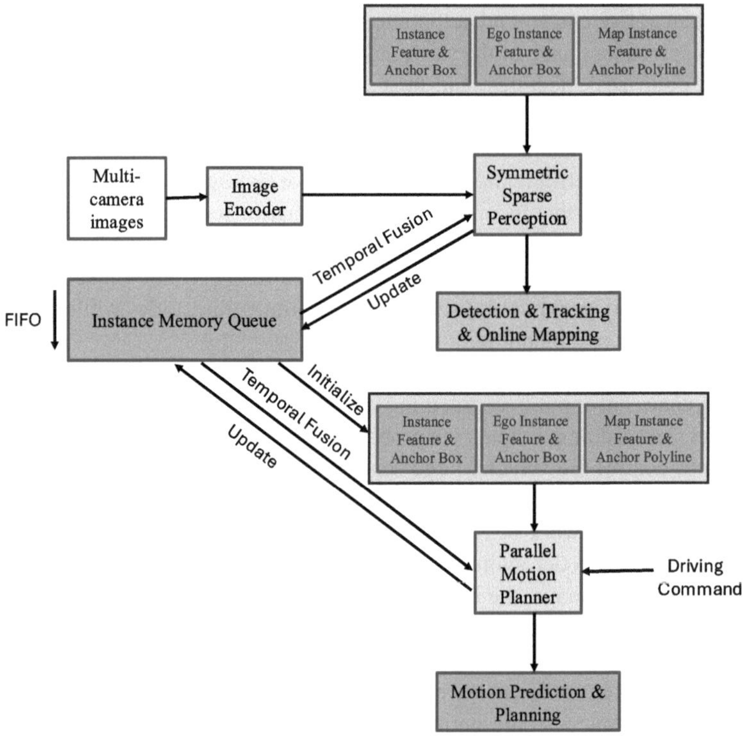

Fig. 8.6 SparseDrive framework

8.5 Summary

This chapter discusses the planning theory of autonomous driving, starting with basic theoretical analysis, then emphasizing the modeling and prediction part of behavior planning, namely vehicle behavior and pedestrian behavior, and finally introducing the current situation of data-driven cloning algorithms for experienced drivers. Chapter 9 will discuss the control part of autonomous driving, that is, the next module where the planning results are sent.

References

1. S D PENDLETON et al. Perception, Planning, Control, and Coordination for Autonomous Vehicles,Machines 2017, 5(1): 6-59.
2. W SCHWARTING, J ALONSO-MORA,D RUS,Planning and Decision-Making for Autonomous Vehicles, Annual Review of Control, Robotics, and Autonomous Systems, 2018. 1(1): 187-210.

3. S. DIXIT et al., Trajectory planning and tracking for autonomous overtaking: State-of-the-art and future prospects,Annual Reviews in Control,2018, 45(2): 76-86.
4. D GONZÁLEZ et al. A Review of Motion Planning Techniques for Automated Vehicles." IEEE Trans. on Intelligent Transportation Systems, 2016, 17(4): 1135-1145.
5. B PADEN et al. A survey of motion planning and control techniques for self-driving urban vehicles. IEEE Transactions on Intelligent Vehicles, 2016, 1(1): 33-55.
6. C KATRAKAZAS et al. Real-time motion planning methods for autonomous on-road driving: State-of-the-art and future research directions. Transportation Research Part C: Emerging Technologies, 2015, 60: 416-442.
7. S LEFÈVRE, D VASQUEZ, C LAUGIER. A survey on motion prediction and risk assessment for intelligent vehicles. Robomech Journal,2014. 1(1): 1-20.
8. W LUO, B YANG, R URTASUM, Fast and Furious: Real Time End-to-End 3D Detection, Tracking and Motion Forecasting with a Single Convolutional Net, IEEE Conference on Computer Vision and Pattern Recognition, Salt Lake city, UT, USA, June 2018: 3567-3577.
9. S CASAS, W LUO, R URTASUM, IntentNet: Learning to Predict Intention from Raw Sensor Data, arXiv preprint arXiv:2101.07907, 2018
10. L SUN et al., A Fast Integrated Planning and Control Framework for Autonomous Driving via Imitation Learning,arXiv preprint arXiv:1707.02515, 2017
11. M BANSAL et al., ChauffeurNet: Learning to Drive by Imitating the Best and Synthesizing the Worst, arXiv preprint arXiv:1812.03079, 2018
12. VASQUEZ D, YU Y, KUMAR S, LAUGIER C,An Open Framework for Human-like Autonomous Driving Using Inverse Reinforcement Learning,IEEE Vehicle Power and Propulsion Conference, Coimbra, Portugal, Oct. 2014: 1-4.
13. XU H,GAO Y, DARELL T, E2E Learning of Driving Models from Large-scale Video Datasets,IEEE Conference on Computer Vision and Pattern Recognition, Honolulu, HI, USA, July 2017: 2174-2182.
14. O OLABIYI et al,Driver Action Prediction Using Bidirectional RNN,arXiv preprint arXiv:1706.02257, 2017
15. L FRIDMAN et al,MIT Autonomous Vehicle Technology Study: Large-Scale Deep Learning Based Analysis of Driver Behavior and Interaction with Automation,arXiv preprint arXiv:1711.06976, 2017
16. L CALTAGIRONE et al,LIDAR-based Driving Path Generation Using Fully Convolutional Neural Networks,arXiv preprint arXiv:1703.08987,2017
17. S MOZAFFARI, O Y. AL-JARRAH, M DIANATI, P JENNINGS, and A MOUZAKITIS, Deep Learning-based Vehicle Behaviour Prediction For Autonomous Driving Applications: A Review,arXiv preprint arXiv:1912.11676,12,2019
18. A RUDENKO, L PALMIERI, M HERMAN, K M. KITANI, D M. GAVRILA and K O. ARRAS, Human Motion Trajectory Prediction: A Survey,arXiv preprint arXiv:1905.06113,12, 2019
19. F LEON, M GAVRILESCU, A Review of Tracking, Prediction and Decision Making Methods for Autonomous Driving, arXiv preprint arXiv:1909.07707,2019
20. W WANG, L WANG, C ZHANG, C LIU, and L SUN, Social Interactions for Autonomous Driving: A Review and Perspective,arXiv preprint arXiv:2208.07541,2022
21. Y CHEN, B IVANOVIC, M PAVONE, ScePT: Scene-consistent, Policy-based Trajectory Predictions for Planning,arXiv preprint arXiv:2206.13387,2022
22. Y. CHAI, B. SAPP, M. BANSAL, and D. ANGUELOV, Multipath: Multiple probabilistic anchor trajectory hypotheses for behavior prediction,Conference on Robot Learning, arXiv preprint arXiv:1910.05449, 2019.
23. D A SU, B DOUILLARD, R AL-RFOU, C PARK, B SAPP,Narrowing the coordinate-frame gap in behavior prediction models: Distillation for efficient and accurate scene-centric motion forecasting,arXiv preprint arXiv:2206.03970, 2022
24. T PHAN-MINH, F HOWINGTON, T-S CHU, S U LEE, M S. TOMOV, N LI, C DICLE, SAMUEL FINDLER, F SUAREZ-RUIZ, R BEAUDOIN, B YANG S OMARI, E M. WOLFF, Driving in Real Life with Inverse Reinforcement Learning, arXiv preprint arXiv:2206.03004, 2022

25. A CUI, S CASAS, A SADAT, R LIAO, R URTASUN, LookOut: Diverse Multi-Future Prediction and Planning for Self-Driving, arXiv preprint arXiv:2101.06547, 2021
26. M LIANG, B YANG, R HU et al.,Learning Lane Graph Representations for Motion Forecasting,arXiv preprint arXiv:2007.13732,2020
27. A MOHAMED, K QIAN,M ELHOSEINY and C CLAUDEL, Social-STGCNN: A Social Spatio-Temporal Graph Convolutional Neural Network for Human Trajectory Prediction, IEEE Conf. on Computer Vision and Pattern Recognition, Virtual, June 2020: 14424-14432.
28. J GAO, C SUN, H ZHAO, Y SHEN, D ANGUELOV, C LI AND and C SCHMID, VectorNet: Encoding HD Maps and Agent Dynamics from Vectorized Representation, IEEE Conf. on Computer Vision and Pattern Recognition, Virtual, June 2020: 11525-11533.
29. M VITELLI, Y CHANG, Y YE et al.,SafetyNet: Safe planning for real-world self-driving vehicles using machine-learned policies,arXiv preprint arXiv:2109.13602,2021
30. Z ZHANG,J GAO, J MAO, Y LIU, D ANGUELOV, C LI, STINet: Spatio-temporal-interactive Network For Pedestrian Detection And Trajectory Prediction, IEEE Conference on Computer Vision and Pattern Recognition, Virtual June 2020: 11346-11355.
31. H ZHAO, J GAO, T LAN, C SUN, B SAPP, B VARADARAJAN, YUE SHEN, YI SHEN, Y CHAI, C SCHMID, C LI, and D ANGUELOV, TNT: Target-driveN Trajectory Prediction, Conference on Robot Learning (CoRL), arXiv preprint arXiv:2008.08294, 2020
32. B VARADARAJAN,A HEFNY,A SRIVASTAVA,K S. REFAAT,N NAYAKANTI,A CORNMAN,K CHEN,B DOUILLARD,C P LAM,D ANGUELOV,and B SAPP,MultiPath++: Efficient Information Fusion And Trajectory Aggregation For Behavior Prediction, arXiv preprint arXiv:2111.14973,2021
33. W ZENG, W LUO, S SUO, A SADAT, B YANG, S CASAS, and R URTASUN. End-to-end interpretable neural motion planner. IEEE Conf. on Computer Vision and Pattern Recognition, Long Beach, CA, USA, June 2019: 8660-8669.
34. J LIANG, L JIANG, J C NIEBLES, A G HAUPTMANN, and F-F LI,Peeking into the Future: Predicting Future Person Activities and Locations in Videos, IEEE Conference on Computer Vision and Pattern Recognition, Long Beach, CA, USA, June 2019: 5725-5734.
35. J HONG,B SAPP,J PHILBIN,Rules of the Road: Predicting Driving Behavior with a Convolutional Model of Semantic Interactions,arXiv preprint arXiv:1906.08945,2019
36. S HU, L CHEN, P WU, H LI, J YAN, and D TAO, ST-P3: End-to-end Vision-based Autonomous Driving via Spatial-Temporal Feature Learning,arXiv preprint arXiv:2207.07601, 2022
37. J FANG, F WANG, P SHEN, Z ZHENG, J XUE, and T CHUA, Behavioral Intention Prediction in Driving Scenes: A Survey,arXiv preprint arXiv:2211.00385,2022
38. H. Fan, F. Zhu, C. Liu, L. Zhang, L. Zhuang, D. Li, W. Zhu, J. Hu, H. Li, and Q. Kong, "Baidu Apollo EM Motion Planner," arXiv 1807.08048,Jul. 2018.
39. Y Hu, J Yang, L Chen, and et al., "Planning-oriented Autonomous Driving" (UniAD), arXiv 2212.10156, 2022
40. B Jiang, S Chen, Q Xu, and et al., "VAD: Vectorized Scene Representation for Efficient Autonomous Driving", arXiv 2303.12077, 2023
41. L Chen et al., "End-to-end Autonomous Driving: Challenges and Frontiers" (Survey), arXiv 2306.16927, 2023
42. T Ye , W Jing, C Hu, and et al., "FusionAD: Multi-modality Fusion for Prediction and Planning Tasks of Autonomous Driving", arXiv 2308.01006, 2023
43. S Hagedorn, M Hallgarten, M Stoll, and et al., "Rethinking Integration of Prediction and Planning in Deep Learning-Based Automated Driving Systems: A Review", arXiv 2308.05731, 2023
44. W Zheng, R Song, X Guo, and et al., "GenAD: Generative End-to-End Autonomous Driving", arXiv 2402.11502, 2024
45. S Chen, B Jiang, H Gao, and et al., "VADv2: End-to-End Vectorized Autonomous Driving via Probabilistic Planning", arXiv 2402.13243, 2024
46. W Sun et al., "SparseDrive: End-to-End Autonomous Driving via Sparse Scene Representation", 2405.19620, 2024

Chapter 9
Control Module for Autonomous Driving

As mentioned earlier, perception is the "eyes" of autonomous driving, planning is the "brain" of autonomous driving, and control is the "limbs" of autonomous driving.

The execution force of the autonomous driving system, also known as motion control, is the process of converting intent into action. Its main purpose is to provide the necessary input to the hardware to execute the planned intent, thereby generating the required motion [18]. The controller maps interactions in the real world based on force, while cognitive navigation and planning algorithms in autonomous driving systems usually focus on the vehicle's speed and position relative to its environment. The measurements within the control system are used to determine the system's performance, so the controller responds, rejects interference, and changes the system's dynamics to the desired state.

Control in autonomous driving can also be seen as motion control [17], which is to generate commands for the vehicle, generally in wire-control mode, i.e., the angle controlled by the steering wheel, the speed or acceleration controlled by the accelerator and brake pedals, and the actual driving trajectory generated under real conditions (road, wind, wheel slip, etc.).

Vehicle motion control can be roughly divided into two tasks [11–14]: Steering control of the vehicle's lateral motion and control of the accelerator and brake pedals for longitudinal motion. The lateral control system aims to control the vehicle's position on the lane and perform other lateral actions, such as lane changes or collision avoidance actions. Longitudinal control can control the vehicle's acceleration, maintain an ideal speed on the road, keep a safe distance from the previous vehicle, and avoid rear-end collisions.

Current traditional vehicle control methods [15] are based on classical optimal control theory, which can be described as a cost function defined on a set of states and control actions. Control inputs are usually defined within a finite time range and are limited to the feasible state space. In this way, rule-based controllers operate autonomous vehicles, where parameters are set by developers and manually adjusted after simulation and field testing.

Y. Huang and Z. Yang, *Autonomous Driving Handbook*,
https://doi.org/10.1007/978-981-95-3725-9_9

Deep learning, on the other hand, allows developers to describe the desired behavior and make the system perform well through learning, and generalize it to new environments, without having to repeatedly adjust each parameter under all foreseeable performance conditions [19, 20]. Learning algorithms can be used as part of a learning controller, used in the motion control module, or as a complete terminal control system that directly maps sensor data to control commands, i.e., end-to-end methods.

Section 9.1 first introduces the kinematics and dynamics models of the vehicle, which are the basis of vehicle control; Sect. 9.2 introduces traditional control algorithms, namely PID and MPC; Sect. 9.3 discusses the control algorithms for path and trajectory stabilization commonly used in autonomous driving, including pure tracking control, rear wheel position feedback, front wheel position feedback, feedback control based on control Lyapunov functions, and feedback control of linearized output. Finally, Sect. 9.4 discusses autonomous driving control algorithms based on deep learning.

9.1 Kinematic and Dynamic Model of the Vehicle

The kinematic model drives the vehicle according to mathematical formulas, forming the vehicle's motion and trajectory. The disadvantage is that this model allows instantaneous direction angle changes, but problems arise when the motion planning module generates solutions in such instantaneous changes.

The dynamic model considers the principles of dynamics and calculates the external forces applied to the vehicle, such as air resistance, vehicle weight, gravity, centrifugal force, and wheel friction, etc.

The efficiency of vehicle automatic control highly depends on the accuracy and complexity of the vehicle model. Due to its simplicity, a planar vehicle model with longitudinal and lateral motion is commonly used in the literature. Accurate high-fidelity models can better represent vehicle motion in different directions. However, this will increase the complexity and computational cost of controller design. Simpler models reduce the complexity of the control system. However, they cannot capture nonlinear effects and bring uncertainty to the system.

Starting from the planar model, a wheeled vehicle similar to a car is assumed to be a 2D rigid body moving in the configuration space. The coordinate system is usually represented by the vehicle's position and direction, describing the vehicle's motion in a special Euclidean group in the 2D configuration space. The planar motion equations of autonomous vehicles are usually represented in two coordinate systems, namely the vehicle coordinate system $B(x, y)$ and the global coordinate system $G(x, y)$. The vehicle's motion is usually represented in the vehicle coordinate system, where x is the vehicle's forward/longitudinal axis and y is the lateral axis facing left forward. These will be converted into the global coordinate system to calculate the vehicle's state in the configuration space.

In the vehicle's kinematic model, motion is replicated regardless of how the dynamic components (such as force, torque, inertial effects) are [9]. In practical applications, the common basic model of a vehicle is the "bicycle model" which includes 2 rigidly connected wheels and is restricted to move within a plane. Sometimes it is also called the "kinematic model", or the "single-track model", where p_r and p_f are the contact points of the front and rear wheels on the road; θ is the forward angle of the vehicle; the time derivatives of p_r and p_f are limited by nonholonomic constraints; δ is the steering angle of the front wheel.

Nonholonomic constraints mean that the vehicle allows the dimension of the velocity space to be smaller than the dimension of the configuration space. Assuming that the wheels roll completely without any sideslip. The motion of points pr and pf must be collinear with the wheel direction to meet the no-slip assumption. This constraint on the rear wheel is expressed as

$$(\dot{p}_r \cdot \dot{e}_y)cos(\theta) - (\dot{p}_r \cdot \dot{e}_x)sin(\theta) = 0 \tag{9.1}$$

And the constraint of the front wheel is

$$(\dot{p}_f \cdot \dot{e}_y)cos(\theta + \delta) - (\dot{p}_f \cdot \dot{e}_x)sin(\theta + \delta) = 0 \tag{9.2}$$

This expression is usually rewritten according to the movement of each point along the basis vector. Along the $\hat{e}_x$ direction, the rear wheel movement is $\dot{x}_r := p_r \cdot \hat{e}_x$; similarly, along the e^y direction, the rear wheel movement is $\dot{y}_r := p_r \cdot \hat{e}_y$. The forward speed of the rear wheel is

$$v_r := \dot{p}_r(p_f - p_r)/\left\|p_f - p_r\right\| \tag{9.3}$$

Thus, taking the point where the rear wheel of the car touches the ground as an example, the obtained differential constraint equation is

$$\dot{x}_r = v_r cos(\theta) \tag{9.4}$$

$$\dot{y}_r = v_r sin(\theta) \tag{9.5}$$

$$\dot{\theta}_r = \frac{v_r}{l} tan(\delta) \tag{9.6}$$

Similarly, taking the point where the front wheel of the car touches the ground as an example, its differential constraint equation is

$$\dot{x}_f = v_f cos(\theta + \delta) \tag{9.7}$$

$$\dot{y}_f = v_f sin(\theta + \delta) \tag{9.8}$$

$$\dot{\theta} = \frac{v_f}{l} sin(\delta) \tag{9.9}$$

where v_f is the forward/longitudinal speed of the front wheel. And the relationship between the speed of the front wheel and the rear wheel is

$$\frac{v_r}{v_f} = cos(\delta) \tag{9.10}$$

Generally speaking, for this model, the planning control problem is how to choose the steering angle $\delta \in [\delta_{min}, \delta_{max}]$ and the forward speed $v \in [v_{min}, v_{max}]$.

A simplified method is to choose the heading rate ω instead of the steering angle δ. The relationship between these quantities is as follows:

$$\delta = arctan(\frac{l\omega}{v_r}) \tag{9.11}$$

This simplifies the heading dynamics, that is

$$\theta = \omega, w \in [\frac{v_r}{l} tan(\delta_{min}), \frac{v_r}{l} tan(\delta_{max})] \tag{9.12}$$

In this case, the model is sometimes referred to as the unicycle model, which can be derived from the movement of a single wheel.

An important variant of this model is the case when v_r is fixed. One is the Dubins car named after Lester Dubins, which is derived from the prescribed tangent to derive the minimum time movement. Another noteworthy change is the Reeds-Shepp car, where the minimum length path is known v_r uses a single forward and reverse speed.

The kinematic model is suitable for planning low-speed routes (such as parking maneuvers and urban driving), and the inertial effects are small compared to the mobility restrictions imposed by the no-slip assumption. The main disadvantage of this model is that it allows instantaneous angle changes. If the motion planning module produces a solution with such instantaneous changes, this may be problematic.

The continuity of the steering angle can be achieved by adding differential constraints, where the steering angle integrates the commanded rate. In this way, the constraint equation of the front wheel becomes

$$\dot{x}_f = v_f cos(\theta + \delta) \tag{9.13}$$

$$\dot{y}_f = v_f sin(\theta + \delta) \tag{9.14}$$

$$\dot{\theta}_f = \frac{v_f}{l} sin(\delta) \tag{9.15}$$

$$\dot{\delta} = v_\delta \tag{9.16}$$

In addition to the range of steering angles, the range of steering rates is $v_\delta \in [\dot{\delta}_{min}, \dot{\delta}_{max}]$. The same problem also occurs in the speed v_r of the car and can be solved in the same way. The disadvantage of this technique is that it increases the size of the model, which may complicate the motion planning and control problem.

Similar to the kinematic model, it is the geometric structure of the dynamic bicycle model [26]. In this model, the mass of the vehicle is m, the moment of inertia at the center of gravity of the body is I_z, perpendicular to the horizontal plane.

The motion of the vehicle is represented by the following Newton-Euler equations:

$$\dot{v}_x = \frac{1}{m}[F_{xf}cos(\delta) + F_{xr} - F_{yf}sin(\delta)] + v_y\dot{\theta} \tag{9.17}$$

$$\dot{v}_y = \frac{1}{m}[F_{yf}cos(\delta) + F_{yr} - F_{xf}sin(\delta)] + v_x\dot{\theta} \tag{9.18}$$

$$\theta = \frac{1}{I_z}[F_{yf}l_r + (F_{xf}sin(\delta) + F_{yf}cos(\delta)l_f) \tag{9.19}$$

where F_x is the longitudinal force; F_y is the lateral force; r and f represent the rear wheel and the front wheel; δ is the steering angle; l_f and l_r are the distances between the front wheel and the rear wheel and the center of gravity; δ is the yaw angle.

The tire slip angles α_f, α_r of the front and rear wheels can be represented as

$$\alpha_f = \delta - \theta_f = \delta - tan^{-1}(\frac{v_y + l_f\dot{\theta}}{v_x}) \tag{9.20}$$

$$\alpha_r = -\theta_r = -tan^{-1}(\frac{v_y - l_r\dot{\theta}}{v_x}) \tag{9.21}$$

An accurate tire model describing tire-road interaction is crucial, as all driving forces are applied to the vehicle through the tire patch. Due to its simplicity, the linear tire model is commonly used in the design of autonomous driving controllers. In this model, the estimation of lateral force is based on the assumption of small slip angles.

If the wheel slip angle is small enough, the lateral tire force is proportional to the tire slip angle:

$$F_{yf} = -C_f\alpha_f \tag{9.22}$$

$$F_{yr} = -C_r\alpha_r \tag{9.23}$$

where C_f and C_r are the cornering stiffness of the front and rear wheels, respectively.

The linear tire model is only applicable to small slip angles and low lateral accelerations. Moreover, when designing controllers using the linear model, it is usually assumed that the longitudinal speed is constant. In this way, the longitudinal and lateral control of the vehicle can be decoupled, with the longitudinal dynamics controlled by a separate controller.

In nonlinear vehicle dynamics models, the forces generated in the wheels at larger slip angles are also considered. These models capture the nonlinearity of vehicle dynamics and provide more accurate solutions at higher wheel slip angles. However, these models significantly increase the complexity of the dynamic vehicle model. One of the most commonly used nonlinear tire models is the Pacejka tire model, where the force on the wheel is calculated according to an empirical model. Here, the formulas for calculating lateral force or longitudinal force are as follows:

$$y = Dsin[C \cdot arctan\{Bx(1 - E) + E \cdot arctan(Bx)\}] \tag{9.24}$$

$$Y(X) = y(x) + S_v \tag{9.25}$$

$$x = X + S_h \tag{9.26}$$

where X represents the input of longitudinal slip κ or lateral slip angle α; Y represents the output of longitudinal force $F_x = y(\kappa)$ or lateral force $F_y = y(\alpha)$. In addition, B is the stiffness factor; C is the shape factor; D is the peak; E is the curvature factor; S_h is the horizontal drift; S_v is the vertical drift.

Another common method for calculating tire force is to use analytical models, such as the Brush model. Here, the tire force is a function of the wheel normal force (F_z), longitudinal slip rate (κ) and lateral slip angle (α). For this model, the tire force is expressed as

$$\begin{cases} F = \mu F_z[3\gamma\sigma - 3(\gamma\sigma)^2 + (\gamma\sigma)^3], \text{if } \sigma \ll \sigma_x \\ F = \mu F_z, \ \text{if } \sigma > \sigma_s \end{cases} \tag{9.27}$$

where

$$\sigma_x = \frac{\kappa}{1 + \kappa} \tag{9.28}$$

$$\sigma_y = \frac{\tan(\alpha)}{1 + \kappa} \tag{9.29}$$

$$\sigma = \sqrt{\sigma_x^2 + \sigma_y^2}. \tag{9.30}$$

Here μ is the tire-road friction coefficient; σ_x and σ_y are the longitudinal and lateral slips, respectively; σ is the total combined slip. In addition, σ_s is the slip at the start of sliding; and γ is the isotropic tire parameter, expressed as

$$\gamma = \frac{2c_p\alpha^2}{3\mu F_z} \tag{9.31}$$

where c_p represents the total stiffness of the tread elements per unit length; a is half the length of the tire contact. Now, the longitudinal and lateral tire forces can be calculated as

$$F_x = \frac{\gamma_x}{\gamma} F \tag{9.32}$$

$$F_y = \frac{\gamma_y}{\gamma} F \tag{9.33}$$

9.2 Traditional Control Algorithms

Traditional control algorithms mainly include two types: PID control and LQR control.

9.2.1 *Classic PID Control*

Proportional-integral-derivative (PID) control [10] is an algorithm that calculates variable values based on the deviation, where the deviation refers to the difference between the planned value and the actual value, generally expressed as

$$u(t) = k_d \dot{e} + k_p e + k_i \int e(t) dt \tag{9.34}$$

where e is the error signal; k_p, k_i, and k_d are the proportional, integral, and derivative gains of the controller, respectively, as shown in Fig. 9.1.

The P term in PID is proportional to the current error e. For example, if the error e is large and positive, considering the gain factor K, the control output will be proportionally large and positive. Using proportional P control alone will result in an error e between the set value and the actual process value, as it requires an error to

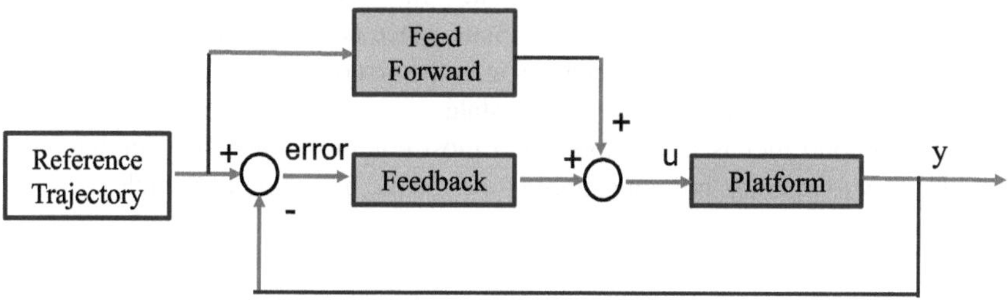

Fig. 9.1 The control flow of PID

generate a proportional response. If there is no error, there is no corrective response. For a given error change, a high proportional gain will result in a large output change. If the proportional gain is too high, the system may become unstable. In contrast, a small gain will result in a small output response to a large input error, as well as a slower or less sensitive controller. If the proportional gain is too low, the control action may be too small when responding to system disturbances. Adjustment theory and industrial practice suggest that the proportional term should contribute most of the output change.

The D term in PID is the best estimate of the future trend of the current change error Δe. It is sometimes referred to as predictive control, as it applies the control effect produced by error changes to effectively seek to reduce the impact of the error. The faster the change, the greater the control or damping effect. The ideal differential is not causal, so the implementation of the PID controller includes additional low-pass filtering of the differential term to limit high-frequency gain and noise. However, the differential term is rarely used in practice.

The I term in PID describes the past values of the error e, and integrates over time to produce the I term. For example, if there is a residual error after proportional P control, the integral term will add the control effect of the accumulated value of the error e to eliminate the residual error. When the error e is eliminated, the integral I term will stop growing. This will cause the proportional P effect to decrease as the error decreases, but it will be compensated by the continuously growing integral effect. The integral term accelerates the process of moving to the set point and eliminates the residual steady-state error that occurs with pure proportional control. However, because the integral term responds to past accumulated errors, it may cause the current value to exceed the set value.

By including a feedforward term in the controller, another degree of freedom can be added to the controller. Adding a feedforward term to the controller helps to overcome the limitations of feedback control. The feedforward term is added to the control signal without considering any measurements of the controlled system. However, the feedforward term may involve measurements of disturbances, etc. Designing feedforward control requires a more complete understanding of the physical system, so model reference is often used for feedforward controllers. The method of combining feedforward and feedback terms in the controller is also called a two-degree-of-freedom (2DOF) controller.

Feedforward is used to generate a reference trajectory, while feedback is used to compensate for disturbances and errors. State space control is often referred to as modern control, which is a technique that attempts to control the entire vector of the system as a unit by examining the system state.

The PID controller is the simplest and most common controller, implemented quickly, easy to operate, but not suitable for dynamic modeling of vehicles.

9.2.2 LQR Control

The Linear Quadratic Regulator (LQR) is an effective method to balance control operations and system responses. In this method, a gain for optimal feedback control is calculated based on a performance index. In the context of autonomous driving control, LQR requires a linear model of a vehicle to calculate the optimal feedback gain. The linearized vehicle dynamics model (bicycle model) can be represented as

$$\dot{x} = Ax + Bu \tag{9.35}$$

where $x = [v_y, \dot{\theta}]$, $u = \delta$.

To implement LQR-based autonomous driving control, an error state space model is usually formulated, thereby treating the path curvature as a disturbance to the system. An error state space model can be represented as

$$\dot{x} = Ax + B_1 u + B_2 \dot{r}_{ref} \tag{9.36}$$

where $x = [e_y, \dot{e}_y, e_\theta, \dot{e}_\theta]$; $u = \delta$; e_y is the cross-track error; e_θ is the heading angle error; r_{ref} is the reference yaw rate. $B_2\dot{r}_{ref}$ is considered to be an external disturbance to the system. In the simplest version of LQR, the task is to find the appropriate gain K for the feedback control law $u(k) = -Kx(k)$ by minimizing the following quadratic cost:

$$J = \sum_{k=0}^{\infty}(x^T(k)Qx(k) + u^T(k)Ru(k)) \tag{9.37}$$

where Q and R are positive definite diagonal weighting matrices.

The optimal solution for the feedback matrix K is $K = R^{-1}B^T P$(9-39)

where P is the solution to the Riccati equation:

$$P = A^T PA - A^T PB(R + B^T PB)^{-1}B^T PA + Q. \tag{9.38}$$

When crossing a curved path, LQR control that treats path information as a disturbance can provide better performance than a PID controller. Compared with a controller based on the main second-order pole (DSOP), LQR shows stronger robustness in the case of unmodeled system dynamics and similar robustness in the case of parameter uncertainty.

When passing through a high curvature path, LQR is not expected to provide steady-state error. To improve the tracking performance of dynamic curvature paths, it is recommended to use an additional feedforward controller. The feedforward controller uses road curvature information, and the feedback controller uses state feedback of cross-track error and yaw error. The integral control with a feedforward control law can be represented as

$$u(k) = -Kx(k) + \delta_{ff}. \tag{9.39}$$

The structure of the LQR controller with feedforward control does not consider future disturbances of the system, which may cause overshoot in the system's response. To avoid this, the optimal rehearsal of this feedforward LQR controller is designed. Based on the path information of a rehearsal range, the feedforward controller can be designed, and the feedback controller can be designed using lateral and yaw errors. Similarly, to compensate for future disturbances of the system, it is recommended to use iterative LQR (ILQR) [29], which turns the linear dynamics of LQR into non-linear, equivalent to the transformation of the differential dynamic programming (DDP) method and the sequential linear quadratic (SLQ) method. Due to the use of a linearized model, LQR cannot handle parameter and non-structural uncertainties. In the presence of uncertainties and external disturbances, LQR cannot provide appropriate robustness.

9.2.3 Model Predictive Control

Model Predictive Control (MPC) [10] is a closed-loop optimization-based control strategy, as shown in Fig. 9.2, which relies on a predictive model of the system. The optimization problem is formulated as:

$$\min_{U(t)} F(x(N|t)) + \sum_{k=0}^{N-1} L(x(k|t), y(k|t), u(k|t)) \tag{9.40}$$

subject to the constraints:

$$x(k+1|t) = f(x(k|t), u(k|t)), \tag{9.41}$$

$$y(k|t) = h(x(k|t), u(k|t)), \tag{9.42}$$

$$x_{\min} \le x(k|t) \le x_{\max}, \quad k = 1, \dots, N_c, \tag{9.43}$$

$$y_{\min} \le y(k|t) \le y_{\max}, \quad k = 1, \dots, N_c, \tag{9.44}$$

$$u_{\min} \le u(k|t) \le u_{\max}, \quad k = 1, \dots, N_c^u, \tag{9.45}$$

$$x(0|t) = x(t), \tag{9.46}$$

$$u(k|t) = K(x(k|t)), \quad k = N_u, \dots, N-1. \tag{9.47}$$

Here, t is the discrete time index. The cost function is composed of the **stage cost** $L(x, y, u)$ and the **terminal cost** $F(x)$. The notation $v(h|t)$ denotes the predicted value of variable v at step h, based on all available information up to time t. The

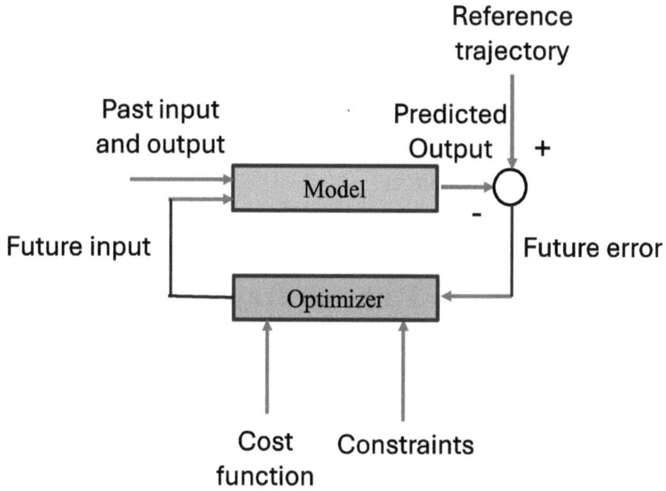

Fig. 9.2 The structure of MPC

system dynamics are modeled by the discrete-time functions $f(x, u)$ and $h(x, u)$, where $x \in \mathbb{R}^n$ represents the state vector, $u \in \mathbb{R}^m$ the control input, and $y \in \mathbb{R}^p$ the system output.

The optimization is carried out over the input sequence:

$$U(t) = (u(0|t), \ldots, u(N-1|t)),$$

where N is the **prediction horizon**. At each control cycle, the MPC procedure operates as follows:

1. Measure the system output and estimate the current state $x(t)$.
2. Initialize $x(0|t)$ with the state estimate.
3. Solve the optimization problem to obtain the optimal input sequence $U^*(t)$.
4. Apply only the first element of this sequence as the control input:

$$u(t) = u^*(0|t).$$

5. At the next cycle, update the state estimate with new measurements and repeat the optimization.

In this way, MPC combines the **performance of optimal control** with the **robustness of feedback control**. Predictions are made over a short time horizon, referred to as the prediction horizon, and the optimization problem is repeatedly solved online to adapt to changing conditions.

A key advantage of MPC lies in its ability to explicitly handle constraints. By predicting future system dynamics, it can incorporate constraints on inputs, outputs, and states directly into the optimization problem, typically formulated as a quadratic program (QP) or nonlinear program (NLP). This explicit treatment of constraints allows MPC to effectively deal with system uncertainty, nonlinearity, and multivariable interactions, providing strong robustness and high control performance.

However, despite its advantages, MPC is computationally demanding and can be challenging to implement in practice, especially for systems with fast dynamics or limited onboard computing resources.

9.3 Control Methods for Path and Trajectory Stability

The purpose of control is to keep the planned reference path or reference trajectory stable under the conditions of environmental modeling errors and uncertainties. The requirements for path stability [16] are ① the initial small tracking error should not increase; ② the tracking error converges to zero; ③ the driving speed along the reference path tends to a given value. Many vehicle control methods, including the ones discussed here, use the feedback rule:

$$u(x) = f(arg \min_{\gamma} \left\| x - x_{ref}(\gamma) \right\|) \tag{9.48}$$

where the feedback is a function of the nearest point on the reference path. An important problem is that the closed-loop vector $f(x, u(x))$ is not continuous. If the path is self-intersecting or non-derivable at a certain point, the discontinuity of $f(x, u(x))$ will fall directly on the path. If the executed trajectory encounters discontinuity, this will lead to unpredictable behavior. In many cases, the stability of the trajectory can be turned into the problem of the stability of the origin in the time-varying system. One issue to note is that controllers are often represented by the asymptotic tracking error as time tends to infinity. In practice, the reference trajectory is finite, so the transient response of the system should also be considered.

9.3.1 *Control of Path Stability*

1. Pure Pursuit Control The control law fits a semicircle to the points on the reference path ahead of the vehicle based on the current configuration of the vehicle, known as the lookahead distance L. This circle is defined as passing through the position of the car and a point on the path ahead of the car at a lookahead distance, tangent to the car's heading. This method is called pure pursuit control, one of the earliest path tracking methods, with a curvature of

$$\mathcal{K} = \frac{2\sin(\alpha)}{L}. \tag{9.49}$$

Given the vehicle speed vr, the forward angular speed is

$$\omega = \frac{2v_r * \sin(\alpha)}{/} L. \tag{9.50}$$

Considering the configuration $(x_r, y_r, \theta)^T$ and the point $(x_{\text{ref}}(s), y_{\text{ref}}(s))$ on the reference path, such that $||(x_{\text{ref}}(s), y_{\text{ref}}(s)) - (x_r, y_r)|| = L$, where s is the reference coefficient. Since there are usually multiple reference points, a point with the largest parameter s value is taken to uniquely define the control. Then, the angle α between the vehicle body (rear wheel) and the lookahead chord L is calculated as

$$\alpha = \arctan[\frac{y_{\text{ref}} - y_r}{x_{\text{ref}} - x_r}] - \theta \tag{9.51}$$

where (x_r, y_r) are the coordinates of the center of the rear wheels of the vehicle, $(x_{\text{ref}}, y_{\text{ref}})$ are the coordinates of the lookahead point, and θ is the yaw angle of the vehicle in the global coordinate system.

The relationship between R and θ, L can be formulated as follows:

$$R = \frac{L}{2\sin(\alpha)}. \tag{9.52}$$

In the case where the distance from the vehicle to the path is greater than L, the controller does not define an output. Another consideration is that changes in the curvature of the reference path may cause the car to deviate from the reference trajectory. This may be acceptable for driving along a road, but may be problematic for tracking parking maneuvers. Finally, as the vehicle speed increases, the heading rate ω becomes increasingly sensitive to the feedback angle α. A common solution to this problem is to scale L relative to the vehicle speed.

2. Rear Wheel Position Feedback Control

Rear wheel position feedback control calculates the control amount through the tracking deviation of the rear wheel center. First find the point on the reference path closest to the rear wheel:

$$s(t) = \arg\min_{\gamma} ||(x_r(t) - y_r(t)) - (x_{\text{ref}}(\gamma), y_{\text{ref}}(\gamma))||. \tag{9.53}$$

Then calculate the unit tangent at this point as follows:

$$\hat{t} = \frac{(\frac{\partial x_{\text{ref}}}{\partial s}|_{s(t)}, \frac{\partial y_{\text{ref}}}{\partial s}|_{s(t)})}{||(\frac{\partial x_{\text{ref}}(s(t))}{\partial s}, \frac{\partial y_{\text{ref}}(s(t))}{\partial s})||}. \tag{9.54}$$

The angle difference between the vehicle and the tangent is

$$\theta_e(t) = \theta - \arctan_2(\frac{\partial y_{\text{ref}}(s(t))}{\partial s}, \frac{\partial x_{\text{ref}}(s(t))}{\partial s}). \tag{9.55}$$

The resulting kinematic equation is

$$\dot{s} = \frac{v_r \cos(\theta)}{1 - \mathcal{K}(s)_e}. \tag{9.56}$$

$$\dot{s} = v_r \sin(\theta_e). \tag{9.57}$$

$$\dot{s} = \omega - \frac{v_r \mathcal{K}(s) \cos(\theta_e)}{1 - \mathcal{K}(s)_e} \tag{9.58}$$

where $\mathcal{K}(s)$ is the curvature of the path at s. Then, the designed heading angle speed controller is as follows:

$$\omega = \frac{v_r \mathcal{K}(s) \cos(\theta_e)}{1 - \mathcal{K}(s)_e} - (k_e |v_r \frac{\sin(\theta_e)}{\theta_e})e. \tag{9.59}$$

3. Front Wheel Position Feedback Control
Front wheel position feedback control, also known as Stanley control (adopted by Stanford University's autonomous driving team), is based on the path tracking deviation of the front wheel center to calculate the control amount. The control variables $s(t)$, $e(t)$ and $\theta_e(t)$ are still used, but $e(t)$ is calculated using the front wheel position instead of the rear wheel position. The time derivative of the lateral position error is

$$\dot{e} = v_f \sin(\theta_e + \delta). \tag{9.60}$$

It can be seen that it can be directly controlled by the front wheel steering angle δ, set $\dot{e} = -ke$, and solve for the front wheel steering angle as

$$\begin{aligned} v_f \sin(\theta_e + \delta) &= -ke \\ \delta &= \arcsin(-\frac{ke}{v_f}) - \theta_e \end{aligned}. \tag{9.61}$$

For the condition of exponential convergence within a finite field, it can be relaxed to local exponential convergence, that is, it becomes:

$$\delta = \arcsin(-\frac{ke}{v_f}) - \theta_e. \tag{9.62}$$

This method from Stanford University, under the condition of variable curvature path and continuous derivability of the path, converges locally exponentially to the path. However, it performs poorly in reverse operations and is not suitable for parking scenarios.

9.3.2 Control of Trajectory Stability

1. Control Based on Lyapunov Function

The control design based on the control Lyapunov function defines the tracking error in the coordinate system fixed to the car [27]. The configuration error can be represented by using the basis change and velocity $(x_{\text{ref}}, y_{\text{ref}}, \theta_{\text{ref}}, v_{\text{ref}}, \omega_{\text{ref}})$ of the inertial coordinate system of the reference trajectory, that is,

$$\begin{pmatrix} x_e \\ y_e \\ \theta_e \end{pmatrix} = \begin{pmatrix} \cos(\theta) & \sin(\theta) & 0 \\ -\sin(\theta) & \cos(\theta) & 0 \\ 0 & 0 & 1 \end{pmatrix} \begin{pmatrix} x_{\text{ref}} - x_r \\ y_{\text{ref}} - y_r \\ \theta_{\text{ref}} - \theta \end{pmatrix}. \tag{9.63}$$

The dynamic change of the configuration error is as follows:

$$\dot{x}_e = \omega y_e - v_r + v_{\text{ref}} \cos(\theta_e) \tag{9.64}$$

$$\dot{y}_e = -\omega x_e - v_r + v_{\text{ref}} \sin(\theta_e) \tag{9.65}$$

$$\dot{\theta}_e = \omega_{\text{ref}} - \omega. \tag{9.66}$$

The control task is

$$v_r = v_{\text{ref}} \cos(\theta_e) - k_1 x_e \tag{9.67}$$

$$\omega = \omega_{\text{ref}} + v_{\text{ref}}(k_2 x_e + k_3 \sin(\theta_e)). \tag{9.68}$$

The closed-loop error dynamics are written as:

$$\dot{x}_e = (\omega_{\text{ref}} + v_{\text{ref}}(k_2 x_e + k_3 \sin(\theta_e)))y_e - k_1 x_e \tag{9.69}$$

$$\dot{y}_e = -(\omega_{\text{ref}} + v_{\text{ref}}(k_2 x_e + k_3 \sin(\theta_e)))x_e - v_r + v_{\text{ref}} \sin(\theta_e) \tag{9.70}$$

$$\dot{\theta}_e = \omega_{\text{ref}} - \omega. \tag{9.71}$$

For $k_{1,2,3} > 0, \dot{\omega}_{\text{ref}} = 0$, $\dot{v}_{\text{ref}} = 0$, its stability is verified by the following Lyapunov equation:

$$V = \frac{1}{2}(x_e^2 + y_e^2) + \frac{1 - \cos(\theta_e)}{k_2}. \tag{9.72}$$

The semi-negative time derivative is

$$\dot{V} = -k_1 x_e^2 - \frac{v_{\text{ref}} k_3 \sin^2(\theta_e)}{k_2}. \tag{9.73}$$

Local analysis shows that the control law provides local exponential stability. However, for time-invariant systems, ω_{ref} and v_{ref} are constants.

2. Linearized output feedback control
For higher vehicle speeds, it is appropriate to limit the steering angle to continuous motion. With state estimation already in place, it becomes more difficult to design controllers considering only simple geometric relationships. In this case, a good choice is to output-linearize the system. This control is not easy to implement using only the front or rear wheel positions.

The system output is denoted as $x_p = x_f + d\cos(\theta+\delta)$ and $y_p = y_f + d\sin(\theta+\delta)$. Replacing the dynamic model of the previous differential equation with the derivative of the output results in:

$$\begin{aligned}\begin{pmatrix}\dot{x}_p\\ \dot{y}_p\end{pmatrix} &= \begin{pmatrix}\cos(\theta+\delta)-\frac{d}{l}\sin(\theta+\delta)\sin(\delta) & -d\sin(\theta+\delta)\\ \sin(\theta+\delta)-\frac{d}{l}\cos(\theta+\delta)\sin(\delta) & d\cos(\theta+\delta)\end{pmatrix}\begin{pmatrix}x_f\\ y_\delta\end{pmatrix}\\ &= A(\theta+\delta)\begin{pmatrix}v_f\\ v_\delta\end{pmatrix}\end{aligned}. \tag{9.74}$$

Define the right side of the formula as the auxiliary control variable, that is

$$\begin{pmatrix}\dot{x}_p\\ \dot{y}_p\end{pmatrix} = \begin{pmatrix}u_x\\ u_y\end{pmatrix}. \tag{9.75}$$

From u_x and u_y, recover the original control v_f and v_δ through the inverse matrix, that is

$$\begin{aligned}&[A(\theta+\delta)]^{-1} =\\ &\begin{pmatrix}\cos(\theta+\delta) & \sin(\theta+\delta)\\ -\frac{1}{d}\sin(\theta+\delta)-\frac{1}{l}\cos(\theta+\delta)\sin(\delta) & \frac{1}{d}\cos(\theta+\delta)-\frac{1}{l}\sin(\theta+\delta)\sin(\delta)\end{pmatrix}\begin{pmatrix}x_f\\ y_\delta\end{pmatrix}.\end{aligned} \tag{9.76}$$

Through the above input-output linear system, the control implementation of local trajectory stability is as follows:

$$u_x = \dot{x}_{p,\text{ref}} + k_x(x_{p,\text{ref}} - x_p) \tag{9.77}$$

$$u_y = \dot{y}_{p,\text{ref}} + k_y(y_{p,\text{ref}} - y_p). \tag{9.78}$$

9.4 Vehicle Control Based on Deep Learning

Physically-based autonomous driving control models usually have two main disadvantages [21]: First, different models have been developed for different driving behaviors, for example, following and lane-changing behaviors are usually modeled separately, and models developed for one behavior must be manually redesigned for different scenarios and tasks. Second, predefined motion heuristic methods usually make strong assumptions about driving behavior with a small set of parameters, which may fail to capture human strategic planning behavior and may not generalize well to various driving scenarios in highly interactive environments.

The manual adjustment of the parameters of this method is very time-consuming and difficult to engineer. In addition, the highly nonlinear characteristics of driving mean that control methods based on vehicle model linearization or other algebraic analysis solutions are often infeasible or not scalable.

Deep learning offers many benefits for vehicle control [22]: It can optimize its behavior based on data and adapt to new scenarios, so it is very suitable for controlling problems in complex and dynamic environments. Deep learning does not need to iteratively adjust parameters for performance in foreseeable scenarios, allowing developers to describe the desired behavior and learn to make the system perform well and generalize to new environments. For this reason, deep learning is becoming increasingly popular for vehicle control.

There are also some differences in the control objectives based on deep learning methods, some work describes the system as a high-level controller, and then implemented by a low-level controller, usually using classic control techniques. The goal of other work is to learn end-to-end driving, mapping observation results directly to low-level vehicle control interface commands.

Controllers based on deep learning can also be divided into lateral and longitudinal directions.

Most lateral control work focuses on a single task. For example, in models trained for lane keeping, these systems do not include any decisions, such as lane changes or turning onto different roads. Moreover, most of these works are trained and evaluated in a simulated environment, requiring further testing to verify their real-world performance. Despite this, there have been significant developments in this field, and these results show the great potential of using deep learning for autonomous vehicle control. While lateral control techniques support supervised learning techniques trained on labeled datasets, longitudinal control techniques support reinforcement learning methods that learn through interaction with the environment.

However, the first challenge is that the reward function in reinforcement learning needs to be carefully designed, where safety, performance, and comfort all need to be considered; if the reward function is poorly designed, it can lead to poor performance or the model failing to converge. Another challenge of reinforcement learning algorithms is the trade-off between exploration and exploitation; during training, the agent must take random actions to explore the environment; however, to better complete the task, the agent should exploit its knowledge to find the best action. Examples of solutions in this area are the ϵ-greedy exploration strategy and the Upper Confidence Bound (UCB) algorithm.

The ϵ-greedy search strategy selects a random action with a probability of ϵ, and after the agent learns its environment, it reduces excessive time overhead. In addition, UCB encourages exploration in states of high uncertainty and exploitation in areas of high confidence.

For end-to-end vehicle control methods, multiple learning strategies are used, but supervised learning is still the preferred method. However, achieving robust and high-performance models here still seems out of reach. For example, compared to techniques that only consider steering, techniques for full vehicle control often perform worse in terms of steering. This can be explained by the significant increase in the complexity of the tasks that the neural network is trained to perform, which still requires a lot of exploration and research.

9.4.1 End-to-End Autonomous Driving Control

Whether it's from sensor data input to control signal output, from perception results input to control signal output, or from planning results input to control signal output, all can be modeled by data-driven machine learning methods, i.e., end-to-end control. Here we introduce methods in this area as an exploration direction for autonomous driving research.

The DeepTest proposed in reference [1] is a system testing tool that can automatically detect erroneous behaviors of autonomous vehicles driven by deep learning models (DNN), prevent fatal collision events from occurring, and systematically explore all parts of the DNN through a large number of generated test inputs. In this exploration, there are some problems to solve in debugging DNN systems.

- How to systematically explore the input-output space of DNN systems?
- How to synthesize the real inputs needed for this exploration?
- How to optimize this exploration?
- How to avoid those detailed manual specifications, can a test library be automatically established to detect erroneous behaviors?

At the conceptual level, the behavior of extreme cases (corner-cases) of software errors based on DNN is similar to that of logic errors (logic bugs) in traditional software.

Reference [2] analyzes that driving models are likely to fail in areas with heavy traffic, complex intersections, poor weather, and lighting conditions. Here, we want to provide a method that can learn how to predict these failures, meaning estimating how difficult a scene is for a driving model, so that we can give early warnings.

This method is to develop a camera-based driving model through real driving data, and the error between the model prediction and the real operation is called the error degree. This defines the scene drivability, the quantified score of which is the degree of safety and danger (safe and hazardous).

The Ref. [3] predicts discrete actions, predicting the probability distribution of all possible actions. However, the limitation of discrete prediction is that it can only predict a finite number of predefined actions. Continuous prediction treats the current state of the vehicle as a regression task. If the driving strategy in the actual state is accurately predicted, the trained model can successfully drive the vehicle. Therefore, treating the driving process as a continuous prediction task, training a model that can predict the correct steering wheel angle and vehicle speed after inputting multiple perception information (including video and point cloud).

Reference [4] is the work of ETH, using a panoramic vision system, a route planner, and a CAN bus reader. The collected driving data includes scattered driving scenes and weather/lighting conditions, integrating information from the panoramic vision system and route planner (with GPS coordinates in the OpenStreetMap map format or TomTom navigator), learning a neural network driving model based on CNN, LSTM, and FCN.

At present, this end-to-end system does not incorporate modules for object detection and tracking, nor additional components such as traffic signal recognition or lane line detection. However, integrating these modules could significantly enhance the overall performance of the system.

Reference [5] is the work of NVIDIA, NVIDIA is a company that started working on end-to-end vehicle control earlier, its method trains a CNN model to complete the mapping from the image pixels of a single forward camera to vehicle control. Its system automatically learns some internal representations of processing steps, such as detecting road features using only the steering angle as a training signal.

It can be seen that this model does not learn the speed adjustment model, such as adaptive cruise control (ACC). That year, the system was demonstrated on Lombard Street in San Francisco, indeed no need to control speed, but obstacles causing braking will also lead to human takeover.

Reference [6] is the work of Comma.ai. As a hacker challenging Tesla and Mobileye's vision methods, Comma.ai is indeed one of the earliest explorers in end-to-end autonomous driving development.

OpenPilot mainly consists of two parts [28]: Software and hardware. The software part includes various algorithms and middleware. The hardware device called EON is the brain of a whole set of devices, responsible for running the NEO system (a highly customized Android system) and software algorithms. Different units on the EON device capture corresponding environmental information, for example, the camera takes forward images of the physical world, and the vehicle interface called Panda extracts vehicle status from the CAN bus. The millimeter-wave radar on the vehicle is also used in this step. Then, using these data, OpenPilot runs the Supercombo autonomous driving model and verifies its output through post-processing software and middleware. Finally, control signals are sent to the vehicle controller through the Panda interface.

The idea of the autonomous driving model Supercombo is to replicate the driving behavior of the driver and simulate the driving planning of the road. The deep learning model used is a variational autoencoder (VAE) under the framework of generative adversarial networks (GAN). An action-conditioned RNN model is used to learn a transition model through 15 frames of video data.

To learn data embedding, a model that forces the latent space probability distribution to be a Gaussian distribution is chosen. The generative adversarial network (GAN) learns the generative model cost function with the generator, alternating training of the generative network and the discriminative network. The generative model transforms samples of the latent space distribution into data from the desired dataset. The discriminator network tries to distinguish samples from the desired dataset from images sampled by the generator.

The VAE method can be simply combined with the learning cost function. VAE can be trained with GAN, where the encoder optimizes the similarity between the Gaussian prior in the latent space and the features extracted by the GAN discriminator network. The generator receives random samples as input from the latent space distribution and the output of the encoder network. The generator optimizes to deceive the discriminator and minimize the similarity between the original image and the decoded image.

The autoencoder is trained for 200 epochs. Each epoch consists of 10,000 gradient updates, with a batch size of 64. Batches are randomly drawn from driving data for processing, and Adam is used for optimization. The generator consists of 4 deconvolution layers, each followed by BN and leaky ReLU activation. The discriminator and encoder consist of convolution layers, each of which is followed by BN except for the first layer. The activation function is ReLU. The output size of the discriminator network is 1, and its cost function is binary cross entropy. The output size of the encoder network is 2048.

There are two training modes for RNN: One is the free-running mode, where the output of the previous state is used as the input of the next state; the other is the teacher-forcing mode, where the training truth is used as the input. In the training of the RNN transition model using 15 frames of video data, the first 5 frames use the teacher-forcing mode, and the remaining 10 frames use the free-running mode. In the relevant literature of RNN, the feedback output as input is informally called RNN hallucination. Note that when the output is fed back as input, the gradient of the output is considered to be zero, and unimportant solutions should be avoided. This method does not consider the separate training of the perception module, and the safety is poor, such as lack of obstacle detection, lane line detection, traffic signal light detection, etc.

9.4.2 Motion Control Based on Machine Learning

The following introduces the work of implementing motion control through deep learning or reinforcement learning. This is different from end-to-end vehicle control in that the input of motion control is not raw sensor data, but intermediate results or representations that have been processed by perception prediction planning.

Reference [7] trains an autonomous driving system by imitating a reinforcement learning coach, namely Roach (RL coach). The end-to-end method of autonomous driving usually relies on expert demonstrations (LfD). For end-to-end algorithms with on-policy intensive supervision, although humans are excellent drivers, they are not very good coaches. On the contrary, autonomous driving experts who provide special information can effectively generate large-scale on-policy and off-policy demonstrations.

However, existing urban driving autonomous driving experts use a lot of manually formulated rules, which perform poorly even on driving simulators with truth values. To solve these problems, a reinforcement learning expert was trained in Ref. [7], mapping BEV images to continuous underlying operations. Under the supervision of this reinforcement learning coach, a benchmark end-to-end agent with a single-camera achieved expert-level performance.

Although imitation learning (IL) methods directly imitate the behavior of experts, reinforcement learning methods usually pre-train part of the model with supervised learning of expert demonstrations to improve sample efficiency. In general, expert demonstrations can be divided into off-policy and on-policy.

1. Off-policy, that is, the expert directly controls the system, and the state-observation distribution follows the expert. The off-policy data of autonomous driving includes some public driving datasets, such as nuScenes, Lyft level 5, Bdd100k.
2. On-policy, the system is controlled by the required agent, and the expert "tags" the data. In this case, the state/observation distribution follows the agent, but can access expert demonstration data. On-policy data is the foundation for mitigating the phenomenon of covariate shift, as it allows the agent to learn from its own mistakes, a feature not present in off-policy data from experts.

However, collecting sufficient on-policy demonstration data from human drivers is challenging. Although expert trajectories and actions can be directly recorded during off-policy data collection, labeling the corresponding targets under given sensor measurements remains a difficult task for humans. In practice, only sparse events–such as human interventions–are typically recorded. These events contain limited information and are therefore difficult to use effectively for training, making them more suitable for reinforcement learning than for imitation learning.

To achieve expert-level performance, autonomous driving experts may rely on extensive computation, expensive sensors, or even ground truth information, making direct deployment undesirable. Although some imitation learning methods do not require on-policy annotation, such as generative adversarial imitation learning (GAIL) and inverse reinforcement learning (IRL), the efficiency of off-policy interaction with the environment is not high.

Conversely, autonomous driving experts can reduce expensive on-policy interactions, enabling imitation learning to successfully apply autonomous driving experts to different aspects of autonomous driving. The "expert" of the autonomous driving simulator CARLA is commonly referred to as Autopilot (or roaming agent). This Autopilot can access the real simulated state, but its driving skills cannot be compared with human experts due to the use of manually formulated rules. Imitation learning can be seen as knowledge transfer, but learning from expert actions alone is not effective enough.

The CARLA Autopilot consists of two PID controllers for trajectory tracking and a hazard detector for emergency braking. Hazards include:

- Pedestrians/vehicles detected ahead.
- Red light/stop signal detected ahead.
- Negative speed of the vehicle, used to handle slopes.

If any hazard appears in the trigger area in front of the vehicle, Autopilot will brake urgently: acceleration = 0, steering = 0, braking = 1. If no danger is detected, the vehicle travels along the desired path through two PID controllers, one for speed control and the other for steering control. The PID controller takes the position, rotation, and speed of the vehicle as inputs, and the specified route is dense (1-meter interval) waypoints. The speed PID generates acceleration, and the steering PID generates steering. By manually adjusting the parameters of the PID controller and hazard detector, CARLA Autopilot can become a powerful benchmark method (target speed is 6 m/s).

From the experiments, after training from scratch for ten million steps, Roach surpasses the rule-based Autopilot, setting a new performance ceiling for CARLA. When learning from Roach experts, you can train imitation learning agents and study more effective training techniques. Given the adoption of neural network policies, Roach can be a better coach for imitation learning agents also based on neural networks. Roach provides many informational goals for imitation learning agents to learn from, far exceeding the deterministic actions provided by other experts.

Roach contains an on-policy supervised learning scheme labeled by Roach on the CARLA simulator. With Roach's output, vehicles can be driven on CARLA, allowing for the recording of on-policy data from Roach. In addition to using 3D detection algorithms and other sensors to synthesize BEV, Roach can also solve the problem of supervision scarcity in the real world where strategies exist.

Roach has the following three features.

- Compared with previous reinforcement learning agents, Roach does not rely on data from other experts.
- Unlike the rule-based CARLA Autopilot, Roach is end-to-end trainable, so it can be extended to new scenarios with a small amount of engineering work.
- High sampling efficiency, based on input/output representation and exploration loss, a single GPU machine can train Roach from scratch in less than a week, achieving top expert performance in 6 LeaderBoard maps in CARLA.

Roach consists of a policy network and a value network. The policy network maps the BEV image and measurement vector to an action distribution. The value network then uses the same input as the policy network to estimate a scalar value output.

Red lights and stop signs are colored at the brightest level, yellow lights at the intermediate level, and green lights at the darker level. If the stop sign is active, it is displayed, i.e., it appears as the vehicle approaches and disappears after the vehicle has completely stopped. The BEV characterizes whether the vehicle has stopped, and the network architecture without a loop structure reduces the size of the Roach model.

To avoid parameter tuning and system identification, Roach directly predicts the action distribution. Its action space is mainly steering and acceleration, with positive acceleration corresponding to acceleration and negative values corresponding to braking. Here, the Beta distribution is used to describe the action.

Compared with the Gaussian distribution often used in model-free RL, the support of the Beta distribution is bounded, avoiding the clipping or squashing operations of forced input constraints. This leads to better behaved learning problems because it does not require a tanh layer and entropy and KL divergence can be explicitly calculated. In addition, the modality of the Beta distribution is also suitable for extreme driving actions, such as emergency braking or sharp turns.

Training uses the Proximal Policy Optimization (PPO) method with clipping to train the policy network and value network. In order to benefit the imitation learning agent from the information supervision generated by Roach, a loss is set for each supervision in Ref. [7], so that Roach's training scheme can be used to improve the performance of existing imitation learning agents.

Learning from all vehicles (LAV) is a framework [23] for training driving policies based on the experience collected by the vehicle and the experience of all vehicles obtained from its observations. The main difficulty of learning from other vehicles is the lack of sensor information. By learning intermediate representations with a set of supervision tasks, it is invariant to the viewpoint of the control vehicle. This not only provides richer signals during training, but also allows for more complex reasoning during the inference process. Understanding the driving style of all vehicles helps predict their behavior during testing and can avoid collisions.

In the agent pipeline, the 3D backbone network fuses lidar measurements and RGB camera semantic segmentation to generate a 2D spatial feature map, which is used as the input to the motion planner. During inference, central cropping is used to predict the vehicle's trajectory; during training, the motion planner is trained from all visible vehicles using the ground truth detection of nearby vehicles. The detection results are characterized by the rotation ROI of the shared feature map. Finally, during inference, the controller aggregates multiple motion predictions into a single steering and acceleration command.

The goal of LAV is to establish a deterministic driving model that maps sensor data, high-level navigation commands, and vehicle status to raw control commands at each moment.

LAV uses an end-to-end differentiable three-step modular pipeline: perception module, motion planner, and low-level controller.

The perception module is trained from large-scale annotated data supervision, with two goals: to create a robust generalization representation of the surrounding world, and to build invariant features of the vehicle, which helps to supervise the motion planner.

PointPillars and PointPainting, two lidar-image detection models, are used as the multimodal 3D perception backbone network. Given RGB images taken by three cameras with external parameters, the semantic segmentation scores are calculated using ERFNet. Five semantic classes are used: background, vehicle, road, lane line marking, and pedestrian. For each lidar point, PointPainting integrates its corresponding semantic class through segmentation scores.

For PointPillars, an FC-64-64 with BatchNorm is used as its PointNet. Pillars are created for lidar points in the $x \in [-10\text{m}, 70\text{m}]$ and $y \in [-40\text{m}, 40\text{m}]$ area. Each pillar represents a spatial area of $0.25\ \text{m} \times 0.25\ \text{m}$. A default 2D CNN with multi-scale features is used to obtain spatial features, with a resolution half that of the original image. Unlike the original PointPillars that directly build dense pillars specified by hyperparameters, here pillars are represented sparsely. In addition, a sparse PointNet is used to process the corresponding sparse pillar features. This allows for efficient processing of all pillars in both spatial and temporal dimensions.

The detection head and map head adopt a branched architecture. For BEV object detection, a simplified single-step CenterPoint model is employed. Specifically, the network predicts two centerness maps—one for vehicles and one for pedestrians—along with class-independent direction and boundary maps. For map prediction, BEV semantic maps are generated for roads, solid lane markings, and dashed lane markings. Each map is produced using a separate 3×3 convolution, followed by a 3×3 transposed convolution with stride 2. All branches share a common backbone network. During testing, a 2D max-pooling layer serves as a simplified alternative to non-maximum suppression (NMS).

The motion planner uses the map perspective features of the perception model to generate a series of waypoints describing the vehicle's future trajectory. Motion planning typically only uses self-vehicle supervision for this prediction. This supervision is very sparse, providing the motion planner with a series of labels for each collected data point. In the LAV framework, motion planning is learned from all the intelligent bodies surrounding the self-vehicle. This is possible because the perception system produces vehicle-invariant features as input; it is also because the output of the motion planner, i.e., the future trajectory, can be easily obtained from real driving logs.

Finally, the low-level controller converts the motion planning into actual steering, braking, and acceleration commands executed on the vehicle. During testing, the low-level controller considers the motion planning of other vehicles and makes emergency stop decisions.

The LAV model training pipeline overview, first, the 3D perception model is trained with detection and semantic mapping as supervision signals. Both tasks help to learn viewpoint-invariant spatial representations, and detection can also predict the posture of other vehicles, predicting their future trajectories during inference. The perception module generates vehicle-independent feature representations for use in motion planning. At the same time, the motion planner is trained with real perception results, using the trajectories of all nearby vehicles to train the model, and their future trajectories as supervision. Finally, the model learned is combined using distillation techniques. This model learns to plan for all vehicles in an end-to-end manner using only the input from the self-vehicle's sensors.

The controller converts motion planning into actual driving commands. In the controller logic uses two PID controllers for lateral (steering) and longitudinal (acceleration) control. Both PID controllers use the modified motion planning to obtain basic statistical information as input and generate continuous output commands. The longitudinal PID controller also uses the current speed as input to calculate acceleration.

The binary brake classifier takes all camera RGB images as input. The telephoto lens image and the other three images are stitched into ResNet-18, followed by a global average pooling layer. This provides two fixed-size embeddings, which are connected together and fed into a linear layer to predict the binary brake.

SHAIL (safety-aware hierarchical adversarial imitation learning) [24] is a method of hierarchical adversarial imitation learning for autonomous driving.

Generative imitation learning is a method of automatically generating policies through real-world and imitation decisions. Previous autonomous driving policies of generative imitation learning focused on learning low-level controllers in simple settings. However, to extend to complex settings, some autonomous driving systems combine fixed, safe, optimization-based low-level controllers with high-level decision logic capable of choosing appropriate tasks and related controllers.

SHAIL is such a learning method, strategically imitating low-level driving data, selecting high-level policies from a set of low-level controller instances.

Many autonomous driving systems combine fast, safe, optimization-based low-level controllers with high-level logic to choose appropriate tasks, controllers, and controller parameters. High-level logic can choose between different options, such as left lane change, accelerate, turn right, easy brake, hard brake, and then pass control to a low-level controller instance with the selected appropriate task and parameters. However, the selection labels for these high-level options are usually inaccessible in expert trajectories, making direct learning difficult.

First, let's review the background knowledge of imitation learning and reinforcement learning.

Optimal decisions are generally made in a Markov Decision Process (MDP), which includes a state space, action space, and state transition function, as well as a reward function, initial state distribution, and discount factor γ. An MDP policy maps states to a distribution through actions, and an optimal policy maximizes the accumulated discounted rewards.

In a reinforcement learning setting, the exact transition and reward functions T and R are unknown, but one can interact with the environment, receiving samples of the next state and reward generated.

In an imitation learning setting, no reward signal is received, but data is provided in the form of trajectory unfolding by experts interacting with the environment. The imitation learning problem can be seen as a moment matching problem between the expert and learner distributions, and methods can be roughly described as off-policy Q-value moments, on-policy Q-value moments, or reward moments.

In an imitation learning setting, the most direct method of learning a policy is behavior cloning (BC), where a supervised learner regresses from states to actions. This method has a long history in autonomous driving systems. Behavior cloning has error accumulation during testing because the agent will eventually enter states not seen during training, a phenomenon commonly referred to as covariance shift.

Under a policy π, the state-action occupancy measure is defined. The occupancy measure under this strategy is the (non-normalized) γ-discounted stationary distribution of states and actions visited. Similarly, the state-action occupancy measure of the expert policy can also be defined.

One perspective formulates imitation learning as a moment matching problem between the occupancy measures of the expert and the learner, achieved by minimizing some f-divergence between the relevant distributions. In the policy reward matching setting, this objective can be cast as a two-player game between a policy generator and an observation-action discriminator.

The optimization of this objective can alternate between discriminator gradient descent for optimizing the discriminator parameters and policy gradient descent for optimizing the stochastic policy, the latter of which can be viewed as reinforcement learning with reward signals. Both of these steps can be estimated using Monte Carlo methods (and a replay buffer) for their expectations.

The specific implementation of SHAIL is discussed below.

First, define an occupancy measure objective that matches a policy capable of hierarchically generating low-level data, then design a safety-aware high-level controller.

First, the occupancy measure-matching objective is defined as a policy that generates states and actions hierarchically, where the occupancy measure is expanded across various high-level options. Then, a hierarchical representation of the occupancy measure is introduced. Here, the discriminator update remains unchanged, while the generator update uses the policy gradient algorithm, where the new "imagined" high-level reward accumulates discounted low-level "imagined" discriminator rewards.

Many practical implementations of policy gradient rely on a fixed-size action space. Based on this, we limit ourselves to a set of high-level options, where any option can be initialized from any state. In terms of safety, this assumption may be very limited. Typically, there is information about restricted options from different states. For example, an accelerate option should not be obtained from a red light. Moreover, it may be possible to predict the safety of different controllers. For example, safety can be strictly predicted through the reachability formula of the controller, or more loosely judged through scene understanding (for example, "turning may be unsafe because there are vehicles crossing the intersection").

SHAIL designs a high-level "option-selection" strategy where option safety includes a sensitivity measure, improving the previous form of hierarchical adversarial imitation learning.

Assuming that the agent can infer the safety or availability of different options from different states, this is incorporated into the safety awareness estimate. SHAIL introduces a binary random variable that predicts the safety or availability of the low-level controller, representing the probability that the option is safe when executed from a high-level state. This way, options can be designed to pass control back to the high-level option selector based on this safety prediction.

Based on the safety of this controller, a high-level controller can be designed that reweights (or masks) the probabilities of options according to the safety prediction of the options. This scheme requires at least one option with a non-zero safety probability (for example, a permanent "safe" controller), otherwise this high-level strategy will not represent a valid distribution in the controller. In addition, to learn a useful option selector, these options should have some semantic meaning that holds under different initial states.

To learn an option selector with this policy through policy gradient, the safety probability seen at option startup needs to be stored in the replay buffer.

GIN is a method of optimizing safe driving strategies based on graph interaction-aware constraints [25]. In this framework, the motion prediction and control modules share a latent representation including a social context and are trained together.

To reflect social interaction, GIN illustrates the motion of the agent in the form of a graph and filters the features with a graph convolution network. This helps to maintain the spatiotemporal position of adjacent nodes. In addition, the feedback loop can effectively combine these two modules. Therefore, GIN encourages the learned controller to be free from dynamic risks, while motion prediction is robust to abnormal motion. Specifically, the motion prediction and control modules are trained simultaneously through a reinforcement learning (RL) framework, sharing a social context vector that can interpret interactions between agents. To learn this context vector, an encoder is used to optimize the dynamic graph features in space and time. With this context vector, two feedback loops can be created to effectively combine these two modules. One method uses geometric interpretation to generate auxiliary costs for predicted motion used as policy optimization constraints; the other method uses samples collected by the policy during training as the ground truth for prediction.

In GIN, it mainly includes three modules: input preprocessing, interaction-aware motion prediction, and interaction-aware control.

Here is a brief description of the input preprocessing module, which mainly consists of two features: path tracking features and social graph features.

The vehicle's location information is obtained in real time through GPS, while road network data is retrieved by querying the database within a specified region. Using this information, the A* algorithm generates a global path from the current position to the target destination on the road network. Based on the planned path and the vehicle's location, the cross-track error and heading angle error are computed at each time step with respect to a preview distance d. Finally, the current vehicle speed is incorporated into the control input.

The motion of dynamic entities on the road is represented in graph form, consisting of nodes and adjacency matrices. Each node vector encodes attributes such as position, orientation angle, relative speed, relative acceleration, relative angular velocity, and a mask, all expressed in the coordinate system centered on the ego vehicle. The mask indicates whether a given entity is within the perception range. Graph edges capture both spatial relationships and temporal continuity across frames, while the adjacency matrix defines the connectivity structure among nodes.

As for the interaction-aware motion prediction module, it consists of an encoder and a decoder, considering the interaction between the vehicle and other vehicles to predict the trajectory. When entities enter or leave the recognition range, the graph structure changes over time. In order to enable the interaction-aware motion prediction module to inductively reason any graph, this module is trained with subgraph samples collected by the reinforcement learning agent.

(1) Encoder: The subgraphs sampled from the replay buffer are used as inputs to the encoder. For the subgraph, a bottleneck layer, i.e., a convolutional layer with a 1×1 kernel, is applied to increase the size of the node features. Then, the features are extracted through three layers of the graph convolution block to encourage

complex driving strategies. Each graph convolution block compresses the latent representation while maintaining spatiotemporal connectivity. In this process, the sum of graph operations extracts spatially related features along the spatial edges. Next, convolution operations are applied along the frame-by-frame edges to understand the temporal relevance between dynamic entities, which helps to keep the learned intent consistent. Finally, a residual connection is used to prevent gradient vanishing. After multiple blocks, a social context is created, and this sequence feature is summarized by the GRU model.

(2) Decoder: Based on the hidden context vector, the decoder generates the future trajectory around and allows the interaction-aware control module to understand long-term risks. This is because it can provide signals for detected risks by geometrically interpreting the predicted trajectory. The initial hidden state of the decoder GRU is the social context vector, and the initial input is the current position of each vehicle. Each GRU unit predicts the vehicle's speed distribution through a skip connection between an input and output. The trajectories sampled from the replay buffer are used as ground truth. The prediction module is trained with a regression loss with p-norm.

Finally, the interaction-aware control module outputs actions by jointly considering path-tracking accuracy and collision-avoidance requirements in dynamic environments within a reinforcement learning framework. In this setting, GIN leverages path-tracking features and social context vectors to model target interactions. To compress social context from the perspective of vehicle control, a shared module is introduced. In addition, auxiliary costs are derived from predicted motion. The interaction-aware motion prediction module and the interaction-aware control module are then optimized simultaneously under a constrained Markov Decision Process (MDP) using the policy gradient (PG) method. (1) Distance Average Pooling: When creating a graph for dynamic targets within the recognition range, the order of nodes changes, so to maintain the permutation invariance of the social context vector, its local features are obtained through a feed-forward network with shared weights. Then, Distance Average Pooling (DAP) is applied to create global features. For the vehicle, the closer the entity, the more its intent is reflected. After DAP, the global features and manually designed path tracking features are connected and passed as inputs to the parameterized policy.

(2) Auxiliary Cost: To avoid obtaining costs only in rare collision situations, auxiliary costs are created with predicted trajectories. Therefore, sparse signals are converted into dense signals to promote learning strategies to avoid long-term risks. Here, risk detection is divided into four situations. To interpret social risks from a geometric perspective, polygons and polylines are used. The polygon is a local frame with a certain safety margin added to its length and width. The polyline represents the trajectory of each vehicle predicted by the decoder. The Separating Axis Theorem (SAT) can check the overlap between the vehicle's polygon and other polygons. Based on the Counter Clockwise (CCW) algorithm, the intersection between two line segments can be detected. To create line segments, connect the start and end points of the polygon. For polygons, each edge is used as a line segment. The auxiliary cost is defined as the overlap between the vehicle's polygon and multiple polylines.

(3) Optimization: The algorithm is called Graph-based Intent-aware constrained Policy Optimization, or GIN. Before the training loop of GIN, a global path τ_g that follows the road network G_n is obtained using the A* algorithm. At each step, the agent interacts with the environment.

During training, samples are continuously generated and stored in a replay buffer. The input preprocessing module constructs either social graph features F_d or path-tracking features F_k, corresponding to ϕ_d and ϕ_k, respectively. Here, F_d encodes the dynamic targets based on the history of observed vehicle positions o_t, while F_k represents the kinematic properties of the system. In addition, a complementary feature $\breve{F}_d$, centered at the midpoint of length $2H$ in the time domain (with H denoting the length of the temporal feature window), is accumulated to predict future trajectories $\hat{\tau}_t$.

Given the state representation processed by the encoder v_η, the agent executes an action and transitions to the next state. To mitigate long-term interaction risks, an auxiliary cost $\hat{c}_\rho$ is introduced, computed from the predicted trajectory $\hat{\tau}_t$ obtained via the decoder $\rho\zeta$. Model parameters are updated from mini-batches sampled from the replay buffer: the encoder and decoder are optimized with prediction loss, while the parameters of the actor, critic, and approximate cost $\langle\theta, \psi, \mu\rangle$ as well as the adaptive weights $\langle\beta, \kappa\rangle$ are updated using the Worst-Case Soft Actor–Critic (WCSAC) algorithm through accumulated samples. Finally, the parameters of the target network $\langle\bar{\theta}, \bar{\psi}, \bar{\mu}\rangle$ are updated via moving averages to ensure training stability.

The pseudocode of the GIN algorithm, where the cost is c, the state is s, the action is a, the reward is r, the state transition is T, the discount factor is γ, the policy distribution is π, the propagation rule of GCN is ψ, and the social context is Z.

Reference [8] proposes a safe autonomous driving system based on Model Predictive Control (MPC) for imitation learning.

On the one hand, many works are developing control systems to imitate the driving behavior demonstrations of drivers. Imitation learning algorithms can be used for this purpose, but it is difficult to provide safety guarantees for the final closed-loop system trajectory. On the other hand, MPC can handle nonlinear systems with safety constraints, but extensive domain knowledge is needed to drive like a human.

Reference [8] suggests using a seamless combination of the two techniques, using MPC as a differentiable control layer in a hierarchical imitation learning strategy, learning a controller from the demonstrations of desired driving behavior. With this strategy, imitation learning can run in a closed-loop and end-to-end manner through the parameters, models, or constraints in the MPC cost.

An MDP includes states s_t, actions a_t, and state transitions $s_{t+1} \sim P(\cdot|s_t, a_t)$. The policy of an agent control system is $\pi(a_t|s_t)$. A policy π induced state occupancy measure $\rho_\pi(s)$ represents the density of state occurrences, while following the policy over an infinite time range adds a discount factor γ:

$$\rho_\pi^S = \sum_{t=0}^{\infty} \gamma^t P(s|t). \tag{9.79}$$

Given the expert policy π_E to be imitated, imitation learning tries to find a policy π that minimizes the occupancy measure distance, i.e.,

$$\min_{\pi} \mathbb{E}_{S\sim\rho_{\pi}^{S}}[\mathcal{L}(\pi_E, \pi)] = \mathcal{L}(\rho_{\pi_E}^{S}, \rho_{\pi}^{S}). \tag{9.80}$$

And MPC as an optimal controller, is defined as

$$a = \arg\min_{x,u} \sum_{k=0}^{N-1} l(\theta, x_k, u_k) + l_N(\theta, x_k, u_k). \tag{9.81}$$

Subject to the following constraints:

$$x_0 = \hat{x} \tag{9.82}$$

$$x_{k+1} = f(\theta, x_k, u_k) \tag{9.83}$$

$$h(\theta, x_k, u_k) \leq 0 \tag{9.84}$$

$$a = u_0. \tag{9.85}$$

Now the control of system state st is through an (unknown) stochastic policy π_H, i.e., $a_t \sim \pi_H(\cdot|s_t)$, and the imitated controller controls with a deterministic policy parameterized by θ, i.e., $a_t = \pi_\theta(s_t)$. The controller part of the policy is represented by MPC, which optimizes its control action a according to system state s and parameter θ (part of the objective function, model, and constraints):

$$\min_{\theta} \mathbb{E}_{S\sim\pi_{\theta}^{S}}[\mathcal{L}(\pi_H, \pi_\theta)] = \mathcal{L}(\rho_{\pi_H}^{S}, \rho_{\pi_\theta}^{S}) \tag{9.86}$$

where

$$\pi \in \arg\min_{x,u} \sum_{k=0}^{N-1} l(\theta, x_k, u_k) + l_N(\theta, x_k, u_k). \tag{9.87}$$

Subject to the following constraints:

$$x_0 = s \tag{9.88}$$

$$x_{k+1} = f(\theta, x_k, u_k) \tag{9.89}$$

$$h(\theta, x_k, u_k) \leq 0 \tag{9.90}$$

$$a = u_0. \tag{9.91}$$

Equations (9.86)–(9.91) describe a two-level optimization problem, where the upper-level problem is the imitation of policy π relative to human policy π_H, and the lower-level problem is the MPC optimization problem given state s.

In gradient-based solution techniques, the lower-level optimization problem is usually replaced by the KKT conditions as constraints. However, in the context of imitation learning and reinforcement learning, directly solving the constrained optimization problem is not easy. Moreover, it requires second-order information (i.e., the Hessian matrix), and when it involves neural networks and large-scale data processing, second-order information does not scale well.

On the contrary, the solution adopted here is based solely on first-order information. The core idea is to calculate the gradient of the MPC parameterized solution with respect to the optimization variables of the upper-level imitation problem, and then solve it as an unconstrained solution, where the gradient descent step size is α:

$$\theta \leftarrow \theta - \alpha \left[\frac{\partial \mathcal{L}}{\partial \theta}\right] \tag{9.92}$$

$$\left[\frac{\partial \mathcal{L}}{\partial \theta}\right]^T = \left[\frac{\partial \pi_\theta}{\partial \theta}\right]^T \left[\frac{\partial \mathcal{L}}{\partial \rho_{\pi_\theta}^S} \frac{\partial \rho_{\pi_\theta}^S}{\partial \pi_\theta}\right]^T . \tag{9.93}$$

It can be seen that the calculation of $\left[\frac{\partial \mathcal{L}}{\partial \theta}\right]$ requires:

- $\frac{\partial \mathcal{L}}{\partial \rho_{\pi_\theta}^S} \frac{\partial \rho_{\pi_\theta}^S}{\partial \pi_\theta}$, depending on the imitation learning algorithm used, the defined distance $\mathcal{L}$, and how to approximate the occupancy measure ρ.
- $\frac{\partial \pi_\theta}{\partial \theta}$, the derivative of the (local) optimal solution of the MPC problem.

The parameter θ refers to the physical properties in the model or the weight matrix in the quadratic regulator cost, so it is static. However, to understand more complex policy behaviors, as an end-to-end control from demonstration, it is necessary to combine MPC control with other differentiable layers.

Regarding some context information of the system at time t, it is analyzed here in the form of latent variable $\mathcal{X}_t$, which can be inferred from the current state of the system, that is, $\mathcal{X}_t = Q(s_t)$. For example, $\mathcal{X}_t$ can be a high-dimensional input (such as an image), from which end-to-end learning can be conveniently carried out. For this, learn a function $\theta = g(\mathcal{X}_t)$ that maps the latent variable $\mathcal{X}_t$ to the MPC parameter θ. For example, θ can represent the target state of the system or the trajectory to be followed, which depends on the current situation at time t. This method is called hierarchical decomposition. MPC is responsible for reactive control behavior, that is, mapping $s_t \rightarrow a_t$, while the previous layers will learn subtasks that indirectly affect the closed-loop behavior of the system, these layers are called indirect controllers. Then, the policy is written as

$a_t = \pi_\theta(\mathcal{X}_t)(s_t)$.

Based on a constrained optimization MPC provides mapping $s_t \to a_t$, the indirect controller outputs MPC parameter θ according to the latent variable $\mathcal{X}_t$. The dashed line shows the gradient flow of the entire policy, which may be due to the differentiability of MPC.

The following introduces how to use observed clone behavior (BCO) and differentiable simulator to close-loop learn hierarchical policy with MPC control through MPC-IL method, referred to as MPC-BCO. Given the state trajectory of length T obtained by policy $\pi_\theta : \tau_{\pi_\theta = S_0, S_1, \ldots, S_T}$, and the state trajectory $\tau_{\pi_H} = S_0^*, S_1^*, \ldots, S_T^*$ sampled from human demonstration. The problem is defined as an optimization problem, that is, starting from the same initial state s_0^*, minimize the $L2$ state error $\mathcal{L}(s_t, s_t^*) = ||s_t - s_t^*||_2^2$ on the state sequence of humans and policies. These sequences are obtained through the simulation model F, which provides state transition $s_{t+1} = F(s_t, a_t)$. Assuming that the first-order derivative of F near some (s_t, a_t) can be calculated, it is suggested to calculate $\left[\frac{\partial \mathcal{L}}{\partial \theta}\right]$ through time backpropagation (BPTT). The error of BPTT future state is caused by the previous operation taken by the policy. Define $J(\theta) = \mathbb{E}_{s_t \sim \rho^s_{\pi_\theta}} \mathbb{E}_{s_t^* \sim \rho^s_{\pi_H}} \sum_{t=0}^{T} \mathcal{L}(s_t, s_t^*)$ is the sum of L2 posture errors along the trajectory. Then, in order to update the policy, use the differentiability of the policy and simulator, and calculate the derivative of J at (s_t, a_t, s_{t+1}) with BPTT. The gradient is calculated by the following recursive formula:

$$\frac{\partial J_t}{\partial s_t} = \frac{\partial \mathcal{L}_t}{\partial s_t} + \frac{\partial J_{t+1}}{\partial s_{t+1}}\left(\frac{\partial F}{\partial s_t} + \frac{\partial F}{\partial a_t}\frac{\partial \pi_\theta}{\partial s_t|\mathcal{X}_t}\right) \tag{9.94}$$

$$\frac{\partial J_t}{\partial \theta} = \frac{\partial J_{t+1}}{\partial s_{t+1}}\left(\frac{\partial F}{\partial a_t}\frac{\partial \pi_\theta}{\partial \theta|\mathcal{X}_t}\right) + \frac{\partial J_{t+1}}{\partial \theta}. \tag{9.95}$$

The final gradient $J(\theta)$ is calculated by recursively applying formulas (9.94) and (9.95), starting from $t = T$ until $t = t_s$. It is suggested not to backpropagate (BP) before $t_s = 0$ to avoid bias. Since the initial state s_0^* belongs to the occupancy measure of the human demonstration state, there may be a transition part in the system before the policy reaches a stable state corresponding to its actual occupancy measure.

9.5 Summary

This chapter discusses the control module of autonomous driving, from the kinematics and dynamics models of vehicles, traditional control algorithms (PID, LQR and MPC) to practical path/trajectory stabilization control algorithms, especially involving end-to-end control systems based on machine learning.

From Chap. 5 to this chapter, we have "browsed" the modular development platform of autonomous driving, that is, perception, mapping, localization, planning and control, etc. Chapter 10 will discuss the simulation module of autonomous driving.

References

1. Y TIAN et al, DeepTest: Auto Testing of DNN-driven Autonomous Cars, arXiv preprint arXiv:1708.08559, 2017.
2. S HECKER, D DAI, L GOOL,Failure Prediction for Autonomous Driving, arXiv preprint arXiv:1805.01811, 2018
3. CHEN Y, WANG J, LI J et al, LiDAR-Video Driving Dataset: Learning Driving Policies Effectively, IEEE Conference on Computer Vision and Pattern Recognition, Salt Lake City, UT, USA, June 2018: 5870–5878
4. S HECKER, D DAI, L GOOL, End-to-End Learning of Driving Models with Surround-View Cameras and Route Planners, European Conference on Computer Vision, Munich, Germany, Oct. 2018: 449–468
5. M BOJARSKI et al., End to End Learning for Self-Driving Cars, arXiv preprint arXiv:1604.07316, 2016.
6. E SANTANA, G HOTZ, Learning a Driving Simulator, arXiv preprint arXiv:1608.01230, 2016.
7. Z ZHANG, A LINIGER, D DAI, F YU and L VAN GOOL, End-to-End Urban Driving by Imitating a Reinforcement Learning Coach, arXiv preprint arXiv:2108.08265, 2021.
8. F S ACERBO, J SWEVERS, T TUYTELAARS, and T D SON, MPC-based Imitation Learning for Safe and Human-like Autonomous Driving, arXiv preprint arXiv:2206.12348, 2022.
9. B PADEN et al. A survey of motion planning and control techniques for self-driving urban vehicles. IEEE Transactions on intelligent vehicles, 2016, 1 (1): 33–55.
10. S D PENDLETON et al., Perception, Planning, Control, and Coordination for Autonomous Vehicles, Machines, 2017, 5(1): 6–59.
11. KHODAYARI A, GHAFFARI A., AMELI S. and FLAHATGAR J., A Historical Review on Lateral and Longitudinal Control of Autonomous Vehicle Motions, Int. Conf. on Mechanical and Electrical Technology, Singapore, Oct. 2010: 421–429.
12. RESCHKA A, BOHMER J., SAUST F., et al., Safe, Dynamic and Comfortable Longitudinal Control for an Autonomous Vehicle, IEEE Intelligent Vehicles Symposium, Madrid, Spain, June 2012: 345–351.
13. R ATTIA, R ORJUELA, and M BASSET, Longitudinal Control for Automated Vehicle Guidance, IFAC Workshop on Engine and Powertrain Control, Simulation and Modeling, Oct. 2012, 45(30): 65–71.
14. R ATTIA, R ORJUELA, and M BASSET, Combined Longitudinal and Lateral Control for Automated Vehicle Guidance, Vehicle System Dynamics, 2014, 52(2): 261–279.
15. S-H LEE, Y SON, C CHUNG, Robust Real-time Optimal Autonomous Highway Driving Control System: Development and Implementation, International Federation of Automatic Control Proceedings Volumes, August 2014. 47(3): 10706–10712.
16. N R. KAPANIA, J. C GERDES, An Autonomous Lane-keeping System for Vehicle Path Tracking and Stability at the Limits of Handling, AVEC Int. Symposium on Advanced Vehicle Control, Yokohama, Japan, Sept. 2014: 720–725.
17. Z GAO, W YAN, A Headway Control Algorithm for ACC Vehicles with the Compensation of the Preceding Vehicle Acceleration, Procedia Engineering, 2016. 137: 669–679.
18. J KONG et al., Kinematic and Dynamic Vehicle Models for Autonomous Driving Control Design, IEEE Intelligent Vehicles Symposium, Seoul, Korea, Aug. 2015: 1094–1099.
19. D LI et al., Reinforcement Learning and Deep Learning based Lateral Control for Autonomous Driving, arXiv preprint arXiv:8101.2778, 2018.
20. J. JIANG, A ASTOLFI, Lateral control of an autonomous vehicle. IEEE Trans. on Intelligent Vehicles, 2018. 3(2): 228–237.
21. X DI, R SHI, A Survey on Autonomous Vehicle Control in the Era of Mixed-Autonomy: From Physics-Based to AI-Guided Driving Policy Learning, arXiv preprint arXiv:2007.05156, 2020.
22. S KUUTTI, R BOWDEN, Y JIN, P BARBER, and S FALLAH, A Survey of Deep Learning Applications to Autonomous Vehicle Control, IEEE Trans. on Intelligent Transportation Systems, 2020. 22(2): 712–733.

23. D CHEN, P KRAHENBUHL, Learning from All Vehicles, arXiv preprint arXiv:2203.11934, 2022.
24. A JAMGOCHIAN, E BUEHRLE, J FISCHER, and M J. KOCHENDERFER, SHAIL: Safety-Aware Hierarchical Adversarial Imitation Learning for Autonomous Driving in Urban Environments, arXiv preprint arXiv:2204.01922, 2022.
25. S-W YOO, C KIM, J-W CHOI, S-W KIM, and S-W SEO. GIN: Graph-based Interaction-aware Constraint Policy Optimization for Autonomous Driving, arXiv preprint arXiv:2206.01488, 2022.
26. M ROKONUZZAMAN, N MOHAJER, S NAHAVANDI, and S MOHAMED, Review and performance evaluation of path tracking controllers of autonomous vehicles, IET Intelligent Transport Systems, March, 2021. 15(1): 646-670.
27. L LI, J LI, S ZHANG, Review article: State-of-art trajectory tracking of autonomous vehicles, Mechanical Sciences, April, 2021, 12(1): 419–432.
28. L CHEN, T TANG, Z CAI, Y LI, P WU, H LI, J SHI, J YAN, and Y QIAO, Level 2 Autonomous Driving on a Single Device: Diving into the Devils of Openpilot, arXiv preprint arXiv:2206.08176, 2022.
29. Y TASSA, T EREZ and E TODOROV, Synthesis and Stabilization of Complex Behaviors through Online Trajectory Optimization, IEEE Int. Conf. on Intelligent Robots and Systems, Vilamoura-Algarve, Portugal, Oct. 2012: 4906–4913.

Chapter 10
Simulation Module of Autonomous Driving

Testing the performance and safety of autonomous driving technology in the real world is a massive undertaking. It is generally believed that approximately 10 billion miles (1 mile $\approx$ 1.609 km) of test driving data are needed to optimize a developing autonomous driving system, and even assembling a fleet of 1000 cars would take about 50 years to complete. Due to the inability to conduct sufficient road tests (road tests require extremely expensive time and cost), it is necessary to rely on computer simulation systems to assist in the development and testing of autonomous driving systems in a virtual environment.

It is well known that Google's simulation system, CarCraft, has run over 15 billion miles (over 10 million miles per day), while actual on-board testing is only 18 million miles (about 1 million miles tested per month). The process of reproducing the problems encountered in actual on-board testing in the simulation system and adding various changes is called "fuzzing." The visualization platform Xview developed at the time is part of the simulation system, serving as a debugging visualization display interface.

The composition of such a simulation system generally includes multiple algorithm modules, such as

- Sensor models. Cameras (similar to game engine functions), lidar, radar, sonar, and even vehicle networking (V2X), GPS, IMU, speedometers, etc.
- High-definition maps and virtual environments (similar to VR). Some high-definition map manufacturers can also generate directly from the real environment, such as San Francisco, New York, and London.
- Vehicle models (OEM data, dynamic models, control models). Pedestrians, motorcycles, and bicycles, etc. Companies that used to do motion capture have good data, and commercial car simulation software is also mature in these areas.
- Road systems (path networks) and road characteristics (geology, slope, risk, utilization, etc.).

Y. Huang and Z. Yang, *Autonomous Driving Handbook*,
https://doi.org/10.1007/978-981-95-3725-9_10

- Environmental models (seasons, climate, weather, time such as day and night). It has a relatively large relationship with sensor simulation.
- Traffic models (traffic rules, speed limits, parking, cross/T-shaped/roundabout, traffic lights, yielding, branching, merging, etc.), that is, the simulation of traffic flow.
- Driving models (driving behavior, navigation rules, collision avoidance, personal characteristics, culture, region, etc.).
- Application related, such as motion planning and simulation training.

This chapter introduces some modules for building a simulation system. Section 10.1 introduces the simulation of sensors, such as the imaging technology of cameras and the data simulation synthesis of lidar and millimeter-wave radar; Sect. 10.2 introduces traffic simulation models based on the open-source software SUMO; Sect. 10.3 introduces vehicle and pedestrian CAD models, and discusses the vehicle dynamics of deep learning modeling, and also introduces some research on rendering human bodies and human motion in the field of computer vision; Sect. 10.4 introduces the visualization technology of autonomous driving with the open-source visualization platform of the American online taxi company Uber as an example; Sect. 10.5 introduces the establishment of a road network simulation environment; Sect. 10.6 analyzes the autonomous driving test technology of the scenario library (including the scenario description language OpenSCENARIO and the scenario library example PEGASUS); finally, Sect. 10.7 discusses the neural network method of constructing safety-critical data for digital twins.

10.1 Sensors

This section mainly introduces cameras, lidar, and millimeter-wave radar, and also introduces inertial navigation systems, navigation systems, ultrasonic radar, and vehicle networking sensors.

10.1.1 Camera

The image processing of the camera refers to Sect. 2.2, and the photoelectric structure of the camera sensor refers to Sect. 3.1.1. Reference [1] simulates the functions of a digital camera system. The simulation includes the conversion from light to digital signals, color processing, and rendering. A spectral image processing algorithm is used to simulate the radiometric characteristics of a digital camera. In this algorithm, the spectral image and transmittance of the light source, lens, filter, and the quantum efficiency of the CMOS image sensor are considered. The characteristics of the optical part lie in the multi-convolution between different point spread function (PSF) optical components, such as the Cooke triplet, aperture, light attenuation, and the optical part of the CMOS sensor. The electrical part includes Bayern sampling, interpolation, dynamic range, and analog-to-digital conversion. To reduce noise, different exposure images are mixed to reconstruct the noise blurred image. Then, the image is filtered, deconvolved, and sharpened to eliminate noise and blur. Next, the interpolation, white balance, color correction, conversion from XYZ color space to LAB color space, then conversion to RGB color space, color saturation, and contrast of the color processing and rendering block are performed.

The camera equation describes the image capture process of the camera. In this equation, the radiometric description of the scenario is converted into an irradiance image on the sensor surface, and then the sensor converts the irradiance image into an electronic signal.

In a camera, the optical device is responsible for illuminating the scenario, capturing reflected light, and focusing the light on the sensor. The image capture system consists of a Cooke triplet objective and an image sensor. The function of optical components can be analyzed from the perspectives of resolution and spectrum. Through the camera equation, the function of the image capture system can be covered from an optical perspective.

The spectral radiance of the reflected target image is obtained by convolving the target's spectral radiance with the point spread function (PSF) of the Cooke triplet. This process captures the color variations introduced during illumination and propagation within the image acquisition system. The point spread function characterizes the optical system in terms of resolution, providing insight into the image quality, sharpness, and distortion of the reflected target as it propagates spatially through the Cooke triplet objective.

The illumination of the target is achieved through a spectral image and an algorithm that converts wavelengths into red, green, and blue (RGB) color space. This conversion process indicates the spectrum of the light source, the spectral response of the lens, the Bayer filter, and the semiconductor quantum efficiency of the sensor manufacturing material. The scenario is represented as a multidimensional pixel array describing the spectral radiance of each pixel in the scenario.

Spectral quantum efficiency is a wavelength-related function that describes the expected fraction of photons that produce electrons. The fill factor is the percentage of pixels occupied by the photodetector. The optical device converts scenario radiance data into an irradiance image on the sensor surface. In addition to the PSF, the optical system in spatial frequency can also be characterized by the modulation transfer function (MTF) and the optical transfer function (OTF). The MTF is a function of the optical path difference (OPD) variance. The root mean square OPD function generates the MTF. The PSF of the entire optical part of the sensor is obtained through multiple convolutions between the single response of the Cooke triplet, the low-pass filter (aperture), light attenuation, and the optical part of the sensor.

The Cooke triplet lens represents the simplest photographic objective, capable of correcting five fourth-order wavefront aberrations, namely, spherical aberration, coma, astigmatism, field curvature and distortion, as well as lateral and transverse chromatic aberrations. The Cooke photographic objective has a circular aperture that controls the amount of light reaching the image sensor. Typically, the aperture stop is placed between the second and third lenses in the Cooke triplet objective.

The irradiance law indicates that even if the lens has no vignetting, the light attenuation in the peripheral area of the image will increase with the increase of the viewing angle. The peripheral area of the image is formed by a set of light rays entering the lens at a specific angle relative to the optical axis.

The sensor converts irradiance into an electronic signal, which is then converted into a digital signal. The functions of the sensor include Bayer sampling, photon noise, fixed pattern noise (FPN) and Gaussian noise, high dynamic image, deconvolution, bilateral filter, analog-to-digital conversion, etc.

Dark current is leakage current, which corresponds to the photocurrent under no light. It fluctuates with temperature and introduces unavoidable photon noise. Photon noise, dark current noise, and thermal noise are signal-related noises. Reset and offset noise are signal-independent noises. Read noise includes all noise sources that are independent of the signal. Noise effects are additive.

The input photon flux is degraded by optical and electronic devices during the radiance capture process. Optical devices are responsible for color fidelity and spatial resolution, while electronic devices introduce temporal and spatial electronic noise. At the output end of the electrical part, the image is destroyed by the combined effects of optical blur and noise. To reduce blur, PSF deconvolution and Laplace filtering techniques of the Cooke triplet objective are used. To reduce the combined effects of photon noise, FPN, and Gaussian noise, six different exposure images are mixed together. Finally, a bilateral filter is used to suppress the remaining noise.

The color processing block includes Bayer sampling, interpolation, white balance, color correction, and color conversion from RGB color space to XYZ color space, then to LAB color space, and reconstruction from LAB color space to RGB color space, color saturation, and contrast.

A single sensor camera needs to use a color filter array (CFA). The Bayer pattern is a periodic array constructed from a 2×2 pixel kernel, including an RGBG arrangement. The data from these four RGBG pixels are combined to form a pixel in the final display. To avoid aliasing artifacts, the brightness must cover the 2×2 pixel kernel.

Once the Bayer in CFA runs sampling, interpolation is required to restore the original image. The color difference space method interpolates pixels in the color difference green-red and green-blue spaces, rather than on the original red, green, and blue channels. The basic assumption is that, due to the correlation between color channels, taking the difference between two channels will produce a color difference channel with lower contrast and less clear edges.

Demosaicing images with lower contrast will produce fewer obvious errors, such as sharp edges, which will cause most of the interpolation errors in the reconstructed image. The color difference space method creates KR (green minus red) and KB (green minus blue) difference channels and interpolates them. Then, this method reconstructs the red, green, and blue channels to obtain a fully demosaiced image. Color balance or white balance is an adjustment to make the colors in the image look the same as the colors in the original scenario. This means that neutral colors are correctly reproduced. In the grayscale world algorithm, the average of the R, G, and B components of the image is a common grayscale value.

The average of the G component is arithmetically averaged with the averages of the other R, G, and B components to obtain grayscale values on each color component separately.

Image rendering engines, such as the popular Unreal Engine or Unity 3D, implement image generation. Some open-source autonomous driving simulation systems, such as Intel CARLA (Car Learning to Act) and Microsoft AirSim, have adopted open-source rendering engines. Some open-source virtual image libraries have been applied in computer vision research, such as Virtual KITTI, the Michigan & Ford Joint Autonomous Driving Research Center (UM Ford Center for Autonomous Vehicles, FCAV), and Synthia. Rendering in computer graphics usually has the following stages:

1. Object materials, including diffuse/Lambertian, glossy, reflective/refractive, microfacet, and anisotropic.
2. Geometric transformations, i.e., the coordinate transformation process from 3D models to 2D images, including distortion models.
3. The rasterization process, mainly shading, including lighting models, shading frequency, and texture models.
    - The lighting model has three parts: diffuse, specular, and ambient.
    - Shading frequency has flat shading, Gouraud shading, and Phong shading models.
    - Textures include diffuse texture, ambient texture, bump/normal texture, and ambient occlusion texture.
4. Image rendering pipeline, including ray tracing, ray casting, texture mapping, voxel shader, and pixel shader/fragment shader methods.
5. Physically based rendering (PBR), i.e., radiance computation models, bidirectional reflectance distribution functions (BRDF), etc.

Computer vision can be seen as the inverse process of computer graphics, but it can also be combined with computer graphics, resulting in two important outputs, one is augmented reality (AR) and the other is image-based rendering (IBR). The idea of AR can also be reflected in simulation systems, such as inserting synthesized vehicles or pedestrians into real street scenarios.

IBR can generate some backgrounds to simplify the actual rendering computation during the process of generating a virtual environment through some captured images. Through machine learning, such as GAN, in the training situation of a large amount of real image data, and the combination of graphics CAD models, new scenario images can also be synthesized, even including image content changes caused by weather changes. Even in the most complex games, the real-time renderings have not yet achieve realistic scenarios people can distinguish simulated from real images at a glance. Intel's work EPE (enhancing photorealistic enhancement) uses the intermediate results G-Buffers produced during the game engine rendering process as additional input signals for training convolutional neural networks, which can further enhance the realism of images in the game. Instead of directly generating data, it enhances already rendered images, integrating scenario information to synthesize geometrically and semantically consistent images.

In EPE, the convolutional network uses the intermediate representation G-buffer generated by the traditional rendering pipeline to provide features such as depth of field, shape, lighting, transparency, and material; it is trained against adversarial objectives, providing supervision signals at multiple perceptual layers, and uses the GAN discriminator to evaluate the realism of the enhanced image; during the training process, it samples image blocks to eliminate image artifacts. The EPE network model has the following two improvements:

1. Ordinary convolution replaces strided convolution.
2. Rendering-aware denormalization (RAD) module, replacing the BN layer.

The overall framework of EPE is shown in Fig. 10.1, the image enhancement network can transform rendered images. In addition to images, there is also rendering information from the traditional graphics pipeline, the G-buffer. The network extracts multi-scale G-buffer feature tensors, which are encoded by the G-buffer encoder network (encoder). The structure of the rendered image is maintained by the LPIPS (learned perceptual image patch similarity) loss, and the perceptual discriminator maximizes the authenticity of the enhanced image. In this way, the two networks are trained together.

The image enhancement network architecture allows the RAD module to replace the BN layer in HRNet, forming a RAD-block (RB). Each branch of HRNet receives a G-buffer feature tensor at a matching scale.

RAD adjusts the image feature tensor through the encoded geometry, material, lighting, and semantic information from the traditional rendering pipeline. The image features are normalized by group normalization (GN), and then scaled and shifted by each element's weight. The weights are learned and adapted to the G-buffer features received by the G-buffer encoder. To better adapt the weights, the G-buffer features are transformed through three residual blocks within each RAD module.

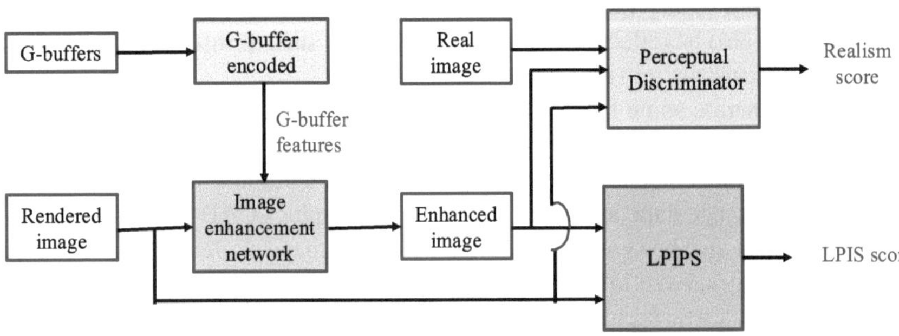

Fig. 10.1 The overall framework of EPE

The G-buffer encoder network architecture considers different data types and different spatial densities of the G-buffer. These features are processed through multiple streams (0 c) and fused into a joint representation consistent with the one-hot encoded target ID. The features are further transformed through residual blocks to match the branch scale in the image enhancement network.

In both the G-buffer encoder and RAD modules, residual blocks are used: composed of convolutional layers (kernel size 3), with spectral normalization and ReLU.

The perceptual discriminator network architecture evaluates the authenticity of the enhanced image. It consists of pre-trained robust segmentation (MSeg) and perceptual (VGG) networks. High-level semantic information is provided by label maps and perceptual feature tensors, which are acquired by the discriminator network to produce a realism score map.

Current image simulation methods either fall short of achieving photorealism or are unable to simulate full 3D environments with dynamic targets, thereby sacrificing high-level control and physical realism. To address this, Uber proposed GeoSim, a geometry-aware image synthesis framework. GeoSim extracts dynamic targets from existing scenarios and re-renders them in new poses, augmenting original images and synthesizing novel autonomous driving scenes.

It builds a diverse 3D target library with realistic geometric shapes and sensor data appearances. During the simulation process, the synthesis method completes the geometric-aware simulation work.

1. Place reasonable, realistic targets in a given scenario.
2. Obtain new views of dynamic targets from the asset library.
3. Synthesize and blend the rendered image.

The generated synthetic images have realism, traffic-awareness, and geometric consistency, allowing for extension to complex use cases. Remote realistic video simulation is performed across multiple camera sensors, and the data generation results are used for data augmentation of downstream tasks.

In a data-driven image processing method for autonomous driving through geometric-aware synthesis of realistic video simulation, it inserts dynamic targets into existing videos. The generated synthetic video clips are highly realistic, with clear layout and geometric consistency, allowing image simulation to extend to complex use cases.

First, the positions of all entities in the scenario are inferred through 3D object detection and tracking. For each new entity to be inserted, both its placement position and the appropriate target model are determined using the HD map and existing traffic detections. The Intelligent Driver Model (IDM) is then applied to govern the motion of the newly placed entity, ensuring realistic behavior by accounting for interactions with other participants and avoiding collisions. The output of this stage defines the updated scenario to be rendered.

Next, new views are generated with 3D occlusion reasoning, rendering all elements in the scenario and incorporating the newly added entities into the image. Finally, neural networks refine the composite by filling in the boundaries of the inserted objects, generating missing textures, and correcting inconsistent lighting conditions. The goal is to insert a new target into an existing image while ensuring that its scale, position, orientation, and motion remain physically and contextually plausible. To achieve this, prior knowledge of urban vehicle behavior is leveraged. Purely 2D target insertion struggles to achieve comparable realism; therefore, high-definition maps with bird's-eye-view (BEV) lane positions are used. The target's position is parameterized by the triplet of its BEV center and orientation, which is then converted into a six-degree-of-freedom (6-DoF) pose using local ground elevation.

The inserted target must exhibit realistic physical interactions with existing participants, conform to traffic flow, and appear within the camera's field of view. To accomplish this, a candidate position is randomly sampled from the lane area within the visible field of view, and the orientation is derived from the lane geometry. Any samples that would result in collisions with other participants or background objects are discarded, yielding a valid placement for the initial frame. For video simulation, the target's trajectory is evolved using an Intelligent Driver Model (IDM), allowing the simulated object to update its state and interact realistically with surrounding traffic.

Finally, the inserted target must respect the occlusion relationships of existing scenario elements. Vegetation, fences, and other dynamic participants may introduce irregular or thin occlusion boundaries that complicate occlusion reasoning. A straightforward approach is to compare the depth of the inserted target with the depth map of the existing 3D scene, thereby determining both the occlusion status of the target and the rendering of its shadow in the composite scenario. To achieve this, a dense depth map of the target image is first estimated through a depth completion network. The input is an RGB image and a sparse depth map obtained by projecting a LiDAR scan onto the image. Using the rendered depth of this target, the depth of the target image is evaluated to be less than the depth of the corresponding target pixel, and the occlusion mask can be calculated.

Even after occlusion inference, the rendered image may still appear unrealistic due to inconsistencies in lighting and color balance between the inserted fragment and the target scenario, visible boundary artifacts, or missing regions from the source view. To address these issues, an image synthesis network (SynNet) is employed to seamlessly blend the source fragment into the target scene.

The network takes as input the background image of the target scenario, the rendered target object, and its outline, and produces as output a final image in which the background and inserted object are naturally synthesized.

The synthesis network architecture is similar to the image completion network, except that the rendered target mask is used as an additional input. The network is trained with instance segmentation mask images in the target scenario.

10.1.2 LiDAR

Next, we introduce LiDAR based on Ref. [2] (for the principle of LiDAR, refer to Sect. 3.1.2).

First, we use the very popular game GTA-V (Grand Theft Auto V) to obtain simulated point clouds and high-fidelity images. Its framework is based on DeepGTAV, where Script Hook V is used as a plugin. To simulate real driving scenarios, an autonomous car (ego vehicle) equipped with a virtual LiDAR is used in the game, and autonomous driving is performed in the virtual world through the AI interface provided by Script Hook V. The system simultaneously collects LiDAR point clouds and captures images. In the virtual environment, the virtual camera and virtual LiDAR are placed in the same position. This has the following two advantages:

1. It is easy to perform a sanity check on the collected data, as the point cloud and the corresponding image must be consistent.
2. The calibration between the game's virtual camera and virtual LiDAR can be automatically completed, and then the collected point cloud and scenario image can be combined as a neural network training dataset for sensor fusion tasks.

Ray casting is used to simulate each laser beam emitted by the virtual LiDAR. The ray casting API takes the 3D coordinates of the ray's start and end points as input and returns the 3D coordinates of the first point hit by the ray. This point will be used to calculate its distance, the category and instance ID of the target hit by the ray, and other data, and to automatically annotate the collected data. LiDAR parameters include vertical field of view (VFOV), vertical resolution, horizontal field of view (HFOV), horizontal resolution, pitch angle, maximum range of the laser beam, and scanning frequency.

This sensor simulation system realizes the automatic calibration of 3D LiDAR point clouds and camera images, and users can choose the desired game scenario and specify and change the eight dimensions of the game scenario: car model, car position, car direction, number of cars, car background, car color, weather, and time. The first five dimensions affect both the lidar point cloud and the game image, while the last three dimensions only affect the game image.

This sensor simulation framework also provides a configurable mode, where users can configure the required scenarios and collect data from them. One advantage of configurable scenarios is the generation of dangerous or rare driving scenarios in the real world as training data; another advantage is that it can systematically sample the modification space of the scenario (such as the number of cars, the position and direction of the cars). Then, these data can be used to test neural networks, expose their vulnerabilities, and improve their performance through retraining. Users can specify and change the eight dimensions of the game scenario: car model, car position, car direction, number of cars, scenario background, car color, weather, and time. The first five dimensions affect both the lidar point cloud and the scenario image, while the last three dimensions only affect the scenario image.

Reference [3] developed a lidar sensor model, which includes accurate modeling of the scanning mode and a complete signal processing tool chain for the lidar sensor. Through the standardized Open Simulation Interface (OSI) and Functional Model Interface (FMI), it is developed as a Functional Model Unit (FMU). A virtual lidar needs to consider the complete signal processing steps and defects of the real lidar sensor to obtain simulation results close to the actual sensor. The defects here are optical losses, inherent detector effects, effects generated by electronic amplifiers, and noise generated by sunlight.

The raw data of the synthetic sensor can be generated through the following rendering techniques: ray tracing, ray casting, rasterization (using Z-buffer technology), and ray path.

In the tool chain and signal processing steps of this lidar model, the sensor model considers the scanning mode and complete signal processing steps of the Blickfeld Cube. The model itself is built as an OSMP FMU and uses CarMaker's virtual environment. It provides a bidirectional reflectance distribution function (BRDF) for the ray tracing framework, considering the direction of the incident light angle θ, material surface, and color properties. The lidar FMU uses CarMaker's ray tracing module. The material library of CarMaker specifies the material characteristics of the simulation target, angle-related spectral reflectance $R_\lambda(\theta)$, and reflection types, including diffuse reflection, specular reflection, reflection, and transmission.

The FMU controller sets the input configuration of the simulation, then calls the lidar simulation library, and passes the ray tracing data for further processing. The central component of the simulation library is the simulation controller. The next module is the link budget module, which calculates the photons that change over time. Capture these photons and convert the captured photons into current signals.

The third module is the circuit module, whose task is to amplify the photoelectric current signal of the detector and convert it into a voltage signal. The last part is the ranging module, which determines the distance and intensity of the target from each reflected scanning point received from the simulated circuit. Finally, the FX engine is an interface that interacts between a series of effects related to the environment or sensor and simulation pipeline blocks. These interactions can include, for example, thermal noise of electrical components, signal attenuation caused by weather phenomena, and backward scattering.

Reference [4] discusses how to generate realistic lidar point clouds. It drives a lidar collection vehicle in the city and establishes a large amount of data of a 3D static map and 3D dynamic targets. Then, select scenarios from the data and "virtually" place the self-driving vehicle (SDV) and a set of dynamic targets from the data in reasonable positions in the scenario to generate new scenarios. To produce realistic simulations, a new simulator, LiDARSim, combines both physics-based and learning-based modeling capabilities. First, use ray casting of the 3D scenario, then use a deep neural network to generate deviations from physics-based simulations, and generate real point clouds.

In an overview of the architecture of LiDARSim, it includes two stages, namely, 3D creation and sensor simulation. In the 3D creation stage, the fleet drives around several cities, building a dataset of a 3D static map and dynamic template grid over time; in the sensor simulation stage, combining physics-based and learning-based simulations, first use ray casting of the 3D scenario to obtain the initial physical rendering, then the deep neural network learns to approximate more complex physics and sensor noise, deviating from physics-based simulations, generating real lidar point clouds.

First, use physics-based simulation to create geometric estimates for generating point clouds. Here we focus on simulating mechanical scanning lidar, namely, Velodyne HDL-64E. This system has 64 emitter–detector pairs, each using light pulses to measure distance. The basic concept is that each emitter emits a light pulse, which propagates until it hits a target, a portion of the light energy is reflected back and received by the detector. The distance can be measured by calculating the travel time. The entire optical assembly rotates on a pedestal, providing a 360-degree panoramic view at about 10Hz, with each complete "scan" providing about 110 k returns.

Given the required six degrees of freedom for attitude and speed, the graphics engine simulates the LiDAR sensor. Based on the internal parameters of the sensor, a set of rays are projected from the center of the virtual LiDAR into the scenario. The rolling shutter effect is simulated by compensating for the relative motion of the vehicle during the LiDAR scanning process. In addition, the motion blur of other vehicles in the scenario during the LiDAR scanning process is also simulated. To balance computational cost and realism, the target attitude in the scan is updated at 360 equal time intervals. Using the Intel Embree ray casting engine, the ray-triangle mesh collisions of all surfaces (surfel) in the scenario are calculated, and the surface closest to the sensor that is hit is found.

This is applied to all rays in the scan, obtaining physically generated point clouds on the constructed scenario. Additionally, a mask is applied to remove rays returned from the vehicle itself.

Real LiDAR typically generates about 10% fewer points than the ray casting version, and some vehicles have more simulated LiDAR points than real ones. One assumption of the physics-based simulation method is that if each ray projected into the virtual world intersects with a physical surface, it will return. However, if the strength of the return signal (intensity value) is not sufficient to be detected by the receiver, the rays emitted by the real sensor may not return (i.e., ray drop). Ray drop is a complex random phenomenon, affected by material reflectivity, angle of incidence, distance value, beam bias, and other environmental factors. Using real-world data, these factors can be captured, despite the interference of noise.

The LiDAR ray casting is set as a binary classification problem, applying neural networks to learn the ray casting characteristics of the sensor, and using machine learning to bridge the gap between simulated and real-world LiDAR data.

To predict the ray drop of LiDAR, the 3D point cloud is converted into a 64×2048 2D polar coordinate image grid, allowing encoding of which rays did not return from the sensor, while also providing a mapping between real and simulated scans. The network is provided with a set of channels as input, channels representing observable factors that may affect the chance of each ray not returning. The network architecture is a standard eight-layer UNet, with the network output being the probability of each element returning or not.

To simulate LiDAR noise, the output LiDAR point cloud is generated by sampling from the probability mask. There are two reasons for sampling from the probability mask rather than directly thresholding: ① ray drop is learned with cross-entropy loss, which means that the estimated probabilities may not be well calibrated, and sampling helps to mitigate this problem compared to thresholding and ② real LiDAR data is uncertain, as current methods may not be able to fully simulate additional noise (atmospheric transmittance, sensor bias).

It is also worth mentioning that Refs. [5–7] discuss the simulation of LiDAR point cloud data in fog, snow, and rain, respectively.

10.1.3 Millimeter-Wave Radar

The principle of millimeter-wave radar is referred to in Sect. 3.1.3. In recent years, many different radar sensor system modeling methods have been developed. There are several different methods for generating synthetic radar data:

1. The ideal sensor model only needs to generate ground-truth (GT) data for targets within the sensor's detection range, thus generating sensor-type-specific target list information for all targets in the sensor's field of view (FOV). This is different from the ideal sensor model.
2. Physics-based sensor models attempt to simulate the physical sensing process of real sensors as accurately as possible. This is a computationally intensive task that requires more computational power and usually sacrifices real-time performance. In addition to the technical expertise required for sensor technology, a detailed description of environmental conditions (material properties, weather conditions, etc.) is also required to accurately simulate the physical characteristics of the sensor. The simulated output data from these models can be raw simulated signals comparable to those of real sensors. Physics-based simulation methods are divided into two categories, one involving the modeling of radar electronic components, including propagation channels modeled by ray tracing technology, and the other considering radar clutter echoes and noise.
3. The probability sensor model can strike a reasonable balance between complexity and computational efficiency. The simplified parameter set and reduced model complexity make the simulation test faster than real time. Although the output data is less realistic compared to real sensors, phenomenological models can be used to test and verify at most stages of the development process.
4. The functional sensor model ignores the sensor hardware architecture and signal processing process. This black-box model only focuses on the detection results of the measured target. This type of sensor model can be achieved by combining certain geometric characteristics and selected physical characteristics. This simplification improves the real-time performance of sensor simulation without completely ignoring the detection limits and characteristics of real sensors.

Function-based modeling methods include two types, namely, geometric models and models dealing with scattering centers.

Radar exhibits some special features, including multipath reflections, interference, ghosting, blurring, clutter, and attenuation. These effects are the result of high-frequency electromagnetic waves propagating and interacting in complex scenarios. Accurate simulation of this phenomenon using physics-based models usually requires extensive use of ray tracing calculations. These methods are currently not fast enough to be practically used in simulations, as they require highly detailed geometric structures (i.e., square millimeter accuracy) to obtain correct results, which requires a lot of computational power.

Existing radar models are usually black box or white box. Black-box models attempt to represent radar phenomena in a random manner. White-box models use ray tracing to estimate electromagnetic path propagation, usually relying on given target radar cross-sectional values or extracting virtual scattering centers. White-box models can achieve faster calculations using modern GPU technology. However, this kind of white-box model requires detailed environmental models to capture important influences related to radar, such as multipath propagation and interference. Detailed models usually do not exist, and the large amount of computation required makes real-time simulation infeasible.

Reference [8] proposes a method for constructing a probabilistic stochastic car radar model, which is based on deep learning and generative adversarial networks (GAN), and the resulting model reflects basic radar effects while maintaining real-time calculation speed. Deep neural networks are used as radar models to learn end-to-end conditional probability distributions from data. The input of the network is a spatial grid and a target list, and the output is the read sensor data.

In the representation of simulating radar data under the deep learning framework, the spatial grid has two main dimensions, namely, the 3D tensor of distance and azimuth, and the third dimension is composed of different types of information layers. This is similar to RGB images, where pixel information is stored in spatial dimensions and color channels. Therefore, this spatial grid is also suitable for CNN models.

The target list is also represented as a 3D tensor. Like the spatial grid, a pre-processing head is used to learn abstract features before connecting with the spatial grid.

The first processing layer of the target list is independent of the target and is designed for parameter sharing. This independence is achieved with 2D convolution, which also reduces the number of parameters and improves robustness. The tensor has the shape nobjects × 1 × nfeature, where the first two dimensions correspond to the length and width of the traditional tensor, and the last feature dimension corresponds to the number of channels. One convolution after another is applied to the tensor, sharing operations among all targets in the list.

The target list representation is universal, allowing for any features to be included, here including the target 2D pose x, y, θ, speed, and one-hot class encoding on the target class. When the target row is not used, there is an additional class, which occurs when the list capacity is greater than the targets in the scenario. When a target row is not used, all classes other than the one-hot class will be zero. The target geometry is not specified because it is given by the target class.

There are two main methods to represent probability distributions with neural networks: Direct parameterization and using noise to generate samples as auxiliary input. These methods are, respectively, called the direct method and the sampling method.

Here, two benchmark radar models that directly parameterize probability distributions are provided: Normal distribution and Gaussian mixture model. Multivariate normal distribution is commonly used in machine learning because it has good mathematical performance. However, the normal distribution is unimodal, and the parameters of the normal distribution increase quadratically with the dimension of the target variable. The output of the CNN model here is a tensor grid with two layers: One mean value and one diagonal log variance.

An important challenge for a stochastic radar model is that the sensor output is multimodal and spatially related. Regression methods will smooth possible solutions, leading to prediction blurring. Variational autoencoders (VAE) allow learning of one-to-many probability distributions without explicitly outputting which distribution.

The architecture of the conditional VAE model is an encoder–decoder network. The encoder takes the raster and target list and generates latent feature representations x, while the decoder uses these feature representations and randomly generated noise values to produce predicted sensor measurements. The encoder consists of two branches, namely, a spatial raster and a target list, which merge together to produce latent feature representations. Both branches are entirely composed of convolutional layers. The output is flattened and concatenated together, then processed with a fully connected layer with ReLU. The decoder uses the encoded features and random noise to generate radar polar coordinate grids of success rates. During VAE reparameterization, random noise is used as input. The noise and latent features are connected together and processed with a fully connected layer activated by ReLU, the output of which is reshaped and then passed through a series of deconvolution layers to output the final radar signal.

10.1.4 Other Sensors

Other sensors, such as GPS, IMU, ultrasonic radar, and V2X sensors, can also simulate their data.

1. GPS simulates GPS location and GPS noise model parameters, outputting the car's latitude and longitude, speed, heading, etc.
2. IMU simulates the car's acceleration and angular velocity, especially the cumulative error of the car's position, speed, and heading when the GPS signal is lost.
3. Ultrasonic radar (mainly for automatic parking) simulates the position, angle, and distance of obstacles of the ultrasonic radar.

4. V2X simulates dynamic traffic flow device data, even reflecting communication delay or packet loss situations.

10.2 Traffic Model

The following introduces the problem of traffic simulation models based on the open-source software SUMo (simulation of urban mobility). Firstly, SUMo road networks can be generated using its own program or by importing digital road maps. The road network importer allows reading networks from other traffic simulators, such as Vissim or MATsim. SUMo is also compatible with common map formats, such as Open Street Map (OSM) and commonly used simulator formats, such as openDRIVE. SUMo can be considered a purely microscopic traffic simulation. Each car is given an identifier (name), departure time, and route in the road network. A macroscopic traffic simulator would view the entire traffic flow as a single unit. SUMo can also define departure and arrival attributes, such as lane, speed, or position. Each car is assigned a type, which describes the physical characteristics of the car and the variables of the motion model.

The simulation is time-discrete and space-continuous, and describes the position of each vehicle internally, i.e., the lane it is in and the distance from the starting point. When a vehicle moves, the car-following model is used to calculate the speed of the car. In addition to traditional traffic measures, SUMo has also extended noise emission and pollutant emission/fuel consumption models. The SUMo traffic modeler defines the total number of traffic groups in a given area and calculates the mobility desire of this group as input to the traffic simulator. Some software can also consider some environmental features, such as weather conditions.

The SUMo ROUTER module reads the departure time, starting point, and destination of a set of virtual groups to be simulated, and then uses the Dijkstra routing algorithm to calculate the route in the traffic network. The SUMo car-driver model uses an extended version of the Gipps model. It can display the main features of traffic, such as free flow and congested flow. At each time step, the speed of the vehicle adapts to the speed of the vehicle in front to avoid collisions in subsequent steps. This speed is called the safe speed, calculated as follows:

$$v_{\text{safe}}(t) = v_l(t) + \frac{g(t) - v_l(t)\tau}{\frac{\overline{v}}{b(\overline{v})} + \tau} \tag{10.1}$$

where $v_l(t)$ is the speed of the car in front, $g(t)$ is the gap to the car in front, τ is the driver's reaction time (generally 1 second), and b is the deceleration function. The "desired" or "expected" speed of the vehicle is the minimum of the following three: the possible maximum speed, the current speed plus the maximum acceleration, and the safe speed calculated above. Therefore, its desired speed is

$$v_{\text{des}}(t) = \min\left(v_{\text{safe}}(t),\ v(t) + a,\ v_{\max}\right) \tag{10.2}$$

where a is the maximum acceleration. If the simulation considers that the driver will not fully adapt to the desired speed due to mistakes, the actual speed is reduced by a random "human error":

$$v(t) = \max[0, \text{rand}[v_{\text{des}}(t) - \epsilon a, v_{\text{des}}(t)]] \tag{10.3}$$

where ϵ is the disturbance coefficient. Since the vehicle must not move backward, the current speed of the vehicle is the maximum of the calculated speed and zero. This model is collision-free, so the simulation does not allow the occurrence of artifacts due to the incompleteness of the model.

SUMo provides the possibility of V2X, and can couple with external communication simulators, such as ns2 or ns3. SUMo provides a complete traffic demand or a set of vehicles with appropriate routes. Its main task is to model the process of traffic participants choosing routes (paths through a given road network) to their destinations. Since the time through the edges of the route map largely depends on the number of traffic participants using this edge, route calculation is a key step in implementing large-scale traffic simulation. In SUMo, this is called user assignment or traffic assignment.

One of the main applications of microscopic traffic flow simulation work is to adapt traffic signals to current traffic. SUMo's fast execution time and open interface for interaction with external applications make it a good candidate for evaluating new traffic control algorithms, including controlling individual intersections and evaluating the entire network. Section 10.7 will introduce the neural network model scenarioGen's insertion of participants into existing real scenarios to generate new traffic scenarios.

10.3 Vehicle and Pedestrian Models

In the simulation environment, some belong to static objects, mainly backgrounds, such as overpasses, buildings, streets, trees, rivers, and hills; others are dynamic objects, such as cars, pedestrians, motorcycles, bicycles, and animals. Static environmental data require 3D models, or IBR models, while dynamic activity models need to provide kinematic and dynamic models.

In the automotive industry, there are many mature simulation companies doing automotive CAD models and vehicle kinematics/dynamics models, such as MSC Software's VIRES, Siemens-acquired company TASS's PreScan, IPG's CarMaker, Mechanical Simulation's CarSim, and DSpace's Automotive Simulation Models (ACM).

Traditional vehicle dynamics models are designed by scientists based on physical principles, and their models mainly take the form of parametric differential equations (for continuous-time dynamic systems). The art of modeling is to determine which physical phenomena are relevant to the intended use and which may be ignored. However, due to the trade-off between modeled and unmodeled physical phenomena, if assumptions about the ignored aspects are not met, analytical models cannot work well. Moreover, if the parametric model is fitted to situations where the unmodeled part stands out, the model will try to explain the unmodeled part with its parameters, thus the model parameters will be changed and no longer have the intended meaning. Another option is models based on machine learning (neural networks/deep learning). Recurrent neural networks (with internal states) can be used to model dynamic systems and are applicable to observations of real-vehicle dynamics. Alternatively, non-recurrent network models can be established, in which case the input to the network is a sufficiently long history of data.

Reference [9] provides the design and experimental results of a neural network model for vehicle longitudinal dynamics. In the RNN model for vehicle longitudinal dynamics, it has inputs consisting of instantaneous speed vk, braking pressure pbk, elevation hk, engine torque Tgk, and gear signal vector Gk, forming a 12-dimensional instantaneous input vector uk, entering the basic recursive layer BR; the basic recursive layer implements state transitions.

Reference [10] discusses the neural network modeling method for vehicle steering control. The model consists of several neural network blocks stacked in the time domain, each neural network block is divided into a known physical function (predicting steering angle output based on the car model) and a stack of fully connected layers. The current input and previous output are stacked together and fed into the fully connected layers and physical function. The outputs of these layers are combined to predict the next state, these predicted variables are fed back into the network along with the new control input, continuing the prediction for the next step.

The vehicle steering control neural network layer predicts the residual dynamics not modeled by the physical function. Therefore, the neural network layer captures the effects of road–tire interaction, power steering logic, and steering system dynamics. The addition of the physical function improves the gradient flow of the network, thus providing better predictive performance.

Reference [11] provides a deep neural network (DNN) modeling method for vehicle lateral-longitudinal dynamics.The input to the DNN is the vehicle's initial longitudinal speed and driving torque, the output is the longitudinal travel distance, lateral travel distance, final longitudinal speed, final lateral speed, and vehicle yaw angle. The DNN model has 4 hidden layers, each involving 28, 24, 20, and 15 neurons. The last activation function is linear, other activation functions are Rectified Linear Unit (ReLU) functions.

Human body models and face models are the basis for simulating pedestrians, and the appearance of clothing also requires 3D modeling. As for more details, such as skin and hair parts, they are not so sensitive to the effect of autonomous driving environment simulation. At the same time, as a moving object, it needs to provide its skeleton model. Models for walking or other actions, such as falling and squatting, need to be determined by kinematic models. Even motorcyclists and cyclists can work on some action details, reflecting their unique action models. Facial expressions, like body movements, also need to provide deformation models (including muscle models and expression unit models) and kinematic equations to determine simulated facial actions, such as surprise, smile, and even crying. Finger movement is similar to body movement, with a unique skeletal model description.

In the past, when it was difficult to understand and model human movement, animation game makers would use motion capture technology, i.e., infrared markers were attached to the moving object, and the movement pattern of the human body model was obtained from the infrared photography data.

In the field of computer vision research, some researchers explore how to model the appearance and movement (including expressions and gestures) of the human body (including the mentioned face, clothing, etc.) using machine learning and probabilistic methods based on real image data. This is combined with the 3D model of computer graphics, i.e., IBR or VBR (video-based rendering) technology.

For the work of reconstructing a 3D human body model based on video data, given a single-eye RGB video depicting a moving person, the goal is to generate a personalized subject 3D model, including the shape of the body, hair, and clothes; personalized texture maps; and the underlying skeleton assembled to the surface. The non-rigid surface deformation of the new posture is entirely skeleton-driven, including three steps: ① posture reconstruction; ② consensus shape estimation; and ③ frame refinement and texture map generation. The consensus shape can be effectively optimized to maximize the explanation of the contour of each frame instance. As time changes, the fabric deforms, and the constructed consensus shape may be slightly misaligned with the frame contour. Therefore, to calculate the texture time-varying details, the deviation of the consensus shape in each frame in the sliding window is optimized. Given the refined frame-by-frame shape, a texture map can be calculated, which depends on the foreground segmentation of the image, i.e., the CNN-based video segmentation method, and manual segmentation is performed three to four times for each sequence.

For a method of obtaining motion conversion based on deep learning, here, the human body motion captured based on vision is transferred to another person. Given a source video of a person dancing, the performance can be transferred to a new (amateur) target after a few minutes of the target subject performing standard actions. The problem is defined as a frame-by-frame "image-to-image" conversion with temporal and spatial smoothness. Pose detection serves as an intermediate representation between the source and the target, learning the mapping from pose images to the target subject's appearance. This method is also used for temporal domain video generation, including realistic facial synthesis.

The entire pipeline is divided into three stages: pose detection, global pose normalization, and mapping from normalized pose bar graphs to the target body. In the pose detection stage, given frames from the source video, a pose bar graph is created using the state of the pre-trained pose detector. The global pose normalization stage considers the differences between the source body and the target body in shape and frame position. A system learns the mapping from the normalized pose bar graph to the adversarially trained target body image.

10.4 Data Visualization Model

Data visualization is a necessary part of the simulation platform and also helps in offline debugging of autonomous driving systems, especially for R&D engineers developing different modules (such as perception, planning, and control modules), who can jointly debug algorithms on the same platform.

The following introduces the visualization model using Uber's open-source visualization system (autonomous visualization system, AVS) as an example.

AVS is a new standard for describing and visualizing autonomous vehicle perception, motion, and planning data, providing a powerful web-based toolkit for building applications, exploring and interacting, and most importantly, developers can use this data to make important development decisions.

Why did Uber choose Web development? There are three reasons:

1. Fast iteration cycle. On the Web, you can quickly and easily develop and deploy features step by step. If users want the latest version of the product, they just need to refresh the browser page instead of downloading and installing a new application.
2. Flexibility and shareability. Since the Web is hardware-independent, anyone can work on the platform using any operating system, in fact, the browser becomes the operating system. Moving the system to the network allows any operating system work within the team to be connected, and there is the possibility of extending the team beyond the ATG headquarters. On the Web, reporting and diagnosing events only require a single click of a URL.
3. Collaboration and customization. As a rapidly developing technology, autonomous vehicles will never stop generating new datasets, metrics, or use cases. New services and endpoints are constantly being added. Each team at ATG has unique visualization and data generation needs, so they need to be able to customize their own experience. HTML5 and JavaScript are tested and trusted tools that can dynamically create custom UIs and can be easily integrated into other infrastructure and task management systems.

To effectively manage data, a new abstract concept needs to be introduced. This will help describe and manipulate the data generated by Web applications. In light of this, there are two key parts to building a visualization system: XVIZ provides the data (including management and standards), while streetscape.gl is the component toolkit that powers Web applications.

10.4.1 XVIZ

There is a need for formal and flexible standardization of data generated by automated systems, so that its format can be integrated with evolving infrastructures, remain consistent with multiple clients, and be close enough to the data source to define necessary controls and bindings for effective management.

XVIZ provides a stream-oriented view of the changing scenario over time and a declarative user interface display system. Like video recording, users can randomly search and understand the environmental conditions at that time. Like an HTML document, it can be focused and structured according to an introspective mode. However, XVIZ can also bind separate stream updates to individual objects for easy exploration and querying of data.

An XVIZ stream is a series of discrete updates that occur at a specific time using a specific primitive type; primitives are objects that can describe information such as LiDAR point clouds, camera images, object boundaries, trajectories, vehicle speeds changing over time, and predictive planning. To simplify user demonstrations, the style of these objects (including stream level) can be set individually or assigned to style classes.

XVIZ organizes streams through hierarchical naming, where a separate metadata section lists streams, stream types, relative transformations, declarative UI panels, and style classes. Then, the user interface bundles the graphical panels of objects with data, providing controls for users to configure a series of layouts and display components through YAML.

10.4.2 streetscape.gl

streetscape.gl is a toolkit for building Web applications that use data defined in the XVIZ protocol. It provides ready-to-deploy components for visualizing XVIZ streams in 3D viewports, charts, tables, and videos. It solves common visualization challenges such as time synchronization across data streams, coordinate systems, cameras, dynamic styles, and interactions with 3D objects and cross-components, so users can spend more time building autonomous vehicle software.

Rendering performance is a primary goal of streetscape.gl. Based on React and Uber's mature WebGL-driven visualization platform, it can support real-time playback and smooth interaction with scenarios supporting hundreds of thousands of geometries.

Composability is also at the forefront and center of this streetscape.gl design. The visualization platform supports a dozen different use cases for classification, labeling, debugging, remote assistance, and scenario editing, from which it can be learned that the components of this design are highly styleable and extensible.

10.5 Road Network Simulation

Taking the commercial software RoadRunner developed by VectorZero as an example (which has been acquired by MathWorks), we introduce the simulation work of road networks.

RoadRunner is a generation software for automotive simulation, urban planning, traffic, roads, and environments. It is a simulation engine based on the Systems Biology Markup Language (SBML), which is a variant on XML used for communicating and storing computational models of biological processes.

RoadRunner can create environments and roads; generate complex road networks containing roundabouts, intersections, and bridges; customize some traffic signs and markings; and then publish to OpenDRIVE, Unreal, Unity, or FBX.

10.5.1 Road and Environment Modeling

Create realistic, complex, interconnected road networks, including intersections, bridges, tunnels, roundabouts, switches, road markings, and signs.

Specifically, there are the following three functions:

1. Real-world road geometry: Create roads with curvature, vertical slopes, and mixed shapes, including cross sections.
2. Complex intersections: Complex intersections with merges, splits, and roundabouts.
3. Surface markings: Detailed road and lane markings, templates, and custom graphics.

10.5.2 Customizable Props and Signs

Customizable and localizable props, such as obstacles, signs, poles, lights, personnel, and obstacles. Similarly, it also includes the following three functions:

1. Prop Library: Use the prop library or import your own FBX and OBJ files to create your own props. Assemble components to build complex components, then use custom tools to place and/or randomize props.
2. Signal Editor: Includes an integrated sign editor that allows you to combine text and graphics to build custom signs. Supports Unicode to create international signs.
3. Profiles: Use the integrated extension tool that complies with road rules to easily form barriers such as fences, walls, and obstacles.

10.5.3 Functional Road Network

Roads, intersections, lanes, and signals (soon to include traffic and pedestrian models) automatically include intelligence when used in the simulation environment.

The functional road network has the following three functions:

1. Complex Road Network: Generate a road network with lane and intersection connections for car and traffic simulation.
2. Traffic Signal Editing: Use the integrated signal timeline and interactive viewport controls to edit the phases and timing of signals.
3. Traffic and Pedestrians: Populated by dynamic traffic (soon to include pedestrian models).

10.6 Construction of Scenario Library and Autonomous Driving Test

The scenario-based autonomous driving car test method is an effective way to achieve accelerated testing and evaluation.

10.6.1 Scenario Definition

A scenario, as a comprehensive manifestation of the driving environment and car driving scenariory, describes the road site of the vehicle's external driving environment, surrounding traffic, meteorology (weather and lighting), and the vehicle's own driving tasks and status information. It is an abstract and mapping of factors affecting and determining intelligent driving functions and performance, with high uncertainty, non-repeatability, unpredictability, and inexhaustibility.

The classification methods of test scenarios vary:

1. According to the abstraction level of the scenario, it can be divided into functional scenarios, logical scenarios, and specific scenarios.
2. According to the data source of the test scenario, it can be divided into natural driving scenarios, dangerous working condition scenarios, standard regulation scenarios, and parameter recombination scenarios.

Based on different generation schemes, test scenarios are divided into two types: Worldsim scenarios and Logsim scenarios. Worldsim scenarios are composed of obstacle behaviors and traffic conditions preset by the simulation platform, with determinism and low complexity. Logsim scenarios are extracted from road test data, including complex moving obstacle behaviors and traffic flow, with random uncertainty.

10.6.2 Classification of Simulation Test Methods

Simulation testing includes the following types: Model-in-the-loop simulation (MIL), Software-in-the-loop simulation (SIL), Hardware-in-the-loop simulation (HIL), and

Vehicle-in-the-loop simulation (VIL). MIL testing is conducted in the model's development environment through simulation, by inputting a series of test cases to verify whether the model meets the design's functional requirements.

SIL testing runs the vehicle model and virtual domain controller model to simulate various required sensor signals. It receives these sensor signals and issues responses from the virtual domain controller to achieve closed-loop simulation and test the control system.

In SIL testing, replace the model in MIL testing, use the same test case input as MIL testing, and record the values of test output, finally comparing whether its output is consistent with the output of MIL testing.

The difference between HIL testing and SIL is that the target hardware needs to be considered, which can additionally evaluate whether the overall performance of the software on the target hardware (running scheduling, memory calling, computing power calling) meets expectations. HIL testing usually connects a test controller with a series of simulation devices, using recorded or simulated raw data as real signal input to complete the testing of the target hardware. The general practice of VIL testing is real road testing. The most important thing about HIL testing is the real time and synchronization of the entire test system. Otherwise, the built environment is not enough to support functional verification, the system step length and real time cannot be guaranteed, and the input signal of the domain controller will have a large deviation from the real-vehicle environment.

10.6.3 Construction of Scenario Library

The construction of the scenario library is basically based on different data sources such as real, virtual, and expert data, and is built into a complete system through scenario mining, scenario classification, and scenario deduction.

A scenario library includes the following dimensions:

1. Scenario: Static part and dynamic part.
2. Traffic: Driving behavior and VRU (pedestrians, bicycles) and other non-motorized behaviors.
3. Weather: Sensors (cameras, radars, lidars) and interference.

The German PEGASUS project (January 2016–May 2019) focused on the research and analysis of highway scenarios, established a scenario database based on accidents and natural driving data, and verified the system based on the scenario database.

The PEGASUS scenario framework is defined, which establishes a three-tier hierarchy of functional-logical-concrete, as well as a process for building a scenario library oriented toward concept-development-testing-calibration, and methods for testing intelligent driving.

PEGASUS attempts to establish a standardized process for simulating, testing, and experimenting with advanced intelligent driving systems in real environments through the development of the OpenScenario interface. The project is divided into four stages: ① scenario analysis and quality assessment, defining a system's scenario generation method and the syntax structure of scenario files, calculating the KPI of scenarios, and defining a set of scenario difficulty (danger) evaluation methods based on expert experience; ② implementation process, designing a sufficiently flexible and robust implementation process for autonomous driving functions based on safety; ③ testing, the output is a set of methods and tool chains for laboratories (simulation software, test benches, etc.) and real traffic scenarios; and ④ result verification and integration, analyzing the results of the first three stages.

PEGASUS establishes three standards for testing scenario formats, namely, OpenCRG, OpenDRIVE, and OpenSCENARIO, and defines a six-layer model for testing scenarios: roads, traffic infrastructure, temporary operations of the first two layers (such as road construction sites), objects, environment, and digital information.

10.6.4 Scenario Description Language OpenSCENARIO

The OpenSCENARIO data format mainly describes dynamic information in autonomous driving simulation tests. The main elements of the scenario are who (the target object), where (the relationship with the road or reference object), what happened (the action description of the scenario), and how long it lasted (the running time of the scenario). The combination of these objective information can effectively cover all test case building, achieve high coverage scenario testing, and enhance the value of the scenario.

Each scenario in OpenSCENARIO has three necessary components:

- Road network.
- Scenario script (storyboard).
- Action.

The road network is composed of entity instances:

- Vehicles.
- Pedestrians.
- Others.

These instances interact through commands contained in the scenario script.

The scenario script covers the complete scenario description. Each scenario script contains an initialization (init) and one or more scenario contents. The scenario script constructs the scenario as a script for virtual dynamic content.

- Initialization is used to set the initial conditions of the scenario.
- The scenario content has a hierarchical structure and is more structured. The scenario content includes actions (acts) that define the action group.
- Triggers (i.e., startTriggers and stopTriggers) control the execution of actions.

- The maneuver group assigns entities as actors to maneuvers. The maneuver is a shared event container.
- Events control the simulation world or the corresponding entity. Through triggering actions, events give user-defined conditions.

The action ultimately taken by the actor is triggered by conditions, where the actor refers to the entity instance participating in the action.

10.7 Safety-Critical Scenario Generalization Based on Digital Twin

Digital twin is to establish a model consistent with the real world in a virtual environment. Digital twin technology will be a great help on the road to the development of autonomous driving simulation testing, and it is an efficiency tool for simulation testing.

Baidu proposed a data-driven method for end-to-end simulation of autonomous driving in Ref. [12]: Enhanced Autonomous Driving Simulation (AADS). This method enhances real-world images by simulating traffic flow, creating realistic simulation scenarios similar to real-world rendering.

Specifically, street scenarios are scanned with lidar and cameras. The input data is decomposed into background, scenario lighting, and foreground targets. With a view synthesis technique, the viewpoint can be changed in the static background. Foreground vehicles are equipped with 3D CG models. With accurately estimated outdoor lighting, 3D vehicle models, computer-generated pedestrians, and other movable targets can be relocated and rendered back into the background image, creating photorealistic street view images. In addition, the simulated traffic flow (such as the placement and movement of synthetic targets) is based on captured real-world vehicle trajectories, which look natural and capture the complexity and diversity of real-world scenarios. Compared with traditional VR-based and game engine-based AV simulation systems, AADS provides more accurate end-to-end simulation capabilities, without the need for expensive CG models or tedious programming to define traffic flow.

Using the semantic information in the ApolloScape dataset, remove specific types of targets, such as cars, bicycles, trucks, and pedestrians. After removing the moving targets, a large number of holes will appear in the RGB images and point clouds, which should be carefully filled in order to generate a complete and clean background for AADS. Use the latest RGB image repair method to fill in the holes, where semantic labels guide the learning-based repair techniques, and then introduce point cloud completion in the depth processing of the new background synthesis. Given the synthesized background image, any 3D CG model can be placed on the ground and then rendered into the image space to create new synthesized simulation images. However, to make the synthesized images realistic, the illumination in the background image is first estimated.

This allows AADS to have consistent shadows on the ground and vehicle-rendered 3D CG models. In addition, to further improve the realism of the synthesized images, AADS also grabs textures from real images to enhance the appearance of the 3D CG models. Specifically, given an RGB image with unmoved vehicles, retrieve the corresponding 3D vehicle models, align these models with the input image, and then transfer and complete the appearance of the 3D CG models from the aligned real images.

Google Waymo's SurfelGAN generates realistic camera images based on collected limited lidar and camera data. It mainly has two steps: first, scan the target environment and reconstruct a scenario composed of a large number of textured surfels; then, render the surfels with the camera trajectory, perform semantic and instance segmentation, and then generate realistic images through GAN.

In Ref. [13], Waymo proposed a method for representing maps with texture-enhanced surfels. Surfel is suitable for dynamic geometric modeling, and a target is represented by a set of dense points or surfels with lighting information. Convert the voxels scanned by the lidar into colored surfels, and discretize them into a $k \times k$ grid.

Due to different lighting conditions and changes in the camera's relative posture, each surfel may have a different appearance in different frames, creating an encoding composed of n different distance $k \times k$ grids to enhance the surfel representation. During the rendering phase, this method decides which $k \times k$ block to create based on the camera posture.

To further improve the geometric shape and texture, train two symmetrical encoding–decoding generators, one from surfel images to real images and the other from real images to surfel images. There are also two discriminators, one for the surfel image domain and the other for the real image domain.

The input data includes paired data and unpaired data. The unpaired data is for two purposes:

- Improve the generalization of the image discriminator.
- Implement cycle consistency constraints.

In addition, since the coverage of the surfel image is limited, the rendered image contains a large area of unknown regions, and the distance between the camera and the surfel introduces uncertainty, an additional distance-weighted loss is added to stabilize the training of the GAN.

VISTA (virtual image synthesis and transformation for autonomy) is an open-source data-driven simulator from MIT, with the following process: ① use a continuous kinematic model to update the vehicle state; ② retrieve the nearest frame related to the current posture from the dataset; ③ project the frame into 3D space to reconstruct the scenario; ④ reproject to the self-driving viewpoint. Unlike traditional geometry-based simulation systems, data-driven simulation directly synthesizes new viewpoints from real data, which has become a method to overcome the gap between photo realism and simulation reality.

From a recent monocular image, a depth map model is estimated using a convolutional neural network with self-supervised learning methods (see Sect. 5.3). Based on the estimated depth map and the camera's internal parameters, the sensor image frame can be back-projected into the 3D world. In this way, the image of the new viewpoint can be projected back into the sensor image coordinate system. VISTA can simulate different local rotations ($\pm 15°$) and lateral and longitudinal translations along the road (± 1.5 m). Since the free lateral space within its lane is usually less than 1m, VISTA can simulate situations beyond the stable driving range of the lane.

The goal of VISTA 2.0 is to extend the simulation to other modalities in a data-driven manner, i.e., to synthesize new sensor measurements of lidar and local event data around the dataset, and to use and publish this platform for robust perception learning. It integrates various sensors for autonomous vehicles, including RGB cameras, 3D lidar, and event-based cameras. VISTA 2.0 quickly generates new viewpoint data in the simulation, enriching the data available for strategy learning, including extreme cases that are difficult to capture in the physical world.

VISTA 2.0's framework is a data-driven simulator that simulates intelligent entities in synthetic scenarios, supporting large-scale learning and testing environments. Here, priority is given to lightweight APIs for processing real-world datasets, which only operate on local viewpoint changes, achieving efficient rendering and low memory overhead.

For VISTA 2.0's multi-sensor simulation schematic, virtual intelligent dynamics unfold trajectories in a data-driven environment, presenting sensor observation data from the new viewpoint at the time.

Lidar sensors have the accuracy of measuring geometric depth information and the robustness to environmental changes (such as lighting). Unlike cameras that return structured grid images, lidar sensors capture sparse point clouds of the environment. Here, each point is represented by a 4-tuple (x, y, z, i), where (x, y, z) is the point's position in 3D Cartesian space, and i is the intensity feature measurement of the point. Given the position of the virtual intelligent entity in the environment, and the relative transformation (rotation R and translation t) to the nearest collected point cloud Ψ, the goal of VISTA is to synthesize a new lidar point cloud Ψ', originating from the relative position of the virtual intelligent entity.

Since Ψ is represented in 3D Cartesian space, a simple solution is to directly apply the relative transformation (R, t) of the intelligent entity as a rigid transformation to $\Psi : \Psi' = R\Psi + t$. However, this method fails for several reasons: the point cloud obtained from the lidar sensor has a specific annular pattern originating from the sensor's optical center; applying a rigid transformation to points not only changes individual points but also transforms and breaks the inherent annular structure that defines the sensor's position; on the contrary, to protect the data space structure of the sensor, the lidar rays from the new sensor position are re-projected into the scenario and the new intensity is estimated; moreover, the result of simply transforming the point cloud is likely to cause points visible in the original scan to be occluded in the new viewpoint, so they need to be rejected to maintain the line-of-sight characteristics of the sensor.

To overcome these problems, VISTA 2.0 has the following measures: ① cull the currently occluded points; ② create a dense representation of the sparse point cloud;

and ③ sample a new sparse point cloud from the dense representation based on sensor-specific prior knowledge.

First, a GPU-accelerated culling algorithm for the transformed sparse point cloud Ψ' is implemented. Ψ' is projected into 2D polar coordinates, i.e.:

$$\alpha = \arctan(\frac{\Psi'_y}{\Psi'_x}), \beta = \arctan(\frac{\Psi'_z}{d}), d = ||\Psi'||_2, \tag{10.4}$$

where (α,β) are the azimuth and elevation angles of the line connecting each point, d is the distance along the line, Ψ' is represented as a sparse image in (α,β) (no information loss), and d is the color or value of each pixel. When culling points in the image, the distance of each pixel is compared with the average distance of the surrounding light "cone," similar to the Z-buffer technique.

Using the sparse and pruned point cloud, it is necessary to construct a dense representation of the scenario, and finally sample new lidar rays and generate new viewpoints. To make the sparse representation dense, a UNet architecture is trained to learn the dense point cloud output of a scenario. The training data for the densification network is generated using 2D linear interpolation. Compared with strict rule-based interpolation algorithms (scipy.interpolate), data-driven methods will produce smoother and more natural qualitative results through data mining. In addition, the obtained model is easy to implement GPU parallelization, achieving significant acceleration (about 100 times).

Finally, sparse points are sampled from the dense representation to form the point cloud of the new viewpoint. To determine the sampling position, a prior Ω can be constructed on the existing light ray projection angle of the dataset sensor. The light ray vector of the sensor is basically fixed over time because it is built into the sensor hardware itself, but there may be some slight changes or fallbacks depending on the environment. Sampling ω from the prior Ω will produce a set of light rays (α_i, β_i) from which point readings are projected and recovered. In addition, the prior Ω will respect several ideal characteristics of the sensor, which can also be specified by the user, such as the number and density of lidar rays. Since operations are still being performed in polar coordinate image space, ω is equivalent to a binary mask image, indicating the positions that should be sampled in the dense image. The newly sampled polar coordinate image can be used to inverse transform the data representation, returning to the desired 3D Cartesian space.

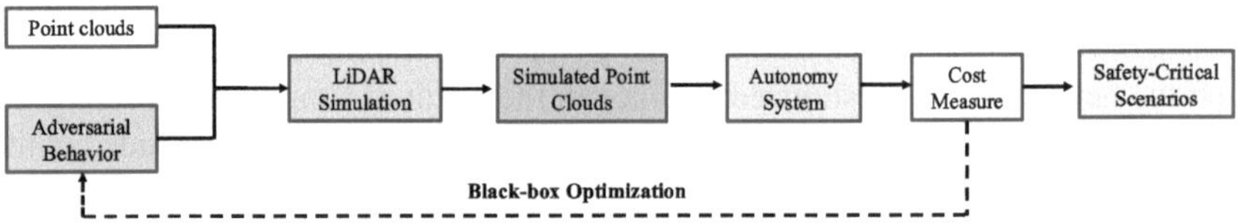

Fig. 10.2 Overview of the adversarial scenario generation pipeline

AdvSim is an adversarial framework proposed by Uber to generate safety-critical scenarios for lidar autonomous driving systems.

Shown in Fig. 10.2, the goal is to interfere with the maneuvers of interacting participants in existing scenarios, adversarial behavior will cause real autonomous driving systems to fail; given the initial traffic scenario, AdvSim modifies the trajectories of participants in a physically reasonable way and updates the lidar sensor data; through sensor data simulation, adversarial scenarios critical to full autonomous driving safety are obtained.

In the real lidar simulation of scenario perturbation, given the scenario perturbation of participant actions, the previously recorded lidar data is modified to accurately reflect the updated scenario configuration; while ensuring sensor authenticity, the original participant lidar observation data is removed and replaced with simulated participant lidar observation data at the perturbed position.

In order to produce physically feasible participant behavior, the trajectory is parameterized as a bicycle model state sequence, including the center position, heading, forward speed and acceleration of the perturbed participant, as well as the curvature of the vehicle path. By setting changes in curvature and acceleration values within boundaries at different time steps and calculating other states with the kinematic bicycle model, candidate adversarial trajectories can be generated. In addition, to expand the space for sampling adversarial behavior, initial states are allowed to be perturbed within set boundaries.

In order to increase the reasonableness of the perturbed trajectory, it is necessary to ensure that it does not collide with other participants or the original expert trajectory of the autonomous driving car. In practice, rejection sampling is first performed to create a set of physically feasible trajectories, and then the generated trajectories are projected onto the physically feasible set, measured by L2 distance. The search space is low-dimensional, which is conducive to query-based black-box optimization, while still allowing fine-grained control of participant motion.

In order to overcome the limitation of modeling the complexity and diversity of real traffic scenarios, Uber proposed a neural autoregressive (AR) model of traffic scenarios, scenarioGen, in Ref. [14], avoiding the need for rules and heuristics. In particular, considering the state of the ego vehicle and the high-definition map of the surrounding area, scenarioGen inserts participants of different categories into the scenario and synthesizes their size, orientation, and speed.

scenarioGen, combined with sensor simulation, can be used to train perception models suitable for real-world solutions.

In the autoregressive traffic scenario generation method, given the state of the ego vehicle and the high-definition map of the surrounding environment, scenarioGen generates traffic scenarios by inserting one participant at a time.

The neural AR model uses a recursive neural network (RNN) to capture long-term correlations in the autoregressive generation process. The basis of this model is the ConvLSTM architecture. In each iteration, the input to the model is a BEV multi-channel image encoding, including the autonomous driving car a_0, the high-definition map, and the other traffic participants a_1, a_2, ..., a_{i-1} generated so far.

The multi-channel image includes lane line polygons (straight lanes, dedicated right lanes, dedicated left lanes, dedicated bus lanes, and dedicated bicycle lanes), lane line centerlines, and dividers (allow crossing, prohibit crossing, and may allow crossing); lane segment (straight lane, dedicated right lane, and dedicated left lane), lane line direction and speed limit, participant occupancy (mask), participant direction and speed, drivable area and road, sidewalk, and the traffic light status (green, yellow, red, flashing yellow, flashing red, and unknown) for each lane, etc. Overall, this generates a 24-channel image.

Each participant has its class label, BEV position, oriented boundary, and speed. To capture the dependencies among these attributes, a factor decomposition of their joint distribution is performed. In the participant probability model, it models the probability for each participant as the product of the distributions of its class, position, boundary, and speed.

This means that the distribution of the participant's position depends on its class; its boundary depends on its class and position; and its speed depends on its class, position, and boundary.

At each generation step, M samples are taken from the output distribution of scenarioGen, and the most likely samples are retained. This helps to avoid degenerate traffic scenarios while maintaining sample diversity. In addition, vehicles and bicycles whose boundaries collide with the boundaries of the currently sampled participants are rejected.

Reference [15] is the previous work of the Lyft Level 5 team, which proposed an end-to-end trainable machine learning system SimNet to realistically simulate driving experiences. The simulation problem is constructed as a Markov process (MP), using deep neural networks to model the state distribution and transition function. It is trained directly from existing raw data, i.e., behavior cloning, without any manual design in the motion model, requiring only a dataset with historical traffic episodes.

For the proposed trainable simulation system, it allows the system to construct unseen scenarios that react realistically to the behavior of self-driving cars; in fact, the system is trained directly with 1000 hours of driving data, and the realism and reactivity of the simulation are two key measured attributes; at the same time, this method can evaluate the performance of the latest machine learning planning system trained with expert driving data, which is prone to causal confusion and difficult to test with powerless simulation or non-reactive simulation methods.

To generate new driving events, an initial state is first selected and sampled, capturing the positions of all traffic participants. Next, the state is forward simulated by the behavior of the neural network-controlled traffic participants and the self-driving control loop. The steps include

1. Choose the initial position of the self-driving car from all allowed map positions.
2. The initial state is obtained from the distribution of all feasible states, capturing the number and initial posture of traffic participants.
3. Driving events are generated by step-by-step forward simulation of participant driving strategies and self-driving control systems.

The features of this simulation include

- Full simulation. Execute all the above steps to generate new, never-before-experienced driving scenarios from all positions.
- Journey simulation. Keep the initial self-driving car position fixed, synthesize many different initial conditions, and get driving events starting from that position.
- Scenario simulation. Existing historical states of interest serve as the initial state, generating many possible future outcomes.
- Behavior simulation. By hard-coding a specific path to follow, the steering angle can be replaced, forcing traffic participants to take specific high-level behaviors, but still leaving some reactive simulation results in execution. This is useful for simulating self-driving car behavior.

In a detailed information about the interactive state unfolding of the simulation system, all the smart entities of this state run a step prediction independently to advance, the self-driving car is controlled by the control algorithm, a new position forms a new state, and then the process is repeated.

Reference [16] introduces STRIVE (stress-test drive), a method for automatically generating challenging scenarios that cause a given planner to exhibit undesirable behavior, such as collisions.

To maintain the reasonableness of the scenario, the key idea is to adopt a learned traffic motion model in the form of a graph-conditioned VAE. Scenario generation is optimized in the latent space of this traffic model, perturbing the initial real scenario to produce trajectories that collide with the given planner. Subsequent optimization is used to find the "solution" to the scenario, ensuring that it helps to improve the given planner.

Further analysis is based on clustering these scenarios by collision type. In the experiment, two planners were attacked, and it was proven that in these two cases, STRIVE successfully generated real and challenging scenarios. In addition, a closed loop was implemented, and these scenarios were used to optimize a rule-based planner's hyperparameters.

STRIVE generates challenging scenarios for a given planner. Adversarial optimization disrupts the real scenarios in the learned traffic model's latent space, leading to collisions with the planner. Subsequent solution optimization will find the planner's trajectory to avoid collisions, and verification scenarios help determine improvements to the planner.

The core idea of STRIVE is to measure the reasonableness of the scenario during the optimization process by learning the likelihood of the generated traffic motion model. This model encourages challenging scenarios while maintaining realism. Therefore, STRIVE does not pre-select specific adversaries, but jointly optimizes all scenario entities, resulting in a variety of scenarios. In addition, to adapt to widely used non-differentiable (or inaccessible) planners in practice, the proposed optimization uses the planner's differentiable entity representation in the learned motion model, allowing for optimization using standard gradient-based methods.

STRIVE does not understand the internal structure of the planner and cannot calculate gradients through it. Unacceptable behaviors include collisions with other vehicles, inability to drive terrain, uncomfortable driving (such as high acceleration), and violation of traffic regulations. Although the formula is general and can theoretically handle other objective function optimizations, the focus is on generating accident-prone scenarios related to vehicle collisions with the planner.

The test architecture of the learned traffic model is used to jointly sample the future trajectories of all entities in the scenario. First, each entity's past motion and local map environment information are processed separately. Then, the conditional prior is calculated, outputting the latent distribution of each node, which can be sampled and fed into the autoregressive (AR) decoder to predict future entity trajectories.

To sample future motion during testing, a conditional prior network and decoder are used, both of which are graph neural networks (GNNs) that run in a fully connected scenario graph of all entities. The prior model includes a set of latent vectors for a group of entities. Each node in the input scenario graph contains context features extracted from the entity's past trajectory, local rasterized map, bounding box size, and semantic class. After message passing, the prior network outputs the Gaussian distribution parameters for each entity in the scenario, forming a "distribution" latent representation that captures possible future changes.

The deterministic decoder operates in the scenario graph, with each node having a sampled latent vector and past trajectory context. Decoding is performed in an autoregressive manner, with a round of message passing resolving interactions before predicting each entity's acceleration at time step t; through the bicycle motion model, acceleration immediately obtains the next state, updating the past trajectory context before continuing to advance. Even when entities are sampled independently, the decoder's determinism and graph structure still support future scenario consistency. Importantly for latent vector optimization, even when input latent vectors are unlikely, the dynamic bicycle model decoder can still ensure reasonable vehicle dynamics.

At each step of adversarial optimization, the planner and non-self-driving car's latent representations are decoded using the learned decoder, and the non-self-driving car's trajectory is provided to the planner to unfold in the scenario. Finally, each loss is calculated.

Reference [17] is a work by NVIDIA, which adopts a data-driven approach and proposes a method that can learn to generate traffic behavior from real driving logs. This method decouples the traffic simulation problem into high-level intent inference and low-level driving behavior simulation, achieving high sampling efficiency and behavior diversity by leveraging the two-level structure of driving behavior.

The method also incorporates a planning model to achieve stable long-term behavior effects. The method, called bi-level imitation learning for traffic simulation (BITS), is empirically validated with two large-scale driving dataset scenarios, demonstrating that BITS achieves balanced traffic simulation performance in terms of realism, diversity, and long-term stability.

The BITS framework's decision context c_t is a tensor that includes a semantic map and rasterized entity history, connected together by channels. Given c_t as input, the spatial target network generates a 2D spatial distribution of short-sighted targets, and the goal-conditioned policy generates a set of actions for each sampled goal. The trajectory prediction model predicts the future motion of adjacent entities, and finally, based on the predicted future state, the framework selects a set of actions that minimize the rule-based cost function.

Traffic simulation can be described as a supervised imitation learning problem. However, the nature of urban driving brings significant technical challenges. It is a partially observed decision-making process, as the model cannot access the demonstrator's latent intentions and other decision-related cues, such as the turn signals of other vehicles. Action supervision is inherently ambiguous and is typically modeled with a probability distribution.

Although this ambiguity complicates training, effectively modeling action distributions can also generate different counterfactual traffic simulations. Secondly, because the behavior of each agent is not explicitly coordinated, their joint behavior generates a combinatorial space of possible future states. This uncertainty makes generating stable traffic simulations very challenging.

The goal of traffic imitation models is to generate a variety of reasonable behaviors by learning from real-world driving logs (as demonstrations). Most existing methods in trajectory prediction use deep latent variable models (such as VAEs) to capture behavior distributions. However, learning to generate stable long-horizon behaviors requires a large amount of training data. In contrast, the method proposed here decomposes the learning problem into training a high-level goal network, capturing the possible short-term goal space distribution; training a deterministic goal-conditioned policy, learning how to reach the predicted goal.

The spatial goal network utilizes the 2D BEV structure of driving motion and efficiently represents the spatial goal distribution with a 2D grid. This decomposition shifts the burden of multimodal trajectory modeling to the high-level goal predictor, allowing the low-level goal-conditioned policy to reuse skills for achieving goals, improving sample efficiency.

Such a two-layer simulation learning method can generate reasonable traffic imitations from limited data. The strategy can sample from the multimodal spatial goal predictor, synthesizing various behaviors. However, the execution of this strategy is still limited by the scale and coverage of the training data.

Driving logs are biased toward normal behavior and hardly contain safety-critical situations such as collisions or off-road driving. The goal of generating diverse behaviors further amplifies this challenge, as it encourages agents to enter unseen areas on the map and create new interactions. Therefore, to achieve stable long-term simulations, agents must generate reasonable behaviors even in states lacking training data guidance.

To this end, it is suggested to use prediction and planning modules to enhance the strategy for stable long-term trajectory unfolding. This method is similar to the motion planning pipeline in a typical modular AV stack, with the important difference that human-like motion trajectory candidates are generated with a learned strategy. The key idea is that the strategy can directly track the data likelihood under distributional states, where most behavior samples follow rules, and accept corrective guidance in states where the most likely action may lead to adverse consequences. In addition, the sampling module allows for flexible adjustments to the simulator without further training (for example, diversity level, emphasis on multiple goals).

Techniques on neural rendering (especially NeRF) will be discussed specifically in Chap. 15.

10.8 Summary

This chapter introduced a very important module in autonomous driving: simulation, emulation, and visualization. Mainly the simulation of sensors, modeling of driving environments (including traffic modeling, road modeling, CAD modeling of vehicles and pedestrians, computer vision modeling, etc.), Uber visualization platform, scenario library construction, scenario-level testing and scenario description language, as well as realistic synthetic data generation methods of digital twins, etc. Chapter 11 will discuss an important issue in autonomous driving development—the safety model.

References

1. F TOADERE, Simulating the functionality of a digital camera pipeline, Optical Engineering, 2013, 52(10): 1–12.
2. X YUE et al, A LiDAR Point Cloud Generator: from a Virtual World to Autonomous Driving, arXiv preprint arXiv:1804.00103, 2018.
3. A HAIDER, M PIGNICZKI, M H. KOHLER, M FINK, M SCHARDT, Y CICHY, T ZEH, L HAAS, T POGUNTKE, M JAKOBI and A W. KOCH, Development of High-Fidelity Automotive LiDAR Sensor Model with Standardized Interfaces, MDPI Sensors, October, 2022. 22(19): 1–23.
4. S MANIVASAGAM, S WANG, M SAZANOVICH et al., LiDARsim: Realistic LiDAR Simulation by Leveraging the Real World, arXiv preprint arXiv:2006.09348,2020
5. M HAHNER, C SAKARIDIS, M BIJELIC et al., LiDAR Snowfall Simulation for Robust 3D Object Detection, IEEE Conference on Computer Vision and Pattern Recognition, New Orleans, USA, June 2022: 16364–16374.
6. M HAHNER, C SAKARIDIS, D DAI, L VAN GOOL, Fog Simulation on Real LiDAR Point Clouds for 3D Object Detection in Adverse Weather, IEEE Int. Conf. on Computer Vision, Virtual, June 2021: 15283–15292.
7. C GOODIN, D CARRUTH, M DOUDE and C HUDSON, Predicting the Influence of Rain on LIDAR in ADAS, MDPI Electronics, 2019. 8(1): 89–97.
8. WHEELER T, HOLDER M, WINNER H et al, Deep Stochastic Radar Models, IEEE Intelligent Vehicles Symposium, Los Angeles, USA, June 2017: 47–53.
9. M D LIO, D BORTOLUZZI and G P R PAPINI, Modeling Longitudinal Vehicle Dynamics with Neural Networks, Vehicle System Dynamics, 2020. 58(11): 1–19.
10. G GARIMELLA, J FUNKE, C WANG, and M KOBILAROV, Neural Network Modeling for Steering Control of an Autonomous Vehicle, IEEE Int. Conf. on Intelligent Robots and Systems, Vancouver, Canada, Sept. 2017: 2609–2615.
11. X NIE, Y PAN, W DAI, F XU, Z LI, A Deep Neural Network-based Nonlinear Dynamics Model for the Prediction of Lateral-longitudinal Vehicle Dynamics, MDPI Sensors, 2022. 22(5): 1–16.
12. W. LI, C. W. PAN et al., AADS: Augmented Autonomous Driving Simulation using Data-driven Algorithms, arXiv preprint arXiv:1901.07849, 2019.

13. Z YANG, Y CHAI, D ANGUELOV, Y ZHOU, P SUN, D ERHAN3, S RAFFERTY, H KRETZSCHMAR, SurfelGAN: Synthesizing Realistic Sensor Data for Autonomous Driving,arXiv preprint arXiv:2005.03844, 2020.
14. S TAN, K WONG, S WANG, S MANIVASAGAM, M REN, R URTASUN, sceneGen: Learning to Generate Realistic Traffic scenarios, arXiv preprint arXiv:2101.06541,2021
15. L. BERGAMINI, Y. YE, O. SCHEEL, L. CHEN, C. HU, L. DEL PERO, B. OSIÅfSKI, H. GRIMMETT, and P. ONDRUSKA, SimNet: Learning reactive self-driving simulations from real-world observations, IEEE Int. Conference on Robotics and Automation, Xi'an, China, 2021: 5119–5125.
16. D. REMPE, J. PHILION, L. J. GUIBAS, S. FIDLER, and O. LITANY, Generating Useful Accident-Prone Driving Scenarios via a Learned Traffic Prior, IEEE Conference on Computer Vision and Pattern Recognition, New Orleans, USA, June 2022: 17305–17315.
17. D XU, Y CHEN, B IVANOVIC, M PAVONE, BITS: Bi-level Imitation for Traffic Simulation, arXiv preprint arXiv:2208.12403, 2022.

Chapter 11
Safety Model

Safety has always been a major concern in the vehicle industry, and the original intention of developing autonomous driving is to reduce traffic accidents caused by human factors. A 2018 report from a U.S. authority shows that more than 100 people die each day in traffic accidents related to non-autonomous vehicles. Similarly, a widely cited study shows that in the past two and a half years, there have been more than 2 million car accidents in the U.S., 94% of which were caused by drivers.

However, as autonomous driving is about to enter the commercial track, it has been found that previous safety standards and testing processes are somewhat unsuitable, hence some new safety standards for testing autonomous driving systems have been revised.

Key modules of autonomous driving, such as perception, planning, and control, use a lot of AI, for example, computer vision, machine learning, and deep neural network algorithms, but these algorithms mostly probabilistically model the problems to be solved, which undoubtedly poses a new challenge to past safety models and testing methods.

Companies developing autonomous driving are also exploring these issues and have proposed their own safety models, such as Mobileye's RSS model and NVIDIA's SFF model. Many autonomous driving companies (mainly Level 4) have published their own safety reports, such as Google WayMo, Ford, GM, NVIDIA, Uber, and AutoX.

Autonomous driving is still plagued by threats from various attacks, which can be divided into physical attacks, network attacks, and learning-based adversarial attacks. Inevitably, the safety & security of autonomous driving systems based on deep learning are seriously challenged by these attacks, so countermeasures should be studied to mitigate potential risks.

If autonomous driving is to be mass-produced or deployed on a large scale, a large amount of verification and validation (V&V) needs to be carried out on it.

Y. Huang and Z. Yang, *Autonomous Driving Handbook*,
https://doi.org/10.1007/978-981-95-3725-9_11

The common strategy for maximizing scenario coverage in verification and validation is to conduct it in virtual simulation and simulate the generation of a large number of scenario samples.

The challenge faced by V&V based on scenario sampling is to determine the sample size required for reasonable coverage, so that the risks generated by improper operations of autonomous driving can be controlled. However, another set of methods based on formal verification handles scenario coverage through specification satisfaction.

The current open challenges in the safety of autonomous driving include the following points.

- Design specifications: The design specifications of machine learning models are often insufficient because the model needs to learn the patterns of the data to identify or generate the distribution of new unknown inputs. Therefore, machine learning algorithms learn the target class through their training data (and regularization constraints), rather than using a formal specification.
- Implementation transparency: Advanced machine learning models trained on high-dimensional data are not transparent. The large number of variables in the model makes the model difficult to understand, or it can be said to be a black box. In practice, interpretability techniques are mainly used by designers to improve network structure and the training process, rather than supporting safety assessment.
- Testing and validation: For DNN, due to the high dimensionality of the data, formally verifying its correctness is challenging (proving NP-hard). Therefore, it is difficult to achieve complete testing and validation within the operation design domain.
- Performance and robustness: The SOTIF standard treats machine learning models as black boxes and suggests improving model performance and robustness. However, improving model performance and robustness itself is a very challenging task. In learning problems, the training of the model usually ends with the error rate on the training set (false positive FP and false negative FN). Empirical error is the prediction error rate of the learning function for its target distribution. Generalization error refers to the difference between the empirical error of the model on the training set and the test set. In addition, operational errors refer to the error rate of the model deployed in the open world, which may be higher than the test set error rate. Domain generalization refers to the ability of the model to learn a general data representation for open-world tasks.
- Running monitoring functions: Designing monitoring functions to predict machine learning failures (such as false positives and false negatives) is essentially different. Machine learning models generate prediction probabilities for input instances, but research shows that prediction probabilities cannot guarantee failure prediction. In fact, DNN and many other machine learning models may produce incorrect outputs in the case of distribution drift and adversarial attacks, but they have high credibility.

Section 11.1 introduces some basic safety concepts, Sect. 11.2 analyzes NHTSA's 12 safety elements, Sect. 11.3 introduces the functional safety standard ISO 26262,

Sect. 11.4 discusses the expected functional safety standard SOTIF, Sect. 11.5 analyzes Mobileye's safety model RSS, Sect. 11.6 introduces network security, and Sect. 11.7 elaborates on the safety hazards of autonomous driving and solutions.

11.1 Basic Concepts

The following terms need to be explained in advance.

(1) Fail-safe: In engineering design, fail-safe means that when a specific type of failure occurs, its own response mode will not cause harm or minimize damage to other equipment, the environment or people; fail-safe means that the device will not endanger life or property safety when a failure occurs; in autonomous driving, L3 is required to be fail-safe.

(2) Fail-secure: Also known as fail-closed, it means that in fail-secure, access or data will not fall into the hands of bad people.

(3) Fail-passive: Disengages after failure, has no further action, and does not interfere with the operation of other systems.

(4) Fault-tolerance: Allows the system to continue to operate normally when certain components fail.

(5) Fail-operational: Continues to operate in the event of a single system failure, so it is necessary to ensure the availability of the required safety components and have a diversified/redundant architecture design; in autonomous driving systems, L4 is required to achieve fail-operational.

Safety is divided into passive safety and active safety; the former refers to the protection measures of the car body, such as airbags and seat restrictions, while the latter is related to autonomous driving, referring to its own accident prevention. Taking the safety report of Google's autonomous driving subsidiary Waymo as an example, we can see some safety issues considered by autonomous driving.

- **Behavioral Safety**: Refers to the driving decisions and behaviors of the vehicle on the road. Similar to a human driver, the system must follow traffic rules and ensure safe navigation in both expected and unexpected situations.
- **Functional Safety**: Ensures that the vehicle remains safe even in the presence of system failures. This is typically achieved through redundancy and backup mechanisms. ISO 26262 primarily addresses functional safety.
- **Crash Safety**: Refers to passive safety, i.e., protecting occupants during collisions.
- **Operational Safety**: Concerns the interaction between the vehicle and its passengers, including the provision of a reliable and intuitive user interface.
- **Non-collision Safety**: Provides physical safety guarantees for pedestrians and other individuals interacting with the vehicle.
- **Cybersecurity/Information Security**: Prevents unauthorized access or malicious attacks on the system, particularly in the context of V2X communication.

Safety involves many aspects, such as the formulation of government regulations, acceptance of moral ethics, and the responsibility of insurance companies in the event of accidents. Therefore, safety is a grand issue that requires consideration of many factors, which will be discussed below.

11.2 NHTSA

What are the duties and mission of the National Highway Traffic Safety Administration (NHTSA) in the United States? The original text is as follows:

Save lives, prevent injuries, and reduce economic costs due to road traffic crashes through education, research, safety standards, and enforcement activity.

For this, NHTSA has given the following 12 safety factors for the automated driving system (ADS), which many departments refer to when making demands on companies and institutions developing autonomous driving.

1. System Safety: Safe design.
2. Operational Design Domain (ODD): Defines the working conditions, time, and place of autonomous driving.

   - Types of roads safe for ADS operation.
   - Geographic area.
   - Speed range.
   - Environmental conditions.
   - Other restrictions.

3. Object and Event Detection and Response: Includes mapping, perception, localization, prediction, planning, and control.
4. Emergency (fallback) or minimal risk conditions.
5. Validation method.
6. Human–machine interface.
7. Vehicle cybersecurity.
8. Crashworthiness.
9. Post-crash ADS behavior.
10. Data recording.
11. Consumer education and training.
12. Federal, State, and Local Laws.

11.3 ISO 26262

International Organization for Standardization (ISO) 26262 has evolved from the International Electrotechnical Commission (IEC) 61508 standard. The goal of this standard is the functional safety of systems composed of electronic and electrical devices and software programming. The purpose of ISO 26262 is also to meet the safety needs of the road vehicle electronic and electrical system field.

What is functional safety? According to the explanation of ISO 26262, "Absence of unreasonable risk due to hazards caused by malfunctioning behavior of E/E systems."

With the increase in software complexity and the degree of mechatronics, the risk of random system and hardware failure also increases. ISO 26262 provides appropriate requirements and processes to avoid such risks. Things that ISO 26262 can do include:

1. Vehicle safety lifecycle (management, development, production, operation, service, scrapping).
2. Vehicle-specific risk assessment, i.e., automotive safety integrity levels (ASIL) method.
3. Based on the ASIL method, obtain operable functional safety requirements to avoid unreasonable risks; provide functional safety requirements.

The basic framework of ISO 26262 (2018 version) is divided into 12 parts:

- Part 1: Vocabulary
- Part 2: Functional Safety Management
- Part 3: Concept Phase
- Part 4: System-level Product Development
- Part 5: Hardware-level Product Development
- Part 6: Software-level Product Development
- Part 7: Production, Operation, Service, and Discontinuation
- Part 8: Support Process
- Part 9: ASIL-oriented and Safety-oriented Analysis
- Part 10: ISO 26262 Guidelines
- Part 11: ISO 26262 Semiconductor Application Guidelines
- Part 12: ISO 26262 Motorcycle Modification Guidelines

ISO 26262-1/2 defines how to avoid unreasonable safety risks caused by electronic and electrical system failures. ISO 26262-3 describes the method of hazard analysis and risk assessment (HARA) to determine vehicle-level hazards. HARA is mainly used to assess potential risks caused by related failures and define high-level safety requirements, such as safety goals and the necessity of risk mitigation.

The remaining sections of ISO 26262 provide safety requirements and recommendations to avoid and control system failures and random hardware failures that affect safety goals.

The division of ASIL safety levels is to define the probability of the system achieving specified safety goals, and each safety function requirement includes two parts: safety goals and ASIL safety levels.

The steps for dividing ASIL safety levels are as follows:

- Determine safety events based on anticipated architecture, functional concepts, operating modes, and system states, etc.

- Hazard analysis and risk assessment to determine the preliminary ASIL safety level.
- Decompose safety requirements and safety levels. The division of ASIL safety levels and the decomposition of ASIL safety levels alternate until they cannot be decomposed.
- Check the rationality of the level allocation based on principles, including coexistence principles, related failure analysis, and safety analysis, etc.
- ASIL defines four levels, namely A, B, C, and D, with the level increasing in order. The specific evaluation is carried out in three dimensions, namely severity, exposure, and controllability.
- Severity (SX). X=0–3. S0 no harm; S1 minor or limited harm; S2 serious or life-threatening harm (survivable); S3 life-threatening harm (possibility of death) or fatal harm.
- Exposure (EX). X=0–4, a total of 5 levels. E0 is almost impossible to be exposed to danger, and E4 is highly likely to be exposed to danger.
- Controllability (CX). X=0–3, a total of 4 levels. C0 is controllable, C3 is almost uncontrollable.

11.4 ISO/PAS 21448 SOTIF

The safety of the intended function (SOTIF) standard is to avoid unreasonable risks caused by performance limitations or misuse.

It can be said that ISO 26262 functional safety and SOTIF are complementary.

The performance limitations of the autonomous driving system are as follows.

- **Perception**: The predefined target scenarios are not comprehensive, leading the system to fail in responding correctly.
- **Decision**: Unreasonable functional logic, arbitration mechanisms, or algorithms may cause decision-making errors.
- **Execution**: The control actions of the execution mechanism may deviate from the expected outcomes, making perfect operation difficult to achieve.

The driving scenarios defined by SOTIF can be divided into the following four parts.

- Known safe.
- Known unsafe.
- Unknown safe.
- Unknown unsafe.

The fourth part is unresolved, and the currently feasible method is to refine the second part and narrow the area of the fourth part.

11.5 Intel Mobileye RSS Safety Model

Intel Mobileye has proposed a "responsibility-sensitive safety model" (RSS), which is a method for determining responsibility in the event of a collision accident for autonomous vehicles.

The construction of RSS is based on the following four common senses.

- Maintain a safe distance from the vehicle in front, so that it can stop in time if it suddenly brakes.
- Maintain a safe distance from vehicles on the side. When making lateral maneuvers and cutting into other vehicles' lanes, sufficient reaction space must be left for other vehicles.
- Respect the right of way, but do not rigidly use it.
- Be careful of obstructions, such as children who may be hidden behind a stopped vehicle.

As a reasonable safety model, Mobileye believes it needs to meet the following two aspects.

- Soundness: When the model believes it is not responsible for the accident, it should clearly conform to human judgment. As long as the model is still useful, it allows for responsibility in ambiguous scenarios, even if it may be overly cautious.
- Usefulness: Effectively formulate policies to ensure that accidents will not occur under normal traffic flow conditions.

Consider two cars driving in the same lane, one following the other in a straight line. If the front car brakes suddenly and the rear car cannot brake in time, a rear-end collision will occur. The safe longitudinal distance that the rear car must maintain from the front car can be calculated, which is actually the safe longitudinal distance.

$$d_{\min} = L + T_f\left[v_r - v_f + \rho(a_a + a_d)\right] - \frac{\rho^2 a_d}{2} + \frac{(T_f - T_r)\left(v_r + \rho a_a - (T_f - \rho)a_d\right)}{2} \tag{11.1}$$

Where, L is the average length of the vehicle; ρ is the response time of the rear car; v_f, v_r are the speeds of the front and rear cars respectively; a_a, a_d are the maximum acceleration and deceleration (braking) of the vehicle; T_f is the time it takes for the current car to come to a complete stop when it is decelerating (braking) at its maximum; T_r is the time it takes for the rear car to switch from maximum acceleration to maximum deceleration (braking) and come to a complete stop.

In another scenario, if the front car recklessly cuts into the lane of another car behind it, and the rear car hits the front car, then it is the front car's fault . RSS can use the same principle to handle this situation. Based on a set of variables, there exists a safety corridor. If a human-driven vehicle violates this corridor before a collision occurs, then it is the fault of that vehicle. Conversely, the safety corridors of other vehicles can be calculated continuously, and they will never be violated.

When an accident occurs with a pedestrian, the car side is almost always at fault, even if the car sees the pedestrian very late, but there are exceptions.

There are also some situations considered by RSS.

- Route priority: In a certain area, multiple roads geometrically overlap, such as highway merge lanes and roundabouts. In these cases, one of the two cars must yield to the other's front corridor. Cutting in can only be merged in the corridor. In this case, vehicles on the priority lane can cut in.
- Two-way traffic: In the case of two-way traffic, the centerline provides a boundary, but it is a violation when a car crosses the centerline to overtake a parked truck or reverse into a parking space.
- Traffic light: The intersection with traffic lights is modeled using dynamic route priority. In other words, according to the condition of the traffic light, some routes have priority over others.
- Unstructured roads: Such as parking lots. In the absence of lane markings, the vehicle's path is predicted based on the current trajectory. If the vehicle deviates from this prediction (by changing course), a sufficient distance is required to allow oncoming cars to adjust.

11.6 Network Security

Another safety task is the cybersecurity of cars, and the corresponding standard is ISO/SAE 21434. Cybersecurity vulnerabilities may impair the system's ability to achieve basic safety objectives, so cybersecurity is an important component to ensure the integrity and safety of autonomous vehicles, which is a priority.

ISO 21434 adopts a risk-oriented approach, clearly distinguishing measures at the product, project, and organizational levels. It does not specify specific technologies, solutions, or remedial methods, nor does it make requirements for autonomous vehicles or road infrastructure. Although ISO 21434 covers cybersecurity principles, it does not provide guidance for implementation.

ISO 21434 can be used to implement a cybersecurity management system, including cybersecurity risk management and governance. Its purpose is to make cybersecurity a mandatory requirement for all organizations involved in the life cycle of road vehicles. Cybersecurity management mainly focuses on different stages of the engineering cycle, the life cycle of the product, and the organizational level.

Autonomous driving cars pose a unique challenge in terms of cybersecurity, combining unprecedented advancements in cars, frequent software updates, and cloud control features. General cybersecurity strategies include not only the vehicle's electronic devices, sensors, and autonomous driving systems, but also any connected functions, such as data ports, mobile applications, and customer service systems.

Cyberattacks on autonomous vehicles are basically divided into three categories: autonomous control systems, autonomous driving system components, and vehicle-to-vehicle communication. Defense methods against such attacks can be divided into security architecture, intrusion detection, and anomaly detection.

Intrusion Detection Systems (IDS) continuously monitor the vehicle network and detect any network attacks. IDS is extremely important and is usually the last line of defense when attackers break through defense mechanisms. Traditional IDS solutions rely on firewalls or rule-based (non-AI) systems to detect network attacks. However, they cannot effectively detect complex attacks because they cannot capture the complex correlations present in time-series network data. Recent AI-based solutions can be used to parse high-dimensional vehicle network data, including monitoring and attack detection at the vehicle network level and the Vehicle Ad hoc Networks (VANET) level.

The cybersecurity industry standard provides a defense-in-depth approach, implementing security layers among multiple overlapping systems. By using component isolation techniques, memory protection, and access control on any embedded system (especially those with external interfaces or security functions), the likelihood of hackers entering the vehicle network is reduced.

Cybersecurity systems also reduce the ability of compromised devices to affect behavior through message authentication, message verification, and credential provisioning. In addition, any violation can be minimized through measures such as network isolation, physical and virtual partitioning. The entire validation technology helps ensure that no single damaged component can endanger the health of the entire system.

11.7 Safety Solutions for Autonomous Driving Systems

The ISO 26262 standard for functional safety of road vehicles provides a set of requirements to ensure system functional safety, but some characteristics of machine learning (including deep learning) methods themselves may affect safety or safety assessment.

- Identifying hazards: The use of machine learning may generate new types of hazards, which are not brought about by component malfunctions. In particular, complex behavioral interactions between humans and high-level functions implemented by machine learning can create hazardous situations that should be mitigated in system design.
- Level of machine learning use: Through the use of end-to-end methods, machine learning can be widely used at the architectural level or limited to use at the component level. However, end-to-end (E2E) methods challenge the assumption that complex systems are defined as a decomposition of layered components with their own functions.

- Required software techniques: ISO 26262 requires the use of many specific techniques at various stages of the software development lifecycle. While some of these are still applicable to machine learning components and others can be easily adjusted, many still particularly favor the assumption of implementing code using imperative programming languages.
- Opacity: All types of machine learning models contain encoded knowledge, but this encoding is more difficult for humans to interpret in some types than others. For example, Bayesian networks used for weather forecasting are easier to interpret because the nodes are random variables representing human-defined concepts, such as precipitation type and temperature. In contrast, neural network models are considered opaque. Increasing the expressive power of machine learning models often comes at the expense of transparency. Opacity is a barrier to security because it makes it harder for evaluators to establish confidence that the model is operating as expected.
- Error rate: Machine learning models usually do not operate perfectly and make mistakes. Therefore, even for test data, the correctness of machine learning components is rarely achieved, and it must be assumed that it makes mistakes periodically. Furthermore, although the estimation of error rates is also a task in the machine learning development process, it only has statistical guarantees of its reliability. Finally, even if the error estimation is accurate, it may not reflect the system's error rate on a finite input set, as the true error is based on an infinite sample set. These characteristics must be considered when designing safety systems using machine learning components.
- Training-based: Machine learning models based on supervised and unsupervised learning are trained with a subset of possible inputs. Therefore, the training set is inevitably incomplete and cannot guarantee or even represent the possible input space. In addition, overfitting may occur in the learning process by capturing details of the training set rather than general inputs. Reinforcement learning is subject to similar limitations, as it typically only explores a subset of behaviors during training. The uncertainty of this machine learning component's performance is a threat to safety. Moreover, even if the training set is representative, it is not sufficient to represent safety-critical cases, as these cases are usually rare in the input space.
- Instability: More powerful machine learning models (such as NN) are usually trained using local optimization algorithms, and there may be multiple optima. Therefore, even if the training set remains unchanged, the training process may produce different results. However, changing the training set may also change the optimal state. Generally speaking, different optimal structures may be far apart in architecture, even if they are similar in performance. This makes it difficult to debug models or reuse parts that have already been safety assessed.

Inherently safe AI is far from being achieved. There are two practical machine learning solutions for safety strategies.

- Safe fail refers to the strategy of maintaining a safe state on the road in the event of a failure. This strategy can mitigate danger in the event of a failure by using monitoring functions and appropriate downgrade plans (such as notifying the driver to control the vehicle). It is recommended to detect erroneous outputs of the vehicle's AI algorithms (such as incorrect classification and error detection) during operation.

- Safety margin in the context of machine learning is described as the performance difference between the model on the training set and the open world. It is recommended to use model robustness techniques to improve the fault tolerance of machine learning components, thereby increasing the safety margin.

In addition, the importance of procedural safeguards for the end user (i.e., the driver and passengers) should also be noted.

The following are some specific methods for implementing AI safety strategies: monitoring functions and algorithm robustness.

1. Monitoring Functions

Monitoring functions use a series of techniques for detecting misclassification errors in machine learning to achieve safe fail behavior. Similarly, various runtime monitoring functions and error detectors can be designed for machine learning components to predict model failures and trigger appropriate warnings.

The uncertainty of probabilistic learning is an important factor in maintaining system failure safety. Even well-trained and calibrated predictors can benefit from uncertainty estimates in detecting domain drift and out-of-distribution (OOD) samples during operation, despite their robustness to noise, deterioration, and disturbances. Quantifying uncertainty can explain what a model does not know in terms of confidence in its predictions (i.e., epistemic or model uncertainty) and uncertainty about unknown samples (aleatoric or data uncertainty).

Quantifying predictive uncertainty in DNNs is a challenging task. Typically, DNN classification models generate normalized prediction scores, which are often overly confident, and regression DNN models do not provide uncertainty characterization in their outputs. Some solutions, such as ensemble and Monte Carlo dropout, estimate predictive uncertainty.

Although uncertainty estimation methods provide effective solutions for DNN failure prediction, the significant computational cost and latency are not ideal for runtime failure prediction in practice.

Misclassification of in-domain samples is often due to weak representation learning capabilities. In recent years, advanced neural networks, regularization techniques, and large training datasets have significantly improved DNN representation learning, thereby improving model performance and robustness. However, in the event of model failure, runtime prediction error detectors are still needed to maintain system safety. Selective classification, also known as classification with a rejection option, is a technique that cautiously provides predictions for high-confidence samples and abstains when in doubt. This credible prediction method can significantly improve model performance, but at the cost of test coverage.

Out-of-distribution (OOD) samples, or outliers, are inputs that fall outside the normal training distribution. OOD errors refer to the misclassification of such samples by machine learning models. In the context of autonomous driving, OOD samples may include rare, unusual, or previously unseen road signs, markings, objects, or scenes that are absent from the training set or insufficiently learned due to issues such as class imbalance.

OOD errors are an inherent problem of the ReLU series of activation functions, which produce arbitrary high confidence as the training distribution deepens. However, various techniques have been proposed, such as OOD detectors, novelty detectors, and outlier detectors, to detect OOD samples. Examples of OOD detector techniques include: modifying network structure to learn prediction confidence, using leaving-out classifier ensembles, and methods for self-supervised representation learning for outlier detection. In addition, a quick and low-cost method in OOD detection is to use class probabilities as a measure for OOD detection.

2. Algorithm Robustness

Robustness techniques in machine learning research have improved the algorithm's tolerance to unknown samples, deterioration and disturbances, adversarial examples, and domain drift. Machine learning literature introduces various techniques, such as dataset augmentation, noise injection, and multi-task learning, to regularize DNN learning of generalizable features. Other techniques, such as transfer learning, which transfers general representations from pre-trained models to new domains, have been proven to improve model robustness. In addition, using unlabeled free data can regularize model training to achieve robustness.

Domain drift (also referred to as distribution drift or dataset drift) describes changes in the input data distribution relative to the training set. Such drift violates the i.i.d. assumption between training and test data, thereby degrading operational performance. Addressing this issue is particularly important in open-world applications such as autonomous driving, where data are collected from uncontrolled and rapidly changing environments.

Domain generalization is therefore a key requirement for machine learning algorithms in these settings. It can be achieved through various approaches. One approach is adversarial domain adaptation, which leverages large amounts of unlabeled data from the target domain to align distributions. Another is multi-task learning, which improves model robustness by training the network to solve two or more tasks simultaneously, thereby enhancing generalization capability.

In open environments, natural data disturbances and deteriorations are common. Robustness benchmark tests for DNN deterioration and disturbances show that machine learning models display unexpected prediction errors on simple disturbances. To achieve model robustness to natural deteriorations (for example, due to snow, rain, and fog appearing on the camera) and disturbances (for example, transient sensor errors, electromagnetic interference of sensors), techniques are needed to improve model robustness to clean datasets. Previously, classic data augmentation was used to enhance robustness to simple image changes (such as rotation and scaling). Other techniques, such as adaptive algorithm selection for augmentation transformations and random image block erasure, have also proven effective for robustness and representation learning.

Recently, advanced augmentation techniques such as style transfer have been proven to improve model robustness to texture biases. Another research direction suggests using larger networks, learning how to improve DNN robustness through multi-scale and redundant feature learning. In addition, adversarial disturbances are small but worst-case disturbances deliberately created by attackers, so disturbed samples can cause the model to make incorrect classifications of samples with high confidence.

11.8 Summary

This chapter discussed the safety model of autonomous driving, introducing each aspect from the basic concept of autonomous driving safety, the safety framework of the renowned institution NHTSA, safety standards ISO 26262 and SOTIF, Mobileye's RSS safety model, network security, and safety issues. Chapter 12 will focus on a very important application scenario of autonomous driving—automatic and autonomous parking, and discuss and analyze it.

References

1. NHTSA. Automated Driving Systems 2.0: A Vision for Safety, US Department of Transportation, 2017.
2. ISO 26262. Road Vehicles – Functional Safety Package, 2018.
3. BERGENHEM C, JOHANSSON R, S ODERBERG A et al., How to Reach Complete Safety Requirement Refinement for Autonomous Vehicles, Critical Automotive applications: Robustness & Safety (CARS), Paris, France, Sep. 2015: 1-5.
4. J E STELLET et al., Testing of advanced driver assistance towards automated driving: A survey and taxonomy on existing approaches and open questions, IEEE Int. Conf. on Intelligent Transportation Systems, Washington DC, USA, Sept. 2015: 1455-1462.
5. ISO/PAS 21448. Road vehicles – Safety of the intended functionality, 2019.
6. S SHWARTZ, S SHAMMAH, A SHASHUA, On a Formal Model of Safe and Scalable Self-driving Cars, arXiv preprint arXiv:1708.06374, 2017.
7. Mobileye RSS. https://www.mobileye.com/responsibility-sensitive-safety
8. R SALAY et al. An Analysis of ISO 26262: Using Machine Learning Safely in Automotive Software, arXiv preprint arXiv:1709.02435, 2017
9. Google WayMo, On the road to fully self-driving, Tech. Report, 2017
10. R MCALLISTER et al. Concrete Problems for Autonomous Vehicle Safety: Advantages of Bayesian Deep Learning, Int. Joint Conference on Artificial Intelligence, Melbourne, Australia, Aug. 2017: 4745-4753.
11. D NISTÉR, H-L LEE, J NG, Y WANG, An Introduction to the Safety Force Field, Nvidia Technical Report, 2019
12. K REN, Q WANG, C WANG, Z QIN and X LIN, The Security of Autonomous Driving: Threats, Defences, and Future Directions, Proceedings of IEEE, 2020. 108(2): 357-372.
13. MOHSENI S et al., Practical Solutions for Machine Learning Safety in Autonomous Vehicles, arXiv preprint arXiv:1912.09630, 2019

14. ISO SOTIF. Road vehicles – Safety of the intended functionality, Jan. 2019.
15. Y DENG, T ZHANG, G LOU, X ZHENG, J JIN and Q HAN, Deep Learning-Based Autonomous Driving Systems: A Survey of Attacks and Defenses, arXiv preprint arXiv:2104.01789, 2021
16. V K KUKKALA, S V THIRULOGA, and S PASRICHA, Roadmap for Cybersecurity in Autonomous Vehicles, arXiv preprint arXiv:2201.10349, 2022

Chapter 12
System Integration, Validation and Verification (V&V)

With the rapid development of autonomous driving technology, the complexity of the system is increasing day by day. The autonomous driving system contains multiple highly collaborative sub-modules, such as perception, decision-making, control, and execution. The seamless integration and collaborative work between these modules are the basis for the regular operation of the system [17]. At the same time, how to verify the safety and reliability of these systems in various complex environments has become a key challenge in the development of autonomous driving.

After the system integration is completed, the V&V process becomes the core step in the implementation of the autonomous driving system [1–4, 11, 12]. V&V requires not only simulation testing and bench testing, but also the functional performance and safety of the V&V system in various scenarios on real roads [24]. For example, how to ensure the safety performance of the system through a large number of scenario tests and how to evaluate the redundant design under extreme working conditions, these are the problems that autonomous driving V&V needs to solve.

This chapter mainly introduces the integration, verification and validation (V&V) of the autonomous driving system, including system architecture design, module integration methods, testing and V&V processes, functional and safety evaluation, and performance metric system. System integration not only solves the interface compatibility problem between hardware and software modules, but also ensures that each module works together to achieve the optimal performance of the system.

12.1 System Integration Methods

System integration is the core link in the development of autonomous driving systems [13]. It integrates multiple functional modules such as perception, localization, decision-making, planning, control, and execution to ensure that the system can

Y. Huang and Z. Yang, *Autonomous Driving Handbook*,
https://doi.org/10.1007/978-981-95-3725-9_12

complete autonomous driving tasks efficiently and reliably [5–7, 14]. This section will introduce the concepts, methods, technical frameworks, and challenges of system integration in detail.

The autonomous driving system is a complex distributed system that includes software, hardware, network communications, etc. [15]. System integration refers to the process of assembling, testing, verifying, and validating each independent module (such as sensors, controllers, and decision-making systems) according to the system architecture requirements [9].

The main objectives of system integration include [8, 15, 20] the following:

- Ensure data flow between modules and real-time performance.
- Ensure the overall functionality, robustness, and stability of the system.
- Solve cross-platform and cross-module interface and protocol coordination issues.
- Provide a standardized, modular, and scalable integration framework.

According to the different objects and stages of integration, system integration can be divided into the following categories [18, 19], as shown in Fig. 12.1.

1. Module-level integration: Develop, debug, and verify individual functional modules (such as perception module and decision-making module). The focus of integration lies in that algorithm implementation within the module and the correctness of data input and output.

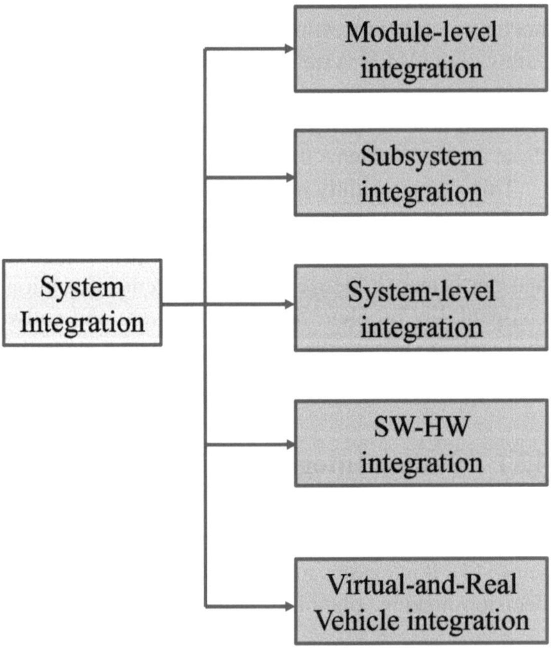

Fig. 12.1 Categories of system integration methods

2. Subsystem integration: Combine modules with similar functions into subsystems for integration, such as sensor fusion systems, path planning and decision-making systems. Ensure data flow and functional coordination between subsystems.
3. System-level integration: Integrate all subsystems into a unified platform to construct a complete autonomous driving system. The focus is on solving problems such as data communication between modules, interface standardization, real-time scheduling, and resource management.
4. SW-HW integration: Deploy software algorithms onto hardware platforms to achieve collaborative work between software and hardware. It involves technologies such as embedded systems, real-time operating systems (RTOS), and hardware acceleration (such as GPU/FPGA).
5. Virtual-and-Real Vehicle integration: Perform virtual integration and testing on a simulation platform to identify potential system problems. Verify the system's performance and reliability in real-world environments through real-vehicle integration.

System integration is generally carried out in two ways [13, 15]: bottom-up and top-down.

1. Bottom-Up Integration: Start from the smallest functional unit, gradually integrate the modules into subsystems, and finally build a complete system.

   - Advantages: Can gradually locate problems and is easy to debug.
   - Disadvantages: The integration cycle is long and it is easy to ignore global optimization.

2. Top-Down Integration: Starting from the overall system architecture, it is gradually decomposed into subsystems and functional modules for integration and verification.

   - Advantages: From a global perspective, system functions can be seen early.
   - Disadvantages: The modules initially may not be completed yet and rely on the simulation environment for testing.

3. Incremental Integration: Divide the system into several incremental functions, integrate some modules at each stage and verify them. It is suitable for iterative development and continuous verification of autonomous driving systems.
4. Continuous Integration and Testing: Adopt the continuous integration/continuous deployment (CI/CD) process to automatically complete code integration, testing and deployment through the tool chain. Use automated testing platforms and simulation environments to discover and resolve integration issues in real time.

The key technologies of system integration include as following [9, 10, 15–17]:

1. The multi-sensor fusion techniques: Integrates data from sensors such as cameras, lidar, millimeter-wave radar, and GPS to provide high-precision environmental perception. Methods such as data layer fusion (low-level fusion), feature layer fusion (mid-level fusion), and decision layer fusion (high-level fusion) are adopted.

2. The sharing techniques of communication and data: Use middleware (such as ROS, DDS) to achieve real-time transmission and sharing of data streams between modules. CAN, Ethernet, and other communication protocols are used to ensure the real time and reliability of data transmission.
3. SW architecture and modular design: Use modular architecture (such as AUTOSAR) to achieve decomposition and integration of system functions. Design standardized interfaces to ensure decoupling and scalability between modules.
4. RTOS and scheduling techniques: The real-time operating system manages system resources to ensure the real-time performance of high-priority tasks. Use multi-threading and task scheduling technology to optimize system performance.
5. HW acceleration and resource optimization: Use hardware accelerators such as GPUs and FPGAs to speed up processing of compute-intensive tasks such as deep learning. Optimize resource allocation and improve overall system performance.
6. Virtual simulation and real-vehicle testing techniques: Use simulation tools (such as CARLA, PreScan, and Apollo simulation platforms) for virtual integration and testing. Through real-vehicle road tests, the actual performance of the system after integration is verified and the collaborative work of the modules is optimized.

The challenges of system integration are [4, 10, 12]:

1. Data communication and interface standardization between heterogeneous modules: Different modules use different communication protocols and data formats, which increases the difficulty of integration.
2. Time synchronization and consistency of multi-sensor data: Sensor data has time delay and noise, which affects the real-time performance and accuracy after system integration.
3. Real-time performance and resource management of the system: How to ensure the real-time response and task scheduling of the system on a hardware platform with limited resources.
4. Software and hardware collaborative optimization: How to efficiently deploy complex algorithms to the hardware platform and achieve stable operation.
5. Complexity of testing and verification: The scenarios of autonomous driving systems are complex, and it is difficult to fully verify the functions and performance of the system after integration.

12.2 System V&V Methods and Processes

System verification and validation are key links to ensure the functional correctness, safety, and performance reliability of the autonomous driving system [16, 21, 24]. Verification focuses on whether the system is correctly implemented according to the design requirements during the development process, while validation emphasizes whether the system ultimately meets user needs and intended use. This section will elaborate on the definition, methods, processes, and key technologies of verification and validation.

Definition of Verification: Verification is the process of evaluating whether a system or its components meet design specifications and functional requirements. Its goal is to ensure that the developed system is consistent with the predetermined design at the technical level. The key issue here is whether the system is "doing the system right." For example, testing whether the perception module can detect obstacles and output correct data as designed.

Definition of Validation: Validation is to evaluate whether the system meets user needs and ensure that the system can operate normally in actual application scenarios. Its goal is to ensure that the overall function of the system meets the real application needs. The key issue is whether the system is doing the right system. For example, in actual road tests, verify whether the system can safely perform autonomous driving.

The difference between verification and validation is

- Verification emphasizes conformance to design requirements and is usually performed during the development phase.
- Validation focuses on actual effects and user needs, and is usually tested and evaluated after system integration.
- The two are complementary processes that jointly ensure the quality and security of the system.

System verification mainly evaluates the technical correctness of the system or module, and the methods include static verification and dynamic verification.

1. Static verification: Verification is performed without running the system, mainly including code review, model analysis, etc. Its methods include code review, i.e., manual or automated code quality checks to identify potential errors; model verification, i.e., logical analysis of system design models to ensure the correctness of system architecture; and static code analysis, i.e., using tools to detect syntax errors, potential vulnerabilities, and consistency issues in the code.
2. Dynamic verification: Verify the correctness of functions and performance by actually running the system. The methods include unit testing, to test the smallest functional unit (such as function or module) in the system; integration testing, to verify whether the interface and data flow after multiple modules are integrated are correct; system testing, to perform comprehensive functional verification of the entire system to ensure that each subsystem works together; and regression testing, to verify whether the original function is affected after the system is upgraded or modified.

The process of system verification is as follows, as shown in Fig. 12.2.

1. Requirement analysis and test case design: Define verification goals and design test cases based on system design specifications and functional requirements.
2. Unit testing: Verify the smallest functional units one by one to ensure that the logic of each module is correct.

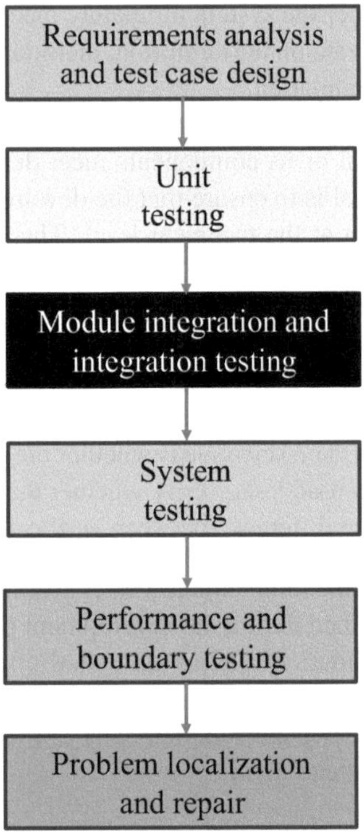

Fig. 12.2 The system verification pipeline

3. Module integration and integration testing: Integrate each module into the subsystem and test the correctness of the interface and data transmission.
4. System testing: Verify the overall system functionality and performance in a simulation platform and laboratory environment.
5. Performance and boundary testing: Perform load testing, pressure testing, and boundary condition testing on the system to evaluate the system's performance under extreme conditions.
6. Problem localization and repair: Find the problem and debug it, and re-verify until the design requirements are met.

System validation mainly evaluates the functionality and safety of the system in actual application scenarios, and the methods used include simulation validation and actual vehicle testing.

1. Scenario-based simulation validation: Validate the functionality and safety of the system in a virtual simulation environment. The methods include: building multiple driving scenarios (such as urban roads, highways, and complex intersections); analyzing and evaluating the behavior of the system in virtual scenarios; and covering normal operating conditions, safety-critical conditions, and failure conditions.

2. Closed-field testing: Validation is carried out in a specially designed test field to validate the performance of the system in a real physical environment. The methods include: testing system functions in specific scenarios, such as obstacle recognition, braking response, lane-keeping, etc.; using test equipment to collect vehicle data and analyze system behavior and performance.
3. Open-road testing: Verify and validate the system on real roads and in actual traffic environments. The methods include: designing test routes to cover different scenarios such as urban roads, highways, and tunnels; testing the system's decision-making ability and performance in dealing with emergencies in a dynamic traffic environment; and focusing on evaluating the system's safety, robustness, and user experience.
4. Corner conditions and safety confirmation: Test the safety performance of the system under corner conditions and boundary scenarios. The methods include: performing fault injection tests on the system to evaluate the system's ability to cope with faults and evaluating the impact of sensor failure and extreme weather (such as heavy rain, snow, and fog) on the system.

The process of system validation is as follows, as shown in Fig. 12.3.

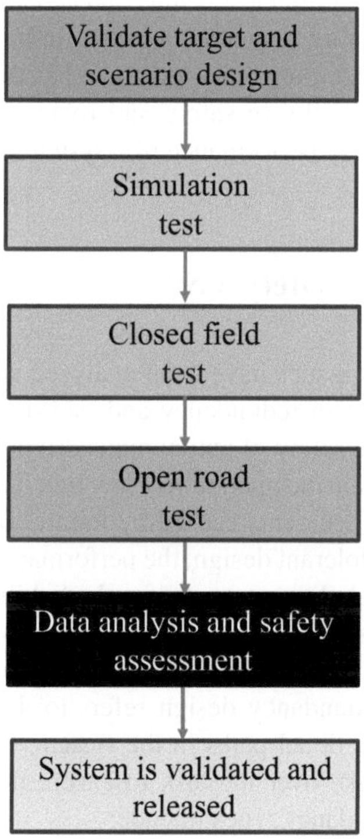

Fig. 12.3 The system validation pipeline

- **Validation of targets and scenario design**: Define the system validation objectives and corresponding test scenarios according to user requirements and intended applications.
- **Simulation testing**: Conduct preliminary validation in a simulation environment to optimize system parameters and functions.
- **Closed-field testing**: Evaluate the system in a controlled environment to identify potential issues.
- **Open-road testing**: Perform long-term, multi-scenario tests and data collection in real-world driving conditions.
- **Data analysis and safety assessment**: Analyze collected test data to evaluate the system's functional integrity, reliability, and safety.
- **System validation and release**: Upon completion of validation tests, prepare a verification and validation report to confirm that the system meets all specified requirements.

Current challenges in verification and validation include:

1. Insufficient scenario complexity and coverage: Real driving scenarios are complex and changeable, making it difficult to cover all edge conditions.
2. Differences between simulation and reality: There may be deviations between simulation test results and actual road performance.
3. Testing resource and cost limitations: Large-scale real-vehicle testing is time-consuming and costly, and has extremely high requirements for system safety.
4. Difficulty in safety and robustness assessment: It is challenging to assess the safety performance of a system under corner conditions and fault states.

12.3 Safety V&V

Safety issues have been analyzed and discussed in Chap. 11. This section focuses on system redundancy and fault-tolerant design, which is an important guarantee for the safety of autonomous driving systems. It aims to ensure that the system can maintain the operation of key functions when some functions or components fail, and avoid accidents caused by failure. By verifying and validating the redundancy and fault-tolerant design, the performance of the system under various fault conditions is evaluated to meet safety and reliability requirements [1–4, 11, 12]. This section will discuss the concepts, verification and validation methods, processes and evaluation metrics of redundancy, and fault-tolerant design.

Redundancy design refers to the introduction of multiple backup components or functional paths in the system. When the main system fails, the backup system can take over the task to ensure the continuity and security of system functions. Redundancy types include:

- Hardware redundancy: such as backup design of sensors, actuators, and controllers.
- Software redundancy: multiple versions of algorithms or codes.

- Time redundancy: verifying the correctness of results by repeating tasks.
- Information redundancy: using checksums and redundant data to ensure information integrity.

Fault-tolerant design means that when a system fails, it uses detection, isolation, recovery, and other strategies to ensure that the system function is not significantly affected and the system can continue to operate. Fault-tolerant mechanisms include:

- Fault detection: Real-time monitoring of system status to detect anomalies or faults.
- Fault isolation: Identify and isolate fault components to prevent the spread of faults.
- Fault recovery: Restore system functions through backup systems or degraded modes.

Redundancy and fault tolerance are closely related. Redundancy design is the basis for realizing fault tolerance, and fault tolerance design is the effective management and utilization of redundant resources. The combination of the two can effectively improve the security, reliability, and robustness of the system.

In order to verify the redundancy and fault-tolerant design of the system, it is usually necessary to evaluate it in simulation environment, laboratory test, and real-vehicle test. The methods used include:

1. Fault injection testing. By artificially introducing faults (such as sensor data loss, controller failure, etc.), the system's fault detection and fault-tolerant recovery capabilities are verified. The methods include: hardware layer fault injection, it physically unplugs or disables hardware components such as sensors and actuators; software layer fault injection, it modifies program logic to simulate incorrect data input or task failure; and network layer fault injection, which simulates network anomalies such as data transmission delay, packet loss, and communication interruption.
2. Simulation verification. Construct fault scenarios in the virtual simulation platform to verify the effectiveness of redundancy and fault-tolerant design. The methods used are: use scenario generation tools to design specific fault conditions (such as single sensor failure, system component failure); verify the system's fault detection capabilities, degraded mode execution, and function recovery time.
3. Boundary conditions and corner working condition testing. Evaluate the redundancy and fault tolerance performance of the system under extreme working conditions. The methods include: simulating extreme weather (such as heavy rain, blizzard), low and high temperatures, complex road scenes, etc.; testing the system's fault handling capabilities under high load and resource constraints.
4. Hardware-in-the-loop (HIL) testing. By connecting the physical hardware to the simulation platform, the system's response capability in a real fault environment is verified. This includes the following methods: using the HIL platform to simulate fault signal input and observe the system's fault-tolerant behavior; verifying the system's backup mechanism and fault recovery time.

In order to validate the redundancy and fault-tolerant design of the system needs to ensure that the system meets the safety and reliability requirements in actual applications, usually following the following process:

1. Requirements Analysis and Design Validation: Clarify the requirements for redundancy and fault-tolerant design, and define metrics such as fault detection time, recovery time, and safe operation mode; evaluate whether the system design meets safety requirements (such as ISO 26262 functional safety standard).
2. Simulation and Fault Scenario Validation: Construct fault scenarios on the simulation platform to validate the system's fault detection, isolation, and recovery mechanisms.
3. Fault Injection and Boundary Testing: Use fault injection tools to simulate hardware and software failures and validate the system's fault tolerance performance under extreme conditions.
4. Closed Site Validation: Introduce faults (such as sensor failure, actuator failure) in a safe and controllable closed environment to evaluate the system's redundancy and fault tolerance capabilities.
5. Open Road Testing: Run for a long time in a real road environment to validate thc stability and reliability of the system's redundancy and fault-tolerant design under complex working conditions.
6. Result Analysis and Validation Report: Analyze the test data and evaluate metric such as system fault recovery time, backup system activation response, and fault coverage; write a redundancy and fault-tolerant design verification and validation report to ensure that the system meets safety and reliability requirements.

12.4 Performance Evaluation and Metrics

System performance evaluation is a key step in measuring the functional integrity, stability, response speed, and reliability of the autonomous driving system. By building a reasonable evaluation method and metric system, the performance of the autonomous driving system can be systematically evaluated to provide data support for system optimization [22, 23]. This section will focus on the concept, method, key metric system, and evaluation process of system performance evaluation.

System performance evaluation is to evaluate the performance of the autonomous driving system in different environments through a series of quantitative and qualitative methods, including function execution, response speed, accuracy, stability, robustness, etc.

The evaluation objectives include:

- Verifying whether the overall functionality of the system meets the design requirements.
- Measuring the stability and security of the system in actual application scenarios.
- Determining the system's corner case performance and boundary conditions.
- Providing data support for system optimization and improvement.

There are several challenges of existing evaluation systems: multi-factor interference in complex scenarios (such as weather, road type, traffic flow); large amounts of data; high difficulty in real-time analysis and evaluation; and the metric system needs to cover multiple dimensions such as function, safety, efficiency, and user experience.

The main methods for evaluating the performance of autonomous driving systems include simulation evaluation, closed-field testing and open-road testing, combined with big data analysis and metric systems to achieve multi-level performance evaluation.

1. Simulation evaluation: Use a virtual platform to simulate real driving scenarios and evaluate the performance of the system in different environments. The main methods are: use simulation tools (such as CARLA, PreScan, LGSVL) to generate a large number of standardized scenarios; evaluate the performance of the system in obstacle avoidance, path planning, acceleration/deceleration, etc. Its advantages are low cost, high efficiency, and coverage of a large number of corner conditions. The disadvantage is that the simulation results may deviate from the real environment.
2. Closed-field test: In a safe and controllable test site, the system performance is evaluated through standardized scenarios. The basic methods are: set up specific driving scenarios (such as obstacle avoidance, lane-keeping, and following the vehicle); monitor the perception, decision-making, and control functions of the system. Its advantages are controllable test conditions and repeatable data acquisition. The limitations are that there are still differences between the test scenarios and the real environment.
3. Open-road test: Evaluate the system performance on real roads and actual traffic environments. The usual operations are: select various environments such as urban roads, highways, and rural roads for long-term testing; record the response performance and safety of the system in complex dynamic environments. The advantage of this method is to verify the adaptability and stability of the system in a real environment. However, the limitations are that the test cycle is long, the cost is high, and there are safety risks.
4. Performance evaluation based on data analysis: Quantitatively evaluate the system performance by collecting and analyzing system operation data. The methods adopted include: using sensor data, log files, driving behavior data, etc. for analysis; evaluating the system's adaptability to complex environments through machine learning algorithms.

System performance evaluation metrics include five categories: functional metrics, efficiency metrics, security metrics, stability metrics, and user experience metrics, as shown in Fig. 12.4.

Functional metrics include: Perception accuracy, i.e., the accuracy of sensor detection of objects, including object recognition rate, false detection rate, missed detection rate, etc.; path planning success rate, i.e., the success rate of the system in generating feasible paths; decision accuracy, i.e., whether the system's decisions in different

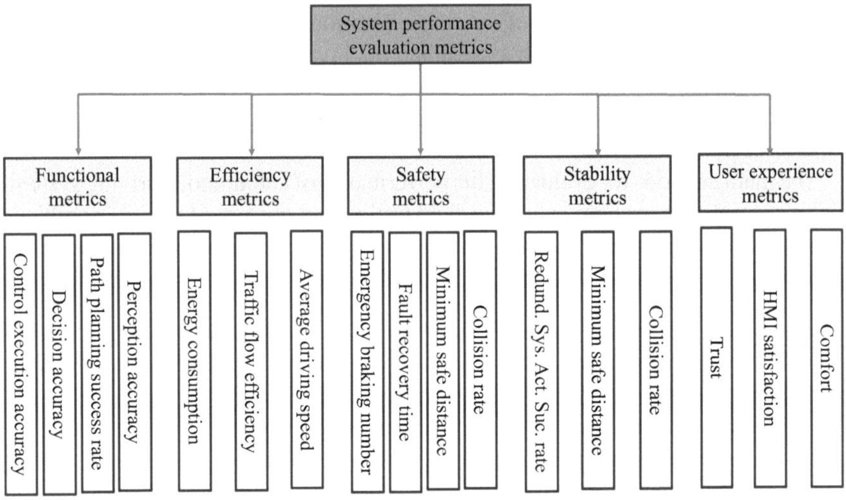

Fig. 12.4 Categories of system performance evaluation metrics

scenarios meet expectations; and control execution accuracy, such as the accuracy of vehicle speed control and direction control.

Efficiency metrics include: Average driving speed, the average speed of the vehicle in different scenarios; traffic flow efficiency, whether the system can effectively deal with congestion or dynamic traffic environment; and energy consumption, the energy consumption performance of the autonomous driving system, such as battery usage efficiency.

Safety metric include: Collision rate, the number or proportion of collisions that occurred during the test; minimum safe distance, the minimum safe distance between the vehicle and other objects; fault recovery time, the time required for the system to restore normal function after a fault occurs; and emergency braking number, the frequency of emergency braking triggered by the system during the test.

Stability metric include: System delay time, the response delay of the system from perception to decision-making and control; system failure rate, the frequency of hardware or software failures in the system; and redundant system activation success rate, the activation of the backup function of the system in a fault state.

User experience metric include: Comfort, whether the vehicle accelerates, decelerates, and turns smoothly to reduce passenger discomfort; human–machine interaction satisfaction, user satisfaction with the ease of use and feedback of the system operation interface; and trust, the level of trust that passengers have in the safety and reliability of the autonomous driving system.

The system performance evaluation process can be listed as follows:

1. Requirements definition and metric determination: Clarify the objectives and scope of system performance evaluation and define key evaluation metrics.

2. Test scenario design: Design standardized scenarios for simulation, closed-field and open-road testing.
3. Data collection and analysis: Record test data through sensors, vehicle systems, log files, etc.
4. Metric calculation and performance evaluation: Perform quantitative analysis based on the collected data according to the defined metric system.
5. Result analysis and performance optimization: Analyze system performance, identify bottleneck problems, and propose optimization and improvement plans.
6. Report writing and confirmation: Prepare system performance evaluation report and provide quantitative data to support system improvement.

12.5 The Formal Method of V&V

Introducing a δ-neighborhood around a discrete scenario and assigning volume to scenario samples is a promising attempt. Mathematical algorithms such as T-wise and Poisson processes can achieve almost full statistical coverage by cleverly selecting samples for finite candidates.

Optimistically assuming that this quantitative coverage-representation problem has been solved and accepted by the entire community, each sampled scenario has a coverage volume. Then, this sample-based full-scenario coverage task will perform a sufficient number of sampling verification tests to ensure that each coverage volume unit is associated with at least one safety test result.

In reality, some parts of the scenario space may not be able to produce safe test results (for example, the reaction distance to obstacles is too small). In this case, extra effort is needed to confirm that these scenarios are indeed unsafe (safety infeasible). In terms of type approval and certification, government agencies need to set an acceptable success threshold to meet public expectations. The simplest threshold is the ratio of the number of tests confirmed to be safe to the minimum number of tests required for ODD scenario testing.

Diversified autonomous driving scenario sampling testing is one of the main methods to strengthen safety control during the development stage. One type of method is based on test sampling, aiming to expand the scenario coverage range to the maximum with minimal effort. Another type of method is based on falsification sampling, aiming to discover extreme security cases that deserve more attention from developers. In terms of achieving the goal of SOTIF to minimize known and unknown unsafe scenarios, sample-based methods have more exploratory power in discovering unknown unsafe scenarios.

According to ISO 21448 SOTIF standard, a scenario is a description of the temporal development between several scenes in a series of scenes. A scene is a snapshot of the environment, including scenery, dynamic elements, self-representation of all participants and observers, and the relationships between these entities. Therefore, ensuring the complete task in the operation domain design (ODD) can be described as conducting safety V&V for all possible scenarios in the ODD.

Sample-based methods do not explicitly possess volume attributes associated with each sample scenario, as each scenario sample is basic and has a small volume. In order to unify the comparison and utilization of the two safety verification methods, the concept of scenario volume must be assigned to sample-based methods.

For simplicity, assume that all N continuous dimensions are orthogonal to each other, and assign an axis-aligned polyhedron volume of an N-dimensional scenario space to a set of scenario samples specified by a group of parameters. In sample-based methods, the safety of simulated scenarios is checked for each scenario sample.

Formal methods usually start with safety specifications before performing any simulations, then convert the specifications into machine-readable language statements, such as Linear Temporal Logic (LTL) or Signal Temporal Logic (STL). Then, the converted specifications can be applied to check the validity of the specifications in simulation/real-world testing, or to convert the specifications into system constraints or other system design features during system synthesis.

Formal methods are a class of techniques that apply mathematical techniques (usually in the form of logical computation) to implement software and engineering design specifications and verification. By definition, formal methods have inherent advantages in safety verification tasks, but the system complexity of continuous dynamic scenarios has always been a major limiting factor for their widespread application in autonomous driving, i.e., difficulty in scaling.

Technically, formal methods can be generalized as a process of implementing or converting abstract specifications into autonomous driving system algorithms/programs, so that the controlled system behavior can meet the stated specifications. (Note: It should be pointed out that the design of autonomous driving algorithms needs to synthesize the specifications into the autonomous driving system architecture, or to perform safety verification on the designed system, with sufficient confidence or conclusive evidence to ensure that it meets the specifications.)

Formal verification mainly implements safety specifications in the autonomous driving system and achieves a certain degree of realism in environmental simulation. Sample-based methods verify scenario changes by adding a large number of scenario samples, due to different safety verification mechanisms. In the safety philosophy of formal methods, safety specifications are either met or violated, and the satisfaction of specifications is synthesized into model-based system design. After the design is completed, the responsibility for implementation is transferred to the online verification of model correctness during operation: as long as the model verification for the autonomous driving system is correct, the system executes according to the comprehensive safety specifications, then the system can be proven to be safe.

During the development phase, it is usually necessary to combine sample-based methods, formal methods, and field operational testing methods to promote prototype design development and discover design errors. Compared with test-based methods, formal methods have the advantage of high reliability of statements, because they have a strict logical basis.

Common formal methods in autonomous driving safety include model checking, reachability analysis, and theorem proving. Model checking originated from software development to ensure that software behavior conforms to design specifications. When safety specifications are represented by axioms and lemmas, theorem proving can be used for safety verification under worst-case assumptions. Reachability analysis holds a special position among the three because of its inherent ability to generate safety statements for dynamic systems, capturing the main features of dynamic driving tasks (DDT).

Real-world road testing or field operational testing (FOT) is the ultimate but also expensive method for autonomous driving verification and confirmation. In a sense, this is the only verification method. However, the disadvantages of FOT are also obvious: it lacks sufficient scenario coverage ability (especially in near-collision and collision scenarios).

The theory of reachability analysis is introduced below. First, a dynamic system is defined as follows:

$$\dot{x} = f(x, u) \tag{12.1}$$

where (x, u) are measurable and continuous state and control variables, and f is a uniformly continuous bounded and Lipschitz continuous function (flow domain).

Given the initial state $X_{\text{init}}(t_0)$ and the time interval $[t_0, t]$, the maximal forward reachable set (maxFRS) is defined as follows:

$$\text{maxFRS:}\{x(t) \in \mathcal{X} \mid \exists u \in \mathcal{U},\ \exists x(t_0) \in \mathcal{X}_{\text{init}}(t_0)\} \tag{12.2}$$

$$x(t) = x(t_0) + \int_{t_0}^{t} f(u, t)dt \tag{12.3}$$

The forward reachable set (FRS) propagates dynamics from the initial time t_0 to all possible reachable states within the future time t.

Conversely, the backward propagation of the backward reachable set (BRS) looks for all possible states from a previous time t that lead to a target state set $\mathcal{X}_{\text{goal}}(t_0)$ at the current time t_0, as defined by the maximal backward reachable set (maximal BRS):

$$\text{maxBRS:}\{x(t) \in \mathcal{X} \mid \exists u \in \mathcal{U},\ \exists x(t_0) \in \mathcal{X}_{\text{goal}}(t_0)\} \tag{12.4}$$

$$x(t_0) = x(t) + \int_{t}^{t_0} f(u, t)dt \tag{12.5}$$

Furthermore, if the selection of states $\forall$ is replaced with $\exists$, the definition of the minimal backward reachable set (minimal BRS) is obtained (other cases are similar):

$$\text{minBRS:}\{x(t) \in \mathcal{X} \mid \forall u \in \mathcal{U},\ \exists x(t_0) \in \mathcal{X}_{\text{goal}}(t_0)\} \tag{12.6}$$

$$x(t_0) = x(t) + \int_{t}^{t_0} f(u, t)dt \tag{12.7}$$

In some cases, the reachability within a defined time range is of more interest. Therefore, the definition is extended to include from the current time to the end of the time range, such as the revised definition of the maximal forward reachable tube (maxFRT) (other cases are similar):

$$\text{maxFRT:}\{x(t) \in \mathcal{X} \mid \exists u \in \mathcal{U},\ \exists s \in [t_0, t_f],\ \exists x(t_0) \in \mathcal{X}_{\text{init}}(t_0)\} \tag{12.8}$$

$$x(s) = x(t_0) + \int_{t_0}^{s} f(u, t)dt \tag{12.9}$$

When interactions between different traffic participants play a key role, the influence of the participants, and sometimes even adversarial entities, must be included in the reachability analysis. In this case, the dynamic system introduces an additional adversarial input d to represent this adversarial control input:

$$\dot{x} = f_a(x, u, d) \tag{12.10}$$

Here, f_a represents the joint or relative dynamics of all participating vehicles. In this case, the influence of adversarial inputs in the system does not depend on the control of the vehicle itself, so conservative modeling must be maintained to provide safety guarantees in the worst-case scenario.

The following further restricts the definition of the maximal forward reachable set under adversarial influence (other cases are similar):

$$\{x(t) \in \mathcal{X} \mid \exists u \in \mathcal{U},\ \forall d \in D,\ \exists x(t_0) \in \mathcal{X}_{\text{init}}(t_0)\} \tag{12.11}$$

$$x(t) = x(t_0) + \int_{t_0}^{t} f_a(u, t, d)dt \tag{12.12}$$

Ideally, for formal methods, if all safety-related specifications are first accurately converted into such machine-readable statements, and if system synthesis fully respects these statements, and the (simulated) environment is perfectly modeled based on the model, then the scenario coverage relative to the simulation layer is 100%.

However, due to the following reasons, practical challenges hinder the 100% ideal conversion of formal safety specifications [21].

- Formal methods depend on viewpoints based on abstractions of the real world, so differences (no matter how small) between abstraction and reality will occur and undermine effective safety guarantees.
- Systems developed in practice often have performance limitations and cannot comply with safety specifications in any case.
- There are practical scalability difficulties when formal safety specifications are converted into safety verification or system synthesis. Formally, the penetration rate of the specification is defined as: the percentage of verifiable scenarios in

the subspace of scenarios stipulated by the formal safety specifications. In other words, the penetration rate of formal methods in real specifications is usually less than 100%.

Please note that through formal methods, the area of safety infeasible scenarios can be determined, and once all scenarios in the ODD (except for safety infeasible scenarios) are verified, no further redesign is needed. Both formal methods and sample-based methods are effective tools to help discover the safety weaknesses of autonomous driving, and autonomous driving developers can choose one or both methods according to their convenience and preference.

Safety verification can be performed by the prediction module during simulation (online), or after simulation (offline), simply by checking the results. For an overview of different verification schemes, a priori verification is performed for white-box control strategies and gray-box control strategies, as shown in the top box of the figure; black-box control strategies can only undergo a posteriori verification, as shown in the bottom box. The difference is that only the final safety result of each episode can be used for a posteriori verification between simulations; as long as environment perception is initiated in the simulation, a priori verification can be performed; when the control strategies of the ego vehicle and other traffic participants are known (white-box control) or partially known (gray-box control), formal methods can utilize this information and narrow down the reachable set, providing more accurate vehicle motion predictions, forming valuable applications such as online safety monitoring and verification. Conversely, if the control strategy is proprietary and completely unknown (black-box control), the safety verification process must involve statistical learning of the black-box controller's behavior after each simulation, which in turn guides how to choose the next scenario to test, better targeting counterexamples.

Below are several types of formal safety verification:

1. Formal safety verification of known autonomous driving strategies. If the autonomous driving strategy is fully understood (usually limited to the ego vehicle), safety verification can be executed very certainly, effectively eliminating the control variable u in the reachability analysis (RA). This usually involves calculating the reachable set of the controlled vehicle and comparing it with the obstacle occupancy set at the relevant time.
2. Formal safety verification of black-box autonomous driving strategies. When the autonomous driving strategy is completely unknown, verification needs to fall back to a more general plan that does not require information about the autonomous driving strategy. Such plans treat autonomous driving and the platform as a whole system and investigate how inputs to the entire system (such as disturbances) lead to undesired outputs (such as safety violations). Methods such as simulated annealing, evolutionary algorithms, Bayesian optimization, or extended ant colony optimization can be used.
3. Formal safety verification of gray-box autonomous driving strategies. If the autonomous driving strategy is partially known, a modified form of reachability analysis can be performed. In the setting of semi-autonomous vehicles, the

behavior of the driver can at most be estimated, and safety verification must be performed with a partially known autonomous driving strategy (referring to human driving strategy). In this setting, the designed specific lane-keeping assist driving system treats the driver's steering strategy as known and immutable, while regarding the driver's acceleration strategy as unknown and variable. This approach allows the reachability analysis to predict the reachable state range of the vehicle within an appropriate time range, and the system can determine whether the driver needs help by comparing the current vehicle state and the backward reachable safety set. When at least the qualitative goal of the autonomous driving strategy or function is known (such as activating intervention, avoiding vehicle collisions due to driver misoperation), reachability analysis can be used to verify whether such autonomous driving strategies or functions are faithful to their qualitative descriptions. For example, it can be used to judge whether the collision avoidance system has missed or abused intervention opportunities.

12.6 Summary

This chapter focuses on the integration and V&V of autonomous driving systems, systematically analyzing the integration and testing, and V&V processes of autonomous driving systems, from system architecture design to V&V methods, and then to safety assessment and performance metric systems.

In the future, with the development of new technologies, such as artificial intelligence, simulation technology, and digital twins, the efficiency of system integration and V&V will be further improved. The establishment of a large-scale scenario library, the application of intelligent testing tools, and the unification of global regulations will accelerate the implementation of autonomous driving systems.

References

1. ISO 26262: Road Vehicles – Functional Safety
2. ISO 21448: Safety of the Intended Functionality (SOTIF)
3. SAE J3016: Levels of Driving Automation
4. UL 4600: Standard for Safety for the Evaluation of Autonomous Products
5. Thrun, S., Montemerlo, M., et al. "Stanley: The Robot that Won the DARPA Grand Challenge". Journal of Field Robotics, 2006.
6. Urmson, C., Anhalt, J., et al. "Autonomous Driving in Urban Environments: Boss and the Urban Challenge". Journal of Field Robotics, 2008.
7. Ziegler, J., et al. "Making Bertha Drive – An Autonomous Journey on a Historic Route". IEEE Intelligent Transportation Systems Magazine, 2014.
8. C Berger and B Rumpe. "Engineering Autonomous Driving Software", arXiv 1409.6579, 2014.
9. Winner, H., Hakuli, S., Lotz, F., & Singer, C. "Handbook of Driver Assistance Systems: Basic Information, Components, and Systems for Active Safety and Comfort". Springer, 2016.

10. Bagloee, S. A., Tavana, M., Asadi, M., & Oliver, T. "Autonomous Vehicles: Challenges, Opportunities, and Future Implications for Transportation Policies". Journal of Modern Transportation, 2016.
11. Kalra, N., & Paddock, S. M. "Driving to Safety: How Many Miles of Driving Would It Take to Demonstrate Autonomous Vehicle Reliability?", RAND Corporation Research, 2016.
12. Koopman, P., & Wagner, M. "Autonomous Vehicle Safety: An Interdisciplinary Challenge". IEEE Intelligent Transportation Systems Magazine, 2016.
13. S Ulbrich, A Reschka, J Rieken, and et al. "Towards a Functional System Architecture for Automated Vehicles", arXiv 1703.08557, 2017.
14. Kato, S., Tokunaga, S., Maruyama, Y., Maeda, S., & Hirabayashi, M. "Autoware on Board: Enabling Autonomous Vehicles with Embedded Systems". ACM/IEEE International Conference on Cyber-Physical Systems (ICCPS), 2018.
15. W Zong, C Zhang, Z Wang, J Zhu, and Q Chen. "Architecture Design and Implementation of an Autonomous Vehicle". IEEE Access, 2018.
16. S Azam, F Munir, A M Sheri, J Kim and M Jeon. "System, Design and Experimental Validation of Autonomous Vehicle in an Unconstrained Environment". MDPI Sensors, 2020.
17. Yurtsever, E., et al. "A Survey of Autonomous Driving: Common Practices and Emerging Technologies". IEEE Access, 2020.
18. M Alcon, H Tabani, J Abella, and F J. Cazorla. "Enabling Unit Testing of Already-Integrated AI Software Systems: The Case of Apollo for Autonomous Driving". EuroMicro Conference on DSD, 2021.
19. R Dona and B Ciuffo. "Virtual Testing of Automated Driving Systems. A Survey on Validation Methods". IEEE Access, 2022.
20. T ZHAO, E YURTSEVER, J A. PAULSON and G RIZZONI, Automated Vehicle Safety Guarantee, Verification and Certification: A Survey, arXiv preprint arXiv:2202.02818, 2022
21. X Zhu, L Wang, C Zhou, and et al. "A survey on deep learning approaches for data integration in autonomous driving system", arXiv 2306.11740, 2023.
22. X Yan, S Feng, D J LeBranc, and et al., "Evaluation of automated driving system safety metrics with logged vehicle trajectory data", arXiv 2401.01501, 2024
23. J Kouril et al., "Performance evaluation of a ROS2 based Automated Driving System", arXiv 2411.11607, 2024
24. S Shetiya, V Vyas, and S Renukuntla. "Verification and Validation of Autonomous Systems". arXiv 2411.13614, 2024.

Chapter 13
Automatic and Autonomous Parking

The product implementation of autonomous driving technology involves application scenarios. Although small, the "sparrow" is complete in every detail. Automatic and autonomous parking is a scenario with practical application value. The previous chapters have analyzed and discussed the internal structure and algorithms of autonomous driving. This chapter will focus on the scenario of automatic parking, and elaborate on the technical modules discussed earlier.

Section 13.1 is a basic introduction to the parking system; Sect. 13.2 discusses how to calibrate the surround-view camera system unique to the parking system; Sect. 13.3 introduces the perception method of fisheye cameras; Sect. 13.4 discusses the method of parking space detection; Sect. 13.5 analyzes some specific motion planning algorithms for parking; Sect. 13.6 analyzes the manifestation of sensor fusion (ultrasound and camera) in automatic parking; Sect. 13.7 analyzes the highest level of autonomous driving method in parking, automatic valet parking; Sect. 13.8 introduces the production and localization technology of semantic maps in parking

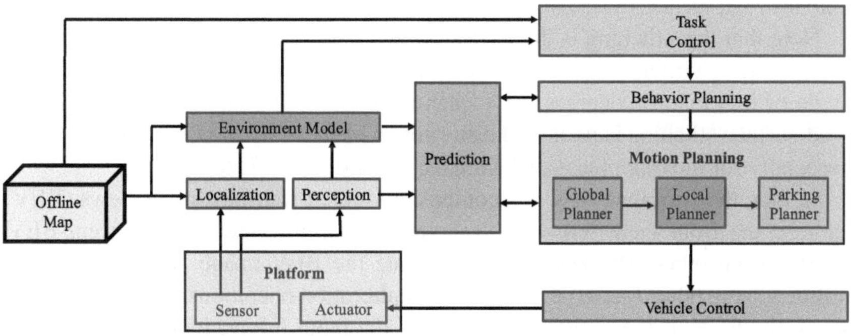

Fig. 13.1 The modular architecture of an automatic parking system

Y. Huang and Z. Yang, *Autonomous Driving Handbook*,
https://doi.org/10.1007/978-981-95-3725-9_13

lots; and Sect. 13.9 introduces the deep learning planning method for automatic parking in parking lots.

13.1 Basic Introduction to the Automatic Parking System

Firstly, the development overview of the automatic parking field can be seen from reference [1], and the modular architecture of an automatic parking system shown in Fig. 13.1 includes perception, localization, mapping, prediction, planning and control. The uniqueness of parking makes the planning module have a small module, the parking planner, at the level of motion planning (as mentioned earlier, planning is divided into task planning, behavior planning, and motion planning) to solve the problem of narrow roads during parking.

In terms of hardware, the requirements for the computing platform and actuator are similar to those of a general autonomous driving system. In terms of sensors, the most basic is ultrasonic radar, but the working distance of this sensor is limited. It can only detect whether parking is possible when the vehicle passes by the parking space, and it cannot determine the exact location of the parking space (especially when there are multiple parking spaces), and it cannot work for inclined parking spaces. Cameras can detect and recognize parking spaces through images, and even create visual parking site maps, which should be upgraded configurations. LiDAR can certainly do the work of ultrasonic radar and cameras, but cost is a major factor. In addition, Vehicle-to-Everything (V2X) will also make parking lot management easier, but this requires the laying of infrastructure, which is not a job that car manufacturers consider unilaterally.

Parking systems based on cameras generally use surround-view camera systems, placing multiple fisheye cameras on the sides and front and rear bumpers of the vehicle. The principle is based on the internal and external parameters of the camera obtained from camera calibration, and through homography, the image is projected onto the road surface (based on IPM), forming a 360-degree surround-view stitching.

Note that this stitching is different from the 360-degree surround-view stitching often used in virtual reality. This stitching is stitched on a cylinder (with a pitch angle of less than 60 degrees) or a sphere centered. Of course, the quality of this road-surface stitching is poor, and distortion is likely to occur at the stitching point, especially for moving objects, which have a "ghosting" effect.

In fact, the surround view is a composite view of four bird's-eye views (BEV), namely, the front view, left view, rear view, and right view. Two adjacent BEVs overlap each other. The key step to generate the BEV image is to construct the offline lookup table $T_{B \to F}$, perform a set of calibration operations, and map the point x_B on the BEV image to the position x_F on the input fisheye image. To determine $T_{B \to F}$, it is necessary to determine the transformation matrix $P_{B \to W}$ from the BEV coordinate system to the world coordinate system, the transformation matrix $P_{W \to U}$ from the world coordinate system to the undistorted input image coordinate system, and the lookup table $T_{U \to F}$ that maps the points on the undistorted input image to the original input fisheye image position. Figure 4 in [2] is an example of stitching of a surround-view fisheye camera, and four surround-view cameras (front, rear, left, and right) are combined into an IPM image.

In the review paper on the perception of autonomous driving fisheye cameras (see reference [18]), the following challenges are pointed out:

1. The most prominent issue is strong radial distortion, which cannot be fully corrected and introduces drawbacks such as reduced field of view and peripheral resampling artifacts. Due to spatial domain deformation, the appearance of objects—especially nearby targets—changes significantly, thereby increasing the learning complexity of convolutional neural networks. Since these networks rely on translation invariance as an inductive bias, the distortion forces the model to learn a wide range of deformed appearances, further complicating training.
2. Another challenge arises in object detection, where bounding boxes provide a poor fit for fisheye-distorted objects. More complex representations are needed beyond rectangular boxes, such as curved frames that explicitly account for the known radial distortion of fisheye cameras.

The primary motivation for using fisheye cameras in surround-view perception systems is to achieve full 360-degree coverage of the near-field area around the vehicle. This is typically accomplished with four fisheye cameras, each providing a horizontal field of view (hFOV) of approximately 190 degrees and a vertical field of view (vFOV) of about 150 degrees. While fisheye cameras offer extensive angular coverage, their angular resolution is relatively low, limiting their ability to detect small objects at greater distances. As a result, surround-view perception systems based on fisheye cameras are primarily employed as near-field sensing solutions.

In contrast, a typical far-field front-facing camera has a horizontal field of view (hFOV) of about 130 degrees and a vertical field of view (vFOV) of about 60 degrees. Although this provides a much smaller angular volume, the higher angular resolution enables the detection of distant targets. Achieving complete 360-degree coverage requires four wide-hFOV fisheye cameras. Their large vertical field of view also allows the capture of regions close to the vehicle, such as traffic signal lights and elevated objects, when parking at intersections.

Most of the current work in surround-view cameras is to independently process each of the four cameras and perform perception algorithms. Joint modeling of four surround-view cameras may be more ideal. First, it helps to detect visible large vehicles (such as transport trucks) through two or three cameras (front, left, and rear); secondly, it eliminates the post-processing of re-identification and single detection of targets seen in multiple cameras, thereby forming a unified output, such as a lane model. Multi-camera models will more effectively gather information and produce more optimized outputs.

13.2 Calibration of Surround-View Vision System

The calibration of the surround-view vision system is similar to that of a general camera. Firstly, a chessboard grid model is used to calibrate the four fisheye cameras, obtaining their intrinsic and extrinsic parameters; secondly, a method based on the polynomial distortion model is used to correct the distortion of each camera; and finally, the BEV image is obtained from the inverse perspective mapping of the image coordinates to the world coordinates. In addition, the nonlinear LM (Levenberg–Marquardt) optimization algorithm can be used to optimize the camera parameters.

The distortion coefficient of the fisheye camera can be estimated by Zhang Zhengyou's calibration method, so the lookup table $T_{U \to F}$ from the undistorted input image to the original input fisheye image can be obtained. The transformation matrix $P_{B \to W}$ from the BEV coordinate system to the world coordinate system is a similarity transformation. If the size of the BEV image (measured in pixels) and the corresponding physical visible range (measured in millimeters) are predetermined, this matrix can be directly determined. As shown in Fig. 4 in [11], assuming the size of the BEV image is $M \times N$ and the height of the corresponding physical area is H_{mm}, it can be easily verified that $P_{B \to W}$ can be represented as follows:

$$P_{B \to W} = \begin{bmatrix} \frac{H}{M} & 0 & -\frac{HN}{2M} \\ 0 & -\frac{H}{M} & \frac{H}{2} \\ 0 & 0 & 1 \end{bmatrix} \tag{13.1}$$

To estimate the homography matrix $P_{W \to U}$ from the world coordinate system to the undistorted input image coordinate system, use a calibration field with known feature point world coordinates in advance. Capture an image of a calibration field, then use the estimated lookup table $T_{U \to F}$ for distortion correction. Then, on the undistorted image, you can manually select N_F feature points (N_F cannot be less than 4, usually set between 10 and 20).

As shown in Fig. 5(a) in [11], it is the original fisheye image of the calibration field, and its undistorted version is shown in Fig. 5(b) in [11]. In Fig. 5(b) in [11], manually select N_F feature points, marked with circles. For the selected feature points, their coordinates x_i^W in the world coordinate system and x_i^U in the undistorted image coordinate system are known.

For each selected feature point i, its coordinates in the world coordinate system are x_W^i, and its coordinates in the undistorted image coordinate system are x_U^i. The relationship between x_W^i and x_U^i is $x_U^i = P_{W \to U} x_W^i$. Therefore, $P_{W \to U}$ can be obtained by solving the least squares problem based on the feature correspondence $\{x_W^i, x_U^i\}_{i=1}^{N_F}$.

When the matrices $P_{B \to W}$, $P_{W \to U}$ and the lookup table $T_{U \to F}$ are ready, the final lookup table $T_{B \to F}$ can be determined. The four fisheye cameras can have four mapping tables $\{T_{B \to F}^i\}_{i=1}^4$, each responsible for generating a BEV. To synthesize the surround view, it is finally necessary to determine the stitching line between each pair of adjacent BEV images. Through four stitching lines, for a point on the 360-degree surround view, you can know which BEV it should come from, and accordingly, which mapping table and associated input fisheye image to use to determine its pixel value.

13.3 Perception System of Fisheye Camera

Before introducing parking spot detection, let's first introduce the overall operation of the fisheye camera perception system through some references.

Reference [19] proposes a multi-task visual perception network OmniDet based on unrectified fisheye images, including the six main tasks required for autonomous driving systems: depth estimation, visual odometry, semantic segmentation, motion segmentation, object detection, and lens contamination detection. It proposes an adaptive mechanism based on camera geometry for encoding the fisheye distortion model during training and inference. Given that target bounding boxes cannot represent deformed fisheye images well, object detection needs to be extended to use a polygon with non-uniform sampling vertices.

Firstly, FisheyeDistanceNet establishes a self-supervised monocular motion recovery structure framework for distance and pose estimation, as shown in Fig. 2 of [22]. It incorporates a polynomial projection model function for view synthesis, and the total loss consists of reconstruction matching items and regularization items. In addition, it uses cross-sequence distance consistency loss and scale recovery techniques; it also considers feature metric loss, using a self-attention autoencoder to learn features; the main purpose of these losses is to prevent the training target from falling into multiple local minima in the homogeneous region, because the homogeneous region of fisheye images is much larger than that of regular camera (rectilinear) images; it is essentially a loss function that penalizes small slopes and emphasizes low-texture areas with image gradients; it uses first-order derivatives to constrain the self-supervised loss landscape, forming an appropriate convergence basin, thereby regularizing the target features.

Due to severe radial distortion, standard bounding box representations are not very successful in fisheye cameras, especially in the peripheral areas of the camera image. For this reason, different output representations of fisheye images are explored, including oriented bounding boxes, curved boxes, ellipses, and polygons. A FisheyeYOLO framework is proposed as shown in Fig. 6 of [23]. Here, the baseline bounding box model is the same as YOLOv3, only the Darknet53 encoder is replaced with a ResNet18 encoder; similar to YOLOv3, object detection is performed at multiple scales. For each grid in each ratio, infer the target width, height, center coordinates, and category. Finally, use non-maximum suppression (NMS) to filter out low-confidence detection results. Use standard category cross-entropy and binary entropy loss, instead of using L2 loss for category and target classification.

Additionally, the direction is discretized into 18 bins, each bin representing a range of 10° with a tolerance of ±5°. To further improve prediction, an IoU loss function is designed to guide the model to minimize the area difference between the prediction box and the ground truth. Calculate the area of the prediction and the ground truth, and apply regression loss to these values. This loss maximizes the overlap area between the prediction and the ground truth, improving the overall results.

Polygons are a universal representation of arbitrary shapes, however, their annotation cost is more expensive than bounding boxes. A PolyYOLO model is proposed, whose target contour can be evenly sampled within a 360-degree range, divided into N equal polygon vertices, each vertex represented by the radial distance from the target centroid in PolyYOLO. Uniform sampling cannot effectively represent high curvature changes in the target contour of fisheye images. Therefore, adaptive sampling based on local contour curvature is used, i.e., vertices are distributed unevenly to best represent the target contour. PolyYOLO uses fewer convolution filters in each layer of the feature extractor part and expands them through SE (squeeze-and-excitation) blocks. The heavy neck net is replaced by a lightweight block with a hypercube, and upsampling is done in a stepwise manner. A single head is used for output instead of three heads, with higher resolution. In summary, PolyYOLO has 40% fewer parameters than YOLOv3, while achieving more accurate predictions.

Motion segmentation uses two frames and outputs a binary dynamic or static mask. During the training process, the network predicts posterior probabilities, using Lovasz Softmax loss and Focal loss instead of cross-entropy loss for supervised optimization, dealing with class imbalance. The posterior probabilities are applied to a per-pixel argmax operation to obtain the final segmentation mask.

The work of SoilingNet is as follows. First, a dataset is created specifically for the development of soiling detection models. In addition, GAN-based data augmentation techniques are applied. In the multi-task network, other tasks and the soiling decoder share the encoder. It includes a simplified ResNet10-like encoder and two decoders. The soiling decoder serves as the third task, with its output being the gridded soiling category output. The number of grids is set during training, one is of size 64×64, and the other outputs at image resolution. The losses of the three models are converted into one loss by weighted averaging, and the weights are optimized by grid search through hyperparameter tuning.

Encoders are designed using vector attention-based pairwise and block self-attention encoders. These networks effectively adjust weights in the spatial dimension and channels. The Siamese network approach is applied to the motion prediction network, where the source frame and target frame features are connected and passed to the super-resolution motion decoder. Since the weights are shared in the Siamese encoder, the encoder of the previous frame can be saved and reused instead of recalculating.

In addition, semantic features are used to guide the detection decoder, while the encoder block is shared across all tasks. The overall framework is composed of multiple processing blocks that handle distance estimation and semantic segmentation, motion segmentation, fisheye object detection based on polygonal representations, and soiling segmentation, all formulated as asynchronous tasks within a self-supervised learning setting. The predicted distance maps are post-processed in

3D space to obtain surround-view geometric information. Furthermore, camera tensors enable OmniDet to generate distance maps from multiple camera viewpoints, ensuring that the network treats cameras independently of one another.

Reference [20] provides a near-field distance estimation model SVDistNet for a vehicle-mounted fisheye camera.

For a given vehicle series, the internal structure of the base camera is not exactly the same. Due to differences in the manufacturing process, each camera instance needs to be calibrated individually. Even after deployment, calibration can change due to rising environmental temperatures. Therefore, a calibration adaptive mechanism is needed in the model. There are four camera instances around the vehicle, each with different internal parameters, forming a surround-view system.

All camera geometric features are converted into a **camera geometry** tensor and used as input to the CNN model during training and inference. From the perspective of distance estimation, the camera's internal parameters are the main adaptive part required by the model. However, the concept of the camera geometry tensor is universal, and here it is extended to include tasks involving camera external parameters and camera motion (visual odometry).

A single model suitable for all cameras has the following advantages: ① lower memory occupancy and better data transfer rate, thus improving efficiency; ② a single neural network model is easier to maintain than multiple models.

Based on the previous FisheyeDistanceNet, SVDistNet is a framework for single-image distance and pose estimation for multiple cameras. The total loss of this distance and pose estimation network consists of image reconstruction loss and regularization loss. The image reconstruction loss minimizes the difference between the reconstructed image and the target frame and enhances edge perception smoothness. In addition, it also utilizes cross-sequence distance consistency loss and scale recovery techniques.

A specific SVDistNet architecture consists of a self-attention encoder based on shared vectors and a task-specific decoder. The encoder is a self-attention network with pair and tile variants, while the decoder uses pixel-adaptive convolution, both of which are supplemented by camera geometry convolution.

The camera geometry adaptive mechanism is the basis of the SVDistNet training process, as the four different cameras installed in the car have different intrinsic parameters and viewpoints. The trained distance and pose estimation network needs to be generalized when deployed on different vehicles, as the multiple viewpoints and intrinsic parameters have changed. For this reason, the camera geometry tensor is included in the mapping from RGB features to 3D information for distance estimation and semantic segmentation. In addition, this tensor is added to the pose encoder.

The tensor can be obtained in three steps. First, the pixel coordinates and the angle of incidence of the image are pre-calculated for more efficient training; then, to incorporate the knowledge encoded in these camera calibration parameters, the normalized coordinates of each pixel are used as additional channels and connected to generate the tensor; finally, these features are passed along with the encoder features to the Significantly Guided Attention Network (SAN) pair and tile model. Therefore, there are six additional channels supplementing the existing decoder channel inputs. Theoretically, this method can be applied to any given fisheye projection model.

In order to better guide distance features, pixel-adaptive convolution is adjusted, distilling the feature knowledge extracted from the segmentation branch into the distance decoder. This follows the intuition that distance estimation can benefit from the specific location knowledge encoded in semantic features, which in turn breaks the spatial invariance of convolution.

Reference [21], a work from the autonomous parking company Zongmu Technology, decomposes the 3D object detection task into multiple subtasks, including vehicle contact point detection, type classification, re-identification, and unit assembly. To address the challenge of acquiring large-scale, high-quality 3D ground-truth data, the authors propose the concept of a multidimensional vector, which integrates information from different dimensions and stages of processing, rather than relying on conventional BEV or 8-point cuboid representations.

In order to maximize the advantages of each algorithm component, a robust mechanism is established, combining geometric constraints to achieve stable acquisition of 3D information under different circumstances; then using a multidimensional vector to improve prediction results, this vector describes the target vehicle through the center point of the target vehicle, heading angle, type, etc.; the system obtains the original image from four surrounding fisheye cameras as input, and only uses 2D marked ground-truth data (including some heading angle ground-truth data) and the required calibration to output the 3D position of the target vehicle.

For the overall architecture of the fisheye camera perception method, first, the four fisheye images are input into the contact point detection module, vehicle type recognition module, and heading angle regression module to obtain the attributes required for the multidimensional vector, including vehicle type, size (l, w, h), front overhang (f_o), rear overhang (r_o), heading angle (φ), and contact point coordinates; then, these attributes are merged and unified by the re-identification (ReID) module to generate a unique identification number for each target vehicle; subsequently, the intermediate results are integrated to generate a vector for the BEV description of the target vehicle, and the final BEV vector diagram is intuitively interpreted.

Using the pixel coordinates of the contact points, their physical coordinates in the real-world coordinate system can be calculated through the fisheye IPM algorithm, and they can be transformed to the distortion plane through the intrinsic parameter matrix. By referring to the distortion removal table, they are transformed to the distortion removal plane. After introducing the re-calibrated intrinsic parameters and pre-calibration table, the coordinates in the camera coordinate system and the target depth value are obtained. Through the external parameter matrix, the coordinates of the contact points in the local coordinate system (self-car coordinate system) are obtained. In summary, some useful vehicle detection information is prepared in this step, including 2D bounding boxes, pixel coordinates, and physical coordinates of the contact points.

In the fisheye surround-view system, the target vehicle can be observed from different cameras and described by one or more vectors. For the convenience of subsequent tasks, it is necessary to describe the target vehicle with a vector with a unique identification number. Therefore, the original vectors from the three branches are merged into a vector with a unique identification number. This process is called target re-identification (ReID), which consists of three different stages:

- The first stage is multi-branch vector fusion. The bounding box information of the target vehicle is contained in the vector of each branch. If two vectors in different branches have the same bounding box, one is appended to the other, and duplicate elements are removed. In this case, the vectors from the vehicle type classification and heading angle regression branches are appended to the contact point detection vector branch.
- The second stage is target vehicle ID generation. Since the same target may be observed in different channels and described by different vectors, the goal of this stage is to assign the same ID to these vectors. This stage is mainly implemented through some manual rules. For example, the same ID is assigned to vectors at physical coordinates less than 50 cm from the ground contact point.
- The third stage is BEV target ID generation, the goal of which is to merge all vectors describing the same target vehicle and assign them a unique identifier. This stage considers two situations: one is the fusion between channels, which is done in a complementary or weighted manner; the other is the fusion between categories, which is done in a complementary manner. Finally, a unique vector representation containing all the information of the BEV vector map is obtained.

13.4 Parking Spot Detection

The following discusses how to check parking slots in IPM images. Common parking slot patterns are vertical, diagonal, and parallel, with two user modes: disabled and non-disabled.

Parking spot detection recognition is essentially a typical image recognition problem. During training, positive and negative samples are input, features are extracted, and an AdaBoost plus decision tree structure is used, finally obtaining a binary classifier. During testing, search from the panoramic view, extract features from each sliding window, and determine whether it is a parking spot in the classifier.

For vehicles with depth sensors, the depth information from lidar or stereo vision can be used in parking spot detection recognition where there are obvious high and low road signs. HM refers to the height map, and the inverse perspective image (IPI) is the result of the image being projected onto the road surface. The HM is obtained based on motion stereo vision, and then spatial segmentation is performed. At the same time, the probability map of road markings is inferred together with salient line extraction, and finally the probability map and segmentation results enter the recognition module based on the Bayesian classifier to judge the parking spot. However, this system also uses a wheel speed sensor to obtain the vehicle's own motion parameters, which are input into the HM estimation module and the probability map inference module.

In indoor/underground parking lots, sometimes pillar information can also be used to detect parking spots for ultrasonic waves.

13.5 Motion Planning for Parking

Once the parking spot is determined, the vehicle can use motion planning to complete the parking action by map localization (autonomous valet) or the driver (automatic) driving to its vicinity.

The EU project V-Charge divides the parking planning into three stages: stage one, connecting an arc between two straight line segments; stage two, connecting two turning points with three circles between two straight line segments; and stage three, mixed planning, using a modified grid search method, can connect any turning points.

The traditional single circular trajectory-based path planning method can be divided into three parts, namely, the first part is a straight line, the second part is a circular arc, and the third part is another straight line. However, this method is always limited by the size of the parking spot, obstacles, parking space, etc., which will cause the automatic parking method to fail. Considering these shortcomings, many improvements have been made to the traditional double circular trajectory-based path planning method. However, this method does not consider the initial position of the vehicle relative to the target parking spot, resulting in a parking trajectory that is not easy to control. In this case, an improved double circular trajectory-based path planning method is proposed in reference [2].

Reference [4] proposes a parking planning framework based on an improved Rapidly-exploring Random Tree (RRT) method. Imagine a parallel parking double circle trajectory planning. A, B, C, D represent the cars parked next to it, points a, b, c, d represent the autonomous car, its width and length are W and L, the overhang lengths at the rear and front of the car are L_{ro} and L_{fo}, the wheelbase is L_{wb}; during the parking process, $R_{\min}$ is the minimum turning radius of the vehicle, l_S is the safety distance between the parking space boundary and the vehicle, the calculation methods of the minimum length and width of the parking space, L_p and W_p, are as follows:

$$L_p = y_B = W + 2l_S \tag{13.2}$$

$$W_p = x_B = \min(x_{B1}, x_{B2}) \tag{13.3}$$

$$x_{B1} = L_{ro} + l_S + \sqrt{(2R_{\min} - l_S) \cdot (W + l_S) + (L - L_{ro})^2} \tag{13.4}$$

$$x_{B2} = L_{ro} + l_S + \sqrt{2}R_{\min} - \sqrt{[(2 - \sqrt{2})R_{\min} - (W + l_S)][\sqrt{2}R_{\min} + l_S]} \tag{13.5}$$

S and E are the starting and ending points of the rear axle center of the autonomous car during the parking process, and their calculation formulas are as follows:

$$x_S = L_{ro} + l_S \tag{13.6}$$

$$y_S = \frac{W_p}{2} \tag{13.7}$$

$$x_E = \sqrt{2}R_{\min} + x_S \tag{13.8}$$

$$y_E = W_p + y_S \tag{13.9}$$

Similarly, for a single circle trajectory planning for vertical parking. Let R_{rr} be the radius of the vertical parking trajectory, L_z be the direct parking distance, then the calculation formulas for the starting point and the ending point are as follows:

$$x_S = \frac{W_p}{2} \tag{13.10}$$

$$y_S = L_{ro} + l_S \tag{13.11}$$

$$x_E = R_{\min} + x_S \tag{13.12}$$

$$y_E = R_{\min} + L_z + y_S \tag{13.13}$$

$$L_z = L_p - l_S - L_{ro} - \frac{\sqrt{2}R_{rr}}{2} \tag{13.14}$$

13.6 Sensor Fusion in Parking

Reference [7] integrates the Around View Monitor (AVM) system and ultrasonic-based automatic parking system sensors in the parking space detection and tracking system. This system consists of three stages: parking space detection, parking space occupancy classification, and parking space tracking. In a sensor fusion scheme for tracking parking spaces, the odometer provides the vehicle's motion information, which is used to predict the next moment position of the parking space. Finally, the parking space position estimation is achieved through Chamfer matching of the image.

The information provided by the AVM image is independent of the previous position, but this information may decrease due to occlusion. In addition, the information provided by the odometer is independent of occlusion, but there is cumulative error. Therefore, by using the advantages of these two sensors, parking space markers can be tracked accurately and stably. The odometer uses wheel speed sensors and steering wheel angle sensors for estimation. In most ultrasonic sensor-based automatic parking systems, the odometer is used to update the vehicle's position. When the vehicle enters the parking space, the image features presented on the ground are less, which makes the odometer method based on image calibration unstable. Here, the odometer is used to predict the current position of the previously estimated parking space. This prediction result has a dual purpose: one is to limit the ROI and the other is to achieve robustness to occlusion. For the first purpose, the ROI is determined based on the current parking position predicted by the odometer, not based on the previous parking position. For the second purpose, to achieve robustness to occlusion, the edges based on the AVM image and the edges based on the odometer are fused here. In the predicted parking position, the located template edge generates the edge based on the odometer, and the ROI edge of the detected AVM image is used to generate the edge based on the AVM. Then, the directional Chamfer matching (DCM) score of the AVM image and the DCM score of the image based on the odometer are calculated. Finally, the weighted sum of the two DCM scores obtains the fused DCM score.

Reference [10] proposes an automatic parking algorithm based on a fuzzy logic controller, with the vehicle posture as input and the steering wheel rotation rate as output. It uses visual sensors and ultrasonic sensors to locate the vehicle.

For example, the image detects the marking points of the parking space, while the ultrasonic detects the empty space, which is suitable for the home garage scene; the image part first extracts the parking space marking candidates, then transforms from the image coordinate system to the world coordinate system, and the extracted parking space markings are sent to the line segment detection module. The ultrasonic part will filter the ultrasonic signal to estimate the parking space, and the result is also sent to the line segment detection module; the final output is an estimate of the vehicle's posture, which enters the fuzzy controller.

If there are other vehicles or garage walls, the vehicle's status can be easily determined through ultrasonic sensors. Because ultrasonic sensor data may contain errors, it needs to be integrated. If the change in the value of the ultrasonic sensor is greater than a certain threshold, the average value of two adjacent sensors in the ultrasonic sensor array can be used to handle this noise. However, if the angle between the vehicle's direction of travel and the empty parking space is large, it is impossible to estimate the vehicle's status. In this case, the system estimates the vehicle's status based on the parking space markings in the image. After extracting the edge and grayscale to get the parking space marking candidates, the extracted parking space markings are converted to the world coordinate system. Then the vehicle's status is estimated by extracting feature points.

13.7 Automatic Valet Parking

Automatic valet parking enables a vehicle to independently search for and navigate to a parking space within a parking lot, eliminating the need for driver intervention and realizing the concept of "one-key parking." This functionality relies on the use of a map, which is primarily a semantic map containing structural and navigational elements such as lane markings, intersections, and pillars (especially in underground parking facilities). The localization method depends on the choice of sensors: for example, vision-based approaches rely on cameras to detect visual features or semantic landmarks; lidar-based methods use 3D point clouds for positioning; and vehicle-to-infrastructure (V2I) approaches leverage connected infrastructure for localization support.

Behavior planning in autonomous parking is more diverse, there are 11 behavior examples: waiting, following the lane, approaching the car, following the car, stopping, emergency braking, preparing to change lanes, changing lanes, turning, parking (entering the parking space), and canceling parking.

The EU project V-Charge implements autonomous parking, which includes a surround-view system composed of a fisheye monocular camera, two stereo vision systems at the front and rear, and ultrasonic radar.

The V-Charge project is basically the development of an autonomous driving system for parking lots. First, motion estimation is performed on all collected images. Since the multi-camera setup of the parking system will adopt a non-perspective camera with a panoramic view (the rays of each camera will not only intersect at a single projection center), it is impossible to use epipolar geometry for motion estimation. Instead, the multi-camera system is modeled as a generalized camera, where all rays are defined relative to a common coordinate system, which brings generalized epipolar geometry for motion estimation. Here, the fact that the multi-camera system is rigidly installed on a car is utilized, so the motion is limited, greatly reducing the computational complexity. After estimating the posture of all cameras, 3D map points are obtained using the triangulation method of feature correspondence. Each 3D point is associated with the feature descriptors of all images observing this point, which can be used for visual localization.

In order to obtain a globally consistent map, a visual location recognizer based on the bag-of-word method can recognize closed loops, so that the 3D map can be optimized. The V-Charge project combines a set of precise sensors with a limited FoV or distance range (stereo vision and ultrasound) with the motion recovery structure of 360-degree panoramic images, providing good metric performance in estimating the space between obstacles and the range of parking spaces. At the same time, the sensor setup completely covers the environment around the vehicle, achieving safe navigation in the parking lot.

Based on the metric map, a semantic layer can be added to further enhance it. This semantic layer includes three different parts to support the planning module, namely: ① road map, detailing the location of lanes, road directions, and intersections; ② parking space location; and ③ speed distribution of vehicles passing through the parking lot.

In order to exploit the potential of the monocular surround-view system, V-Charge has developed a system capable of detecting and tracking pedestrians and vehicles around the car. A traditional Soft-Cascade+ACF classifier is run on each camera image to detect vehicles and pedestrians. Here, a cylindrical camera model is used to minimize distortion outside the image center while maintaining a large FoV. For the multi-camera tracking problem, a correlation algorithm between cameras is first applied, and then the UKF is used to track obstacles.

The overlapping FoV between cameras can improve detection performance, especially for crossing or temporarily occluded targets. As shown in Fig. 6 of [24], in the panoramic tracking of dynamic targets, the available multi-viewpoints allow overcoming occlusions (the distant pedestrian is only visible in the left image).

To achieve "natural" parking maneuvers, that is, maneuvers that contain only the minimum number of cusps and occupy very little space, V-Charge uses a three-level planning cascade of increasing complexity. The first two stages are composed of deterministic methods, as humans tend to park in similar behaviors (i.e., single moves or two direction changes). The third stage uses a hybrid A* planner, which provides the highest flexibility. The parking planner goes through these three stages until it finds a collision-free maneuver.

13.8 Semantic Maps and Localization in Parking Lots

In parking scenarios, vehicles need to navigate in narrow, crowded parking lots where GPS is not available. Accurate localization capability is crucial. Here we introduce a work that builds maps and locates based on semantic features in parking lots.

In parking lots, common semantic features include directional signs, parking lines, speed bumps, etc. Compared to traditional features, these semantic features have long-term stability and robustness against changes in viewpoint and illumination.

The parking lot map construction and localization system uses four fisheye cameras for perception. With the assistance of an IMU (inertial measurement unit) and a wheel speedometer, the system generates a global visual semantic map that can be used for centimeter-level vehicle localization.

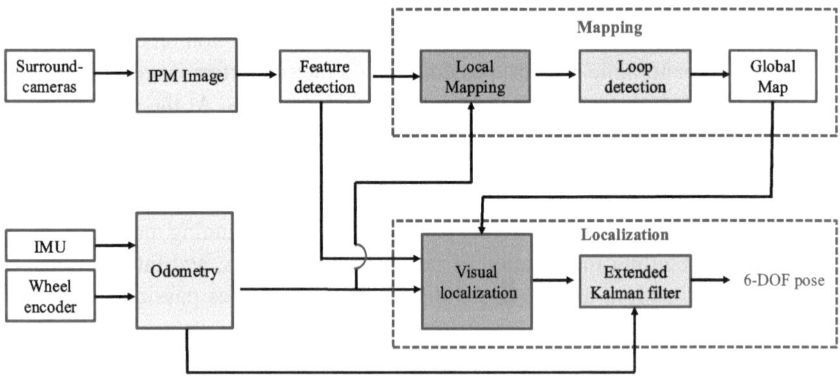

Fig. 13.2 The whole system pipeline of AVP-SLAM

Next, we introduce a work that builds maps and locates based on semantic features in parking lots, called AVP-SLAM. For a complete pipeline of the AVP-SLAM system shown in Fig. 13.2, it starts with four omnidirectional body cameras, an IMU, and a wheel speedometer. The four images from the surround-view cameras are projected into BEV and combined into an omnidirectional image. A neural network detects semantic features, including lanes, parking lines, directional signs, and speed bumps. Based on the odometry method, semantic features are projected into global coordinates. Due to long-term drift of the odometry, loop closure detection is needed through local map matching to reduce cumulative error. During the mapping process, semantic features are extracted from the IPM image. The semantic features are matched with the previously built map to locate the vehicle. Finally, an Extended Kalman Filter (EKF) fuses the visual localization results with the odometry to ensure the system outputs a smooth trajectory.

A modified UNet is used to segment image pixels into different categories. The network is trained on parking lot BEV images, segmenting pixels into lanes, parking lines, guide signs, speed bumps, free space, obstacles, and walls. Among these categories, parking lines, directional signs, and speed bumps are unique, stable features used for localization. Parking lines are also used for parking spot detection. Free space and obstacles are used for planning.

Due to the long drift time of the odometry, loop detection is run to eliminate drift. For the latest local map, it is compared with other surrounding local maps. Two local maps are matched using the ICP method. If the two local maps match successfully, the relative pose between the two local maps is obtained. This relative pose will be used for global pose graph optimization to correct drift.

After loop detection is completed, global pose graph optimization is performed to eliminate cumulative drift and maintain the consistency of the entire map. In this pose graph, nodes are the poses of each local map. It also includes two types of edges: one is the odometry edge, which constrains two consecutive local maps through odometry measurements and the other is the loop closure edge, which constrains the local maps of the loop closure.

The optimization of the pose graph uses the Gauss–Newton method. After global pose graph optimization, the updated poses overlay the local maps together. This generates a globally consistent map.

Based on this semantic map, when the vehicle comes to this parking lot again, it can be located. Similar to the mapping process, the surround-view images are combined into a BEV image. Semantic features are detected on the image and transformed into vehicle coordinates. Then, the current pose of the vehicle is estimated by matching the current feature points with the map.

It should be noted that a good initial value is crucial for the ICP method. Initially, there are two initialization strategies: one strategy is to mark the entrance of the parking lot on the map, so the vehicle is directly initialized at the entrance of the parking lot; the second strategy is to use GPS as the initial pose before entering the underground parking lot. After locating the vehicle on the map, GPS is no longer used. After initialization, the prediction of the odometry is used as the initial value.

There are some areas in the parking lot where the texture is extremely sparse. To overcome this problem, an EKF framework is adopted, which fuses the odometer with the visual localization results. In this filter, the odometer is used for prediction, and the visual localization results are used for updates. This filter not only improves the robustness of the system but also smooths the estimated trajectory.

13.9 Deep Learning Planning Techniques for Autonomous Parking

Parking lots are a very complex environment, with less structure than highways and streets. The following analyzes how to learn autonomous parking planning techniques through deep learning.

Reference [15] proposes a neural motion planning algorithm that combines deep generative networks with traditional motion planning methods. Because the planning algorithms A* and Hybrid A* have limitations in terms of efficiency, hence the Conditional Variational Autoencoder (CVAE) is applied to guide the search algorithm, and to learn planning space information given the parking environment. Based on the feasible trajectory distribution learned from expert demonstrations, an effective expansion strategy is obtained.

State transition model: The state $x_k = (x_k, y_k, \theta_k)^T$ represents the state of planning step k, where x_k, y_k, and θ_k, respectively, represent the x-axis position, y-axis position, and heading angle. The discrete form of the state transition model is represented as follows:

$$x_{k+1} = x_k + d\cos(\theta_k)dir \tag{13.15}$$

$$y_{k+1} = y_k + d\sin(\theta_k)dir \tag{13.16}$$

$$\theta_{k+1} = \theta_k + \frac{d}{L}\tan(\delta_k)dir \tag{13.17}$$

where δ_k belongs to the steering angle candidate values of the discrete steering angle set D; dir represents the vehicle's direction of movement, the value of dir is -1 or 1, representing moving backward and forward, respectively; and d is an extension amount in a search step. These two control actions include steering angle and direction. D includes steering angles set between $-40° - 40°$, with an interval of 10°.

Hybrid A*: The Hybrid A* algorithm uses the transition model to generate kinodynamic paths, Hybrid A* achieves a resolution of 2 meters in the X-Y dimension and a 15° heading angle. This information is used to determine whether candidate nodes are located in a certain grid unit, where only the nodes with the lowest cost are retained.

Algorithm 1: Hybrid A* Algorithm

Input: Start state x_s, goal state g, heuristic function $h(x)$, obstacle set O
Output: Trajectory from x_s to g

```
C ← ∅;
N ← (x_s, 0, 0, 0, ∅);   // Node with state, cost, heuristic, total cost, and parent
O ← {N};
KeepSearching ← 1;
while KeepSearching is 1 do
    if O is not empty then
        N ← Extract node with smallest c + h from O;
        Remove N from O, add N to C;
        if N is goal g then
            trajectory ← Backtracking(N);
            KeepSearching ← 0;
        end
        else
            foreach action a in AvailableActions(N) do
                (x_n, c_n) ← transition(N.x, a, N.c);
                if isCollide(x_n, O) is 0 then
                    N_n ← (x_n, c_n, h(x_n), a, N);
                    if N_n ∈ C then
                        continue;
                    end
                    else if there exists n ∈ O with same state and lower cost then
                        replace n with N_n in O;
                    end
                    else
                        add N_n to O;
                    end
                end
            end
        end
    end
end
return trajectory
```

Algorithm 2: Neuro-Hybrid A* Algorithm

Input: Start state x_s, goal state g, heuristic function $h(x)$, obstacle set O, distribution map D_{map}

Output: Trajectory from x_s to g

```
C ← Ø;
N ← (x_s, 0, 0, 0, Ø);
O ← {N};
KeepSearching ← 1;
while KeepSearching is 1 do
    if O is not empty then
        N ← Extract node with smallest c + h from O;
        Remove N from O, add N to C;
        if N is goal g then
            trajectory ← Backtracking(N);
            KeepSearching ← 0;
        end
        else
            foreach action a in AvailableActions(N) do
                (x_n, c_n) ← transition(N.x, a, N.c);
                if Rand > 0.2 then
                    if CheckDistMap(x_n, D_map) then
                        continue;
                    end
                end
                if isCollide(x_n, O) is 0 then
                    N_n ← (x_n, c_n, h(x_n), a, N);
                    if N_n ∈ C then
                        continue;
                    end
                    else if there exists n ∈ O with same state and lower cost then
                        replace n with N_n in O;
                    end
                    else
                        add N_n to O;
                    end
                end
            end
        end
    end
end
return trajectory
```

To understand the feasible trajectory distribution of automatic parking, a CVAE architecture is designed, including an encoder–decoder network with conditional input. The conditional input characterizes map information, including the position and direction of the initial and target points, obstacles, and free space. In its overall structure, the 2D image represents map information c, with a width and height of 250 and 150, respectively, and a resolution of 0.1 m. In the image, each pixel's specific value 0 represents free space, 1 obstacles, 2 the starting position, 3 the target position, and the arrow indicates heading information. The real trajectory ξ is the planned trajectory obtained by the Hybrid A* algorithm. A total of five trajectories are generated, and the corresponding pixels of the 2D image are assigned a value 1. The dashed line representing the real trajectory ξ is only used in the training phase.

By inputting different conditions, such as map information, various predicted trajectory distributions can be generated. For example, by changing the starting and target positions and heading values, different trajectory distributions can be generated.

The Neural Hybrid A* algorithm adopts the trajectory distribution map information obtained from the CVAE network. Algo 3 and Algo 4 show the pseudocode for Hybrid A*and Neural Hybrid A*, where N represents a node composed of {state, action, path cost, heuristic, parent node}. O and C, respectively, represent the open list and closed list.

Algo 3: The current state and zero-cost state are first added to the open list. At each iteration, the state x_k with the lowest cost is removed from the open list and added to the closed list. If x_k corresponds to the target state, the search terminates. Otherwise, the available actions (e.g., steering angle and direction) are applied through the transition model to generate the next state x_{k+1}. If x_{k+1} is collision-free and already exists in the open list, but the sum of its step cost and heuristic is lower than the current total cost of that state, the entry in the open list is updated with x_{k+1}. If x_{k+1} is not in the open list, it is added. This process is repeated for all states in the open list until the target state is selected. At that point, the best trajectory can be reconstructed by tracing back through the closed list.

The collision check is performed by the `IsCollide` function, which returns 0 when the vehicle is collision-free, allowing the transition model to propagate the next state under the given control input.

Algo 4: The requirements include new inputs, lines 15–17 are newly added parts; D_{map} represents the distribution map of the predicted trajectory; line 15 checks a random number Rand() $\in (0, 1)$ to determine whether to use D_{map}; 80% of the node expansions are carried out with the help of the neural network model; if the current state value of D_{map} is greater than a certain threshold, the *CheckDistMap* process will return 0; as shown in lines 16 and 17, if the value of D_{map} at state x_n is below the threshold, it is considered that the operation selected in line 13 is unnecessary; otherwise, continue to the next step. Therefore, under this state expansion strategy, the main expansion is based on the learned trajectory distribution, which can effectively select control actions.

In an example of this planning strategy, the expansion of the predefined action set is carried out in the traditional method in the top figure, and the effective expansion method based on the CVAE predicted distribution is carried out at the bottom.

Reference [16] is based on imitation learning, using expert driving data to train deep neural networks, to obtain a planning method for drivable areas in parking lots based on visual occupancy grid maps.

There are two methods for collecting datasets, reinforcement learning and imitation learning. For reinforcement learning, data is collected to get the maximum reward. However, to achieve the goal, heuristic methods are needed to model the reward function. In addition, the agent collects data through repeated trials of random operations, so it takes a long time to complete the training. Imitation learning collects successful driving data from experts, which is used to train deep neural networks in a similar way to supervised learning, to simulate the pattern of expert driving. Compared with reinforcement learning, the required training time is less.

Traditional motion planning algorithms include **optimization**, **graph search**, and **incremental tree search**. These algorithms have three problems. first, if the local area is large or complex, generating a path requires a long computation time, and it may not be possible to find a solution in the control loop (i.e., non-real time); second, selecting a target in the global path to search for a local path is heuristic; finally, when implementing the algorithm, due to inaccurate localization data, it is difficult to identify whether obstacles are close to the global path and to track the path without collision. In unstructured environments, various types of obstacles exist in the drivable area, and it is difficult to obtain accurate localization data for each point.

In addition to searching and tracking a path, other methods can be used, including **candidate path selection** (dynamic window method, curvature speed method, and constraint algorithm) and **artificial field generation** (virtual force field, artificial potential field, and velocity vector field algorithm). However, these two types of methods have some problems and are difficult to use in unstructured environments. First, the parameters (weights) of the objective function or field model may vary, and it is not easy to find specific parameters that can handle all these situations; second, inaccurate localization data makes it difficult to know the exact location of the global path in practice; third, if local obstacle information is difficult to accurately identify, especially in areas with road boundaries or shadows (i.e., noisy states), the vehicle may not be able to drive smoothly, may drive out of the drivable space, or drive toward obstacles.

To solve these problems, the method proposed in reference [16] is to use a spatial representation based on a visual occupancy grid map, which is obtained by segmenting images captured by a camera through deep learning. In motion planning, imitation learning selects a look-ahead point with the occupancy grid map as input, and the pure pursuit algorithm calculates the steering angle to reach the look-ahead point, as well as the speed determined by the longitudinal distance between the look-ahead point and the vehicle. The Dataset Aggregation (DAgger) algorithm is used to address unsafe and near-collision situations that arise from the training network strategy.

An occupancy grid map is a 2D map that divides a region into grids. There are two advantages to a driving strategy based on an occupancy grid map: first, segmenting the image can ignore information irrelevant to driving, such as the type of obstacles and differences in road surfaces in the drivable area. Therefore, the driving strategy can achieve similar performance in untrained environments, enhancing the universality of driving performance; second, near and far information can be clearly distinguished because the occupancy grid map is a 2D map (i.e., BEV). Therefore, depending on the situation, the vehicle can prioritize avoiding nearby obstacles or consider obstacles in the distance in advance.

Through semantic segmentation, the deep neural network decomposes the image into drivable and non-drivable areas. Roads, crosswalks, and road markings are marked as drivable areas, while road boundary lines, crosswalks, parking spaces (including parking lines), pedestrians, and vehicles are also marked as non-drivable areas.

Imitation learning refers to imitating the behavior of experts in specific states. When human experts drive, state-action data pairs are collected. The policy π_{net} (i.e., the deep neural network) uses these data for training in a behavior cloning process. To overcome the limitations of behavior cloning, DAgger executes the training strategy, retrains π_{net}, and collects additional data. This process is repeated until the optimal strategy is obtained.

This driving strategy network is an encoder composed of CNN and fully connected layers. The encoder consists of two pairs of convolutional layers and max pooling layers, with flat layer nodes connected together.

The behavior cloning driving strategy network can learn various complex driving patterns suitable for the existing training set, including learning patterns of all information in the occupancy grid map, not just information around the candidate path or artificial field.

13.10 Summary

This chapter is analysis and discussion of an important landing scenario in autonomous driving—the automatic parking system—by introducing its system calibration, parking space detection, motion planning, sensor fusion, and valet parking technology. Chapter 14 will introduce the application of vehicle networking in autonomous driving.

References

1. ANZHAF H, NIENHÜSER D., KNOOP S. and ZÖLLNER J. M., The Future of Parking: A Survey on Automated Valet Parking with an Outlook on High Density Parking, IEEE Intelligent Vehicles Symposium, Los Angeles, USA, June 2017: 1827-1834.
2. C WANG et al., Automatic Parking Based on a Bird's Eye View Vision System, Advances in Mechanical Engineering, Jan. 2014. https://journals.sagepub.com/doi/10.1155/2014/847406

3. L ZHANG et al., Vision-Based Parking-Slot Detection: A DCNN-Based Approach and a Large-Scale Benchmark Dataset, IEEE Trans. on Image Processing, 2018. 27(11): 5350-5364.
4. B LEE, Y WEI, I GUO,Automatic Parking of Self-driving Car based on Lidar, Int. Arch. Photogramm. Remote Sens. Spatial Inf. Sci. (ISPRS), XLII-2/W7, Sept. 2017: 241–246
5. U SCHWESINGER et al., Automated Valet Parking and Charging for e-Mobility, IEEE Intelligent Vehicles Symposium, Gothenburg, Sweden, June 2016: 157-164.
6. J K SUHR, H G JUNG, Automatic Parking Space Detection and Tracking for Underground and Indoor Environments, IEEE Trans. on Industrial Electronics, 2016. 63(9): 5687-5698.
7. J K SUHR, H G JUNG, Sensor Fusion-Based Vacant Parking Slot Detection and Tracking, IEEE Trans. on Intelligent Transportation Systems, 2014. 15(1): 21-36.
8. C LOEPER et al., Automated Valet Parking as Part of an Integrated Travel Assistance, IEEE International Conference on Intelligent Transportation Systems, The Hague, Netherlands, June 2013: 1827-1834.
9. S-J HAN, J CHOI, Parking Space Recognition for Autonomous Valet Parking Using Height and Salient-Line Probability Maps, ETRI (Electronics and Telecommunications Research Institute) Journal,2015. 37(6): 1220-1230.
10. Y RYU, S OH, S KIM, Robust Automatic Parking without Odometry using an Evolutionary Fuzzy Logic Controller, Int. J. Control, Automation, and Systems, 2008. 6(3): 434-443.
11. LI L, ZHANG L, Li X et al., Vision-Based Parking-Slot Detection: A Benchmark and A Learning-Based Approach, IEEE Int. Conf. on Multimedia and Expo, Hong Kong, China, July 2017: 5350-5364.
12. T QIN, T CHEN, Y CHEN, and Q SU, AVP-SLAM: Semantic Visual Mapping and Localization for Autonomous Vehicles in the Parking Lot,arXiv preprint arXiv:2007.01813, 2020
13. M HEIMBERGER, J HORGAN, C HUGHES, J MCDONALD, S YOGAMANI, Computer Vision in Automated Parking Systems: Design, Implementation and Challenges, arXiv preprint arXiv:2104.12537, 2021
14. Y G JO, S H HONG, S HWANG, and J M HA, Fisheye-Lens-Camera-based Autonomous Valet Parking System,arXiv preprint arXiv:2104.13119, 2021
15. D KIM and K HUH, Neural Motion Planning for Autonomous Parking,arXiv preprint arXiv:2111.06739, 2021
16. J AHN, M KIM and J PARK, Vision-based Autonomous Driving for Unstructured Environments Using Imitation Learning, arXiv preprint arXiv:2202.10002, 2022
17. J LEU, Y WANG, M TOMIZUKA, and S D CAIRANO, Autonomous Vehicle Parking in Dynamic Environments: An Integrated System with Prediction and Motion Planning, arXiv preprint arXiv:2204.10383, 2022
18. V R KUMAR, C EISING, C WITT, and S YOGAMANI, Surround-view Fisheye Camera Perception for Automated Driving: Overview, Survey & Challenges, arXiv preprint arXiv:2205.13281, 2022
19. V R KUMAR, S YOGAMANI, H RASHED, G SITSU, C WITT, I LEANG, S MILZ and P MÄDER, OmniDet: Surround View Cameras based Multi-task Visual Perception Network for Autonomous Driving, arXiv preprint arXiv:2102.07448, 2021
20. V R KUMAR, M KLINGNER, S YOGAMANI, M BACH, S MILZ, T FINGSCHEIDT, and P MÄDER, SVDistNet: Self-Supervised Near-Field Distance Estimation on Surround View Fisheye Cameras, arXiv preprint arXiv:2104.04420, 2021
21. Z WU, W ZHANG, J WANG, M WANG, Y GAN, X GOU, M FANG and J SONG, Disentangling and Vectorization: A 3D Visual Perception Approach Autonomous Driving Based on Surround-View Fisheye Cameras, arXiv preprint arXiv:2107.08862, 2021
22. Kumar, Varun Ravi, et al. "Fisheyedistancenet: Self-supervised scale-aware distance estimation using monocular fisheye camera for autonomous driving." 2020 IEEE international conference on robotics and automation (ICRA). IEEE, 2020
23. Rashed, Hazem, et al. "Generalized object detection on fisheye cameras for autonomous driving: Dataset, representations and baseline." Proceedings of the IEEE/CVF Winter Conference on Applications of Computer Vision. 2021.
24. Bertozzi, Massimo, et al. "360 detection and tracking algorithm of both pedestrian and vehicle using fisheye images." 2015 IEEE intelligent vehicles symposium (iv). IEEE, 2015.

Chapter 14
Vehicle Networking

Section 3.1.7 has briefly introduced vehicle networking, i.e., DSRC and Cellular-V2X communication technologies. Vehicle networking includes vehicles that communicate with the public network through V2V (vehicle-to-vehicle), V2I (vehicle-to-infrastructure), and V2P (vehicle-to-pedestrian) interactions, collecting and sharing key information about road network conditions in real time.

Using onboard units, vehicles can communicate with each other (V2V) and roadside units (V2I). This can enable various other forms of communication, such as vehicle-to-broadband cloud (V2B) (where the vehicle communicates with the surveillance data center), vehicle-to-human (V2H), i.e., communication with pedestrians or bicycles, or vehicle-to-sensor (V2S), i.e., communication with sensors embedded in the environment.

For autonomous vehicles, like high-definition maps, vehicle networking itself is an extension of sensors and can be seen as a sensor. Compared with commonly used cameras or lidar in autonomous driving technology, V2X technology has the ability to break through visual blind spots and obtain information across obstacles, and can also share real-time driving status information with other vehicles and facilities, and generate predictive information through judgment algorithms. In addition, V2X is the only sensor technology that is not affected by weather conditions, and rain, fog, or strong light will not affect its normal operation. Therefore, V2X technology is widely used in the field of transportation, especially in the field of autonomous driving.

This chapter introduces the technology of vehicle networking in autonomous driving. Section 14.1 discusses the development of intelligent connected vehicles; Sect. 14.2 introduces the social characteristics of vehicle networking and its AI elements; Sect. 14.3 analyzes various edge computing; Sect. 14.4 discusses the collaborative methods of vehicle-roadside; Sect. 14.5 introduces the technology of vehicle collaborative perception through vehicle networking; Sect. 14.6 discusses the planning and control of vehicle platooning.

Y. Huang and Z. Yang, *Autonomous Driving Handbook*,
https://doi.org/10.1007/978-981-95-3725-9_14

14.1 Intelligent Connected Vehicles

Intelligent Connected Vehicles (CAV) can improve safety by reducing traffic accidents. They can also provide opportunities to reduce transportation energy consumption and emissions by improving traffic flow. Communication between vehicles and traffic structures and traffic signals can optimize the operation of individual vehicles and respond to unpredictable changes. Reference [1] summarizes the development and research trends in the field of CAV collaboration and discusses other challenges and potential future research directions.

Clearly, V2V communication has the potential to enable vehicles to consider changes in their mutual environment more quickly, reducing traffic accidents and alleviating traffic congestion. Similarly, V2I communication, such as communication with traffic structures, nearby buildings, and traffic lights, allows individual vehicle control systems to consider unpredictable changes in local infrastructure.

A large amount of research using centralized or distributed collaborative methods focuses on coordinating CAV at intersections and merging on highway ramps. If at least one task in the system is globally decided by a single central controller for all vehicles, it is classified as centralized. In decentralized methods, vehicles are viewed as autonomous agents, trying to maximize their collaborative efficiency through strategic interaction. Under this framework, each vehicle obtains information from other vehicles and roadside infrastructure, optimizes specific performance criteria (such as efficiency and travel time), while satisfying the physical constraints of the transportation system (such as stop signs and traffic signals).

Ramp metering is a common method used to regulate the flow of vehicles converging onto highways to reduce traffic congestion. Although it has been shown that it can help improve the overall traffic flow and safety of highways, some problems may arise due to the short length of the ramp, such as causing interference to traffic on adjacent roads. Different strategies to address these challenges have been explored in the past, including feedback control theory, optimal control methods, and heuristic algorithms.

Given recent technological developments, some teams have made efforts in several ways to solve traffic congestion problems caused by road convergence. In these efforts, it is assumed that vehicles on the road are connected and have a certain degree of autonomy. This assumption helps to design strategies to achieve safe and effective coordination of merging actions, thus avoiding unnecessary stop-and-go operations of vehicles.

Furthermore, for intersections, traffic lights are considered one of the most effective methods of controlling traffic, and attempts are still being made to improve their effectiveness. Please note that the intersection control problem and the merging control problem are essentially very similar, and most methods proposed for intersection control can easily be used for merging coordination, and vice versa.

Typically, the order of passage at intersections is controlled by traffic lights or stop signs. In the case of merging highways, ramp metering is a common method used to regulate the flow of vehicles converging onto highways, but this also means that vehicles on the secondary road (the other being the main road) must stop to reduce traffic congestion. Figures 14.1 and 14.2 illustrate these two situations, the area in the center of the intersection, or the confluence of roads, is called the merging zone, with a length of S. In addition, there is a control zone, within which vehicles can communicate with each other; the distance between the entrance to the control zone and the entrance to the merging zone is L.

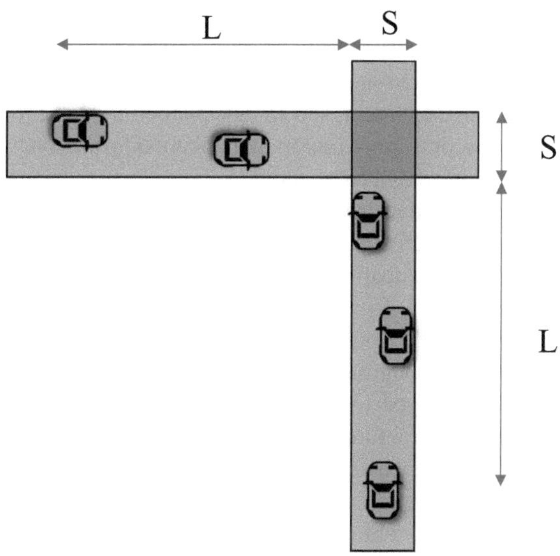

Fig. 14.1 Intersection of CAV

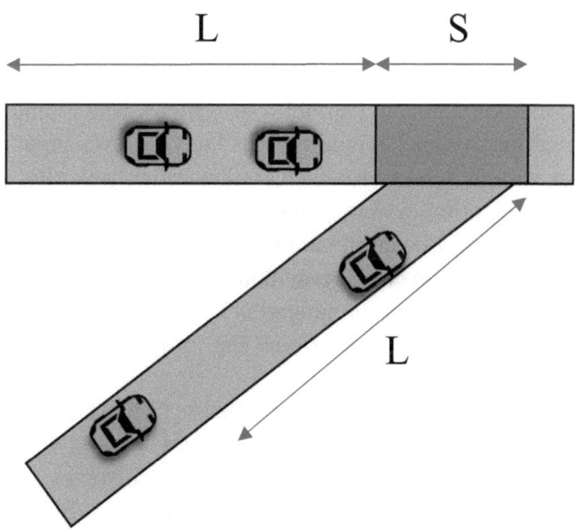

Fig. 14.2 CAV's converging road scene

The goal of the work here is to coordinate vehicles through intersections (or merging), and there can be no rear-end or lateral collisions in the merging area. There are generally two methods to solve this problem: centralized methods and decentralized methods.

In centralized methods, the system has at least one task globally decided by a single central controller for all vehicles, including heuristic rule methods and optimization control theory methods. In decentralized methods, each vehicle determines its control strategy based on information received from other vehicles on the road or a coordinator. One of the main challenges faced by decentralized methods is that solutions may deadlock due to the use of local information.

Reference [2] reviews current collaborative solutions for autonomous vehicles based on various applications (i.e., smart parking, lane changing and merging, intersection management, and vehicle platooning). The role and function of this collaboration become more important in platoon formation, where one challenge is to select a leader in advanced platooning.

Collaborative driving is a key subcomponent of intelligent transportation systems, dedicated to creating cooperative autonomous vehicles and navigating on highways using communication technology. It then adopts the ideal control law to achieve a common goal. To achieve this, the **Collaborative Driving System** (CDS) should consider 4 key components: ① Vehicle dynamics, defined by a complete vehicle model (i.e., longitudinal and lateral); ② Actuator lag, describing the information to be exchanged between vehicles; ③ Communication topology, determining the connection structure of the vehicle network; ④ Control law, defining the control law executed on each vehicle.

The motivation behind CDS is to solve two main problems: ① managing the decentralized control of vehicles; ② coordinating the actions of each vehicle controller. The solution to the first problem is to use longitudinal and lateral control. To solve the second problem, CDS assumes that vehicles are grouped in a platoon. In collaborative driving, a platoon is usually formed.

The vehicle control layer is responsible for two functions: ① detecting the condition and state variables of vehicles behind and adjacent; ② activating lateral and longitudinal actuators. This layer receives inputs from the sensor system (such as various sensors for acceleration, radar, yaw rate, vehicle speed, and vision) and the execution system (lateral and longitudinal control), and outputs to the vehicle management layer, obtaining steering and vehicle speed queries.

The vehicle management layer uses data from the control layer, adjacent vehicles in inter-vehicle communication (IVC), and the traffic control layer in road vehicle communication to determine the torque of each vehicle under the constraints of cooperative driving.

The traffic control layer consists of two components: physical components and logical components. The physical part consists of traffic lights, signs, and roadside communication, while the logical part involves social laws, rules, etiquette, and other social ethics. Finally, this layer passes traffic information to each vehicle in the management layer to maintain coordination and execute actions.

Common cooperative driving scenarios include:

1. Cooperative intelligent parking. Intelligent parking systems can be divided into six types: parking reservation systems, parking guidance and information systems, crowdsourcing in intelligent transportation systems, centralized assisted parking search, guidance systems based on smart bodies, and electric vehicle parking systems.
2. Vehicle lane changing or merging. Using intelligent driving technology for cooperative and automatic lane changing can greatly improve road efficiency and safety; the rapid development of V2V communication makes cooperative automatic lane changing possible; similarly, lane merging can minimize vehicle travel time and maximize the number of merging vehicles as the goal, solving an optimization problem.
3. Cooperative Intersection Management (CIM). At signal intersections, cooperative intersection management allows vehicles to communicate and exchange various information with the infrastructure, such as traffic perception, negotiation through intersections, vehicle signs, etc.; in non-signal intersections, CIM helps drivers make the best decisions by providing a global view of intersection management.

Vehicle platooning is an important use case for implementing cooperative automatic driving. Due to its potential advantages such as improving road safety, reducing fuel consumption, reducing carbon dioxide emissions, increasing road throughput, improving asset utilization, and optimizing labor costs, it has received significant attention.

Vehicle platooning, also known as **convoy driving**, refers to driving a group of two or more consecutive vehicles from head to tail (nose-to-tail) at the same speed, traveling in the same lane with a smaller vehicle spacing (usually less than 1s at the same speed). In the platoon, vehicles are virtually connected through lidar (or other sensors) and communicate with each other through wireless communication technology (such as CACC).

In a vehicle platooning, the leading vehicle drives in the first position and is followed by a batch of following vehicles, which means that the following vehicle (FV) can automatically brake, steer, and decelerate according to the actions of the leading vehicle (LV). The vehicles in the platoon coordinate their lateral and longitudinal control to maintain the required headway with the leading vehicle.

The leading vehicle controls the driving direction, and the following vehicles use advanced technologies to follow the leading vehicle, such as Adaptive Cruise Control (ACC), Automatic Emergency Braking (AEB), Blind Spot Warning (BSW), Forward Collision System (FCA), Lane Departure Warning (LDW), etc. The vehicles behind automatically start to follow the vehicle in front without driver intervention. V2V communication allows vehicles to communicate back and forth, so the leading vehicle can also adjust its speed and position based on the response of the following vehicle. The wireless coupling and decoupling of the platoon make it easy for vehicles to join or leave the platoon without having to stop.

In many applications, a leader or coordinating vehicle is needed to synchronize and complete the cooperative goals of a group of vehicles. Some of these applications are intersection coordination, virtual traffic lights (VTL), and highway merging, and vehicle platooning is one of them. These applications allow vehicles to communicate in a V2V manner and eventually unanimously elect a leader.

The leading vehicle plays an important role in the platoon, in addition to transmitting information to subsequent vehicles, it is also responsible for maneuvering tasks such as platooning, merging, and splitting.

14.2 Vehicle Networking and Social Internet of Things

The Social Internet of Things (SIoT) introduces social relationships between targets, creating a social network where the participants are not humans but intelligent targets. Reference [3] discusses the concept of Social Internet of Vehicles (SIoV), which makes social interaction between vehicles and drivers possible; it also discusses the technology and components of SIoV, possible applications, and potential security, privacy, and trust issues.

Context-aware technology and wireless vehicle communication technology, such as Dedicated Short Range Communication (DSRC), Long Term Evolution (LTE), IEEE 802.11p, and Global Interoperability for Microwave Access (WiMax), their latest developments have promoted the design, development, and deployment of vehicle networks. More and more proposals want to apply social networks to vehicle networks, which leads to the transformation of traditional vehicle networks to SIoV.

The Vehicular Ad-hoc Network (VANET) is a type of mobile ad hoc network that has been proposed to enhance traffic safety and provide comfortable applications for drivers.

DSRC makes VANET possible. VANET does not require any infrastructure for vehicle communication, which is key to ensuring safety in remote and underdeveloped areas, especially in preventing accidents on foggy highways. In DSRC, vehicles transmit Public Awareness Messages (CAM) and Basic Safety Messages (BSM) with a delay of less than 100 ms. DSRC is a slight modification of the IEEE 802.11 protocol, which can be easily used to deploy VANET.

A key aspect of SIoV is to make vehicles context-aware, that is, to understand the surrounding situation, especially those particularly relevant to them. Context-aware systems are those that can adapt their behavior to their current context environment. Vehicle context awareness can be provided through three main subsystems: perception, reasoning, and execution.

- The perception subsystem collects context information from different sensors integrated with the vehicle's OBU. The types of these sensors vary according to the requirements of the vehicle, such as location, infrared, or ultrasonic. In other words, this representation is a way to collect context data.
- The reasoning subsystem processes raw data to extract high-level context information, such as the driver's situation. Context information can be extracted from

a single sensor that defines specific context information, or from multiple sensors that define uncertain context information.
- The execution subsystem represents the application executor, which provides services to users or other drivers, disseminates warning messages, and vehicle alarm systems and intelligent parking assistance systems are examples of deploying advanced applications that can prevent accidents and reduce road congestion.

Vehicle clustering is a method widely used in various networks, from mobile ad hoc networks to sensor networks. Clusters exist in different forms and can use a wide range of network characteristics to execute, bringing significant benefits to the network. For example, clustering can alleviate the problem of a large number of broadcasts in dense networks (broadcast storms), increase system throughput, and reduce the bit error rate. It can significantly reduce packet delay and provide better spectrum utilization in time and space.

Smartly dividing the network's clusters can also allow data aggregation and extend network life. By creating groups of nodes sharing the same characteristics, the network appears smaller and more stable. To form clusters, vehicles can combine LTE and DSRC communication capabilities, as shown in Fig. 1 of [4]. The cluster head can perform special operations within the cluster, such as channel usage, data aggregation, scheduling, and packet routing.

A special category of clustering is vehicle platooning, as discussed earlier, the leading vehicle can act as the cluster head and take over most of the communication between the platoon and the external network (see Fig. 1 of [3]).

The security of vehicle networks is an important aspect of SIoV, as if vehicles do not meet the requirements, life-threatening situations may occur, and SIoV itself is also part of the smart city infrastructure. Unlike standard computer networks, vehicle networks provide computing capabilities for a large number of distributed heterogeneous resources. The connectivity between vehicles in different geographical locations makes security a more complex issue, as does the heterogeneity of ownership, manufacturers, and users. Social vehicle network security considers social characteristics and human behavior.

Autonomous vehicles can be regarded as vehicles equipped with artificial intelligence capabilities, enabling autonomous driving through functions such as perception, recognition of the surrounding environment and objects, and the execution of reasoning and decision-making. Connected and Autonomous Vehicles (CAVs) form the foundation of the Internet of Vehicles (IoV), which is expected to transform transportation systems and urban life on a global scale.

Vehicles need to interact not only with other vehicles but also possibly with motorcycles, bicycles, pedestrians, and other road users through DSRC or 5G-V2X networks, adopting IoT services (including through roadside units). Through network protocols, vehicles have the opportunity to exchange application layer messages and cooperate to improve safety and increase their effectiveness, thereby creating a cooperation layer above the vehicular network layer.

The social vehicle concept envisions **the social brain** of CAV, CAV is defined as a software module that determines how vehicles cooperate with other vehicles, how vehicles cooperate with pedestrians and services on the vehicle network, how vehicles reason social behavior, how vehicles behave when they receive specific messages, and how to use cooperation between vehicles under road conditions.

The social brain of CAV can infer the social situation for other vehicles and pedestrians on the road, remember past interactions to provide information for future cooperation, work within the social norms of the road, and have context-aware attention (similar to human social cognition).

The social brain not only reasons situation-awareness but also determines the behavior of the vehicle (how the vehicle behaves relative to other vehicles in social activities) and the way the vehicle interacts with other entities (such as which messages to send).

The social brain of a CAV enables it to predict or model the behavior of other vehicles and pedestrians, infer their intentions, and exhibit situational awareness across diverse traffic scenarios.

The social brain of a vehicle can implement diverse cooperation protocols tailored to different road conditions, such as collision-warning protocols for intersections, merging protocols for roundabouts, lane-merging protocols for highways, overtaking protocols, and yielding protocols at intersections. All of these are integrated into the vehicle's social brain module. Beyond rules, policies, and robotic laws that govern general road traffic, it is also essential to ensure trustworthy and ethical interactions—both among CAVs and between CAVs and human road users.

In addition to perceiving the direct environment and basic network technology, as a form of **social AI** function of autonomous vehicles, it also needs to have the ability to behave cooperatively. The social AI of the vehicle is a major component supporting various cooperative behaviors.

Traditionally, markings on roads or traffic lights are used to coordinate vehicles to drive in an orderly manner, but they may cause delays or reduce road usage. Traffic flow can be coordinated without physical signals through cooperation and coordination. Cooperation between vehicles can also form flexible collective vehicle behavior. For example, on highways, lanes are fixed equally on both sides (for example, there are 5 lanes in each direction); but at certain times, the traffic flow in one direction may be larger than the other.

One situation is with the help of CAV, when the traffic flow in one direction is high, vehicles can cooperate on a large scale and notify other vehicles, allowing now one direction to have 8 lanes, and the opposite direction to have 2 lanes. Then, rebalance the lanes at other times. In addition, some "lanes" may become narrower at certain times (cars are close to each other but slower), and at other times, lanes will become wider (cars move faster but are farther apart from each other). This idea is called "traffic shaping".

Another situation is to give way to emergency vehicles—CAVs receiving emergency vehicle notifications can create paths through coordination. All nearby vehicles can receive the same information from the emergency vehicle, but each vehicle needs to decide what is the best action to take based on its own and the context knowledge of the surrounding vehicles in order to create a path. Vehicles must also be able to detect that the emergency vehicle has passed in order to be able to coordinate and resume normal driving.

14.3 Edge Computing

Edge Computing (EC) is a viable solution that can address the challenges of latency, scalability, and privacy, and will become an important supplement to the cloud, and even replace the role of the cloud in some scenarios. In edge computing, a fine grid of computing resources provides computing capabilities close to terminal devices. For example, edge computing nodes can be located in the same place as cellular base stations and IoT gateways or campus networks.

To solve the latency challenge, edge computing, close to the data source on terminal devices, can reduce end-to-end latency, thereby achieving real-time services. To solve the scalability challenge, edge computing supports a hierarchical architecture of terminal devices, edge computing nodes, and cloud data centers, can provide computing resources and expand with the increase in the number of clients, thereby avoiding network bottlenecks at the central location. To address the privacy challenge, edge computing allows data to be analyzed close to the source, possibly by a locally trusted edge server, thereby avoiding traversing the public internet and reducing the risk of privacy and security attacks.

In the development of edge computing, various new technologies aimed at working at the network edge have emerged, with the same principles but different focuses, such as Cloudlet, Micro Data Center (MDC), Fog Computing, and Mobile Edge Computing (now Multi-Access Edge Computing). However, the edge computing community has not yet reached a consensus on the standardized definition, architecture, and protocols of edge computing. The following introduces the concepts and differences of various edge computing.

1. **Cloudlet and Micro Data Center** (MDC). Cloudlet is a network architecture element that combines mobile computing and cloud computing. It represents the middle layer of the three-layer architecture, namely mobile devices, micro clouds, and clouds. Its highlights are: ① defining the system and creating algorithms that support low-latency edge cloud computing; ② implementing related functions in open source code, as an extension of Open Stack cloud management software. Similar to Cloudlet, MDC is also designed to supplement the cloud. The idea is that all the computing, storage, and network equipment required to run client applications are encapsulated in a chassis, as an independent secure computing environment, for applications with lower latency or terminal devices with limited battery life and computing power.
2. **Fog Computing**. One of the highlights of fog computing is the assumption of a fully distributed multi-layer cloud computing architecture, with billions of devices and large cloud data centers. Although cloud and fog paradigms share a set of

similar services, such as computing, storage, and networking, fog deployment is targeted at specific geographical areas. Moreover, fog is designed for applications that require less latency for real-time responses, such as interactive and Internet of Things applications. Unlike Cloudlet, MDC, and MEC, fog computing focuses more on the Internet of Things.

3. **Mobile/Multi-access Edge Computing** (MEC). Mobile edge computing places computing capabilities and service environments at the edge of cellular networks. It aims to provide lower latency, more accurate context and location awareness, and higher bandwidth. Deploying edge servers on Base Stations (BSs) allows users to deploy new applications and services flexibly and quickly. The European Telecommunications Standards Institute (ETSI) has further extended the term MEC from mobile edge computing to multi-access edge computing to accommodate more wireless communication technologies, such as Wi-Fi.
4. **Edge Computing Terminology**. Common edge devices are further divided into terminal devices and edge nodes. Terminal devices (end devices, end level) refer to mobile edge devices (including smartphones, smart vehicles, etc.) and various Internet of Things devices, while edge nodes (edge level) include Cloudlet, Roadside Units (RSUs), fog nodes, edge servers, MEC servers, etc., i.e., servers deployed at the network edge.
5. **Collaborative End-Edge-Cloud Computing**. Although cloud computing was created to handle compute-intensive tasks (such as deep learning), it cannot meet the latency requirements from data generation to transmission to execution. Moreover, the independent processing capability on terminal or edge devices is limited by their computing power, power consumption, and cost bottlenecks. Therefore, collaborative end-edge-cloud computing is becoming an important trend. In this new computing paradigm, compute-intensive tasks generated by terminal devices can be executed directly on the terminal devices or offloaded to the edge, thus avoiding the latency caused by sending data to the cloud. For compute-intensive tasks, they will be reasonably decomposed and allocated to terminal, edge, and cloud devices for execution, reducing task execution latency while ensuring result accuracy. The focus of this collaborative mode is not only on successfully completing tasks but also on achieving the optimal balance of device energy consumption, server load, transmission, and execution latency.

14.4 Vehicle-Roadside Collaboration Technology

The emergence of CAV has also paved the way for significant improvements in transportation, including improving safety (for example, by detecting vehicles that obstruct the line of sight) and reducing reliance on traffic signals.

Reference [5] summarizes the research work of **automatic intersection management** (AIM) in various disciplines (such as traffic engineering, control engineering) from three levels: corridor coordination layer, intersection management layer, and vehicle control layer. Compared with **signal intersection management** (SIM), AIM coordinates intersections at the single-vehicle level, providing more flexibility. AIM can potentially eliminate stops at intersections due to traffic signals while maintaining safe separation between conflicting traffic flows. Reference [5] discusses in detail the key aspects of AIM design, including conflict detection schemes, priority rules, centralized control, computational complexity, etc.

For example, three hierarchical structures can be extracted from existing intersection control practices in traffic engineering, namely the corridor coordination layer, intersection management layer (also called trajectory planning layer), and vehicle control layer. The layered structure framework is also applicable to evaluating AIM. The connections between the layers are implemented by communication networks. In SIM, loop detectors collect passing traffic conditions for signal controllers carrying intersection management protocols (such as signal phase and timing SPaT schedules). Traffic signals inform drivers whether they can pass. AIM replaces the above procedures with V2X communication. In addition, it is expected that human-driven vehicles will be replaced by ADAS and eventually by AIM's autonomous driving systems (ADS).

The corridor coordination layer deals with the coordination of multiple intersections at the road level. This coordination is common in arterial traffic under SIM. Maximizing the green band is usually used to ensure the main straight-through progress across multiple intersections. The fixed time SPaT (signal phase and timing) schedule between intersections is the most commonly used method to achieve coordination.

The intersection management layer assigns intersection traffic order for AIM vehicles or SIM vehicles. For SIM, signal phases with conflicting active behaviors cycle according to a predefined phase sequence. For AIM, the intersection manager is responsible for allocating limited time-space resources. This aspect of AIM is often referred to as trajectory planning because AIM separates each conflicting activity at the single-vehicle level.

The vehicle control layer focuses on single-vehicle longitudinal and lateral motion control. Typically, vehicles are driven by drivers, sometimes with the help of vehicle subsystems, such as power steering and auxiliary braking. In the CAV environment, autonomous driving systems can supplement and eventually replace human driving input.

At the intersection management level, priority allocation and reservation systems are two key aspects. Vehicle trajectory allocation is similar to aircraft separation planning in the time-space dimension. Separation in AIM is done at the vehicle level using a reservation-based system, while separation in SIM is done at the vehicle group level using traffic signals.

AIM has 4 types of reservation systems for cutting conflicts: Intersection-based reservation, tile-based reservation, conflict point-based reservation, and vehicle-based reservation.

Intersection-based reservation allows only one vehicle in the intersection. In tile-based reservation, space is discretized into block grids. If two vehicles occupy the same tile at the same time, the reservation is rejected. Tiles can also be clustered into larger areas to reduce the computational complexity of reservations. Conflict point-based reservation can make full use of intersection space. Vehicle-based reservation is the least restrictive reservation system. However, due to high-dimensional collision avoidance constraints, solving nonlinear programming (NLP) problems requires a lot of computational resources.

Priority policies dictate the allocation of intersection resources and are independent of the reservation framework. The main goal of the reservation framework is to separate conflicting activities. In SIM, the priority between vehicle groups is determined by operational requirements (such as queue length or delay) and implemented through SPaT planning. Priority can be increased by extending the green signal timing of the interested activity group.

Unlike SIM, AIM's priority allocation is done at the single vehicle level. The only vehicle-level priority allocation in SIM is signal preemption by buses or emergency vehicles. Most AIMs adopt a fair, first-come-first-served (FCFS) strategy. System optimal strategy is the second most commonly used priority strategy, where traffic order is determined based on system-level performance indicators such as overall delay, throughput, travel time, etc. Other known priority strategies include longest queue first strategy, vehicle type-based strategy, strategy based on defined priority scores, and auction-based strategy.

Centralized, decentralized/distributed, and distributed are terms used to describe the communication structure of AIM systems.

Centralized AIM has single-point contact between nodes (vehicles) for information sharing and decision-making. Therefore, single-point failure is a major problem of centralized AIM. The implementation cost of centralized interruption management strategies is high, and its scalability is also problematic. The current state of V2X wireless communication may technically not guarantee such performance for thousands of vehicles near the intersection.

Decentralized/distributed AIM contains several central hubs within the system. Note that processing in a decentralized system is shared across multiple nodes. For example, communication within a platoon is assumed to be local communication between platoon members.

Distributed AIM is the extreme case of decentralization. In a distributed system, there is no single point making decisions, each node only makes decisions for its own behavior. System behavior is the result of the aggregated response of each node within the system. Scheduling allocation between vehicles can become truly fault-tolerant.

From a traffic engineering perspective, AIM generally assumes the availability of vehicle control and emphasizes the coordination of conflicting intersection activities. Roadside units (RSUs) take over vehicle control and guide them safely through the intersection. With vehicle automation, driving functions are expected to be replaced by vehicle controllers.

Reference [6] proposes a framework for lane merging coordination for CAVs using a centralized system. Connected vehicles on the road can provide trajectory suggestions, which are implemented by a **traffic orchestrator** (TO) and **data fusion** (DF) module as the main components. The DF works in conjunction with a centralized microservice-oriented architecture to provide descriptors of connected and unconnected vehicles to the TO. Here, deep reinforcement learning (DRL) and data analysis are used to predict trajectory recommendations for connected vehicles, taking into account unconnected vehicles.

Connected vehicles will attempt to merge onto the main lane where there are connected and unconnected vehicles. Through edge cloud methods, customized trajectory recommendations are determined and sent to connected vehicles by the centralized coordination mechanism. This architecture is able to collect data for lane merging algorithms, while being able to respond to sudden road changes. This architecture also allows for easy scalability in vehicle coordination. Five different components contribute to lane coordination merging: V2X gateway, image recognition (IR) system, **global dynamic map** (GDM), DF, and TO.

The V2X gateway is responsible for forwarding messages to various applications and interfaces in the architecture. The V2X gateway serves as a communication medium, connecting interfaces and applications to connected vehicles based on message exchange methods.

The image recognition system collects all vehicle information on the specified area of the road, including the specific road user localization and trajectory-based parameters given by the Road User Description (RUD). It collects and processes information about connected and unconnected vehicles, and sends all information to the V2X gateway, which forwards the messages to the Global Dynamic Map (GDM).

GDM stores the environmental information of connected and unconnected vehicles in the database. This information is provided by the V2X gateway. GDM ensures that the stored RUD is up-to-date. The Data Fusion (DF) provides a synchronization mechanism for RUDs from different sources (for example, from the image recognition system and connected vehicles in precise localization time coordinates). DF sends information to applications that subscribe to specific location boundaries. In addition, it includes a monitoring and evaluation platform for assessing communication KPIs.

The Trajectory Optimizer (TO) will store and process environmental factors related to connected and unconnected vehicles, generating trajectories for connected vehicles. TO needs to consider time-critical variables, such as timestamps of vehicle positions, vehicle speed, and specific vehicle dimensions. Once TO provides coordinated trajectory recommendations for a single or a group of road users, the recommendation will be sent to the V2X gateway and forwarded to the connected vehicles. Connected vehicles can choose to accept, reject, or abort the recommendation. This feedback information is provided by the connected vehicles to the V2X gateway and can be used to recalculate trajectory recommendations.

Unconnected vehicles cannot communicate with TO, nor can they interpret or use trajectory recommendations. However, the lane merging coordination module can identify unconnected vehicles through the image recognition system. This image recognition system provides RUD to GDM for storage. TO will request road user information to create trajectory recommendations. For this, a set of messages needs to be defined to communicate all components of the lane merging coordination module.

The novelty of TO lies in its use of a deep learning model, which can be more widely applied to different lane merging scenarios. The TO architecture is minimal, relying on lower-level architectures (such as image recognition) to provide the status of roads and vehicles.

The architecture diagram of TO, the main function of the Detection Interface (DI) is to receive any data being sent from the V2X gateway through the Transmission Control Protocol (TCP) connection; the Detection Interface also acts as an intermediate filter, reading JSON strings and processing JSON messages into a more compact and efficient structure for TO to use; similarly, the Network Interface (NI) converts information within TO into JSON messages, inputting them into the V2X gateway.

A knowledge base has been designed to store information sent to TO. The Exchange Interface (EI) is designed with two responsibilities: executing the TO application, and mediating the information flow between all interfaces in TO.

Two reinforcement learning models are used: Deep Q-Network (DQN) and Dueling Deep Q-Network (dueling DQN). The dueling DQN showed the most ideal results, providing human-like trajectories with extremely low bias.

DF is responsible for updating the GDM data with the latest Road User Description, avoiding duplicate road users from connected vehicles and the image recognition system. It is also responsible for combining the most accurate data from each source to improve the accuracy of RUD, i.e., the acceleration accuracy from the camera system is not as accurate as the acceleration provided by the connected vehicles.

DF consists of four different components: Network Interface, Data Synchronization, Data Association, and Fusion History Map.

The Network Interface is responsible for interacting with other components. It receives RUD from the V2X gateway, deserializes the message and forwards it to Data Synchronization. It also handles the fusion and corrected descriptions from Data Association, serializing them and sending them to GDM. DF also receives the fusion and corrected descriptions from Data Association, serializing them and sending them to GDM. Data Synchronization aims to synchronize the received RUD in a timely manner. It collects all descriptions received within a specific time period (100ms), then extrapolates each description to the same time reference. Considering the speed of the target, a uniform acceleration linear motion model is sufficient under the assumption that the cycle is small. It updates the coordinates and timestamps of each target. Data Association matches the targets detected by the camera system with the connected vehicles. For each target detected by the camera system, it first checks whether the target has already been matched with another target in the Fusion History Map. Then, it increases or decreases confidence based on Euclidean distance and angle. The Fusion History Map stores the history of matched targets.

The goal of the KPI evaluation platform is, first, to aggregate the logs of each component in a single, easy-to-search platform, simplifying real-time monitoring of the entire system; second, to evaluate software and network KPIs, such as latency and reliability.

The KPI evaluation platform consists of three components: Collector, Database, and Data Visualization.

The Collector receives logs from different components through a public network interface, analyzing, formatting, and enriching them before forwarding the messages to the database. The Database is used to store messages collected by the Collector and provides a query language to browse data and calculate KPIs. Data Visualization is a GUI that allows users to monitor database data in real time. It provides the ability to explore raw data as well as create charts and dashboards.

The Elastic Stack is a set of well-integrated open-source components, with Logstash for data collection, Elasticsearch for storing and querying data, and Kibana for visualization. The Micro Services Manager supports connections between independent components in scalable solutions, providing a central logging system for all monitorable components for further operation and analysis.

The reference [7] proposes a V2I scheme based on Visible Light Communication (VLC), which increases the redundancy of V2I communication for Autonomous Valet Parking (AVP) applications through the interference-free characteristics and location-based characteristics of VLC. Its architecture, through vehicle data transmission and location-based map propagation methods, uses bidirectional VLC communication to achieve online map updates in indoor parking lots. The proposed system achieves error-free LoS transmission through DCO-OFDM, with a distance of up to 33 m between the transmitter and receiver, and can share vehicle CAN bus data, camera video, and lidar point cloud data in indoor parking lots.

The VLC-based V2I communication system is based on VLC On-Board Unit (OBU), VLC Roadside Unit (RSU), optical detector, and LED.

The vehicle transmits on-board data through the LED driven by VLC-OBU. In addition, infrastructure information is captured by the vehicle's VLC-OBU and transmitted to the vehicle's mobility planner.

The online map of the parking lot is created by cameras, lidar, and vehicle information. The map server of the infrastructure is responsible for integrating and propagating online map updates through the LED driven by VLC-RSU. The VLC-RSU captures vehicle information through an optical receiver.

VLC provides a unique advantage for location-based transmission due to its LoS characteristics. Therefore, only the relevant parts of the online map are transmitted from the infrastructure LED. In addition, even if the vehicle continuously transmits information through its LED, due to the limitations of the detector's LoS and Field of View (FoV), the deployed infrastructure optical receiver only captures vehicle data at specific locations. Therefore, VLC-based V2I communication realizes the exchange of specific location information between vehicles and infrastructure, reducing the need for on-board computing while providing secure communication without radio frequency interference.

The vehicle transmits deceleration, acceleration, speed, and braking status information through VLC, and receives lidar point cloud maps and videos from the infrastructure. In addition, optical detectors deployed in the parking lot receive vehicle data and aggregate this information into the lidar point cloud map. The continuously updated aggregated map is propagated to the vehicle through the infrastructure LED.

On the vehicle side, the VLC-OBU is responsible for converting vehicle CAN bus information into DCO-OFDM frame information, and driving the vehicle LED with the baseband DCO-OFM signal and the required DC bias; more importantly, it combines an optical receiver to capture information from the infrastructure LED. On the infrastructure side, the VLC-RSU converts lidar point cloud data into DCO-OFDM frame information, drives the infrastructure LED with the baseband DCO-OFM signal and the required DC bias; the VLC-RSU integrates the same hardware as the VLC-OBU, and after optimization, continuously transmits map information instead of periodic vehicle information.

14.5 Collaborative Perception in Autonomous Driving

The reference [8] reviews the related work of collaborative perception technology, including basic concepts, collaborative modes, key elements, and applications.

Two important issues of single-vehicle perception are long-distance occlusion and sparse data. The solutions to these problems are that vehicles in the same area share collective perception information (CPM), collaboratively perceive the environment, which is called collaborative perception or cooperative perception.

Thanks to the construction of communication infrastructure and the development of communication technologies such as V2X, vehicles can exchange information in a reliable manner, thus achieving collaboration. Recent work shows that collaborative perception between vehicles can improve the accuracy of environmental perception as well as the robustness and safety of the traffic system.

Perception is one of the key modules of autonomous driving systems, but the limited field of view of a single vehicle becomes a bottleneck for improving perception performance. To break through the limitations of single perception, collaborative perception is proposed, enabling vehicles to share information and perceive the environment beyond the line of sight and field of view.

In addition, autonomous vehicles are usually equipped with high-fidelity sensors to achieve reliable perception, resulting in expensive cost overhead. Collaborative perception can alleviate the stringent requirements of single vehicles for perception equipment.

Collaborative perception shares information with nearby vehicles and infrastructure, enabling autonomous vehicles to overcome certain perception limitations, such as occlusion and short field of view. However, achieving real-time and robust collaborative perception requires solving some challenges brought by communication capacity and noise. Recently, some work has studied the strategies of collaborative perception, including what is collaboration, when to collaborate, how to collaborate, alignment of shared information, etc.

Similar to fusion, there are also 4 types of collaboration.

1. **Early collaboration** takes place in the input space, sharing raw sensory data between vehicles and infrastructure. It aggregates the raw measurements of all vehicles and infrastructure to get an overall view. Therefore, each vehicle can carry out downstream processing and complete perception based on the over-

all perspective, which can fundamentally solve the occlusion and long-distance problems that occur in individual perception. However, sharing raw sensory data requires a lot of communication and can easily congest the communication network due to excessive data load, which hinders its practical application in most cases.

2. **Late collaboration** takes place in the output space, promoting the fusion of the perception results output by each intelligent body, achieving refinement. Although late collaboration is bandwidth economical, it is very sensitive to the localization error of the intelligent body and suffers from high estimation errors and noise due to incomplete local observations.
3. **Intermediate collaboration** takes place in the intermediate feature space. It can transmit the intermediate features generated by the prediction model of a single intelligent body. After fusing these features, each intelligent body decodes the fused features and produces perception results. Conceptually, representative information can be compressed into these features, saving communication bandwidth compared to early collaboration, and improving perception capabilities compared to late collaboration. In practice, the design of this collaboration strategy poses two algorithmic challenges: ① How to select the most effective and compact features from the original measurements for transmission; ② How to maximize the fusion of features from other intelligent bodies to enhance the perception ability of each intelligent body.
4. **Hybrid collaboration** mentioned above, each collaboration mode has its advantages and disadvantages. Therefore, some work has adopted hybrid collaboration, combining two or more collaboration modes to optimize the collaboration strategy.

Reference [9] uses V2V communication to improve the perception and motion prediction performance of autonomous vehicles, proposing V2VNet. By aggregating information received from multiple nearby vehicles, the same scene can be observed from different angles. This allows for detection of participants through obstructions and at long distances, even in cases where observations are very sparse or non-existent.

V2VNet consists of three stages: ① Feature extraction stage, where a convolutional network block processes raw sensor data and creates compressible intermediate representations; ② cross-vehicle aggregation stage, where information received from multiple vehicles is aggregated with the vehicle's internal state (calculated from its own sensor data) to compute updated intermediate representations; ③ an output network that computes the final perception-prediction.

Convolutional Network Block: Extracts features from LiDAR data and converts them into BEV. Specifically, the past five LiDAR point cloud scans are voxelized, outputting feature maps.

Compression: Uses a variational image compression algorithm to compress the intermediate representation. Then, it is quantized through entropy coding and losslessly encoded with very few bits. Note: The compression module is differentiable, so it can be trained.

Cross-vehicle aggregation: Decodes the intermediate representations received from other vehicles. Specifically, entropy decoding is applied to the bitstream, and a decoder CNN is used to extract the decompressed feature map. Then, the information received from other vehicles is aggregated to generate the latest intermediate representation.

Each vehicle uses a fully connected Graph Neural Network (GNN) as an aggregation module, where each node of the GNN is a state representation of an autonomous vehicle in the scene, including itself. Each autonomous vehicle maintains its own local graph within a certain range (i.e., within 70 m). For the receiver information of the autonomous vehicle, the GNN can perform temporal warping and spatial transformation to the receiver coordinate system. After the autonomous vehicle sends a message, each receiving autonomous vehicle compensates for the time delay of the received message, and the GNN aggregates spatial messages to compute the final intermediate representation.

First, **the time delay** between vehicles is compensated for, creating an initial state for each node of the GNN. Based on universal GPS time, the time delay is the time difference between the start times of each vehicle's LiDAR scan. The time-delay-compensated intermediate representation is taken, zeros are concatenated to increase the capacity of the node state, and the GNN aggregates the information propagated after receiving from other vehicles.

Then, GNN message passing is performed. Since other autonomous vehicles are in the same local area, the node representations have overlapping fields of view. First, a relative **spatial transformation** is applied to warp the intermediate state, and then a CNN is used for joint inference on the spatially aligned feature maps of the two nodes.

Next, at each node, the received messages are aggregated through a mask-perception permutation invariant function, and the node state is updated with a Convolutional GRU (ConvGRU). After the final iteration, a multilayer perceptron outputs the updated intermediate representation.

Output Network: After message passing, a set of 4 Inception-like convolutional blocks are applied to efficiently capture multi-scale context, which is crucial for prediction. Finally, the feature map is obtained and two network branches are used to output detection and motion prediction estimates respectively. A large-scale V2V communication dataset, called V2V-Sim, is generated using a high-fidelity LiDAR simulator, LiDARsim. LiDARsim applies ray casting and then uses a deep neural network to generate real LiDAR point clouds for each frame in the scene.

A large-scale cooperative perception dataset, V2X-Sim, is presented in reference [10]. V2X-Sim provides: ① Synchronized records of multiple vehicles at roadside infrastructure and intersections, achieving cooperative perception; ② multi-modal sensor streams, promoting multi-modal perception; ③ various annotated ground truths, supporting various downstream tasks, including detection, tracking, and segmentation.

For the sensor setup and coordinate system, the sensors consist of RGB cameras, depth sensors, semantic segmentation sensors, BEV semantic segmentation sensors, and LiDAR. Meanwhile, the road infrastructure is equipped with RGB cameras, BEV semantic segmentation sensors, LiDAR, and semantic LiDAR, etc.

On electric vehicles and roadside infrastructure, cameras and LiDAR cover a 360° range, thus achieving full-view perception. Specifically, each vehicle is equipped with 6 RGB cameras, following the nuScenes configuration; roadside infrastructure is equipped with 4 RGB cameras in 4 directions at intersections. Please note that the BEV semantic segmentation sensor is based on orthogonal projection, while the ego-car semantic segmentation sensor uses perspective projection.

On the ego-car and roadside infrastructure, LiDAR and semantic LiDAR, RGB, depth, or semantic sensors are all placed in the same position to obtain depth or semantic ground truth. The BEV semantic segmentation sensor shares the same BEV position with the LiDAR, but is positioned higher to ensure a certain field of view. As for roadside infrastructure, sensors are placed at random heights within a suitable range to enhance diversity.

To address the latency issue of multi-agent cooperative perception, a latency-aware system is proposed from a machine learning perspective in reference [8], which actively adapts the asynchronous perception features of multiple agents to the same timestamp, improving the robustness and effectiveness of cooperation.

The cooperative perception system follows the middle representation cooperation framework. It consists of five components: ① Encoding module, extracting perception features from raw data; ② communication module, transmitting perception features across agents under different communication latencies; ③ latency compensation module, synchronizing the features of multiple agents to the same timestamp; ④ fusion module, aggregating all synchronized features and producing fused features; ⑤ decoding module, using the fused features to obtain the final perception output.

The key component of this system is the latency compensation module, which aims to achieve feature-level synchronization. To achieve this, a SyncNet is proposed in reference [8], which uses historical cooperative information to simultaneously estimate the current feature and the corresponding cooperative attention, both of which are uncertain due to latency. As the attention weight between the two cooperating agents, cooperative attention has the same spatial resolution as the feature and can indicate the information level of each spatial domain in the feature.

Compared with common temporal sequence prediction methods, SyncNet has two main differences: ① Feature-level estimation, rather than output-level prediction; ② estimation of coupled features and associated cooperative attention, rather than predicting a single output.

The function of SyncNet is to use historical information to achieve latency compensation, which includes two parts: Feature Attention Symbiotic Estimation (FASE), which uses a dual-branch pyramid LSTM to simultaneously estimate real-time features and cooperative attention; temporal modulation, which adaptively adjusts the final estimation of cooperative features with delay time.

In the FASE part, each branch is implemented by a pyramid LSTM, which models a series of historical cooperative information and estimates the current state. The pyramid LSTM is specifically used to capture spatially related cooperative features.

The feature estimation branch aims to obtain the feature with the maximum amount of cooperative information at the current time. To achieve this, the feature estimation branch should be attention-aware. The goal of the attention estimation branch is to find the most informative cooperative area at the current time, and it must also suppress areas with large estimation errors. Similarly, the attention estimation branch should be feature-aware. To allow the feature estimation and the corresponding attention estimation to be aware of each other, the estimated feature map and the cooperative attention from the previous timestamp are repeatedly used as the input for the next timestamp of each branch.

In the multi-resolution spatial convolution of FASE the multi-scale convolution architecture replaces the matrix multiplication of LSTM. The original LSTM does not specifically consider extracting spatial features, convLSTM extracts spatial features at a single scale, while pyramid LSTM captures local-global features at multiple scales.

When the delay is low, the performance degradation it causes is relatively less than the estimation noise caused by FASE. To solve this problem, the temporal modulation module focuses on merging the original features (which work well under low delay) and the estimated features (which work normally under high delay) under the condition of delay time, generating more comprehensive and reliable estimates.

14.6 Cooperative Planning and Control of Vehicle Platooning

Reference [11] mainly summarizes the cooperative methods between autonomous vehicles, including vehicle formation control or vehicle platooning.

To achieve formation control, sharing information with other vehicles is a key task in multi-vehicle systems. Therefore, the implementation of the topology structure in vehicle platooning solves the problem of information sharing. These topologies are responsible for the information exchange process, which describes how vehicles in the platoon exchange information with each other.

In multi-agent systems, achieving consensus can be considered one of the important requirements, in which case, autonomous vehicles can reach an agreement to form a formation by locally sharing information with their neighboring vehicles.

Generally, converging to a common value is called consensus, which depends on the communication between autonomous vehicles. According to some designed distributed protocols, the key problems to be solved are local interactions between mobile agents, neighboring agents sharing information, and the analysis of consensus behavior.

Leader-follower control is a widely adopted platooning method. Leadership can be achieved in three ways: static leadership (unchanging leader), virtual leadership (software leader), and dynamic leadership (changing according to the situation). In addition, the communication topology structure responsible for information exchange between vehicles plays a crucial role in the leader-follower method. Once the platoon is formed, string stability analysis is used to analyze the stability of the platoon. In the leader-follower method, the sum of each position vector and vehicle separation vector reaches a consensus, and the information flow itself is a directed spanning tree.

Behavioral control is used to achieve cooperative control of multi-agent systems in unknown or dynamic environments. When it is impossible to accurately model or characterize the real world, behavior-based methods are most effective. Avoiding offline path planning and using sensors to obtain real-time information about the environment can provide the system with autonomy to navigate in complex or chaotic environments.

The environment of autonomous vehicles is full of uncertainties and is unpredictable, noisy, and dynamic. Behavior-based architecture provides answers to overcome these difficulties, enabling real-time processing methods that heavily rely on sensors, avoiding the construction of global world models with potential errors. Behavior-based agents have high autonomy and mechanical precision, almost no computational resources, can be improved through learning, can be programmed through software reuse, and integrated with the environment.

Motor schemas used for mobile agents are a series of actions to complete goal-directed behavior. Schemas and motion units do not represent the simplest basic actions available to agents (such as simple commands for actuators), but represent higher levels of abstraction of agent actions, such as avoiding obstacles, avoiding other agents, maintaining formation, and heading to the goal. These schemas and motion units define control strategies encoded with only a few parameters and serve as the basic set or motion vocabulary for agents.

Such units are sufficient to generate all the actions of the agent through the combination of schemas or primitives. By implementing several motion schemas, such as moving to the goal, avoiding static obstacles, avoiding other agents, and maintaining formation, many formation shapes have been considered, such as straight lines, cylinders, diamonds, and wedges.

In **the Artificial Potential Field** (APF) method, a mobile intelligent body has two fields generated by the target and obstacles in the search space. These two fields are the repulsive field generated by obstacles and the attractive field generated by the target. These forces are stronger when close to obstacles or targets and have less impact at a distance. In this method, the target location gains attraction, while obstacles generate repulsion for the intelligent body. The resultant force (the sum of all forces) of the intelligent body is used to determine its motion, speed, and direction of travel, while avoiding collisions. It has been proven that APF is a good obstacle avoidance algorithm that can be used for orchestration problems.

Flocking describes the behavior of birds, the release of fish, or the swarm behavior of insects. Flocking control mainly includes three behaviors: collision avoidance (also known as separation), velocity matching (also known as alignment), and flock centering (also known as cohesion). Velocity matching is a vector, which refers to the combination of heading and speed. Collision avoidance is a separation behavior that can prevent overcrowding and collisions with each other, while flock centering places the intelligent body near the center of the flock or its flock mates.

The flocking problem can be seen as a sub-case of the formation control problem, which requires intelligent bodies to move together along certain paths of the flock, but the requirements for the paths adopted by specific intelligent bodies are very low. Compared with flocking, formation is more strict, requiring intelligent bodies to maintain a certain relative position when moving in the environment. Therefore, multiple mobile intelligent body systems consider both flocking and formation control simultaneously.

The virtual structure method aims to solve the problem of maintaining geometric configurations during the coordinated movement of intelligent bodies. A virtual structure is a collection of intelligent body elements that maintain (semi) rigid geometric relationships with each other and with the reference system.

The advantages of the virtual structure method are as follows.

- The ability to achieve high-precision control.
- Inherent fault tolerance during intelligent body failure by maintaining formation.
- No need to elect a leading intelligent body.
- Can be reconfigured for different types of virtual structures without modification.
- Can be implemented in a distributed manner without increasing communication overhead like centralized implementation.
- No explicit functional decomposition.

Reference [12] introduces a specific truck platooning path planning algorithm.

During truck platooning, the front truck is manually driven, the rear truck is automatically driven, and the distance between the trucks is short. To successfully platoon in various situations, each truck must maintain dynamic stability, and the entire system must maintain string stability. However, due to the short forward-looking range, the path planning ability of subsequent vehicles will be significantly reduced.

In addition, during the platooning of **articulated cargo trucks**, the off-tracking phenomenon that occurs on bends makes it difficult for the rear truck to follow the trajectory of the front truck. Moreover, without global coordinate system information, the local coordinate systems used by each truck to perceive the environment and dynamic signals are difficult to correlate with each other.

To solve these problems, reference [12] uses a Kalman filter, V2V communication, and an update-and-conversion method to allow each following vehicle to accurately calculate the trajectory of the front part of the leading vehicle and use it as the target path. This algorithm can provide lateral string stability and robustness for truck platooning.

For the architecture of truck platooning, the leading vehicle (LV) is manually driven by an experienced driver, and the following vehicle (FV) is operated by an automatic driving system.

The following vehicle uses environmental sensors such as radar and cameras to perceive the vehicle and lane ahead, and performs automatic driving through vehicle longitudinal and lateral control. The automatic driving algorithm of the following vehicle does not rely on GPS, because the vehicle cannot receive the correct GPS signal in some cases, such as when driving in a tunnel. In the case of large truck platooning, the length of the convoy can easily reach 100 m. Therefore, considering the safety of nearby vehicles, the number of trucks in a platoon is usually limited to 3 or 4.

The longitudinal control of the following vehicle aims to maintain a short distance from the front vehicle, which is basically achieved by adopting a mature ACC algorithm that relies on radar. A shorter vehicle spacing is important in truck platooning because reducing air resistance provides fuel economy for the following vehicle.

The lateral control of the following vehicle aims to follow the driving path of the leading vehicle while maintaining its own lane. The following vehicle uses a camera for lane keeping, but mature LKS algorithms cannot be used because the forward view of the camera is severely limited by the truck in front. The path-following control of the following vehicle sets the trajectory passed by the leading vehicle as its target. Therefore, in truck platooning, a higher level of lateral control method than general autonomous driving is required. There are two main platooning lateral control methods, namely direct vehicle-following and vehicle path-following.

In **the direct vehicle-following** method, the following vehicle operates with the relative longitudinal and lateral distances to the leading vehicle, and calculates the steering angle based on geometric principles. The relative position and relative angle between the center of gravity (CG) of the subject vehicle and the rear center of the leading vehicle can be used to calculate its virtual bending path to the rear of the leading vehicle. However, the relative position information behind the leading vehicle is used, not the trajectory of the leading vehicle's steering wheel, so there may be a problem of driving inside the actual trajectory of the leading vehicle when turning. In addition, when driving on a high-speed highway with small curvature, the reliability of following the virtual bending path of the leading vehicle cannot be guaranteed if the relative yaw angle with the leading vehicle is small and the perception accuracy is low or the measurement resolution is small.

In addition, **vehicle path-following** is a method of following the trajectory of the leading vehicle. The trajectory of the leading vehicle can be obtained by using the motion parameters of the vehicle and storing the position coordinates of the rear of the leading vehicle. Since the look-ahead distance within the trajectory of the leading vehicle can be controlled, the performance of path tracking is improved. However, because the distance between vehicles is short during platooning, there is a problem that sufficient look-ahead distance cannot be obtained at high speeds. If it is a semi-trailer truck, the off-tracking phenomenon during steering, that is, the path between the steering axle of the tractor and the rear bumper of the trailer is different, will cause tracking errors when following the leading vehicle.

This is the **off-tracking** phenomenon. At low speeds, the trajectory of the rear bumper of the trailer is formed inside the trajectory of the tractor; at high speeds, due to the increase in lateral acceleration, the trajectory of the trailer is outward compared to the trajectory of the tractor. Off-tracking is a major factor damaging the lateral dynamics stability of the platoon, and as it propagates to the tail of the platoon, the stability worsens.

To overcome the problem of off-tracking, the following vehicle needs to determine its target path with the tractor of the leading vehicle instead of the trailer. However, the following vehicle cannot perceive the position of the leading vehicle's tractor only through the camera. One study proposed using the curvature of the position trajectory of the leading vehicle's trailer, but it is only effective when the yaw angle between the two vehicles is very small. Another study proposed using DRTK and V2V to access the global position of the leading vehicle's tractor, but because the GPS signal cannot be recognized, platooning trucks usually do not use GPS, because under certain conditions, such as when driving through a tunnel, the GPS signal cannot be received.

For example, a monocular camera and radar are installed on the dashboard and front bumper respectively to perceive the center point of the front vehicle's bumper; the V2V module uses dual antennas, installed inside the left and right rearview mirrors, to minimize the area of the communication blind zone; the controller is MicroAutoBox II, which serves as the CPU to implement platooning control logic and path planning algorithms.

For the truck platooning lateral controller, all vehicles run V2V communication; the leading vehicle controller creates its driving trajectory and transmits it to the following vehicle through V2V communication; using the trajectory of the leading vehicle, the following vehicle performs path planning, that is, calculates the target path it needs to follow; finally, the following vehicle performs path tracking, following the target path; similarly, a target path is created between any adjacent following vehicles, so essentially all following vehicles can follow the trajectory of the leading vehicle. Here we only discuss the path planning algorithm.

For the main vehicle driver trajectory generation, including steps such as state estimation Kalman filter, rear bumper center calculation, front-end and back-end trajectory caching, and trajectory curve fitting. The inputs are speed, yaw rate, steering wheel angle, and kingpin angle, and the outputs are front trajectory, rear trajectory, and current rear point, which are sent to the following vehicle via V2V.

For a 3DoF articulated vehicle model, all variables and parameters of the model are defined according to the local coordinate system of each truck. The origin of this coordinate system is at the center of gravity of the tractor, the x-axis points forward, and the y-axis points to the left.

Design a Kalman filter that estimates the state every 10 ms, which is equivalent to the CAN communication cycle of each car. In the path planning algorithm, the leading car must generate its own driving trajectory, as this trajectory allows the following car to create its target path. This trajectory is composed of two parts, each storing 300 samples in the cache.

Suppose there are two sets of points P and Q representing the same scene in different coordinate systems, then coordinate matching is equivalent to point matching between the two data sets P and Q. First, select a reference point from P and Q respectively, these points must represent the same point in the scene; then, based on their position differences, you can find the translation vector and accordingly translate all points in Q; finally, find the rotation matrix that makes P and Q consistent.

Applying the above algorithm to the two sets of rear trajectories in the coordinate systems of the leading and following cars can achieve coordinate matching between the following car and the leading car. This will be used for the platoon truck path planning discussed below.

The path planning process is as follows: first, generate TF.LV (the leading trajectory in the leading car's coordinate system) and TR.LV (the following trajectory in the leading car's coordinate system), and at the same time, the following car generates TR.FV (the following trajectory in the following car's coordinate system); through V2V communication, the following car receives TF.LV and TR.LV in the form of third-order polynomial coefficients. Use TR.LV and TR.FV, the following car performs coordinate transformation, because they represent the same trajectory; the coordinate transformation is to find the translation and rotation relationship between the leading car and the following car.

Reference [13] discusses cooperative driving under the 6G network, which is a driving mode through information sharing. In the context of the sixth-generation cellular network (6G), connected autonomous vehicles (CAVs) are a key vertical field with great potential to improve road safety, road and energy efficiency. However, the requirements of CAVs for reliability, latency, and high-speed communication will pose great challenges to the 6G network. CAVs supported by 6G need new channel access algorithms and intelligent control schemes for connected vehicles.

Roadside units (RSUs) are not available in all scenarios, here we consider the worst case of the cooperative driving network, distributed V2V communication for communication. As stipulated by DSRC and 3GPP LTE-V2V technology, vehicles compete for access to the broadcast channel in a random manner for regular data transmission.

For V2V communication with relatively low radio frequency, all vehicle data transmission is through random channel access. And high-frequency (for example, millimeter wave and terahertz frequency band) 6G communication, it is assumed that vehicles can only communicate with adjacent vehicles. Without loss of generality, take millimeter wave communication as an example. To distinguish between low-frequency communication and high-frequency communication, they are called **cellular V2V** and **6G communication**, and V2V networks with both low-frequency and high-frequency communication are called **hybrid networks**.

In good weather conditions, cellular V2V and 6G communication channels can be established for short safety messages and large messages containing sensor data, respectively. However, if the environment is not suitable for establishing a 6G channel, cellular V2V can also maintain the basic functions of cooperative driving. According to the characteristics of cellular V2V and 6G communication, cellular V2V communication sends wide-range short packets, and 6G communication sends large sensing data packets to nearby neighbors. If the connectivity status of 6G communication is not good, cellular V2V communication is needed to transmit part of the sensor data.

For hybrid vehicle networks, each vehicle has a message queue for cellular V2V communication and 6G communication. For the same type of communication, vehicles have the same queue capacity. The queue length of cellular V2V communication is represented by L. Only when the distance is less than the required threshold can a vehicle establish a 6G connection, such as a mmWave link. To simplify performance analysis, assume that the topology of the 6G communication network is a connected graph. 6G communication is assumed to be a fluid-flow discrete-time queuing system with infinite buffering. In addition, the longest flow in the 6G network is a flow with a starting node S, n-2 full-duplex relays, and a destination node D, where n is the number of vehicles.

In order to improve the level of traffic safety and improve road tests, it is necessary to predict the communication performance of vehicle networks. The communication delay upper limit of the V2V network is analyzed and modeled through network calculus (SNC), and a deep learning neural network developed and trained for fast prediction of delay boundaries in real-time operation. The analytical results obtained from SNC are used to create training samples for the deep learning neural network.

Based on the upper limit of delay, an intelligent inter-vehicle distance control strategy for connected vehicles can be developed. This strategy must meet the requirements of safety, traffic efficiency, and energy efficiency. Here, deep learning modeling is used, and the entire strategy is divided into two parts: **safety distance** and **decision-making**. In addition, an algorithm based on the Markov chain predicts parameters, and there is a safety distance mapping method for implementing smooth speed changes.

14.7 Summary

This chapter discusses the application of vehicular networks in autonomous driving, including Connected Autonomous Vehicles (CAV), the social characteristics of vehicular networks and their AI elements, edge computing, vehicle-road collaboration, collaborative perception, and collaborative planning control, etc.

References

1. J RIOS-TORRES and A A. MALIKOPOULOS, A Survey on the Coordination of Connected and Automated Vehicles at Intersections and Merging at Highway On-Ramps, IEEE Trans. on Intelligent Transportation Systems, 2017. 18(5): 1066–1077.
2. S MALIK, M AHMED KHAN and H EL-SAYED, Collaborative Autonomous Driving–A Survey of Solution Approaches and Future Challenges, MDPI Sensors, 2021. 21(11): 1–24.
3. L A. MAGLARAS, A H. AL-BAYATTI, Y HE, I WAGNER and H JANICKE, Social Internet of Vehicles for Smart Cities, J. Sensor and Actuator Network, 2016. 5(1): 1–23.

4. Mehmood, Amjad, et al. "ANTSC: An intelligent Naïve Bayesian probabilistic estimation practice for traffic flow to form stable clustering in VANET". IEEE Access 6 (2017): 4452–4461.
5. Z ZHONG, M NEJAD, E E LEE, Autonomous and Semi-Autonomous Intersection Management: A Survey. arXiv preprint arXiv:2006.13133, 2020.
6. O NASSEF, L SEQUEIRA, E SALAM, T MAHMOODI, Deep Reinforcement Learning in Lane Merge Coordination for Connected Vehicles, arXiv preprint arXiv:2010.10567, 2020
7. B TURAN, A UYRUS, O N KOC, E KAR, and S COLERI, Vehicular Visible Light Communications for Automated Valet Parking, arXiv preprint arXiv:2202.02453, 2021
8. S REN, S CHEN, W ZHANG, Collaborative Perception for Autonomous Driving: Current Status and Future Trend, arXiv preprint arXiv:2208.10371, 2022
9. T-H WANG, S MANIVASAGAM, M LIANG, B YANG, W ZENG, J TU, and R URTASUN, V2VNet: Vehicle-to-Vehicle Communication for Joint Perception and Prediction, arXiv preprint arXiv:2008.07519, 2020
10. Y LI, Z AN, Z WANG, Y ZHONG, S CHEN, C FENG, V2X-Sim: A Virtual Collaborative Perception Dataset for Autonomous Driving, arXiv preprint arXiv:2202.08449, 2022
11. A SONI and H HU, Formation Control for a Fleet of Autonomous Ground Vehicles: A Survey, MDPI Robotics, 2018. 7(4): 67–91.
12. S MALIK, M A KHAN and H EL-SAYED, A Novel Path Planning Algorithm for Truck Platooning Using V2V Communication, MDPI Sensors, 2020. 20(24): 1–26.
13. X CHEN, S LENG, J HE and L ZHOU, Deep Learning Based Intelligent Inter-Vehicle Distance Control for 6G Enabled Cooperative Autonomous Driving, arXiv preprint arXiv:2012.13817, 2020

rendered view of the scene representation. Minimizing this error across multiple views specifies the volume density and color for the scene content, encouraging the network to predict a consistent model of the scene.

First, let's introduce the scene representation of NeRF. The continuous scene is represented as a 5D vector-valued function, whose input is a 3D position $\mathbf{x} = (x, y, z)$ and a 2D observation direction (θ, ϕ), and its output is radiance color $\mathbf{c} = (r, g, b)$ and volume density σ. In fact, the direction is represented as a 3D Cartesian unit vector $\mathbf{d}$. An MLP network $F_\theta : (\mathbf{x}, \mathbf{d}) \rightarrow (\mathbf{c}, \sigma)$ is used to approximate this continuous 5D scene representation and optimize its weight θ, to map the input 5D coordinates to volume density and color.

To restrict the network to predict volume density σ (only as a function of position $\mathbf{x}$), while allowing RGB color $\mathbf{c}$ (as a function of position and observation direction) to predict, thus making the representation have multi-view consistency. To achieve this, the MLP network F_θ first passes the input 3D coordinates $\mathbf{x}$ through 8 fully connected layers (activation function ReLU, channel number is 256), and outputs volume density σ and a 256-dimensional feature vector. Then, this feature vector is connected with the observation direction of the camera ray, and passed to another fully connected layer (activation function ReLU, channel number is 128), which outputs the view-related RGB color $\mathbf{c}$.

Next, the volume rendering method of the radiation field is elaborated.

The 5D neural radiation field represents the scene as the volume density and directed emission radiance of any point in space. Then, the color of any ray passing through the scene can be rendered using the principles of classical volume rendering, i.e., the volume rendering equation (VRE). The volume density $\sigma(\mathbf{x})$ can be interpreted as the differential probability of the ray terminating at an infinitesimal particle at position $\mathbf{x}$. The expected color $C(\mathbf{r})$ of a camera ray $\mathbf{r}(t) = \mathbf{o} + t\mathbf{d}$ with near and far boundaries t_n and t_f is

$$C(\mathbf{r}) = \int_{t_n}^{t_f} T(t)\sigma(\mathbf{r}(t))\mathbf{c}(\mathbf{r}(t), \mathbf{d})\, dt \tag{15.1}$$

where

$$T(t) = \exp\left(-\int_{t_n}^{t} \sigma(\mathbf{r}(s))\right) ds \tag{15.2}$$

The function $T(t)$ represents the cumulative transmittance along the ray from t_n to t, i.e., the probability that the ray propagates from t_n to t without hitting any other particles. To render a view from a continuous neural radiation field, it is necessary to estimate the integral $C(\mathbf{r})$ of each pixel of the virtual camera traced back along a camera ray.

Numerical integration (quadrature) is used to estimate this continuous integral. Because the MLP can only query at a fixed set of discrete positions, deterministic integration is usually used for rendering discretized voxel grids, which limits the resolution of the representation. Instead, using stratified sampling, $[t_n, t_f]$ is divided into N evenly spaced bins, and then a sample is uniformly randomly drawn from each bin:

$$t_i \sim \mathcal{U}[t_n + \frac{i-1}{N}(t_f - t_n), t_n + (t_f - t_n)] \tag{15.3}$$

Although a set of discrete samples is used to estimate the integral, stratified sampling can represent a continuous scene representation, because the MLP is evaluated at continuous positions during the optimization process. The numerical integration rule of volume rendering is used to estimate these sample colors $C(\mathbf{r})$:

$$\hat{C}(\mathbf{r}) = \sum_{i=1}^{N} T_i(1 - \exp(-\sigma_i \delta_i))c_i \tag{15.4}$$

where

$$T_i = \exp\left(-\sum_{j=1}^{i-1} \sigma_j \delta_j\right) \tag{15.5}$$

And $\delta_i = t_{i+1} - t_i$ is the distance between adjacent samples. This function is used to calculate $\hat{C}(\mathbf{r})$ from a set of (c_i, σ_i) values, it is a differentiable function, which can be simplified to traditional alpha synthesis with $\alpha_i = 1 - \exp(-\sigma_i \delta_i)$ as parameters.

It can be seen that, in contrast to methods that use neural networks to learn rendering functions, NeRF more explicitly uses knowledge of computer graphics, and due to (physical) inductive bias, this 3D structured intermediate representation of scene density and radiance can better generalize to new views. Therefore, NeRF learns physically meaningful color and density values in 3D space, and physically excited ray casting and volume integration can be continuously rendered to new views.

Finally, how to optimize a NeRF is discussed.

Two improvements are mainly adopted to achieve high-resolution complex scene representation. The first is the position encoding of input coordinates, which helps the MLP represent high-frequency functions (a kind of Fourier transform feature); the second is hierarchical sampling, which effectively samples this high-frequency representation.

1. Position Encoding

Although neural networks are universal function approximators, letting the network F_θ directly operate on the input coordinates $xyz\theta\phi$ leads to poor rendering performance in terms of high-frequency changes in color and geometry representation. Then, redefining F_θ as the combination of two functions $F_\theta = F_\theta \circ \gamma$, one learned and one unlearned, can significantly improve performance. Here γ is a mapping from $\mathbf{R}$ to high-dimensional space $\mathbf{R}^{2L}$, and F_θ is still just a regular MLP. Formally, the encoding function is

$$\gamma(p) = (\sin(2^0\pi p), \cos(2^0\pi p), \ldots, \sin(2^{L-1}\pi p), \cos(2^{L-1}\pi p)) \tag{15.6}$$

The function $\gamma(\cdot)$ is applied separately to the three coordinate values in x (normalized to $[-1, 1]$) and the three components of the Cartesian observation direction unit vector $\mathbf{d}$ (located in $[-1, 1]$). In the experiment, $L = 10$ for $\gamma(x)$ and $L = 4$ for $\gamma(d)$.

In the popular Transformer architecture, a similar mapping called positional encoding is used. However, the Transformer uses it for a different purpose, namely, to provide the discrete positions of tokens in the sequence as input to an architecture that does not contain any concept of order. In contrast, these functions are used here to map continuous input coordinates to a higher-dimensional space, making it easier for the MLP to approximate higher-frequency functions.

2. Hierarchical volume sampling

The rendering strategy of densely evaluating the NeRF network at N query points along each camera ray is inefficient—free space and occluded areas that do not contribute to the rendered image are still repeatedly sampled. Therefore, a hierarchical representation is proposed, which scales the samples according to the expected effect on the final rendering, improving rendering efficiency.

Not only a single network is used to represent the scene, but two networks are optimized simultaneously: a coarse network and a fine network. First, hierarchical sampling is used to sample at a set of N_c positions, and the coarse network at these positions is evaluated, as described in Eqs. 15.3 to 15.5. Given the output of this coarse network, a more insightful point sampling is then generated along each ray, where the samples are biased toward the relevant parts of the body.

To this end, the alpha composite color is first rewritten from the coarse network $\hat{C}_c(\mathbf{r})$, i.e., the weighted sum of all sampled colors c_i along the ray:

$$\hat{C}_c(\mathbf{r}) = \sum_{i=1}^{N_c} w_i c_i\,, \quad w_i = T_i\,(1 - \exp(-\sigma_i \delta_i)) \tag{15.7}$$

where the weights are normalized to $w^i = w_i / \sum_{i=1}^{N_c} w_j$, resulting in a piecewise linear probability distribution function (PDF) along the ray.

Inverse transform sampling is used to sample the second set of N_f positions from this distribution, the fine network is evaluated at the union of the first and second set of samples, and the final rendered color of the ray $\hat{C}_c(\mathbf{r})$ is calculated with all $N_c + N_f$ samples using Eqs. 15.4 and 15.5. This process allocates more samples to areas expected to contain visible content. This is similar to the goal of importance sampling, but here the entire integration domain is discretized non-uniformly with the sample values, rather than treating each sample as an independent probability estimate of the entire integration.

In the specific implementation, a separate neural continuous body representation network is optimized for each scene. This only requires collecting an RGB image dataset for a scene, obtaining the corresponding camera pose, intrinsic parameters, and scene boundaries (in the experiment, the true pose, intrinsic parameters, and synthetic data boundaries are used, and the open-source software COLMAP is used to estimate the parameters of real data). In each optimization iteration, a batch of camera rays is randomly sampled from the set of all pixels in the dataset, and then N_c samples are queried from the coarse network and $N_c + N_f$ samples are queried from the fine network according to hierarchical sampling. Then, the color of each ray from the two sets of samples is rendered using the volume rendering process. The loss calculation is simply the total square error between the coarse rendering, fine rendering, and real pixel color:

$$L = \sum_{\mathbf{r}\in\mathcal{R}} \left[\left\| \hat{C}_c(\mathbf{r}) - C(\mathbf{r}) \right\|_2^2 + \left\| \hat{C}_f(\mathbf{r}) - C(\mathbf{r}) \right\|_2^2 \right] \tag{15.8}$$

It can be said that the impact of the NeRF method on this field lies in its astonishing simplicity: just an MLP outputting density and color under 5D coordinates, plus positional encoding and hierarchical sampling improvements. Despite this, the original NeRF leaves many areas for improvement.

- Both training and rendering speeds are slow.
- Can only represent static scenes.
- Fixed lighting.
- Trained NeRF representations do not generalize to other scenes/targets.

The following discusses improvements to address these weaknesses.

15.2 Acceleration Methods for NeRF

NeRF needs to directly calculate the color and density of individual points in space, rather than querying a simple data structure, which requires evaluating the entire neural network (hundreds of thousands of floating-point operations), making its rendering speed very slow. Some acceleration work is based on MLP representation of volume rendering methods, such as Neural Sparse Voxel Fields. Other methods cache various quantities learned by the MLP on a sparse 3D grid, so that real-time rendering can be performed after training is completed, such as PlenOctrees and non-neural network Plenoxels. Another way to accelerate rendering is to train the MLP representation itself, effectively precomputing part or all of the volume integral along the ray, such as AutoInt. Many new methods use classic data structures such as grids, sparse grids, trees, and hashes to accelerate rendering speed and achieve faster training times, such as Instant Neural Graphics Primitives.

Here, AutoInt, PlenOctree, and Plenoxels are used as examples to discuss NeRF acceleration schemes.

15.2.1 AutoInt

AutoInt is the automatic integration proposed in reference [1]. The method of integration is used to learn effective closed-form integral solutions in coordinate-based neural networks (also known as implicit neural representations). Assuming an MLP is used as the network, its derivative is obtained to get a gradient network, which can be trained according to the signal to be integrated. If the parameters of the gradient network are recombined into the original MLP, an indefinite integral (antiderivative) of the signal represented by the neural network is constructed. According to the fundamental theorem of calculus, any definite integral can be calculated between two estimates of this network. This method is applied to neural rendering, and the rendering time is increased by more than 10 times under the premise of slightly reducing the image quality.

The mathematical definition of automatic integration is as follows: The coordinate-based network, which maps low-dimensional input coordinates to low-dimensional output, is a neural network F_θ with parameters θ. Suppose this network accepts a (sub) gradient relative to the input x, represented by $G_\theta^i = \frac{\partial F_\theta}{\partial x_i}$ for the derivative with respect to the coordinate x_i. Then, according to the fundamental theorem of calculus, we get

$$F_\theta(\mathbf{x}) = \int \frac{\partial F_\theta}{\partial x_i}(\mathbf{x})\, d\mathbf{x}_i = \int G_\theta^i(\mathbf{x})\, d\mathbf{x}_i \tag{15.9}$$

Here, the coordinate-based network F_θ and its derivative G_θ^i are linked, so F_θ is the indefinite integral (antiderivative) of G_θ^i.

G_θ^i is a different neural network, which shares parameters θ with F_θ while satisfying (15–9). Now, instead of optimizing the coordinate-based network F_θ, we optimize G_θ^i to represent a target signal, reorganize the optimized parameters (i.e., weights and biases) θ to form F_θ, which represents the network of the indefinite integral of G_θ^i, the whole process is the automatic integration of the indefinite integral.

The coordinate-based network of the MLP or fully connected architecture is defined as

$$F_\theta(\mathbf{x}) = W_n\left(\phi_{n-1} \circ \phi_{n-2} \circ \cdots \circ \phi_0\right)(\mathbf{x}) \tag{15.10}$$

where ϕ_k is the kth layer of the neural network, defined as $\phi_k(\mathbf{y}) = \mathrm{NL}_k(W_k\mathbf{k} + \mathbf{b}_k)$, parameters $\theta = \{W_k, \mathbf{b}_k\}$, where NL is a nonlinear function.

Then the derivative of this MLP is obtained according to the chain rule:

$$G_\theta^i(x) = \hat{\phi}_{n-1} \circ (\phi_{n-2} \circ \cdots \circ \phi_0)(x) \odot \cdots \odot \hat{\phi}_1(x)\phi_0(x) \odot W_0\mathbf{e}_i \tag{15.11}$$

where $\odot$ represents the Hadamard product, $\hat{\phi}_k(\mathbf{y}) = W_k^T \mathrm{NL}'_{k-1}(W_{k-1}\mathbf{y} + \mathbf{b}_{k-1})$, and $\mathbf{e}_i$ is a vector with only the ith element being non-zero.

It can be said that although the two networks have different architectures (the number of nodes is also very different), they share the same parameters. The network associated with F_θ is called the integration network, and the neural network associated with G_θ^i is called the gradient network.

In order to calculate the antiderivative and definite integral of function f in the AutoInt framework, first select the details of the MLP architecture for the integration network F_θ. Then instantiate the gradient network G_θ^i from this integration network according to AutoDiff. In practice, a custom AutoDiff framework is used to record the integration network and explicitly instantiate the corresponding gradient network, while maintaining shared parameters.

The implementation of AutoInt relies on the evaluation of the computation graph, where dependencies are modeled with a directed acyclic graph (DAG). Using this graph-based representation, an automatic pipeline is created to instantiate the gradient network from the integration network, and an efficient program is provided to evaluate the gradient network during training.

Most nodes in the DAG represent computational operators, with two types of leaf nodes: differentiable input nodes and constant input nodes. Directed edges represent dependencies between nodes and point to dependencies (i.e., other nodes that need to be computed first). Therefore, zero-degree nodes (in-degree zero) are the final results of the computation graph.

To instantiate a gradient network, AutoInt performs automatic differentiation on the DAG of the integration network. Each node is called in topological order and provides its own derivative. This recursive call chain constructs the computation graph of the gradient network.

Once the gradient network is established, it can be evaluated in a reverse topological order of its nodes: starting from the leaves and tracing the computation back to the root. Note that this process is different from backpropagation (BP), which stores the intermediate results of the forward pass for evaluating the graph associated with the backward pass.

In AutoInt, if the gradient network is an entity separate from the integration network, the intermediate results of the forward pass cannot be obtained. However, the gradient network can still be efficiently computed by reusing the computations of network legs, as these nodes share weights and perform the same computations.

The gradient network has a tree-like structure and calculates the partial derivative of the integration network with respect to one of its inputs. Nodes within the "branches" of the gradient network appear multiple times, so these computations can be improved in performance by reusing them during the forward pass. This is achieved by maintaining a dictionary order between nodes of the same level in topological order.

This dictionary order is defined by the order in which nodes are created during the differentiation process of the integration network, with the last created node appearing first in the order. Therefore, in forward mode, the evaluation of the gradient network calls each node in this dictionary topological order. Nodes save their computations in a cumulative manner; when computing the branches of the network, the final result is retained for reuse in other branches.

The training of gradient network weights uses the BP algorithm. During the evaluation of the gradient network, the calculation results of the BP algorithm's backward propagation are saved, and then the weights of the gradient network are updated with Adam (a variant of SGD).

Once instantiated, the parameters of the gradient network will be optimized using traditional AutoDiff and optimization tools to adapt to the signal of interest. Specifically, the optimization loss function is as follows:

$$\theta^* = \arg\min_{\theta} \mathcal{L}(G^i_\theta, f(x)) \tag{15.12}$$

where $\mathcal{L}$ is a cost function that penalizes the difference between the target signal $f(x)$ and the output of the coordinate network G^i_θ.

Once trained, the gradient network approximates the signal, i.e., $G^i_{\theta*} \approx f(x), \forall x$. Therefore, the indefinite integral of $f(x)$ is calculated as follows:

$$\int f(x)\,dx_i \approx \int G^i_{\theta*}(x)\,dx_i = F_{\theta*}(x) \tag{15.13}$$

Then, according to the Newton–Leibniz formula, any definite integral of the signal $f(x)$ can be calculated by two evaluations of F_θ:

$$\int_a^b f(x)\,dx_i = F_\theta(b) - F_\theta(a) \tag{15.14}$$

Automatic integration can effectively calculate definite integrals with closed-form solutions of derivatives. However, the volume rendering equation cannot be directly evaluated with AutoInt because it contains multiple nested integrals, such as the radiative integral along the ray and the integral weighted by accumulated transmittance. Therefore, we choose to approximate this integral in segments, each of which can be effectively calculated with AutoInt. For the N segment cross sections along the ray, the approximate VRE and transmittance are given as follows:

$$\tilde{C}(\mathbf{r}) = \sum_{i=1}^{N} \bar{\sigma}_i \bar{c}_i \bar{T}_i \delta_i, \quad \bar{T}_i = \exp\left(-\sum_{j=1}^{i-1} \bar{\sigma}_j \delta_j\right) \tag{15.15}$$

where

$$\bar{\sigma}_l = \delta_i^{-1} \int_{t_{i-1}}^{t_i} \sigma(t)\,dt, \quad \bar{c}_l = \delta_i^{-1} \int_{t_{i-1}}^{t_i} c(t)\,dt \tag{15.16}$$

Equation 15.15 can be seen as a repeated alpha blending operation, where the alpha value is $\bar{\sigma}_j \delta_j$. After simplification, the following segmented VRE can be obtained:

$$\tilde{C}(\mathbf{r}) = \sum_{i=1}^{N} \delta_i^{-1} \int_{t_{i-1}}^{t_i} \sigma(t)\,dt \int_{t_{i-1}}^{t_i} c(t)\,dt \prod_{j=1}^{i-1} \exp\left(-\int_{t_{j-1}}^{t_j} \sigma(s)\,ds\right) \tag{15.17}$$

Although this segmented expression is only an approximation of the entire VRE, it can effectively calculate the integrals of absorption and radiation in each segment with AutoInt. In practice, as N decreases, a trade-off is made between improving computational efficiency and reducing approximation accuracy.

Subsequently, stratified sampling is also correspondingly corrected, dividing the interval δ_i into M grids and calculating the samples as follows:

$$t_{i,j} \sim \mathcal{U}\left(t_{i-1} + \frac{j-1}{M}\delta_i, t_{i-1} + \frac{j}{M}\delta_i\right) \tag{15.18}$$

15.2.2 PlenOctree

PlenOctree is an efficient data structure that can reproduce non-Lambertian effects in scenes. Specifically, it is an octree that stores spherical harmonic (SH) coefficients at the leaves, encoding view-dependent radiance.

To more directly convert NeRF to PlenOctree, NeRF-SH, as a variant of the NeRF network, directly outputs SH coefficients, eliminating the need to input viewing direction to the network. With this change, model evaluations can be performed on a unified grid and then thresholded.

The NeRF-SH model uses the same optimization program and volume rendering method as NeRF. However, this network does not directly predict RGB color c, but predicts spherical harmonic coefficients k. The color c is calculated by summing the weighted spherical harmonic basis for the corresponding ray direction (θ, ϕ). The representation of spherical harmonic coefficients can model view-dependent appearance. To build PlenOctree, the NeRF-SH model is densely sampled in the volume around the target object, and the density and SH coefficients are tabulated, which can then be directly optimized with training images to improve its quality.

SH has become a popular low-dimensional representation of spherical functions and has been used to model Lambertian surfaces and even smooth surfaces. Specifically, adjust the NeRF network f to output spherical harmonic coefficients k, rather than RGB values.

$$f(x) = (k, \sigma), \quad \text{where } k = (k_l^m)_{0 \le l \le L \le l_{\max}}^{m, -l \le m \le l} \tag{15.19}$$

Each k_l^m is a set of three parameters corresponding to RGB color. In such a setup, the view-dependent color c at position $\mathbf{x}$ can be determined by querying the SH function at the desired view angle $\mathbf{d}$:

$$c(d; k) = S\left(\sum_{l=0}^{l_{\max}} \sum_{m=-l}^{l} k_l^m Y_l^m(d)\right) \tag{15.20}$$

where $S : x \to (1 + \exp(-x))^{-1}$ is the sigmoid function for normalizing color.

In the absence of any regularization, this model can freely generate arbitrary geometry in unobserved areas. Although this does not directly degrade image quality, it adversely affects the conversion process because the extra geometry occupies a large amount of voxel space.

To address this issue, a sparsity prior is introduced during NeRF training. Intuitively, when both blank and pure color are possible solutions, this prior encourages NeRF to choose blank space. Formally speaking,

$$\mathcal{L}_{\text{sparsity}} = \frac{1}{K}\sum_{k=1}^{K} |1 - \exp(-\lambda\sigma_k)| \tag{15.21}$$

where, $\{\sigma_k\}_{k=1}^{K}$ are the density values estimated at K uniformly random points within the bounding box, and λ is a hyperparameter.

The trained NeRF-SH model can be converted into a sparse octree representation for real-time rendering. PlenOctree stores density and SH coefficients, simulating the view-dependent appearance of each leaf.

To render PlenOctree, for each ray, the ray-voxel intersections in the octree structure are first determined. This generates a series of segments of length $\{\delta_i\}_{i=1}^{N}$, with constant density and color, between voxel boundaries. Then, NeRF's volume rendering model is applied to assign color to the ray. This method allows skipping large voxels in one step without missing small voxels.

During testing, early stopping is applied when the accumulated transmittance T_i of the ray is less than $\gamma = 0.01$, which can further speed up this rendering process.

The conversion process from NeRF-SH to PlenOctree can be divided into three steps: first, evaluate the network on a grid at a higher level, retaining only the density values; then, filter the voxels through a threshold. Finally, sample random points within each remaining voxel, average them, and obtain the SH coefficients stored in the octree leaves. Details are as follows:

- Evaluate the NeRF-SH model to obtain σ values on a uniformly spaced 3D grid. The grid will automatically scale to fit the scene content tightly.
- Filter this grid to obtain a sparse set of voxels centered on the grid points, which is sufficient to represent the scene. Specifically, render the alpha map of all training views with this voxel grid, tracking the maximum ray weight $1 - \exp(-\sigma_i\delta_i)$ for each voxel. Then, eliminate voxels with weights below the threshold τ_w. When constructing the octree, the remaining voxels are placed in the deepest leaves, while elsewhere is empty. Compared with the simple method of dividing at each point with σ as the threshold, this method eliminates invisible voxels.
- Sample 256 random points in each remaining voxel, and set the corresponding leaves of the octree to the average of these values to reduce aliasing. Now, each leaf contains the density σ and spherical harmonic coefficient vector for each RGB color channel.

Since this volume rendering process is fully differentiable with respect to these tree values, the octree generated from the original training images can be fine-tuned directly with the NeRF loss function of SGD, improving image quality. Note: The tree structure is obtained from NeRF.

15.2.3 Plenoxel

The Plenoxel (plenoptic voxel) model extends PlenOctree, optimizing the sparse voxel representation with spherical harmonics end to end, providing faster training speed (two orders of magnitude faster than NeRF). The Plenoxel model is a generalization of PlenOctree, supporting sparse plenoptic voxel grids of arbitrary resolution (not necessarily quadratic), and has the ability of trilinear interpolation, making this sparse voxel structure easier to implement.

For the Plenoxel model, given a set of images of a target or scene, it reconstructs a sparse voxel (Plenoxel) grid with the density and spherical harmonic coefficients of each voxel. To render a ray, it calculates the color and opacity of each sample point through trilinear interpolation of adjacent voxel coefficients; like NeRF, it uses differentiable volume rendering to integrate these sample colors and opacities; and it optimizes voxel coefficients with a standard MSE reconstruction loss relative to the training images and a total variation (TV) regularizer.

Similar to PlenOctree, the geometric model uses a sparse voxel grid. However, for easy implementation of trilinear interpolation, Plenoxel does not use an octree as the data structure. Instead, a dense 3D index array with pointers is stored in a separate array, containing only the values of occupied voxels. Like PlenOctree, each occupied voxel stores the scalar opacity σ and spherical harmonic coefficient vector for each color channel.

Spherical harmonics form an orthogonal basis for functions defined on the sphere, with low-order harmonics encoding smooth (more Lambertian) color variations and high-order harmonics encoding higher-frequency (more specular reflection) effects. The color value of sample c_i is the sum of these harmonic basis functions for each color channel, weighted by the corresponding optimized coefficients, and evaluated in the appropriate viewing direction. Here, second-order spherical harmonics are used, which require 9 coefficients for each color channel, a total of 27 spherical harmonic coefficients per voxel. The reason for using second-order spherical harmonics is that PlenOctree found that the benefits of higher-order harmonics are minimal.

The PlenOctree assumes that opacity and spherical harmonic coefficients remain constant within each voxel. The Plenoxel grid uses trilinear interpolation to define a continuous plenoptic function overall. At each sampling point along each ray, the calculation of opacity and color is a trilinear interpolation of the opacity and harmonic coefficients stored in the 8-voxel neighborhood.

Plenoxel adopts a coarse-to-fine strategy to achieve high resolution, starting with a low-resolution dense grid, optimizing and pruning unnecessary voxels, dividing each voxel in half in each dimension, refining the remaining voxels and continuing to optimize. After each voxel subdivision step, trilinear interpolation is used to initialize the grid values. Voxel pruning still uses the method mentioned in PlenOctree. In addition, there is a dilation operation, that is, only when the voxel itself and its adjacent voxels are considered unoccupied, is it pruned.

Finally, through total variation (TV) regularization, the voxel opacity and spherical harmonic coefficients are optimized for the mean square error (MSE) of the rendered pixel color. Specifically, the basic loss function is

$$\mathcal{L} = \mathcal{L}_{\text{recon}} + \lambda_{\text{TV}}\mathcal{L}_{\text{TV}} \tag{15.22}$$

where the MSE reconstruction loss and TV regularizer are calculated as follows:

$$\mathcal{L}_{\text{recon}} = \frac{1}{|\mathcal{R}|}\sum_{r\in\mathcal{R}} ||C(r) - \hat{C}(r)||_2^2 \tag{15.23}$$

$$\mathcal{L}_{\text{TV}} = \frac{1}{|\mathcal{V}|}\sum_{\substack{\mathbf{v}\in\mathcal{V}\\ d\in D}} \sqrt{\Delta_x^2(\mathbf{v}, h) + \Delta_y^2(\mathbf{v}, h) + \Delta_z^2(\mathbf{v}, h)} \tag{15.24}$$

where $\Delta_x^2(\mathbf{v}, h)$ represents the square difference of the h-th value normalized by the resolution between voxel (i, j, k) and $(i + 1, j, k)$, and $\Delta_y^2(\mathbf{v}, h)$ and $\Delta_z^2(\mathbf{v}, h)$ are similar.

As can be seen, Plenoxel does not require the weights of the neural network to extract features, maintains the consistency of features between different scenes, and facilitates fast rendering and transfer.

With minor modifications, Plenoxel can be extended to real unbounded scenes, including forward and 360° panoramic views. For forward scenes, the same sparse voxel grid structure and normalized device coordinates are used as defined by the original NeRF. For 360° scenes, a multi-spherical image (MSI) background model is used to enhance the sparse voxel grid foreground representation, while also using learned voxel color and opacity, as well as trilinear interpolation within and between spheres. Note: The regularizer will be appropriately adjusted to adapt, such as sparse prior knowledge and foreground-background segmentation.

15.3 Rendering of Dynamic Scenes

The original NeRF is used to represent static scenes and targets, but there have been some methods that can handle dynamically changing content. These methods can be classified into two types: one is time-varying representation methods, which synthesize new viewpoints of dynamically changing scenes into unmodified playback (for example, producing "bullet-time" slow-motion effects); the other is techniques that control deformation states, synthesizing and editing content from new viewpoints.

Deformation of NeRF can be implicitly or explicitly implemented, implicitly adjusting the radiance field in deformation, and explicitly using a separate deformation MLP to warp space, regressing the offset from the deformed space to the standard space.

Time-varying NeRF representation allows the playback of videos with new viewpoints. Since these methods give up state control, they do not depend on a specific motion model, so they can handle general targets and scenes.

On the other hand, in order to control the deformation state of NeRF, another type of method uses a specific category of motion model as the basic representation of the deformation state (for example, the deformation model of the face or the skeletal deformation map of the human body).

Next, we will introduce the extension of NeRF technology for dynamic scenes with NR-NeRF as an example.

A simple way to model deformation in the NeRF framework is to adjust the body in deformation (for example, adjust in the time domain or a deformation latent code). However, besides providing hard correspondence, this simple method only produces satisfactory results when reconstructing the input camera path, but the results are very poor in new view synthesis.

Non-rigid Neural Radiance Field (NR-NeRF) is a method for reconstructing and synthesizing new views of general non-rigid dynamic scenes. It inputs RGB images of dynamic scenes (such as monocular videos), creating high-quality spatio-temporal geometry and appearance representations. NR-NeRF decomposes the dynamic scene into a standard body and its deformation. The scene deformation is implemented as a ray bending network, where straight-line rays undergo non-rigid deformation.

In addition, NR-NeRF also proposes a rigidity network to better constrain the rigid areas of the scene, thus obtaining more stable results. The ray bending network and rigidity network can be trained without explicit supervision. In NR-NeRF, the ray bending network b' related to deformation and the rigidity network ω unrelated to deformation bend the straight-line rays in the deformed body into a single static NeRF body $\mathbf{v}$.

Simply put, the NR-NeRF method takes a set of N RGB images $\{\hat{c}_l\}_{i=0}^{N-1}$ and their intrinsic and extrinsic $\{R_i, t_i\}$ and $\{K_i\}$ as inputs for a non-rigid scene. Then, NF-NeRF finds a single canonical neural radiance field that can correctly render each $\{\hat{c}_l\}$ by bending light rays. Specifically, appearance and geometric information are collected from the static canonical body v parameterized by weight θ. The direct light rays emitted by the camera are bent to simulate deformation, obtaining the deformed rendering of v.

This ray bending is implemented as a ray bending network b with weight ψ. Given the current deformation, it maps 3D points (for example, sampled from direct light rays) to the 3D position v. For each image i, this deformation condition appears in the form of automatically decoded latent codes $\{l_i\}$.

Specifically, ray bending is implemented as a ray bending network $b(x, l_i)$. For point x, such as a point on a straight line, the network regresses the offset under the deformation represented by l_i. The offset is then added to x to bend the light ray. Finally, the new bent ray point is passed to the canonical body, i.e., $(c, o) =$

$v(x + b(x, l_i))$. Please note that v is not conditioned on l_i, which results in the separation of deformation (b and l_i) and geometric appearance (v). The bent version of the straight line $\bar{r}$ is represented as $\tilde{r}(j) = \bar{r}(j) + b(\bar{r}(j), l_i)$.

However, the rigid part of the scene is not sufficiently constrained by this formula. Recalculate $b(x, l_i)$ as a product of an original offset $b'(x, l_i)$ and a rigid mask $\omega(x) \in [0, 1]$, i.e., $b(x, l_i) = \omega(x)b'(x, l_i)$. For rigid targets, we want to prevent deformation, so $\omega(\mathbf{x}) = 0$; for non-rigid targets, $\omega(x) > 0$. This makes b' easier to focus on the non-rigid parts of the scene that change over time, as the rigid parts may be masked by the jointly trained rigid network ω.

Since the rigid network is not constrained by the latent code l_i, it is forced to share knowledge of the rigidity of the area in the time domain, which also ensures that part of the rigid background can be reconstructed at all times, even if it does not exist at some times, without any deformation.

Next, we will correct the loss of NeRF for non-rigid settings:

$$L_{\text{data}} = \left\| c_c\left(\tilde{C}\right) - \hat{c}(r)\right\|_2^2 + \left\| c_f\left(\tilde{C} \cup \tilde{F}\right) - \hat{c}(r)\right\|_2^2 \tag{15.25}$$

where the uniformly sampled points are set as $j \in [j_n, j_f]$, given the latent code l, the bent coarse ray points $\tilde{C} = \{\tilde{r}_l(j)\}_{j \in C}$, the fine ray points $\tilde{F} = \{\tilde{r}_l(j)\}_{j \in F}$, $\hat{c}(r)$ is the true color of the pixel, and $c(S)$ is the estimated ray color at the discrete ray point set S.

In addition, a transmittance loss is used to regularize the offset. Since we hope that visually unoccupied space (i.e., air) is compressible and will not hinder optimization, opacity is used to weight the loss of each point. However, this still applies high weights to fully occluded points in the ray, which can cause artifacts when rendering new views. Therefore, it is weighted by transmittance:

$$L_{\text{naive offsets}} = \frac{1}{|C|} \sum_{j \in C} \alpha_j \cdot \|b(\bar{r}(j), l)\|_2^{2-\omega(\bar{r}(j))} \tag{15.26}$$

where each point is weighted together with occupancy by transmittance $\alpha_j = V(j) \cdot o(\tilde{r}(j))$. In fact, the loss used for regression of the rigid mask and the original offset would be better, i.e.:

$$L_{\text{offsets}} = \frac{1}{|C|} \sum_{j \in C} \alpha_j \cdot \left(\left\| b'(\bar{r}(j), l) \right\|_2^{2-\omega(\bar{r}(j))} + \omega_{\text{rigidity}} \omega(\bar{r}(j)) \right) \tag{15.27}$$

where b' is penalized instead of b.

Since the offset loss only restricts the visible area, additional regularization of the hidden area is introduced. Inspired by local isometric shape preservation in computer graphics, such as regularization of surfaces to be as rigid as possible or body preservation, we seek to maintain local shape after deformation. For this, it is proposed to regularize the absolute value of the divergence of the offset field.

Using Helmholtz decomposition, any twice differentiable 3D vector field on a bounded domain can be decomposed into a rotation-free vector field and a divergence-free vector field. Therefore, by penalizing divergence, the vector field can be encouraged to be mainly composed of translations and rotations, effectively maintaining the body. The divergence loss is defined as

$$L_{\text{divergence}} = \frac{1}{|C|} \sum_{j \in C} \omega'_j \cdot |\text{div}(b(\bar{r}(j), l))|^2 \tag{15.28}$$

where

$$\text{div}(b(x)) = \text{Tr}\left(\frac{\text{d}b(x)}{\text{d}x}\right) = \frac{\partial b(x)_x}{\partial x} + \frac{\partial b(x)_y}{\partial y} + \frac{\partial b(x)_z}{\partial z} \tag{15.29}$$

where $b(x)k$ is the kth component of $b(x)$ and $db(x)/dx$ is the 3×3 Jacobian matrix. And the calculation of $Tr(\cdot)$ follows the Hutchinson formula as follows:

$$\text{Tr}(A) = \mathbb{E}_e\left[e^{\top} A e\right] \tag{15.30}$$

where e is a Gaussian distribution.

15.4 NeRF Relighting Techniques

General NeRF applications are based on a simplified absorption–emission volume rendering model, where the scene is modeled as a volume of particles that block and emit light. Although this model is sufficient to render images of the scene from new viewpoints, it cannot render images of the scene under different lighting conditions. Enabling relighting requires a scene representation that simulates the transmission of light within the volume, including the scattering of light by particles with various material properties.

The following introduces how to relight in NeRF using the neural reflectance field as an example.

The neural reflectance field is a deep scene representation that encodes the volume density, normals, and reflectance properties of any 3D point in the scene using a fully connected neural network. This representation, combined with a physically based differentiable **ray marching** framework, can render images from NeRF under any viewpoint and light. Images are captured from a simple camera light co-located setup, and the neural reflectance field is estimated, accurately modeling the appearance of the real scene with complex geometry and reflectance. Once estimated, it can be used to render photorealistic images from new viewpoints and (non-co-located) lighting conditions, accurately reproducing challenging effects such as specular reflections, shadows, and occlusions.

Let's start with the reflectance field. In general, the physical rendering of a non-emitting, non-absorbing volume is determined by the volume rendering equation (VRE), which is rewritten here for convenience to calculate the radiance $L(c, \omega_0)$ at point c in direction ω_0:

$$L(c, \omega_0) = \int_0^{\infty} \tau_c(x)\sigma(x)L_s(x, \omega_0)\, dt \tag{15.31}$$

where

$$\tau_c(x) = e^{-\int_0^x \sigma(c-u\omega_0)\, du} \tag{15.32}$$

Here, t represents the 1D position along a ray traced within the volume, $x = c - t\omega_0$ represents the 3D position of t, and point c typically represents the camera position. σ is the extinction coefficient, representing the probability density of medium particles, i.e., volume density. $\tau_c(x)$ represents the transmission coefficient, which determines the loss of light from c to x. And $L_s(x, \omega_0)$ represents the scattered light at x in direction ω_0 as follows:

$$L_s(x, \omega_0) = \int_S f_p(x, \omega_0, \omega_i)L_i(x, \omega_i)\, d\omega_i \tag{15.33}$$

where S is the unit sphere, s represents scattering, f_p is the phase function determining light scattering, and $L_i(x, \omega_i)$ is the incident radiance arriving at position x from direction ω_i.

Here, L_s is approximated by single bounce direct illumination under a point light source, calculated as follows:

$$L_s(x, \omega_0) = f_r(x, \omega_0, \omega_i, n(x), R(x))L_i(x, \omega_i) \tag{15.34}$$

where f_r represents a differentiable reflection model with parameters R and n represents the local surface shading normal. When considering direct illumination from a point light source only, L_i is determined by the light source intensity and the extinction loss traversing the volume:

$$L_i(x, \omega_i) = \tau_l(x)L_l(x) \tag{15.35}$$

where τ_l is the transmittance from the light to the shading point; L_l represents the light intensity considering distance decay; l is the position of the point light source; and ω_i is the direction of the vector $l - x$.

Finally, a reflectance-aware volume rendering equation (VRE) takes the following form:

$$L(c, \omega_0) = \int \tau_c(x)\tau_l(x)\sigma(t)f_r(x, \omega_0, \omega_i, n(x), R(x))L_l(x)\, dt \tag{15.36}$$

This VRE equation considers the complete single bounce camera-volume-light path in light transport. Unlike previous work that only considered view transmittance or opacity between the shading point and the camera, it explicitly represents the transmittance from the point to the light (τ_l), allowing for the rendering of realistic shadows under different point light sources. Essentially, it does not model scene radiance L_s like previous view synthesis work, it decouples multiple factors (τ_l, f_r, L_l) embedded in L_s , and explicitly models scene reflectance parameters in f_r, thus allowing for reflectance-aware volume rendering for view synthesis and relighting, with realistic shading and shadow effects.

Now let's talk about the ray marching framework. This method can numerically estimate the continuous integral in Eq. 15.36, similar to the approach previously used in volume rendering. Specifically, ray marching starts from the camera center, passes through each pixel on the image plane, and samples a sequence including N shading points x_j on each ray. This rendering equation is calculated as follows:

$$L(c, \omega_0) = \sum_{j=0}^{N} \tau_c(x_j)\tau_l(x_j)(1 - \exp(-\sigma(x_j)\Delta t_j)) f_r(x_j) L_l(x_j) \tag{15.37}$$

where Δt_j represents the ray step length at point x_j, here ignoring other parameters of f_r (ω_0, ω_i, n, R), etc. $\tau_c(x_j)$ is an integral of Eq. 15.2, which can be numerically estimated as follows:

$$\tau_c(x_j) = \exp\left(-\sum_{k=0}^{j} \sigma(x_k)\Delta t_k\right) \tag{15.38}$$

The transmittance $\tau_l(x_j)$ can be estimated in the same way, but requires sampling an additional sequence of points x'_p from the light source to the shading point x_j using another ray:

$$\tau_l(x_j) = \exp\left(-\sum_{p} \sigma(x'_p)\Delta' t_p\right) \tag{15.39}$$

Equations 15.37 to 15.39 represent the framework of light propagation in reflection and perception. Given a camera c and a point light source l, this framework calculates the radiance of any propagating light ray passing through the scene based on the volume density σ, normal n, and reflection characteristics R of each point in the scene.

Furthermore, this light propagation framework supports any differentiable reflection model f_r. Based on the classical analytical BRDF based on f_r, this model uses the diffuse reflection coefficient (albedo) and specular roughness to model the reflection on opaque surfaces.

The following introduces the use of neural network regression for these necessary rendering characteristics. For the neural reflection field and the light propagation framework, the light ray passes through the pixels of each camera and samples a series of shading points on the ray. The light ray radiation under the point light source is calculated by the volume density σ, normal n, and reflection characteristics R through the differentiable light propagation equation. At each point x_j, the shading is calculated according to the specified reflection model f_r, the normal $n(x_j)$, the reflection characteristics $R(x_j)$, and the corresponding light and viewing directions (ω_i and ω_o). The volume density $\sigma(x_j)$ weakens the shading like opacity ($\alpha = 1 - \exp(-\sigma\Delta t)$); it is also used to calculate the viewpoint transmittance τ_c along the camera light ray, and the additional light transmittance τ_l along the direction of the light. Here, a neural reflection field represented by an MLP is used to regress the required rendering characteristics (σ, n, R) from the 3D position $x_j = (x, y, z)$ for light propagation.

Given a reflection model f_r with m parameters, the neural reflection field outputs a $(4 + m)$-dimensional vector at any 3D position x in the scene, which includes the volume density σ(1D), normal n(3D), and reflection characteristics R(mD). In fact, a microfacet BRDF model is used, where R includes the diffuse reflection coefficient and specular roughness, but a hair reflection model is also used as an extension. The parameterization of the neural reflection field uses an MLP with 15 fully connected layers and ReLU activation layers.

Given the 3D position x, a frequency-based position encoding is applied. Specifically, for each dimension of the known 3D point x, the scalar value v is mapped to

$$\gamma(v) = (\sin(2^0\pi v), \cos(2^0\pi u), \ldots, \sin(2^{W-1}\pi v), \cos(2^{W-1}\pi u)) \tag{15.40}$$

where W represents the highest frequency level. These are the inputs to the MLP that regresses scene characteristics in the encoding $\gamma(x)$, $\gamma(y)$, and $\gamma(z)$.

This framework only uses the 3D position as input, inferring the scene appearance attributes that are independent of the view (and light).

Since the entire rendering process (characterization and light propagation) is differentiable, the neural reflection field network is trained to minimize the error between the rendered image and the scene capture image. Flash images captured with a combined light and view (such as a mobile phone with a camera and flash) are used to train the network. To optimize point sampling in light propagation, two networks are trained for each scene—coarse and fine neural reflection field networks—and rendering is performed with a coarse-to-fine adaptive sampling process. Hierarchical sampling is also used to sample the sparse point set on each propagating light ray, so the distribution function is calculated with the coarse network, and then the dense point set is sampled from the distribution function, and the final radiance brightness value is calculated with the fine network.

Specifically, each complete ray segment is divided into N_1 units, and a point is randomly drawn from each unit to obtain stratified sampling points. Starting from these points, the coarse network radiance of the light ray can be calculated using Eq. 15.37 as a product, and the corresponding point-by-point contribution weights can also be generated:

$$a(x, x_j) = \tau_c(x_j)\left(1 - \exp(-\sigma(x_j)\Delta t_j)\right) \tag{15.41}$$

The weight $a(x_j)$ essentially describes the visibility of the point at x to the camera. Normalize the weight $a(x_j)$ of each point to construct a piecewise constant probability distribution, and then draw N_2 points (importance sampling) from this distribution. According to the visibility information collected from the coarse neural reflection field, this distribution adaptively selects new samples.

Then, using all $N = N_1 + N_2$ sampling points and the rendering parameters of the fine neural reflection field, the final radiance is calculated according to Eq. 15.37. This coarse-to-fine adaptive sampling effectively distributes more sampling points in the areas that contribute the most to the rendering integral, thus allowing precise shading calculations of high-frequency details.

It is suggested to pre-calculate an adaptive transmittance body to effectively approximate τ_l. The learned neural reflection field can quickly calculate the transmittance of the transmittance body. Specifically, a virtual image plane is placed in front of the point light source pointing to the scene, and the light propagation passes through each pixel, similar to the classic camera practice. Similarly, adaptive sampling is performed using two trained networks (coarse network and fine network). First, the visibility-aware distribution function is calculated with sparse points sampled in stratified sampling bins using the coarse representation method.

Then, dense points are drawn from the distribution, samples from the two processes are merged, and their transmittance is calculated to generate a transmittance body that adapts to the visibility information inferred from the coarse network. This transmittance body adaptively distributes the sampling points around the visible scene structure along each light ray, thereby achieving an efficient transmittance representation.

Using the learned network and pre-calculated adaptive transmittance body, ray tracing is started from the viewpoint, and an image is rendered from any viewpoint under any point light source. At any given shading point, locate the nearest sampling point, then linearly interpolate the transmittance body to obtain the required transmittance. This method can well recover the real shadow effect when relighting.

Furthermore, during inference rendering, the rays are sampled from coarse to fine. Basically, during inference, the rays and the camera are adaptively sampled from coarse to fine during ray tracing, thereby achieving efficient transmittance calculation and final image synthesis.

L2 loss is used to supervise the regression radiance values of the coarse network and the fine network, as well as the true radiance values from the acquired image. Since opaque targets are considered, it is also necessary to regularize the ray transmittance (from the fine network) to make it close to 0 or 1, which helps to obtain a clean background.

15.5 Generalization Techniques of NeRF

The method of overfitting a single MLP in a single scene requires a large amount of image observation data. The core goal of generalization in scene representation is to synthesize new views when given very few or possibly only a single input view.

Different scenes can be generalized through various key methods. One type of method follows a method similar to Image-Based Rendering (IBR), where multiple input views are warped and blended to synthesize new viewpoints; in the context of MLP-based scene representation, this is usually achieved through local adjustment, where the coordinate input of the scene representation MLP is connected with a local variation feature vector, stored in a discrete scene representation, such as a voxel grid.

Another type of method aims to learn the overall, global representation of the scene, rather than relying on images or other discrete spatial data structures. Given a set of observation data, this method uses the scene representation MLP to infer a set of weights to describe the entire scene. Some work encodes the scene in a single low-dimensional latent code, which is then used to adjust the MLP of the scene representation.

There is also a similar method, which does not infer a low-dimensional latent code based on a set of observations of the 3D scene, but learns an unconditional generative model. A 3D scene representation embedded in a Generative Adversarial Network (GAN) with a neural renderer. There is no need to infer a low-dimensional latent code from a set of observations, but the distribution of the latent code is directly defined.

The following introduces generalization techniques using GRAF and GIRAFFE as examples.

15.5.1 GRAF

GRAF (generative radiance field) is a generative model of radiance fields that achieves the synthesis of high-resolution 3D perception images by introducing a discriminator based on multi-scale patches, while the model training only requires 2D images taken by cameras with unknown poses.

The goal is to learn a model that synthesizes new scenes by training on raw images. More specifically, an adversarial framework is used to train a generative model of radiance fields (GRAF).

In GRAF model, the generator G_θ takes the camera matrix K, camera pose ξ, 2D sampling pattern ν, and shape/appearance code z_s/z_a as input and predicts an image patch P'; the discriminator D_ϕ compares the synthesized patch P' with the patch P extracted from the real image I; during inference, a color value is predicted for each image pixel; however, this operation is too expensive at training time, so the pixel sampling module Γ predicts a fixed patch P of size $M \times M$ pixels, which is randomly scaled and rotated, providing gradients for the entire radiance field.

Let $\nu = (u, s)$ determine the center $u = (u, v)$ and scale s of the virtual $M \times M$ patch $P(u, s)$ to be generated. The random patch center u comes from a uniform distribution $u \sim U(\Omega)$ of an image domain Ω, while the patch scale s comes from a uniform distribution $s \sim U([1, S])$, where $S = \min(W, H)/M$, and W and H represent the width and height of the target image. The shape and appearance variables z_s and z_a are sampled from the shape and appearance distributions $z_s \sim p_s$ and $z_a \sim p_a$, respectively. In experiments, both p_s and p_a use standard Gaussian distributions.

The $M \times M$ patch $P(u, s)$ depends on a set of 2D image coordinates:

$$P(u, s) = \left\{ (sx + u, sy + v) \mid x, y \in \left\{ -\frac{K}{2}, \ldots, \frac{K}{2} - 1 \right\} \right\} \tag{15.42}$$

The corresponding 3D ray uniquely depends on $P(u, s)$, camera intrinsic parameters K, and camera pose ξ. Below, we define the pixel/ray index r, the normalized 3D ray d_r, and the number of rays R (training time $R = K^2$, inference time $R = WH$). For the numerical integration of the radiance field, the stratified sampling method is still used, and N points $\{x_r^i\}$ are sampled along each ray r.

The radiation field is represented by a deep connected neural network, where the parameter θ maps the position encoding of 3D position x and observation direction d to RGB color value c and volume density σ:

$$g_\theta : (\gamma(x), \gamma(d), z_s, z_a) \rightarrow (c, \sigma) \tag{15.43}$$

where g_θ depends on two additional latent codes: one is the shape code z_s determining the target shape and the other is the appearance code z_a determining the appearance. Here, g_θ is referred to as the conditional radiation field. First, the shape encoding h is calculated based on the position encoding of x and the shape code z_s; the density head σ_θ converts this encoding into volume density σ; to predict the color c at the 3D position x, h is connected with the position encoding of d and the appearance code z_a, and the resulting vector is passed to the color head c_θ; σ is calculated independently of the viewpoint d and appearance code z_a, encouraging multi-view consistency while separating shape and appearance; this encourages the network to model shape and appearance separately with latent codes z_s and z_a, and allows them to be handled separately during the inference process.

A more formal representation is as follows:

$$h_\theta : (\gamma(x), z_s) \rightarrow h \tag{15.44}$$

$$c_\theta : (h(x, z_s), \gamma(d), z_a) \rightarrow (c, \sigma) \tag{15.45}$$

$$\sigma_\theta : h(x, z_s) \rightarrow \sigma \tag{15.46}$$

The implementation of $(h_\theta, c_\theta, \sigma_\theta)$ uses a fully connected network with ReLU activation. To avoid confusion, the same symbol θ is used to record the parameters of each network.

Given the color and volume density of all points on the ray $\{(c_r^i, \sigma_r^i)\}$, volume rendering yields the pixel color c_r corresponding to the ray r. Combining the results of all R rays, the predicted patch is represented as P'.

The discriminator D_ϕ is implemented as a convolutional neural network, comparing the predicted patch P' with the patch P extracted from the real image I extracted from the distribution p_D. To extract an $M \times M$ patch from the real image I, first extract $\nu = (u, s)$ from the same distribution p_υ used to extract the aforementioned generator patch; then, query I at the 2D image coordinates $P(u, s)$ through bilinear interpolation, sampling the real patch P. This bilinear sampling operation is represented by $\Gamma(I, \nu)$.

Experiments have found that a single discriminator with shared weights is sufficient for all patches, even if these patches are sampled at random positions at different scales. Note: The scale determines the receptive field of the patch. Therefore, to facilitate training, start capturing the global context from patches with a larger receptive field. Then, gradually sample patches with smaller receptive fields to refine local details.

Finally, the training model is a GAN objective function with R_1 regularization as follows:

$$\begin{aligned} V(\theta, \phi) = & \mathbb{E}\left[f\left(D_\phi(G_\theta(z_s, z_a, \xi, \nu))\right)\right] \\ & + \mathbb{E}\left[f\left(-D_\phi(\Gamma(I, \nu))\right) - \lambda \left\|\nabla D_\nabla(\Gamma(I, \nu))\right\|^2\right] \end{aligned} \tag{15.47}$$

where

$$f(t) = -\log(1 + \exp(-t)) \tag{15.48}$$

15.5.2 GIRAFFE

GIRAFFE is used to generate scenes in a controllable and realistic manner when training on original unstructured images. The main contributions are twofold: ① incorporating a composite 3D scene representation directly into the generative model, achieving more controllable image synthesis; ② combining this explicit 3D representation with a neural rendering pipeline, achieving faster inference and more realistic images. For this, the scene representation is a composite generative neural feature field. For a randomly sampled camera, a feature image of the scene is volume rendered based on a single feature field; a 2D neural rendering network converts the feature image into an RGB image; training only uses original images, and during testing, the image formation process can be controlled, including camera pose, target pose, and the shape and appearance of the target; in addition, the model can be extended beyond the range of training data, for example, it can synthesize scenes containing more targets than the training images.

Rendering the scene into a relatively low-resolution feature image can save computation time. The neural renderer processes these feature images and outputs the final rendering. In this way, this method can obtain high-quality images and scale to real scenes. When trained on a collection of original unstructured images, this method allows for controllable image synthesis of single-target and multi-target scenes. The following discusses the details, including the separate target modeling of the neural feature field, the combination of scenes with multiple separate targets through the feature field, the effective combination of volume rendering and neural rendering, model training, etc.

This method replaces the 3D color output of GRAF with a more general M_f-dimensional feature f, and the target is represented as a generative neural feature field:

$$h_\theta : (\gamma(x), \gamma(d), z_s, z_a) \rightarrow (\sigma, f) \tag{15.49}$$

GIRAFFE represents each target with a separate feature field and the following affine transformation:

$$T = \{s, t, R\} \tag{15.50}$$

where s, t are scale and translation parameters and R is a rotation matrix. Based on this representation, the spatial domain transforms the target point into the scene space, that is:

$$k(x) = R \cdot \begin{bmatrix} s_1 & 0 & 0 \\ 0 & s_2 & 0 \\ 0 & 0 & s_3 \end{bmatrix} \cdot x + t$$

In practice, volumetric rendering is performed on the scene space, and the feature field is evaluated in a normalized target space:

$$(\sigma, f) = h_\theta \left(\gamma \left(k^{-1}(x)\right), \gamma \left(k^{-1}(d)\right), z_s, z_a\right) \tag{15.51}$$

This allows multiple targets to be placed in the scene. All target feature fields share weights, and the affine transformation is sampled from a data-related distribution.

When combining scenes, two situations need to be considered: N is fixed and N changes (where the last one is the background). In practice, the background is represented in the same way as the target, but the scale and translation parameters s_N, t_N are fixed across the entire scene, with the origin of the scene space as the center.

A single body $h^i_{\theta_i}$, for a given point x and observation direction d, predicts a density σ_i and a feature vector f_i. Considering non-rigid targets, the total density at point x is calculated by summing the individual densities and taking the density-weighted average of all its features in direction d. All features:

$$C(x, d) = \left(\sigma, \frac{1}{\sigma} \sum_{i=1}^{N} \sigma_i f_i\right), \quad \sigma = \sum_{i=1}^{N} \sigma_i \tag{15.52}$$

Start with 3D volumetric rendering. For a given camera extrinsic parameter ξ, assume $\{x_j\}_{j=1}^{N_s}$ is a set of sample points along the camera ray d, and $(\sigma_j, f_j) = C(x_j, d)$ is the volumetric density and feature vector. The volumetric rendering operator π_{vol} maps these evaluations to the final feature vector f:

$$\pi_{\text{vol}} : \{\sigma_j, f_j\}_{j=1}^{N_s} \rightarrow f \tag{15.53}$$

Using numerical integration, f can be obtained as follows:

$$f = \sum_{j=1}^{N_s} \tau_j \sigma_j f_j, \quad \tau_j = \prod_{k=1}^{j-1} (1 - \alpha_k), \quad \alpha_j = 1 - e^{-\sigma_j \delta_j} \tag{15.54}$$

where τ_j is the transmittance; α_j is the alpha value of x_j; and $\delta_j = \|x_{j+1} - x_j\|_2$ is the distance between adjacent samples. The entire feature image is estimated by π_{vol} for each pixel. For efficiency, the feature image is rendered at a resolution of 162 (lower than the output resolutions of 642 and 2562). Then, the low-resolution feature image is upsampled to a high-resolution RGB image through 2D rendering.

The weights θ of the 2D rendering operator π_{neural} map the feature image I_V to the final synthesized image $\hat{I}$, which can be parameterized as a 2D CNN with leaky ReLU activation, followed by a combination of 3×3 convolution and nearest neighbor (NN) upsampling Up_{NN}. At each resolution, the feature image is mapped to an RGB image, then summed with the previous output through a bilinear upsampling block Up_{BL}, thereby increasing the spatial resolution. The skip connections ensure strong gradient flow in the feature field. The final layer applies a sigmoid operation to obtain the final image prediction.

The generator is defined as

$$G_\theta \left(\left\{ z_s^i, z_a^i, T_i \right\}_{i=1}^{N}, \xi \right) = \pi_\theta^{\text{neutral}} (I_V) \tag{15.55}$$

where

$$I_V = \{\pi_{\text{vol}}\{C(x_{jk} \cdot d_k)\}_{k=1}^{N_s}\}\}_{k=1}^{H_V \times W_V} \tag{15.56}$$

where N is the number of individuals in the scene; N_s is the number of sample points along each ray; $\mathbf{d}_k$ is the ray of the kth pixel; and x_{jk} is the jth sample of the kth pixel/ray.

The discriminator is a CNN with leaky ReLU activation. During network training, the number of individuals sampled from the scene is $N \sim p_N$, the latent code is $\{z_s^i, z_a^i\} \sim N(0, I)$, the camera pose is $\xi \sim p_\xi$, and the target-level transformation is $T_i \sim p_T$. In practice, p_ξ and p_T are defined as uniform distributions of data-related camera pitch and effective target transformation, respectively.

The model training uses a GAN objective function and R1-gradient penalty as follows:

$$V(\theta, \phi) = \mathbb{E}\left[f(D_\phi(G_\theta(z_s^i, z_a^i, T_i)_i, \xi))\right] + \mathbb{E}\left[f(-D_\phi(I)) - \lambda \left\|\nabla D_\phi(I)\right\|^2\right] \quad (15.57)$$

where

$$f(t) = -\log(1 + \exp(-t)) \quad (15.58)$$

15.6 Techniques for Quality Improvement

There are some improvements, including adding supervised data (DONeRF), optimizing camera pose (bundle-adjusting neural radiance field, BARF), hybrid surface/volume representation (NeuS), robustness and quality improvement (NeRF++, NeRF in the wild, Mip-NeRF, Mip-NeRF 360, NeRF-in-the-dark) and combination with standard computational imaging methods (NeRF-in-the-dark), large-scale scenes (urban radiance field, URF, CityNeRF, Block-NeRF) and NeRF from text (CLIP-NeRF), etc.

The following introduces recent improvement works with Mip-NeRF and NeRF-in-the-dark as examples.

15.6.1 Mip-NeRF

When observing scene content at different resolutions, NeRF may generate overly blurred or jagged renderings. The solution provided by Mip-NeRF extends NeRF, enabling it to represent scenes at a continuous scale. Mip-NeRF reduces the annoying jaggedness by effectively rendering anti-aliasing conical frustums instead of rays, significantly enhancing NeRF's ability to represent details, while also being 7% faster than NeRF, and the model is only half the size. Mip-NeRF matches the accuracy of brute-force oversampling NeRF on multi-scale datasets, while being two times faster.

The input to Mip-NeRF is a 3D Gaussian function, representing the region of the radiance field integral. Mip-NeRF is queried at intervals along a cone to render prefiltered pixels, approximating the conical frustum corresponding to the pixel with a Gaussian function. To encode the 3D position and its surrounding Gaussian region, a feature representation, namely, Integrated Position Encoding (IPE), is proposed. This is a generalization of NeRF's Position Encoding (PE), allowing a region to be compactly characterized, rather than a single point in space.

Mip-NeRF does not perform point sampling along each ray, but divides the projected cone into a series of conical frustums (cones cut perpendicular to their axis); it does not construct Position Encoding (PE) features from an infinitesimal point in space, but constructs the Integrated Position Encoding (IPE) representation for each conical frustum. These changes enable the MLP to infer the size and shape of each conical frustum, not just its centroid. This conical frustum and IPE feature allow NeRF's two separate "coarse" and "fine" MLPs to be reduced to a single multi-scale MLP, thereby improving training and evaluation speed and reducing the model size by 50%.

Like NeRF, the image in Mip-NeRF is rendered one pixel at a time, so the process can be described based on the single pixel to be rendered. For this pixel, a cone is projected from the projection center o of the camera along the direction d passing through the center of the pixel. The vertex of this cone is located at o, and the radius of the cone at the image plane $o + d$ is parameterized as $\dot{r}$. Setting $\dot{r}$ to the width of the pixel in world coordinates, scaled by $2/\sqrt{12}$, will generate a cone whose part on the image plane matches the variance of the pixel area in x and y. The set of positions x within the conical frustum between the two thresholds $[t_1, t_0]$ is

$$F(x, o, d, \dot{r}, t_0, t_1) = \mathbb{I}\left\{ \left(t_0 < \frac{d^\top (x - o)}{\|d\|_2^2} < t_1 \right) \wedge \left(\frac{d^\top (x - o)}{\|d\|_2 \cdot \|x - o\|_2} > \frac{1}{\sqrt{1 + \left(\frac{\dot{r}}{\|d\|_2} \right)^2}} \right) \right\} \tag{15.59}$$

where $\mathbb{I}\{\cdot\}$ is an indicator function. $F(x, \cdot) = 1$, when and only when x is in the conical frustum defined by $(o, d, \dot{r}, t_0, t_1)$.

Now construct the body feature representation within this conical frustum. The simplest and most effective solution is to simply calculate the expected position encoding of all coordinates within the conical frustum:

$$\gamma^*(o, d, \dot{r}, t_0, t_1) = \frac{\int \gamma(x) F(x, o, d, \dot{r}, t_0, t_1)\, dx}{\int F(x, o, d, \dot{r}, t_0, t_1)\, dx} \tag{15.60}$$

Approximate the conical frustum with a multivariate Gaussian to effectively approximate the required feature, called **Integrated Position Encoding** (IPE).

To approximate the conical frustum with a multivariate Gaussian, the mean and covariance of $F(x, \cdot)$ must be calculated. Since each conical frustum is assumed to be circular and symmetric about the cone axis, this Gaussian distribution is fully represented by three values (excluding o and d): the average distance μ_t along the ray, the variance σ_t^2 along the ray, and the variance σ_r^2 perpendicular to the ray.

$$\mu_t = t_\mu + \frac{2t_\mu t_\delta^2}{3t_\mu^2 + t_\delta^2},$$
$$\sigma_t^2 = \frac{t_\delta^2}{3} - \frac{4t_\delta^4(12t_\mu^2 - t_\delta^2)}{15(3t_\mu^2 + t_\delta^2)^2}, \tag{15.61}$$
$$\sigma_r^2 = \dot{r}^2\left(\frac{t_\mu^2}{4} + \frac{5t_\delta^2}{12} - \frac{4t_\delta^4}{15(3t_\mu^2 + t_\delta^2)}\right)$$

These quantities are parameterized according to the midpoint $t_\mu = (t_0 + t_1)/2$ and half-width $t_\delta = (t_1 - t_0)/2$. Convert this Gaussian function from the conical frustum coordinate system to world coordinates as follows:

$$\mu = o + \mu_t d, \quad \Sigma = \sigma_t^2(dd^\top) + \sigma_r^2\left(I - \frac{dd^\top}{\|d\|_2^2}\right) \tag{15.62}$$

to get the final multivariate Gaussian function.

Next, derive the Integrated Position Encoding (IPE), which is the expectation of a position encoding coordinate according to the above Gaussian function distribution. To achieve this, first rewrite the Position Encoding (PE) as a Fourier feature:

$$P = \begin{bmatrix} 1\,0\,0\,2\,0\,0 & & 2^{L-1} & 0 & 0 \\ 0\,1\,0\,0\,2\,0 & \cdots & 0 & 2^{L-1} & 0 \\ 0\,0\,1\,0\,0\,2 & & 0 & 0 & 2^{L-1} \end{bmatrix}^T, \quad \gamma(x) = \begin{bmatrix} \sin(Px) \\ \cos(Px) \end{bmatrix} \tag{15.63}$$

This reparameterization allows for the derivation of a closed form for IPE. Using the fact that the covariance of a linear transformation of variables is a linear transformation of the variable covariance ($\mathrm{Cov}[Ax, By] = A\mathrm{Cov}[x, y]B^T$), after lifting it to the IPE base P, the mean and covariance of the conical truncated Gaussian can be determined:

$$\mu_\gamma = P\mu, \quad \Sigma_\gamma = P\Sigma P^T \tag{15.64}$$

The final step in generating IPE features is to calculate the expectation of this lifted multivariate Gaussian function, modulated by position sines and cosines. These expectations have simple closed-form expressions:

$$E[\sin(x)] = \sin(\mu)\exp\left(-\left(\frac{1}{2}\right)\sigma^2\right) \tag{15.65}$$

$$E[\cos(x)] = \cos(\mu)\exp\left(-\left(\frac{1}{2}\right)\sigma^2\right) \tag{15.66}$$

This expectation of sine or cosine is just the mean sine or cosine attenuated by the variance Gaussian function. By this method, the final IPE feature can be calculated as the expectation of the mean sine, cosine, and the diagonal elements of the covariance matrix as follows:

$$\gamma(\mu, \Sigma) = E[\gamma(x)] = \begin{bmatrix} \sin(\mu_y) \circ \exp\left(-\left(\frac{1}{2}\right) \operatorname{diag}(\Sigma_\gamma)\right) \\ \cos(\mu_y) \circ \exp\left(-\left(\frac{1}{2}\right) \operatorname{diag}(\Sigma_\gamma)\right) \end{bmatrix} \tag{15.67}$$

Because position encoding encodes each dimension independently, this expectation encoding only depends on the marginal distribution of $\gamma(x)$, only needing the diagonal of the covariance matrix (a vector of per-dimension variances). Since the size of Σ_γ is relatively large and the computation cost is high, the diagonal elements of Σ_γ are calculated directly:

$$\operatorname{diag}(\Sigma_\gamma) = \left[\operatorname{diag}(\Sigma), 4\operatorname{diag}(\Sigma), \cdots, 4^{L-1}\operatorname{diag}(\Sigma)\right]^T \tag{15.68}$$

This vector depends on the diagonal elements of the 3D position covariance matrix Σ, calculated as follows:

$$\operatorname{diag}(\Sigma) = \sigma_t^2(d \circ d) + \sigma_r^2\left(1 - \frac{d \circ d}{\|d\|_2^2}\right) \tag{15.69}$$

If these diagonal elements are calculated directly, the construction cost of the IPE feature is roughly the same as that of the PE feature.

In addition to cone tracing and IPE features, Mip-NeRF behaves similar to NeRF. Cone projection and IPE features allow explicit encoding of scale into input features, enabling the MLP to learn multi-scale representations of the scene. Therefore, Mip-NeRF uses an MLP with a parameter θ, repeatedly queried in a hierarchical sampling strategy. This has several benefits: the model size is halved, rendering is more accurate, sampling is more efficient, and the overall algorithm is simpler. The optimization problem is

$$\min_\theta \sum_{r \in R}\left(\gamma \left\|C^*(r) - C(r; \theta, t^c)\right\|_2^2 + \left\|C^*(r) - C(r; \theta, t^f)\right\|_2^2\right) \tag{15.70}$$

Because there is only one MLP, the "coarse" loss must be balanced with the "fine" loss, which is achieved by using the hyperparameter λ (set $\lambda = 0.1$). Coarse samples t^c are generated by hierarchical sampling, while fine samples t^f are sampled by inverse transform sampling of the alpha composite weights ω. Unlike NeRF, in Mip-NeRF, 128 samples are sampled for the coarse model and 128 samples for the fine model. Before sampling t^f, the weights ω are slightly modified:

$$\omega_k' = \frac{\max(\omega_{k-1}, \omega_k) + \max(\omega_{k+1}, \omega_k)}{2} + \alpha \tag{15.71}$$

Mip-NeRF is implemented in JaxNeRF, following the training procedure of NeRF: 1 million Adam iterations, batch size of 4096, learning rate annealed from 5×10^{-4} to 5×10^{-6}.

15.6.2 *NeRF-in-the-Dark*

Like most view synthesis methods, NeRF uses tone mapping low dynamic range (LDR) as input. These images have already been processed by a lossy camera pipeline that smooths details, clips highlights, and distorts the single noise distribution of the original sensor data. NeRF-in-the-dark modifies NeRF to train directly on linear raw images, preserving the full dynamic range of the scene. By rendering raw output images from the generated NeRF, new high dynamic range (HDR) view synthesis tasks can be performed. In addition to changing the camera viewpoint, focus, exposure, and tone mapping can be manipulated post hoc.

Although the noise of a single raw image is significantly higher than that of a post-processed image, NeRF has strong robustness to the zero-mean distribution of raw noise. When optimized with many noisy raw input images, the scene representation generated by NeRF is very accurate, and its rendering of new views surpasses dedicated image denoising algorithms. This method, called RawNeRF, reconstructs scenes from almost dark noisy images.

The comparison of RawNeRF and original NeRF is given below. The original NeRF training pipeline receives LDR images sent through the camera processing pipeline, reconstructs the scene, and renders new views in the LDR color space. Therefore, its rendering has been effectively processed and cannot be significantly adjusted. In contrast, RawNeRF modifies NeRF to train directly on linear raw HDR input data. The generated scene representation generates new views that can be edited like any raw photo.

Firstly, this method applies a loss that more severely penalizes errors in dark areas, consistent with the human perception of compressing dynamic range. One approach is to pass the rendering estimates $\hat{y}$ and noise observation intensity y through a tone mapping curve ψ before applying this loss:

$$L_{\psi} = \sum_i \left(\psi(\hat{y}_i) - \psi(y_i)\right)^2 \tag{15.72}$$

To avoid bias in tone mapping under noise conditions, a weighted L2 loss is used here. After linearizing the tone curve ψ, we get

$$L_{\psi}(\hat{y}, y) = \sum_i \left[\psi'(\mathrm{sg}(\hat{y}_i))(\hat{y}_i - y_i)\right]^2 \tag{15.73}$$

where $sg(\cdot)$ records a cutoff gradient, which considers the input as a constant with zero derivative, preventing the loss gradient from affecting BP. A gradient-supervised tone curve $\psi(y) = \log(y + \epsilon)$, where $\epsilon = 10^{-3}$, results in high-quality results with minimal perceptual distortion, meaning the loss weighting term is $\psi'(sg(\hat{y}_i)) = (sg(\hat{y}_i) + \epsilon)^{-1}$. The final form of the loss is

$$L_\psi(\hat{y}, y) = \sum_i \left(\frac{\hat{y}_i - y_i}{g(\hat{y}_i) + \epsilon} \right)^2 \tag{15.74}$$

In scenes with a high dynamic range, even 10–15 bit original images may not be sufficient to capture both bright and dark areas in a single exposure. This has been solved in many digital cameras' bracketing mode, where multiple images with different shutter speeds are taken in succession, then merged to utilize the bright highlights preserved in shorter exposures and the more detailed dark areas captured in faster exposures.

Variable exposure in RawNeRF can also be utilized. Given a series of images I_i with exposure time t_i (and all other capture parameters kept constant), the linear space color output of RawNeRF is exposed by the recorded shutter speed t_i, thus matching the brightness of image I_i.

In practice, due to sensor calibration errors, shutter speed alone cannot precisely align different exposures. For this, a learned per-color channel scale factor is added for each unique shutter speed in the image set, optimized together with the NeRF network. Given a network's output color y_i, the final exposure of RawNeRF is to minimize $(\hat{y}_i^c \cdot t_i \cdot \alpha_{t_i}^c, 1)$, where c is the color channel, and $\alpha_{t_i}^c$ is the learned scale factor for shutter speed t_i and channel c ($\alpha_{t_{\max}}^c = 1$ in long exposure situations).

The above estimate is clipped at 1, which can explain the fact that pixels saturate in overexposed areas. This scaling and clipping value is passed to the loss described earlier.

The specific implementation is based on the Mip-NeRF codebase. The only network architecture change is modifying the activation function of the MLP outputting color from a sigmoid to an exponential function, better parameterizing linear radiance values. The Adam optimizer is used for batch 16k random ray sampling of all training images, with the learning rate decaying from 10^{-3} to 10^{-5} over more than 500k optimization steps.

Scenes with extreme noise interference benefit from a regularization loss on volume density, preventing partially transparent "floating" artifacts. A loss term also needs to be added for the variance of the weight distribution that accumulates color values along the ray during volume rendering. The weight variance regularizer is a function that computes the combined weights of colors, given the MLP output, i.e., the lengths of the corresponding ray segments, these weights are computed as

$$\omega_i = (1 - \exp(-\Delta_i \sigma_i)) \exp\left(-\sum_{j<i} \Delta_j \sigma_j \right) \tag{15.75}$$

If these weights define a piecewise constant probability distribution p_w of a ray segment, then the variance regularizer equals:

$$L_\omega = \mathrm{Var}(X) = E\left[(X - E[X])^2\right] \tag{15.76}$$

The mean (expected depth) is calculated as

$$\bar{t} = E[X] = \sum_i \int_{t_{i-1}}^{t_i} \frac{\omega_i}{\Delta_i} t\, dt = \sum_i \frac{\omega_i}{\Delta_i} \frac{t_i^2 - t_{i-1}^2}{2} = \sum_i \omega_i \frac{t_i + t_{i-1}}{2} \tag{15.77}$$

The regularizer is calculated as follows:

$$\mathrm{Var}(X) = \sum_i \omega_i \frac{(t_i - \bar{t})^2 + (t_i - \bar{t})(t_{i-1} - \bar{t}) + (t_{i-1} - \bar{t})^2}{3} \tag{15.78}$$

Regularization with large weights results in a small loss of sharpness, which can be improved by annealing during training.

15.7 Gaussian Splatting Techniques

The emergence of Neural Radiance Fields (NeRF) marks a major milestone in the field of computer graphics and 3D scene reconstruction, revolutionizing the way of dealing with new view synthesis. Based on deep learning and computer vision, NeRF is able to render photorealistic scenes from a set of sparse input views, establishing a new paradigm for image synthesis. However, NeRF also encounters challenges and limitations, especially in terms of computational efficiency and controllability. In this context, 3D Gaussian Splatting (3D GS) [2–5] came into being. It is not only an incremental improvement, but also a paradigm shift approach that redefines the boundaries of scene representation and rendering.

3D GS introduces a scene representation technique using millions of 3D Gaussians, as shown in Fig. 15.2. Unlike implicit coordinate-based models, 3D GS adopts explicit representation and highly parallelized workflows, which promotes more efficient computation and rendering. The innovation of 3D GS lies in its unique fusion of the advantages of differentiable pipelines and point-based rendering techniques. By using a learnable 3D Gaussian representation for the scene, it preserves the desirable properties of the continuous radiation field, which is critical for high-quality image synthesis, while avoiding the computational overhead associated with rendering in empty space, a common drawback of traditional NeRF approaches. The GS radiance field [6], also known as 3D Gaussian Splatting (3DGS), is a scene representation based on explicit radiance fields that uses a large number of 3D anisotropic spheres

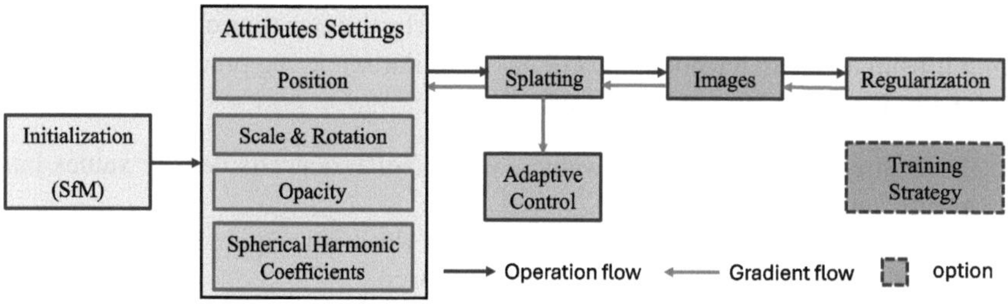

Fig. 15.2 The illustration of 3DGS

to represent the radiance field, each of which is modeled using a 3D Gaussian distribution as follows:

$$G(X) = M^T \Sigma^{-1} M,$$

where each anisotropic sphere has a mean M and covariance Σ for modeling view-dependent color. For regularization optimization, the covariance matrix is further decomposed into a rotation matrix R and a scaling matrix S as follows:

$$\Sigma = RSS^T R^T.$$

These matrices can be further represented as a quaternion r and a scale factor s.

For this scene representation, view rendering is performed via point splatting. Specifically, all Gaussian spheres in the scene are first projected onto a 2D image plane and their colors are computed based on the spherical harmonic parameters. Then, for each 16 × 16 pixel patch in the final image, the projected Gaussian functions that intersect it are sorted by depth. For each pixel in the patch, its color is alpha composited with the opacity and color of all Gaussian functions covering that pixel in depth order, computed as follows: $C = \sum_{i \in N_{cov}} c_i \alpha_i \prod_{j=1}^{i-1} (1 - \alpha_j)$, where N_{cov} represents the splats covering the pixel, α_i represents the opacity of the Gaussian splat multiplied by the projected 2D Gaussian distribution density at the pixel location, and c_i represents the calculated color.

From the perspective of point-based rendering technology, we can see the progress brought by Gaussian splatting. Traditional point-based rendering methods focus more on how to use given geometry to produce high-quality rendering results. But with the recent development of implicit representation, neural implicit representation is used for point-based rendering without any given geometry to complete 3D reconstruction tasks. NeRF uses an implicit density field to model geometric shapes and another appearance field to predict view-related colors c_i. To ensure high-quality rendering, NeRF usually needs to sample 128 points on a single ray, which inevitably requires longer training and rendering time. In order to speed up training and rendering, 3D Gaussian splatting abandons neural networks and directly optimizes a set of

Gaussian ellipsoids. The pixel color is determined by the Gaussian ellipsoid projected onto it from a given viewpoint. 3D GS has a similar rendering process to NeRF, but there are two main differences between them:

- 3DGS directly models opacity values, while NeRF converts density values into opacity values.
- 3DGS uses rasterization-based rendering and does not require sampling points, while NeRF requires dense sampling in 3D space.

The pipeline is roughly divided into the following stages, i.e., initialization, attribute optimization, splatting, regularization, training strategy, and adaptive control.

3DGS is usually initialized using structure from motion (SfM) or randomly generating sparse points. However, these methods are often unreliable, especially under weak supervision signals. Thus, techniques that incorporate pre-training or structured information can be used, as high-fidelity initialization can alleviate training instabilities, especially in underdetermined scenes. It is feasible to determine the initialization protocol based on existing prior knowledge.

The number of initial Gaussian primitives may not be sufficient to achieve high-quality scene reconstruction; therefore, 3DGS provides a method to adaptively control the Gaussian primitives. The method evaluates whether the primitive is "under-reconstructed" or "over-reconstructed" by observing the gradient of the position attributes of each Gaussian primitive in view space. Based on this evaluation, the method increases the number of Gaussian primitives by cloning or splitting the primitives to enhance the scene representation capability. In addition, the opacity of all Gaussian primitives is periodically reset to zero to mitigate artifacts during the optimization process. This adaptive process allows 3D GS to start optimization with a smaller initial set of Gaussians, thereby alleviating the dependence on previous point-differentiable rendering methods on dense point clouds.

The original attributes of 3DGS include position, scale, rotation, spherical harmonics (SH) coefficients, and opacity. Modifying Gaussian attributes helps to perform a wider range of downstream tasks and provides an efficient method because it does not require additional structural elements. In addition, combining the new attributes with supplementary information constraints may also significantly enhance the representation effect of the original 3DGS. For example, in some cases, semantic attributes can produce more precise object boundaries.

The role of splatting is to efficiently convert 3D Gaussian data into high-quality 2D images, ensuring smooth and continuous projection and significantly improving rendering efficiency. As a core technology of traditional computer graphics, splatting is also working on improving it from the perspective of efficiency and performance.

Regularization can be divided into two types: two dimensional and three dimensional. Three-dimensional regularization directly constrains 3DGS, while two-dimensional regularization imposes constraints on the rendered image, thereby affecting the optimization of attributes. Introducing regularization terms is an effective way to improve the reconstruction performance of 3DGS. These regularization terms can impose constraints on various attributes of 3DGS according to specific task requirements, including geometric shape, spatial distribution, etc. In addition, in underdetermined scenarios, the optimization process can be further constrained by introducing additional prior information. For specific tasks, multiple different constraints can be introduced at the same time.

Training strategies can be divided into multi-stage strategies and end-to-end strategies. Using different 3D representations in different training stages is a typical multi-stage training paradigm. In dynamic 3DGS, two-stage (static and dynamic) reconstruction can be adopted. Generally speaking, it trains a time-independent static 3DGS in the first stage and then trains a time-dependent deformation field in the second stage to characterize the dynamic Gaussian function. In addition, incremental reconstruction of dynamic scenes frame by frame is also an idea. In the multi-task optimization, the multi-stage training paradigm can enhance stability and performance.

The end-to-end strategy includes four typical methods: (1) the progressive optimization strategy helps 3DGS prioritize learning global representations before optimizing local details; (2) the block optimization strategy is often used for large-scale scene reconstruction, which can improve efficiency and alleviate catastrophic forgetting; (3) the robust optimization strategy introduces noise perturbations, especially for camera poses, time steps, and images, which can be regarded as a form of data enhancement to prevent overfitting; (4) the distillation-based strategy uses a teacher–student model to further compress model parameters. Adaptive control is mainly used to adjust the number of Gaussian primitives, usually including cloning, splitting, and pruning. This technique helps improve rendering fidelity and computational efficiency. However, excessive densification or pruning can have an adverse effect on the efficiency and performance of 3D Gaussian distribution.

It is worth mentioning that the post-processing strategy of the pre-trained Gaussian model can improve the efficiency and performance of the original model. Common post-processing usually enhances Gaussian representation through various optimization strategies.

In addition, the transformable representation is beneficial to the integration of 3DGS with other representations, and its advantages are used to improve the original 3DGS. The implicit representation can be used to alleviate the conditional difficulty and surface artifacts of 3DGS. Finally, when dealing with underdetermined problems such as sparse view settings, introducing additional priors is a direct way to improve the performance of 3DGS. Although some priors may increase computational overhead and cost, they can still significantly enhance the representation capabilities of 3DGS.

15.8 NeRF and 3DGS for Autonomous Driving

There have been a few recent surveys discussing the applications of neural rendering techniques in robotics or autonomous driving [7–10]. Besides basic view synthesis, scene-conditioned neural fields can also modify environmental lighting conditions such as camera exposure, weather, or time of day to further enhance simulated scenes.

Neural fields have become an important framework in the field of autonomous driving due to their ability to generate realistic 3D environments from RGB images. These environments are valuable for building immersive simulation systems with several key properties [9]: (1) Neural fields offer extensive operability and composability, allowing for seamless integration and manipulation of objects in the scene; this helps simulate complex scenarios, such as collisions, which are difficult to replicate in physical environments. (2) The scenes generated by neural fields have impressive photorealism, allowing realistic simulations from visual data. (3) The strong generalization ability of neural fields from sparse inputs allows the creation of accurate and scalable environments, enhancing research on embodied intelligence.

The following examples illustrate the application of NeRF and 3D GS in autonomous driving.

MARS [11] proposes a NeRF-based autonomous driving simulator to address special situations in autonomous driving. It adopts instance-aware modeling and proposes a modular design for foreground instances and background environments, respectively, which can flexibly switch between different NeRF-related components and achieve photorealistic effects. Note: MARS is open source.

UniSim [12] is a neural sensor simulator that enables closed-loop evaluation of autonomous vehicles by converting recorded driving logs into realistic multi-sensor simulations. UniSim reconstructs scene elements using neural feature grids and dynamically simulates LiDAR and camera data, which helps to accurately evaluate autonomous driving systems in safety-critical scenarios.

NeuRAD [13] is a view synthesis method for dynamic AD data. It is designed to handle large-scale automotive scenes. Its network design is simple, where static and dynamic elements are identified only by their position embeddings, and modeled for camera and LiDAR sensors.

DrivingGaussian [14] reconstructs dynamic driving scenes using static 3D Gaussian functions and dynamic Gaussian maps, and applies LiDAR priors for Gaussian superposition, thereby achieving photorealistic synthesis and multi-camera consistency, surpassing the existing technology.

Street Gaussians [15] introduces a dynamic urban scene model that uses point clouds with semantic logic vectors and 3D Gaussian functions for vehicles and backgrounds to achieve efficient scene editing and fast rendering.

S^3 Gaussians [16] is a self-supervised Street Gaussian method that decomposes dynamic and static elements in the scene based on 4D consistency. Each scene is represented by a 3D Gaussian to maintain explicitness, and a spatio-temporal field network is used to further compactly model the 4D dynamics.

15.9 Summary

This chapter introduces the technology of the recent neural rendering branch NeRF. It mainly discusses its basic algorithms, acceleration methods, dynamic scenes, relighting, and generalization issues. Besides, Gaussian splatting, a latest technique with the combination of point rendering, is also analyzed. Neural rendering technology is still immature and is an emerging field with many open challenges to solve. Due to space limitations, it is not possible to cover all aspects of its cutting-edge technology here.

References

1. D. LINDELL, J. MARTEL, and G. WETZSTEIN, AutoInt: Automatic integration for fast neural volume rendering. arXiv preprint arXiv:2012.01714, 2020
2. G Chen, W Wang, "A Survey on 3D Gaussian Splatting", arXiv:2401.03890, 2024
3. B Fei, J Xu, R Zhang, et al., "3D Gaussian as a New Vision Era: a Survey", arXiv:2402.07181, 2024
4. T Wu et al., "Recent Advances in 3D Gaussian Splatting", arXiv:2403.11134, 2024
5. Y Bao et al., "3D Gaussian Splatting: Survey, Technologies, Challenges, and Opportunities", arXiv:2407.17418, 2024
6. B Kerbl, G Kopanas, T Leimkuhler, and G Drettakis. "3D Gaussian Splatting for real-time radiance field rendering". ACM TOG, 42(4), 2023.
7. L He, L Li, W Sun, and et al., "Neural Radiance Field in Autonomous Driving: A Survey", arXiv:2404.13816, 2024
8. S Zhu, G Wang, X Kong and et al., "3D Gaussian Splatting in Robotics: A Survey", arXiv:2410.12262, 2024
9. M Z Irshad, M Comi, Y-C Lin, and et al., "Neural Fields in Robotics: A Survey", arXiv:2410.20220, 2024
10. H Zhu, Z Zhang, J Zhao, and et al., "Scene reconstruction techniques for autonomous driving: a review of 3D Gaussian splatting", AI Review, 58(1), Nov. 2024
11. C Wu et al., "MARS: An Instance-aware, Modular and Realistic Simulator for Autonomous Driving", arXiv:2307.15058, 2023
12. Z. Yang, Y. Chen, J. Wang, et al, "UniSIM: A neural closed-loop sensor simulator", IEEE CVPR, arXiv:2308.01898, 2023
13. A Tonderski, Carl Lindstroom, G Hess, et al., "NeuRAD: Neural Rendering for Autonomous Driving", arXiv:2311.15260, 2023
14. X Zhou, Z Lin, X Shan, et al., "DrivingGaussian: Composite Gaussian Splatting for Surrounding Dynamic Autonomous Driving Scenes", arXiv:2312.07920, 2023
15. Y. Yan, H. Lin, C. Zhou, and et al., "Street Gaussians for modeling dynamic urban scenes," arXiv:2401.01339, 2024.
16. N Huang et al., "S^3 Gaussian: Self-Supervised Street Gaussians for Autonomous Driving", arXiv:2405.20323, 2024

Chapter 16
Diffusion Models

Given observation samples x of the distribution of interest, the goal of a generative model is to learn to model its true data distribution $p(x)$. Once learning is complete, new samples can be generated at will according to this approximate model. Moreover, under certain formulas, the learned model can be used to evaluate the likelihood of observed or sampled data.

Generative Adversarial Networks (GANs) model the sampling process of complex distributions, learning in an adversarial manner. Another class of generative models, known as likelihood-based, attempt to learn a model that assigns a high likelihood to observed data samples, including autoregressive (AR) models, normalizing flows (NFs), and variational autoencoders (VAEs). A similar approach is energy-based modeling (EBM), where a distribution is learned as an arbitrarily flexible energy function, then normalized.

Generative models in deep learning generate observed results based on hidden parameters. However, current generative models face four major problems: slow generation process (low sampling efficiency), single type of data processing (weak generalization ability), suboptimal likelihood, and the need for model dimension reduction.

Recently, Diffusion Models, with their powerful generative capabilities, have become one of the hot topics in generative models. Application areas include computer vision, speech generation, bioinformatics, and natural language processing.

So far, in the field of computer vision, diffusion models have been applied to various generative modeling tasks, such as image generation, image super-resolution, image inpainting, image editing, and image-to-image translation. In addition, the latent representations learned by diffusion models are also useful in discriminative tasks, such as image segmentation, classification, and anomaly detection.

Diffusion probabilistic models were initially proposed as latent variable generative models inspired by non-equilibrium thermodynamics. These models consist of

Y. Huang and Z. Yang, *Autonomous Driving Handbook*,
https://doi.org/10.1007/978-981-95-3725-9_16

two processes: the forward process, which adds noise at multiple scales, gradually disturbing the data distribution; and the reverse process, which learns to recover the data structure.

This chapter covers the theory and application of diffusion models. Section 16.1 discusses fraction-based generative networks with Langevin dynamics; Sect. 16.2 introduces denoising diffusion probabilistic models; Sect. 16.3 analyzes denoising diffusion implicit models (DDIMs); Sect. 16.4 discusses the SDE framework that encapsulates previous methods; Sect. 16.5 introduces the application of diffusion models in image and video synthesis; Sect. 16.6 discusses another application, image-to-image translation; Sect. 16.7 introduces the application of diffusion models in text-to-image/video generation; Sect. 16.8 introduces the application of diffusion models in policy and planning learning, Sect. 16.9 summarizes some improvements to diffusion models.

16.1 Score-Based Generative Networks

The core idea of diffusion models is to reconstruct a target data distribution by gradually transforming random noise into structured samples. To achieve this, the distribution of the generated samples must closely match that of the original data. Score-based generative models follow this principle: rather than directly modeling the energy function, they learn the score function—the gradient of the log-probability density—using a neural network. This learned score guides the iterative denoising process that underpins diffusion models.

The score matching model aims to solve the original data distribution estimation problem by approximating the gradient of the data density (called the score). The main method of score matching is to train a score network to predict this score, obtained by perturbing the data at different noise levels.

Assume that the dataset consists of i.i.d. samples $\{x_i\}_{i=1}^N$ from an unknown data distribution $p_{\text{data}}(x)$. The score of the probability density $p(x)$ is defined as $\nabla_x \log p(x)$. The score network s_θ is a neural network parameterized by θ, trained to output an approximation of the score of the $p_{\text{data}}(x)$ distribution. The goal of generative modeling is to learn a model from the dataset to generate new samples from $p_{\text{data}}(x)$.

With score matching, the score network $s_\theta(x)$ can be trained directly to estimate $\nabla_x \log p_{\text{data}}(x)$, without first training a model to estimate $p_{\text{data}}(x)$. Unlike the typical use of score matching, here we choose not to use the gradient of the energy-based model as the score network, to avoid additional computations due to high-order gradients. The goal is to minimize $\frac{1}{2}\mathbb{E}_{p_{\text{data}}}\left[\|s_\theta(x) - \nabla_x \log p_{\text{data}}(x)\|_2^2\right]$, which can be expressed as equivalent to the following:

$$\mathbb{E}_{p_{\text{data}}}\left[\operatorname{tr}\left(\nabla_x s_\theta(x)\right) + \frac{1}{2}\left\|s_\theta(x)\right\|_2^2\right] \tag{16.1}$$

where $\nabla_x s_\theta(x)$ represents the Jacobian matrix of $s_\theta(x)$.

For large-scale score matching, to avoid the computation of $\operatorname{tr}(\nabla_x s_\theta(x))$, the following two methods are used.

1. Denoising score matching. First, a predetermined noise $q_\sigma(\tilde{x}|x)$ distribution is used to disturb the data x, then score matching is performed to estimate the score of the disturbed data distribution $q_\sigma(\tilde{x}|x) \triangleq \int q_\sigma(\tilde{x}|x) p_{\text{data}}(x)dx$. Its goal is equivalent to

$$\frac{1}{2}\mathbb{E}_{q_\sigma(\tilde{x}|x)p_{\text{data}}(x)}\left[\|s_\theta(\tilde{x}) - \nabla_{\tilde{x}} \log q_\sigma(\tilde{x} \mid x)\|_2^2\right] \tag{16.2}$$

2. Piecewise score matching. Random projection approximates $\operatorname{tr}(\nabla_x s_\theta(x))$ in score matching, and its goal is

$$\mathbb{E}_{p_v}\mathbb{E}_{p_{\text{data}}}\left[v^\top \nabla_x s_\theta(x) v + \frac{1}{2}\|s_\theta(x)\|_2^2\right] \tag{16.3}$$

   where p_v is a simple distribution of random vectors. Piecewise score matching provides a score estimate of the original undisturbed data distribution, but it requires quadrupling the computation.

Langevin dynamics generate samples from the probability distribution $p(x)$ using the score function $\nabla_x \log p(x)$. Given a fixed step size $\varepsilon > 0$, and an initial value $\tilde{x}_0$ from the initial distribution $\pi(x)$, the Langevin method can be recursively calculated as follows:

$$\bar{x}_t = \bar{x}_{t-1} + \frac{\varepsilon}{2}\nabla_x \log p(\bar{x}_{t-1}) + \sqrt{\varepsilon} z_t \tag{16.4}$$

where z_t follows a standard normal distribution.

It can be observed that in order to obtain samples from $p_{\text{data}}(x)$, the score network is first trained so that $s_\theta(x) \approx \nabla_x \log p_{\text{data}}(x)$, and then samples are approximated using Langevin dynamics based on $s_\theta(x)$, which is the so-called score-based generative modeling method.

Observations show that data perturbed with random Gaussian noise makes the data distribution more suitable for score-based generative modeling. Firstly, because the Gaussian noise distribution supports the entire space, the perturbed data will not be confined to the low-dimensional manifold, which eliminates the difficulty of the manifold hypothesis and makes the score estimation clearly defined. Secondly, large Gaussian noise fills the low-density areas in the original unperturbed data distribution, so score matching can obtain more training signals to improve score estimation. In addition, by disturbing with multiple noise levels, a sequence of noise perturbation distributions converging to the real data distribution can be obtained. These intermediate distributions can be used to improve the mixing rate of Langevin dynamics under multimodal distribution through simulated annealing and annealed importance sampling.

Based on this, the score-based generative modeling is improved in the following ways: ① Disturb the data with noise at different levels; ② Estimate the scores corresponding to all noise levels simultaneously by training a single conditional score network. After training, when using Langevin dynamics to generate samples, the score corresponding to large noise is first used, and then the noise level is gradually reduced. This helps to smoothly transfer the benefits of high noise levels to low noise levels. At low noise levels, the perturbed data is almost indistinguishable from the original data. The final result is $s_\theta(x, \sigma)$, a Noise Conditional Score Network (NCSN).

Similar to likelihood-based generative models and GANs, the design of the model architecture plays an important role in generating high-quality samples. Since the main focus is on useful architectures for image generation, and the output of the noise conditional scoring network has the same shape as the input image, inspiration was drawn from successful image dense prediction model architectures (such as semantic segmentation). In the experiments, the model combines the architecture design of UNet with dilated convolution/deconvolution. In addition, Instance Normalization (IN), which is actually a modified version of conditional instance normalization, is adopted in the score network.

16.2 Denoising Diffusion Probabilistic Model

Based on the connection between the diffusion model and the denoising score matching of Langevin dynamics, a weighted variational bound is designed. The Denoising Diffusion Probabilistic Model (DDPM) acknowledges a progressive lossy decompression scheme, which can be interpreted as a generalization of autoregressive (AR) decoding.

DDPM consists of two parameterized Markov chains and uses variational inference to generate samples matching the original data after a finite time. The forward chain gradually adds Gaussian noise to disturb the data distribution using a predesigned schedule until the data distribution converges to a given prior, i.e., the standard Gaussian distribution. The reverse chain starts from the given prior, using a parameterized Gaussian transformation kernel, learning to gradually restore the undisturbed data structure.

The diffusion model can be formalized as the following latent variable model:

$$p_\theta\left(x_0\right) := \int p_\theta\left(x_{0:T}\right) dx_{1:T} \tag{16.5}$$

where $x_1, x_2, \ldots, x_T$ are latent variables of the same dimension as the data $x_0 \sim q(x_0)$. The joint distribution $p_\theta(x_{0:T})$, known as the reverse process, is defined as a Markov chain starting from $p(x_T) = \mathcal{N}(x_T; 0, I)$ with a learned Gaussian transformation as follows:

$$p_\theta(x_{0:T}) := p(x_T)\prod_{t=1}^{T} p_\theta(x_{t-1} \mid x_t),$$
$$p_\theta(x_{t-1} \mid x_t) := \mathcal{N}(x_{t-1}; \mu_\theta(x_t, t), \Sigma_\theta(x_t, t)) \tag{16.6}$$

The diffusion model differs from other latent variable models in that the approximate inference posterior $q(x_{1:T} \mid x_0)$, known as the forward process or diffusion process, is fixed as a Markov chain that gradually adds Gaussian noise to the data according to the variance level table $(\beta_1, \beta_2, \ldots, \beta_T)$:

$$q(x_{1:T} \mid x_0) := \prod_{t=1}^{T} q(x_t \mid x_{t-1}), \quad q(x_t \mid x_{t-1}) := \mathcal{N}(x_t; \sqrt{1-\beta_t}x_{t-1}, \beta_t I) \tag{16.7}$$

During training, the usual negative log-likelihood variational bound is optimized:

$$\begin{aligned}\mathbb{E}\left[-\log p_\theta(x_0)\right] &\leqslant \mathbb{E}_q\left[-\log \frac{p_\theta(x_{0:T})}{q(x_{1:T} \mid x_0)}\right] \\ &= \mathbb{E}\left[-\log p(x_T) - \sum_{t\geqslant 1} \log \frac{p_\theta(x_{t-1} \mid x_t)}{q(x_t \mid x_{t-1})}\right] \\ &:= L\end{aligned} \tag{16.8}$$

A noteworthy feature of the forward process is that it allows sampling of x_t at any time t in closed form, i.e., using the symbols $\alpha_t := 1 - \beta_t$, $\bar{\alpha}_t := \prod_{s=1}^{t} \alpha_s$, the following holds

$$q(x_t \mid x_0) = \mathcal{N}(x_t; \sqrt{\bar{\alpha}_t}x_0, (1-\bar{\alpha}_t)I) \tag{16.9}$$

To improve training efficiency, the random terms in L need to be optimized. The L formula in Eq. 16.8 is rewritten as follows:

$$\mathbb{E}_q\left[D_{KL}(q(x_T \mid x_0)|p(x_T))\right] + \sum_{i>1} \mathbb{E}_{p_{\text{data}}}\left[v^\top \nabla_x s_\theta(x) v + \frac{1}{2} s_\theta(x)^2\right] \tag{16.10}$$

Among them, the KL divergence is used to directly compare $p_\theta(x_{t-1} \mid x_t)$ and the forward process posterior, the latter of which can be controlled under the condition of x_0:

$$q(x_{t-1} \mid x_t, x_0) = \mathcal{N}(x_{t-1}; \tilde{\mu}_t(x_t, x_0) \cdot \tilde{\beta}_t I) \tag{16.11}$$

where $\tilde{\mu}_t$ is the posterior mean of the forward process:

$$\tilde{\mu}_t(x_t, x_0) := \frac{\sqrt{\bar{a}_{t-1}}\beta_t}{1-\bar{a}_t}x_0 + \frac{\sqrt{a_t}(1-\bar{a}_t)}{1-\bar{a}_t}x_t, \quad \tilde{\beta}_t := \frac{1-\bar{a}_{t-1}}{1-\bar{a}_t}\beta_t \tag{16.12}$$

The forward process variance β_t can be learned or kept constant as a hyperparameter in the reparameterization, and the expressiveness of the backward process partly comes from the Gaussian condition chosen in $p_\theta(x_{t-1} \mid x_t)$, because when β_t is small, the two processes have the same function form.

The diffusion probability model allows a great deal of freedom in implementation. It is necessary to choose the forward process variance β_t and model structure, as well as the Gaussian distribution parameterization of the backward process. To facilitate selection, an explicit connection has been established between the diffusion model and denoising score matching, which can bring a simplified weighted variational boundary objective of the diffusion model.

Ignoring the forward process variance β_t can be learned by reparameterization, and it is fixed as a constant. In implementation, the approximate posterior q has no learnable parameters, so L_T is a constant during training and can be ignored.

In the backward process, $\Sigma_\theta(x_t, t) = \sigma_t^2 I$ is set as a time-related constant, then it can be written as

$$L_{t-1} = \mathbb{E}_q \left[\frac{1}{2\sigma_t^2} \| \tilde{\mu}_t(x_t, x_0) - \mu_\theta(x_t, t) \|^2 \right] \tag{16.13}$$

It can be seen that the most direct μ_θ parameterization is a model predicting $\tilde{\mu}_t$.

Further parameterize $q(x_t|x_0)$ as $x_t(x_0, \epsilon) = \sqrt{\bar{\alpha}_t} x_0 + \sqrt{1 - \bar{\alpha}_t} \epsilon$, where $\epsilon \sim \mathcal{N}(0, I)$, and apply the forward process posterior formula, it can be obtained:

$$\begin{aligned} L_{t-1} - C &= \mathbb{E}_{x_0,\epsilon} \left[\frac{1}{2\sigma_t^2} \left\| \tilde{\mu}_t \left(x_t(x_0, \epsilon), \frac{1}{\sqrt{\bar{\alpha}_t}} \left(x_t(x_0, \epsilon) - \sqrt{1 - \bar{\alpha}_t}\, \epsilon \right) \right) - \mu_\theta(x_t(x_0, \epsilon), t) \right\|^2 \right] \\ &= \mathbb{E}_{x_0,\epsilon} \left[\frac{1}{2\sigma_t^2} \left\| \left(x_t(x_0, \epsilon) - \frac{\beta_t}{\sqrt{1 - \bar{\alpha}_t}} \epsilon \right) - \mu_\theta(x_t(x_0, \epsilon), t) \right\|^2 \right] \end{aligned} \tag{16.14}$$

Equation 16.14 indicates that μ_θ must predict $\frac{1}{\sqrt{\bar{\alpha}_t}} \left(x_t - \frac{\beta_t}{1 - \bar{\alpha}_t} \epsilon \right)$ under the given x_t. Therefore, the following parameterization can be chosen:

$$\mu_\theta(x_t, t) = \tilde{\mu}_t \left(x_t, \frac{1}{\sqrt{\bar{\alpha}_t}} (x_t - \sqrt{1 - \bar{\alpha}_t} \epsilon_\theta(x_t)) \right) = \frac{1}{\sqrt{\bar{\alpha}_t}} \left(x_t - \frac{\beta_t}{\sqrt{1 - \bar{\alpha}_t}} \epsilon_\theta(x_t, t) \right) \tag{16.15}$$

where ϵ_θ is a function approximating ϵ predicted from x_t. Based on the above parameterization, L_{t-1} can be further simplified:

$$\mathbb{E}_{x_0,\epsilon} \left[\frac{\beta_t^2}{2\sigma_t^2 \alpha_t (1 - \bar{\alpha}_t)} \left\| \epsilon - \epsilon_\theta \left(\sqrt{\bar{\alpha}_t} x_0 + \sqrt{1 - \bar{\alpha}_t} \epsilon, t \right) \right\|^2 \right] \tag{16.16}$$

This is similar to denoising score matching with multiple noise scales (t).

In summary, the backward process mean function approximator μ_θ can be trained to predict $\tilde{\mu}_t$, or its parameterization can be corrected, and it can be trained to predict ϵ.

It can be seen that the ϵ-prediction parameterization is similar to Langevin dynamics and simplifies the variational boundary of the diffusion model to a target similar to denoising score matching.

16.3 DDIM

In order to accelerate sampling, the denoising diffusion implicit model (DDIM) uses non-Markov diffusion processes. These non-Markov processes correspond to deterministic generation processes, thus obtaining implicit models that generate high-quality samples faster.

As analyzed in Sect. 16.2 for DDPM, if all conditions are modeled as Gaussian distributions with trainable mean functions and fixed variances, then the goal of DDPM to maximize the variational bound can be simplified as follows:

$$L_\gamma(\epsilon_\theta) = \sum_{t=1}^{T} \gamma_t \mathbb{E}_{x_0 \sim q(x_0), \epsilon_t \sim \mathcal{N}(0,I)} \left[\left\| \epsilon_\theta^{(t)} \left(\sqrt{\bar{\alpha}_t} x_0 + \sqrt{1-\bar{\alpha}_t} \epsilon_t \right) - \epsilon_t \right\|_2^2 \right] \quad (16.17)$$

where $\epsilon_\theta := \{\epsilon_\theta^{(t)}\}_{t=1}^T$ is a set of T functions, each $\epsilon_\theta^{(t)}$ is a function with trainable parameters $\theta^{(t)}$, and $\gamma := [\gamma_1, \gamma_2, \ldots, \gamma_T]$ is a positive coefficient vector depending on the $\bar{\alpha}_{1:T}$ target.

In Ref. [3], the goal of $\gamma = 1$ optimization is to maximize the generative performance of the training model, which is also the same goal in the noise-conditioned score network based on score matching. First, sample x_T from the prior $p_\theta(x_T)$ in a training model, and then iteratively sample x_{t-1} in the generation process.

The length T of the forward process is an important hyperparameter in DDPM. From the perspective of parameter changes, a large T allows the backward process to approach the Gaussian distribution, so the generation process modeled with Gaussian conditional distributions becomes a good approximation, which prompts the choice of large T values, such as $T = 1000$ in Ref. [5]. However, because all T iterations must be performed sequentially rather than in parallel to obtain samples x_0, DDPM sampling is much slower than other deep generative model sampling, which is not suitable for tasks with limited computation and tight latency.

The key observation is that the DDPM target of the L_γ form only depends on the margin $q(x_t|x_0)$, not the joint distribution $q(x_{1:T}|x_0)$. As there are many inference distributions (joint) with the same margin, non-Markov alternative inference processes can bring new generation processes, as shown in Fig. 16.1.

The following will analyze these non-Markov inference processes that have resulted in alternative objective functions identical to DDPM (Note: The non-Markov perspective is also applicable to situations beyond Gaussian distribution).

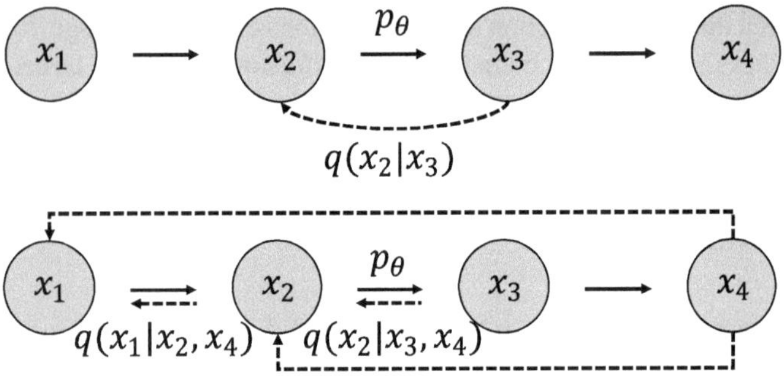

Fig. 16.1 Graph models of diffusion inference model (top) and non-Markov inference model (down)

Consider a set of inference distributions Q, marked with a real vector σ:

$$q_\sigma(x_{1:T} \mid x_0) := q_\sigma(x_T \mid x_0) \prod_{t=2}^{T} q_\sigma(x_{t-1} \mid x_t, x_0) \tag{16.18}$$

where $q_\sigma(x_T \mid x_0) = \mathcal{N}(\sqrt{\bar{\alpha}_T} x_0, \sqrt{1-\bar{\alpha}_T} I)$, and for all $t > 1$:

$$q_\sigma(x_{t-1} \mid x_t, x_0) = \mathcal{N}(\sqrt{\bar{\alpha}_{t-1}} x_0 + \sqrt{1-\bar{\alpha}_{t-1}-\sigma_t^2} \cdot \frac{x_t - \sqrt{\bar{\alpha}_t} x_0}{\sqrt{1-\bar{\alpha}_t}}, \sigma_t^2 I) \tag{16.19}$$

This average function is chosen to ensure that for all t, $q_\sigma(x_t \mid x_0) = \mathcal{N}(\sqrt{\bar{\alpha}_T} x_0, \sqrt{1-\bar{\alpha}_t} I)$. This defines a joint inference distribution and the expected "marginal distribution" match. The forward process is derived from the Bayesian principle as follows:

$$q_\sigma(x_t \mid x_{t-1}, x_0) = \frac{q_\sigma(x_{t-1} \mid x_t, x_0) q_\sigma(x_t \mid x_0)}{q_\sigma(x_{t-1} \mid x_0)} \tag{16.20}$$

Unlike the diffusion process, this forward process is no longer a Markov chain, as each x_t depends on both x_{t-1} and x_0. Here, the magnitude of σ controls the randomness of this forward process.

The next step is to define a trainable generation process $p_\theta(x_{0:T})$, where each $p_\theta^{(t)}(x_{t-1} \mid x_t)$ utilizes the knowledge of $q_\sigma(x_{t-1} \mid x_t, x_0)$. Intuitively, given a noisy observation x_t, first predict the corresponding x_0, then use it to obtain a sample x_{t-1} from the defined reverse conditional distribution $q_\sigma(x_{t-1} \mid x_t, x_0)$.

For some $x_0 \sim q(x_0)$, $\epsilon_t \sim \mathcal{N}(0, I)$, x_t can be obtained without the knowledge of x_0, and the model $\epsilon_\theta^{(t)}(x_t)$ tries to predict ϵ_t from x_t. This way, a denoised observation can be predicted, given the prediction of x_0 given x_t:

$$f_\theta^{(t)}(x_t) := (x_t - \sqrt{1-\bar{\alpha}_t} \cdot \epsilon_\theta^{(t)}(x_t))/\sqrt{\bar{\alpha}_t} \tag{16.21}$$

This generation process is defined in this way, through a fixed prior $p_\theta^{(t)}(x_T) = \mathcal{N}(0, I)$ and the following conditional distribution:

$$p_\theta^{(t)}(x_{t-1} \mid x_t) = \begin{cases} \mathcal{N}(f_\theta^{(1)}(x_1), \sigma_1^2 I), & t = 1 \\ q_\sigma(x_{t-1} \mid x_t, f_\theta^{(t)}(x_t)), & \text{others} \end{cases} \tag{16.22}$$

where $q_\sigma(x_{t-1} \mid x_t, f_\theta^{(t)}(x_t))$, $f_\theta^{(t)}(x_t)$ replaces x_0. For the case $t = 1$, adding Gaussian noise ensures that the generation process is supported everywhere.

Optimize θ from the following variational inference objective:

$$J_\theta(\epsilon_\theta) := \mathbb{E}_{x_{0:T} \sim q_\sigma(x_{0:T})} \left[\log q_\sigma(x_{1:T} \mid x_0) - \log p_\theta(x_{0:T})\right] \tag{16.23}$$

$$= \mathbb{E}_{x_{0:T} \sim q_\sigma(x_{0:T})}[\log q_\sigma(x_T \mid x_0) + \sum_{t=2}^{T} \log q_\sigma(x_{t-1} \mid x_t, x_0) -$$

$$\sum_{t=1}^{T} \log p_\theta^{(t)}(x_{t-1} \mid x_t) - \log p_\theta(x_T)]$$

where, decompose $q_\sigma(x_{1:T} \mid x_0)$.

From the definition of J_σ, it seems that for each choice of σ, a different model must be trained, as it corresponds to a different variational objective (and a different generation process).

The variational objective L_γ is special, because if the parameters θ of the model $\epsilon_\theta^{(t)}$ are not shared between different t, then the optimal solution of ϵ_θ will not depend on the weight γ (because the global optimum is achieved by maximizing each item in the sum separately). This property of γ has two implications: on the one hand, it proves that using L_1 as an alternative objective function for the DDPM variational lower bound is reasonable; on the other hand, since J_σ is equivalent to some L_γ, the optimal solution of J_σ is also the same as L_1. Therefore, if the parameters in the model ϵ_θ are not shared between t, then the L_1 objective of DDPM can also be used as an alternative to the variational objective J_σ.

Using L_1 as the target, not only learning the generation process of the Markov inference process, but also learning the generation process of the non-Markov forward process parameterized by σ. Therefore, a pre-trained DDPM model can basically be used as a solution for the new target, and focus on finding a generation process that generates better samples by changing σ.

When all t's $\sigma_t = 0$, given x_{t-1} and x_0, the forward process becomes deterministic, except for $t = 1$; in the generation process, the coefficient before the random noise ϵ_t becomes zero. The resulting model becomes an implicit probability model, where samples are generated from latent variables through a fixed procedure (from x_T to x_0), which can be named as Denoising Diffusion Implicit Model (DDIM), because it is an implicit probability model trained with the DDPM target (although the forward process is no longer a diffusion process).

16.4 SDE

Based on score-based generative modeling and diffusion probabilistic modeling, a unified Stochastic Differential Equation (SDE) framework was proposed in Ref. [7]. In this framework, the forward SDE gradually perturbs complex data distributions into tractable prior distributions by continuously injecting noise, while the corresponding reverse-time SDE removes the noise to transform the prior distribution back into the data distribution. Crucially, the reverse-time SDE depends only on the time-dependent gradient field (i.e., the score) of the perturbed data distribution. Leveraging score-based generative modeling, these scores can be efficiently estimated with neural networks, and new samples can be generated through numerical SDE solvers.

This diffusion process is defined as a solution to an SDE, i.e., the forward process:

$$dx = f(x,t) + g(t)dw \tag{16.24}$$

where w is a standard Wiener process (Brownian motion); $f(\cdot, t)$ is a vector-valued function, known as the drift coefficient of $x(t)$; $g(t)$ is a scalar function, i.e., the diffusion coefficient of $x(t)$.

Let $p_t(x)$ be the probability density of $x(t)$, and we start the backward SDE from the sample $x(T) \sim p_T$ to get the sample $x(0) \sim p_0$. The reverse of a diffusion process is also a diffusion process, i.e.,

$$dx = \left[f(x,t) - g^2(t)\nabla_x \log p_t(x)\right]dt + g(t)d\bar{w} \tag{16.25}$$

where $\bar{w}$ is a standard Wiener process, and dt is an infinitesimal negative time step. Once each marginal distribution score $\nabla_x \log p_t(x)$ is known, the backward diffusion process can be derived, simulating sampling from p_0.

As shown in Fig. 16.2, solving the backward time SDE generates a schematic diagram of the score-based generative model. A continuous-time SDE can complete the transformation of data into a simple noise distribution. If the distribution score $\nabla_x \log p_t(x)$ of each intermediate time step is known, the SDE can be reversed.

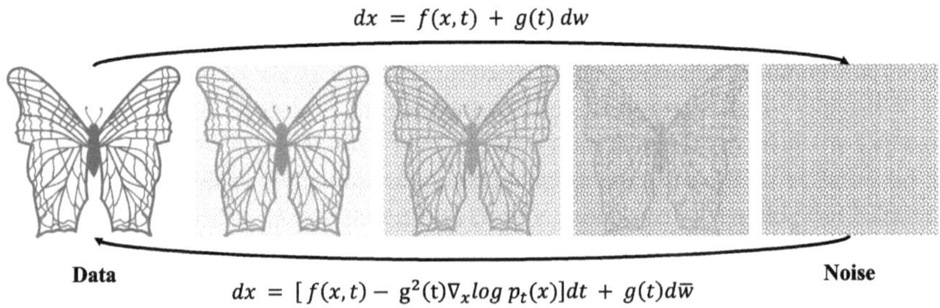

Fig. 16.2 Schematic diagram of the SDE generating a score-based generative model

Under this framework, score-based generative modeling methods can be described as a variance-exploding (VE) SDE with an increasing noise scale $\sigma(t)$:$dx = d\left[\sigma^2(t)/dt\right]dw$ In contrast, diffusion probability modeling is a variance-preserving (VP) SDE with another noise sequence $\beta(t)$: $dx = -[\frac{|\beta(t)|}{2}]xdt + \sqrt{\beta(t)}dw$ Note: VP SDE can be reparameterized into an equivalent VE SDE.

Specifically, Ref. [7] also provides a predictive–corrective framework to correct errors in the evolution of discrete backward time SDEs, as well as an equivalent neural ordinary differential equation (ODE), which samples from the same distribution as the SDE, but relies on accurate likelihood calculations to improve sampling efficiency.

Any diffusion process can be represented by a deterministic ODE, which carries the same marginal density as the diffusion process throughout the evolution process, known as the Probability Flow (PF) ODE. The PF-ODE supports the unique identifiable encoding of data and is the core of DDIB. For the forward SDE in Eq. 16.24, the equivalent PF-ODE has the following form:

$$dx = \left[f(x,t) - \frac{1}{2}g(t)^2\nabla_x \log p_t(x)\right]dt \tag{16.26}$$

When the score function is approximated by a time-based score model (usually a neural network), a neural ODE example is obtained.

In addition, Ref. [17] introduces a set of diffusion-based generative models, which add a set of Fourier features to the input of the noise prediction model to capture the fine-scale details of the data. Including these features in the denoising model input greatly improves the likelihood, especially when combined with a learnable signal-to-noise ratio (SNR) function. By analyzing the variational lower bound (VLB) of the diffusion model, a very simple expression for the SNR of a diffusion process is derived, which helps to understand the theoretical understanding of density modeling of diffusion models.

For the continuous-time (infinite-depth) setting, Ref. [17] proves that its VLB is invariant to noise scheduling. This allows learning a noise scheduling scheme that minimizes the variance of the VLB estimator, thereby achieving faster optimization.

16.5 Image/Video Synthesis

16.5.1 Image

Image super-resolution aims to recover high-resolution (HR) images from low-resolution (LR) images, while image inpainting is to reconstruct missing or damaged areas in images.

Reference [11] proposes a method for achieving image super-resolution by repeated refinement, SR3. SR3 transforms the standard normal distribution into the empirical data distribution through a series of refinement steps, similar to Langevin dynamics. SR3 adapts DDPM to the scenario of conditional image generation and performs super-resolution through a stochastic iterative denoising process. The generated output starts from pure Gaussian noise and iteratively refines the noisy output using a UNet model trained to denoise at different noise levels.

SR3 replaces the original DDPM residual block with the residual block in BigGAN, scales the residual connection by a factor of $1/\sqrt{2}$, and increases the number of residual blocks and channel multipliers at different resolutions. To adapt the model to the input, the low-resolution image is upsampled to the target resolution using bicubic interpolation. The result is concatenated with the output along the channel dimension.

The Ref. [12] specifically studied the application of image generation, and believed that the gap between diffusion models and GANs stems from at least two factors: ① The GAN model architecture has been extensively explored and improved; ② GANs can exchange diversity for fidelity, producing high-quality samples, but cannot cover the entire distribution. For this reason, Ref. [12] improved the diffusion model architecture and designed a scheme to exchange diversity for fidelity.

The improvements made are as follows.

- Increase depth and width, keeping the model size relatively unchanged.
- Increase the number of attention heads.
- Use attention mechanism at 32×32, 16×16 and 8×8 resolutions.
- Use BigGAN residual blocks for up-sampling and down-sampling of activations.
- Scale residual connections by a factor of $1/\sqrt{2}$.

In addition, Ref. [12] also studied other attention configurations that better match the Transformer architecture, and conducted experiments on the Adaptive Group Normalization (AdaGN) operation. Similar to Adaptive Instance Normalization, the AdaGN layer introduces time steps and class embeddings into each residual block after the Group Normalization (GN) operation. This layer is defined as $AdaGN(h, y) = y_s GroupNorm(h) + y_b$, where h is the intermediate activation of the first convolution residual module, and $y = [y_s, y_b]$ is obtained from a linear embedding of time steps and class embeddings.

GANs used for conditional image synthesis make extensive use of class labels, usually adopting class conditional normalization statistics and the form of classifier $p(y \mid x)$ head discriminator. For this reason, Ref. [12] explored a different method: using the classifier $p(y \mid x)$ to improve a diffusion generator, that is, the classifier-guided method.

The classifier-guided method uses the gradient of a classifier to adjust a pre-trained diffusion model. Specifically, a classifier $p_\phi(y \mid x_t, t)$ can be trained on the noise image x_t, and then the diffusion sampling process can be guided toward an arbitrary class label y using the gradient $\nabla_{x_t} \log p_\phi(y \mid x_t, t)$.

Starting with a diffusion model with an unconditional noisy reverse process $p_\theta(x_t \mid x_{t+1})$, in order to adjust a label, each transformation is sampled as follows:

$$p_{\theta,\phi}(x_t \mid x_{t+1}, y) = Zp_\theta(x_t \mid x_{t+1})p_\phi(y \mid x_t) \tag{16.27}$$

where Z is the normalization constant. It is difficult to sample directly from this distribution, but it can be approximated as a perturbed Gaussian distribution.

Recall that the diffusion model predicts x_t from x_{t+1} using a Gaussian distribution:

$$p_\theta(x_t \mid x_{t+1}) = \mathcal{N}(\mu_t, \Sigma) \tag{16.28}$$

$$\log p_\theta(x_t \mid x_{t+1}) = -\frac{1}{2}(x_t - \mu_t)^\top \Sigma^{-1}(x_t - \mu_t) + C \tag{16.29}$$

An approximation can be made by Taylor expansion at $x_t = \mu$:

$$p_\phi(y \mid x_t) = (x_t - \mu_t)g + C_1 \tag{16.30}$$

where $g = \nabla_{x_t} \log p_\phi(y \mid x_t)\big|_{x_t=\mu}$, C_1 is a constant. Thus we have:

$$\begin{aligned}\log p_\theta(x_t \mid x_{t+1})p_\phi(y \mid x_t) &\approx -\frac{1}{2}(x_t - \mu)^\top \Sigma^{-1}(x_t - \mu_t) + (x_t - \mu_t)g + C_2 \\ &= \log p(z) + D, z \sim \mathcal{N}(\mu + \Sigma g, \Sigma)\end{aligned} \tag{16.31}$$

The constant D can be directly ignored.

However, the above derivation of the conditional sampling formula is only valid for the random diffusion sampling process, and cannot be used for deterministic sampling methods, such as DDIM. If there is a model $\epsilon_\theta(x_t)$ that predicts sample noise, a score function can be directly derived:

$$\nabla_{x_t} \log p_\phi(x_t) = -\frac{1}{\sqrt{1-\bar{\alpha}_t}}\epsilon_\theta(x_t) \tag{16.32}$$

Substitute into the score function of $p(x_t)p(y|x_t)$:

$$\begin{aligned}\nabla_x \log p_\theta(x_t)p_\phi(y \mid x_t) &= \nabla_{x_t} \log p_\phi(x_t) + \nabla_{x_t} \log p_\phi(y \mid x_t) \\ &= -\frac{1}{\sqrt{1-\bar{\alpha}_t}}\epsilon_\theta(x_t) + \nabla_{x_t} \log p_\phi(y \mid x_t)\end{aligned} \tag{16.33}$$

Finally, define a new prediction model $\hat{\epsilon}(x_t)$, corresponding to the score function of this joint distribution:

$$\hat{\epsilon}(x_t) := \epsilon_\theta(x_t) - \frac{1}{\sqrt{1-\bar{\alpha}_t}}\nabla_x \log p_\phi(y \mid x_t) \tag{16.34}$$

In this way, apply the same sampling procedure as DDIM, where a corrected noise prediction model $\hat{\epsilon}(x_t)$ is used.

Since we mentioned the classifier-guided diffusion model, we also need to introduce the classifier-free guided diffusion model. Pure generative models without classifiers can indeed be guided, but they do not sample in the gradient direction of the image classifier, but jointly train a conditional diffusion model and an unconditional diffusion model, and combine the obtained conditional scores and unconditional score estimates. Through mixing weights, a FID/IS trade-off similar to that obtained through classifier guidance is achieved.

At this time, the label y of the class conditional diffusion model $\epsilon_\theta(x_t \mid y)$ is replaced with a null label $\emptyset$ at a fixed probability during training. During sampling, the output of the model is further extrapolated along the direction of $\epsilon_\theta(x_t \mid y)$ and away from $\epsilon_\theta(x_t \mid \emptyset)$, as follows:

$$\hat{\epsilon}_\theta(x_t \mid y) = \epsilon_\theta(x_t \mid \emptyset) + s \cdot (\epsilon_\theta(x_t \mid y) - \epsilon_\theta(x_t \mid \emptyset)) \tag{16.35}$$

where $s \geq 1$ is the guiding scale factor. In fact, the form of this function is inspired by the implicit classifier, that is,

$$p^i(y \mid x_t) \propto \frac{p(x_t \mid y)}{p(x_t)} \tag{16.36}$$

Its gradient can be written in the form of real scores ϵ^*:

$$\begin{aligned} \nabla_x p^i(x_t \mid y) &\propto \nabla_x \log p(x_t \mid y) - \nabla_x \log p(x_t) \\ &\propto \epsilon^*(x_t \mid y) - \epsilon^*(x_t) \end{aligned} \tag{16.37}$$

As proposed in Ref. [13], the cascaded diffusion model (CDM) can generate high-fidelity images on the class conditional ImageNet generation benchmark without any auxiliary image classifier to help improve sample quality. A cascaded diffusion model includes a pipeline of multiple diffusion models generating images of increasing resolution, starting from the standard diffusion model at the lowest resolution, followed by one or more super-resolution diffusion models, and then image up-sampling and adding more details.

The sample quality of the cascading pipeline largely depends on conditioning augmentation, that is, data augmentation for the low-resolution conditioning input of the super-resolution model.

Cascading is a technique that learns a pipeline of models trained separately at multiple resolutions to model high-resolution data; the base model generates low-resolution samples, and the super-resolution model upsamples the low-resolution to high-resolution. Sampling from a cascading pipeline is sequential, first sampling from the low-resolution base model, and then sampling from the super-resolution model to increase the resolution.

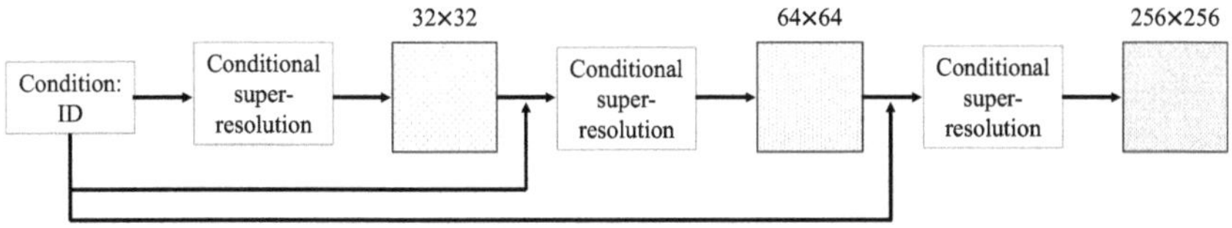

Fig. 16.3 Schematic diagram of CDM generating class conditional 256 × 256 images

Assume x_0 is high-resolution data, and z_0 is low-resolution data. At low resolution, there is a diffusion model $p_\theta(z_0)$; at high resolution, there is a super-resolution diffusion model $p_\theta(x_0 \mid z_0)$. The cascading pipeline forms a latent variable model for high-resolution data, for example, $p_\theta(x_0) = \int p_\theta(x_0 \mid z_0) p_\theta(z_0) dz_0$. It is also simple to adjust an entire cascading pipeline based on class information or other conditioning information c, where the model uses $p_\theta(z_0 \mid c)$ and $p_\theta(x_0 \mid z_0, c)$. An example of a cascading pipeline, generating a detailed CDM of class conditional 256 × 256 images. As shown in Fig. 16.3, the first model is a class conditional diffusion model, followed by a sequence of two class conditional super-resolution diffusion models.

The most effective technique to improve the sample quality of the cascading pipeline is to use data augmentation for training on the low-resolution input of each super-resolution model, that is, conditioning augmentation. The most effective at low resolution is to add Gaussian noise (forward process noise), and for high resolution, to randomly apply Gaussian blur to z. In some cases, a more practical method is to train the super-resolution model and select the best intensity in the post-training hyperparameter search to achieve the best sample quality.

In order to achieve diffusion model training on limited computational resources while maintaining its quality and flexibility, the paper reporting high-resolution image synthesis (see Ref. [16]) applies the diffusion model to the latent space of pre-trained autoencoders, that is, the latent diffusion model (LDM). It introduces cross-attention layers in the model architecture, converting the diffusion model into a generator for general conditional inputs such as text or borders, and achieving high-resolution synthesis in a convolutional manner. At the same time, it significantly reduces computational requirements compared to pixel-based diffusion models.

In the training of the encoder, to avoid a high-variance latent space, two different regularizations are added. One is KL regression, which imposes a slight KL penalty on the learned latent variables, similar to VAE; the other is VQ regression, with a vector quantization (VQ) layer inside the decoder. This model can be interpreted as a VQGAN, but the quantization layer is absorbed by the decoder.

Through the perceptual compression model composed of trained encoders and decoders, you can enter a high-efficiency, low-dimensional latent space, where high-frequency imperceptible details are abstracted. Compared with the high-dimensional pixel space, this space is more suitable for likelihood-based generative models because it: ① focuses on the important semantic bits of data; ② trains in a low-dimensional, computationally more efficient space.

16.5.2 Video

Due to the spatio-temporal continuity and complexity of video frames, high-quality video generation still poses a challenge. The Ref. [20] proposes a diffusion model for video generation, which is a natural extension of the standard image diffusion architecture, using a 3D UNet diffusion model architecture to train the model to generate a fixed number of video frames.

This 3D UNet diffusion model architecture supports joint training from image and video data, which can reduce minibatch gradient variance and accelerate optimization. To generate long-duration and high-resolution videos, the Ref. [20] introduces a conditional sampling technique for spatial and temporal video expansion.

The spatio-temporal decomposition of the 3D UNet architecture extends the image diffusion model to video data. First, each 2D convolution is changed to a spatial-only 3D convolution. The attention of each spatial attention block remains as spatial attention, with the first axis as the batch axis. Secondly, a temporal attention block is inserted after each spatial attention block, which pays attention on the first axis and treats the spatial axis as the batch axis. Relative position embedding is used in each temporal attention block so that the network can distinguish the order of video frames without needing absolute video time.

The 3D UNet architecture is shown in Fig. 16.4, where each block represents a 4D tensor, its axes are marked as frame × height × width × channel; the input is noisy video z_t, condition c, and log SNR λ_t. Each of the K blocks of down-sampling/up-sampling blocks adjusts the spatial height × width resolution by factor2. Channel multipliers $M_1, M_2, ..., M_K$ specify the number of channels, and the up-sampling channels have cascaded skip connections with the down-sampling channels.

16.5.3 New View Synthesis

The Ref. [26] proposes a diffusion model for 3D new view synthesis, 3DiM, which transforms a single input view into a consistent and clear completion result across

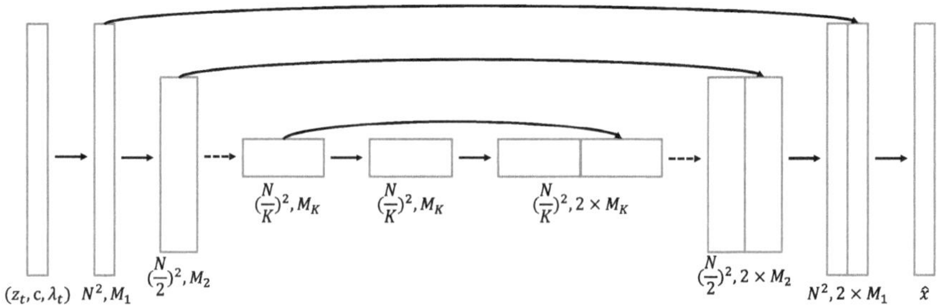

Fig. 16.4 Diffusion model's 3D UNet architecture

multiple views. The core component of 3DiM is a pose-conditioned image-to-image diffusion model, which takes the source view and its pose as input and generates a new view for the target pose as output.

Compared to GAN, diffusion not only provides a simpler architecture but also offers a simpler hyperparameter tuning experience. 3DiM uses stochastic conditioning to generate multiple 3D consistent views. The output view is autoregressively generated, and during the process of generating a new view, each denoising step selects a random condition view from the available view set.

3DiM is an image-to-image diffusion model trained on paired images of the same scene, where it is assumed that the poses of the two images are known. After training, 3DiM can construct a conditional generative model of a view given another view and its pose. This image-to-image model can be transformed into a model that generates a set of 3D consistent frames through autoregressive generation.

Given a complete description of a 3D scene S, for any pose p, its view $x(p)$ is completely determined by S, that is, given S, the views are conditionally independent. However, the point of interest is the modeling of the distribution $q(x_1, \ldots, x_m \mid x_{m+1}, \ldots, x_n)$ in the absence of S, where the views are no longer conditionally independent. An image-to-image model, if only back view sampling is done, should indeed produce different outputs for each front view, but cannot guarantee their consistency, especially in the case of perfect learning data distribution.

This contrasts with the NeRF method. Given a 3D representation S, query rays are conditionally independent—this condition is even stronger than the conditional independence constraint between frames. The NeRF method attempts to learn the richest possible representation of a single scene S, while 3DiM avoids the difficulty of learning a generative model for the entire S.

Given a scene view pair data distribution $q(x_1, x_2)$ for poses p_1, p_2, a homoscedastic Gaussian process can be defined, which adds a certain amount of noise to the data samples as the signal-to-noise ratio λ decreases, i.e.,

$$q(z_k^{(\lambda)} \mid x_k) := \mathcal{N}(z_k^{(\lambda)}; \sigma(\lambda)^{\frac{1}{2}} x_k, \sigma(-\lambda) I) \tag{16.38}$$

where $\sigma(\cdot)$ is the sigmoid function. Apply the reparameterization method, and sample from the marginal distribution through the following formula:

$$z_k^{(\lambda)} = \sigma(\lambda)^{\frac{1}{2}} x_k + \sigma(-\lambda)^{\frac{1}{2}} \epsilon, \epsilon \sim \mathcal{N}(0, I) \tag{16.39}$$

Then given a pair of views, learn to do the reverse process in one of the pair, producing better sample quality than maximizing the Evidence Lower BOund (ELBO):

$$\mathbb{E}_{q(x_1,x_2)} \mathbb{E}_{\lambda,\epsilon} \left\| \epsilon_\theta(z_2^{(\lambda)}, x_1, \lambda, p_1, p_2) - \epsilon \right\|_2^2 \tag{16.40}$$

where ϵ_θ is a neural network that denoises frame $z_2^{(\lambda)}$ given different (clean) frame x_1, and λ is the log signal-to-noise ratio. For convenience, it is abbreviated below as $\epsilon_\theta(z_2^{(\lambda)}, x_1)$.

Ideally, 3D scene frames can be modeled as a chain rule decomposition as follows:

$$p(\mathbf{x}) = \prod_i p(x_i \mid x_{<i}) \tag{16.41}$$

However, this factor decomposition performs poorly. The following introduces a random conditional sampler, which uses a random conditional sampling program to generate 3D consistent samples from a 3DiM. First, start with a set of conditional views $\chi = \{x_1, \ldots, x_k\}$ of a group of static scenes, where k is relatively small. Then for $\lambda_{\min} = \lambda_T < \lambda_{T-1} < \ldots < \lambda_0 = \lambda_{\max}$, run the following standard denoising diffusion reverse process revised version to produce a new frame:

$$\hat{x}_{k+1} = \frac{1}{\sigma(\lambda_t)^{\frac{1}{2}}}(z_{k+1}^{(\lambda_t)} - \sigma(-\lambda_t)^{\frac{1}{2}}\epsilon_\theta(z_{k+1}^{(\lambda_t)}, x_i)) \tag{16.42}$$

$$z_{k+1}^{(\lambda_{t-1})} \sim q(z_{k+1}^{(\lambda_{t-1})} \mid z_{k+1}^{(\lambda_t)}, \hat{x}_{k+1}) \tag{16.43}$$

where x_i is resampled in each denoising step, $i \sim \mathcal{U}(1, \ldots, k)$. In other words, each individual denoising step depends on a random different view in χ. Once this sampling chain is completed, the final x_{k+1} is obtained, which is simply added to χ, and this process can be repeated if necessary. Given enough denoising steps, the random conditional method allows each generated frame to be guided by the previous frame.

The structure of the random conditional sampler, its sampling process consists of two main components: ① Multi-frame autoregressive generation; ② The denoising process generates each frame. When generating a new frame, a previous frame is randomly selected as the conditional frame in each denoising step.

In practice, 256 denoising steps are used, which is sufficient to achieve high sample quality and approximate 3D consistency. Note: The first (most noisy) sample is a Gaussian, i.e., $z_i^{(\lambda_T)} \sim \mathcal{N}(0, I)$, and the last step λ_0, the sample is basically noiseless.

The random conditional method can be regarded as a simple approximation to true autoregressive (AR) sampling. Exact AR sampling requires a score-based model, such as

$$\nabla_{z_{k+1}^{(\lambda)}} \log q\left(z_{k+1}^{(\lambda)} \mid x_1, \ldots, x_k\right),$$

but this formulation strictly depends on multi-view training data. In practice, however, the goal is to train with as few views as possible for each scene.

The 3DiM model requires a neural network architecture that takes the conditional frame and noise frame as inputs. A natural way is to simply concatenate the two images along the channel dimension and use the standard UNet architecture. However, the results produced by this Concat-UNet method are very poor, with serious

3D inconsistencies and lack of alignment with the conditional image. It is assumed that given the limited model capacity and training data, it is difficult to learn complex nonlinear image transformations that rely solely on self-attention. Therefore, X-UNet is introduced, whose core changes are: ① Shared parameters handle each of the two views; ② Cross-attention mechanism between the two views.

X-UNet allows each frame to have its own noise level (remember, the input to the DDPM residual block is the feature map and a position encoding of the noise level), and uses λ_{max} position encoding for clean frames.

16.6 Image-to-Image Translation

A unified image-to-image translation framework Palette based on the conditional diffusion model was developed in Ref. [14], which was evaluated for 4 challenging image-to-image translation tasks (i.e., coloring, restoration, decropping, and JPEG recovery) and found to outperform GAN and regression baseline methods on all tasks, without the need for specific task hyperparameter tuning, architecture customization, any auxiliary loss or complex new techniques. The study proves the importance of self-attention in neural structures.

The tasks of image-to-image translation include super-resolution, coloring, restoration, and pixel-level image understanding tasks, such as instance segmentation and depth estimation. A natural method of image-to-image translation is to learn the conditional distribution of the output image given the input, using deep generative models to capture the multimodal distribution of the high-dimensional space of images.

Given a training output image y, given an input image x and noise level γ, generate a noisy version $\tilde{y}$, and train a neural network f_θ to denoise $\tilde{y}$. Palette uses a UNet architecture, inspired by Ref. [7], with several modifications. This network architecture is based on the class conditional 256×256 UNet model in Ref. [12]. The two main differences from previous methods are: ① no class condition; ② As in Ref. [11], conditions are added to the source image through cascading.

A major contribution in Ref. [14] is the evaluation of image-to-image translation models. Considering scale, diversity, and public availability, it proposes a unified evaluation protocol for tasks such as restoration, decropping, and JPEG recovery on the ImageNet dataset. It mainly includes four automatic quantitative measures of image-to-image translation sample quality: ① Inception Score (IS); Fréchet Inception Distance (FID); ② Classification Accuracy (CA) of pre-trained ResNet-50 classifier; ③ Simple measure Perceptual Distance (PD), i.e., Euclidean distance in Inception-v1 feature space. However, the final evaluation is by human eye observation.

Training details: The minimum batch size for all models is 1024, and the number of training steps is 1M; The standard Adam optimizer is used, with a fixed learning rate of 1e-4 and a 10k linear learning rate warm-up schedule. No task-specific hyperparameter tuning or architectural modifications are made.

The specific training details for each task are as follows.

- Coloring: Coloring is done using RGB parameterization, with grayscale images as the source images, and a palette is trained to predict the complete RGB image; During training, the largest square crop is randomly selected from the image and resized to 256×256.
- Repair: A combination of free masks and rectangular masks is used to train the Palette; The total area covered by the rectangular mask is kept between 10% and 40% of the image; A random combination of 60% free masks and 40% rectangular masks is drawn; No additional mask channels are provided, the mask area is simply filled with random Gaussian noise; During training, the loss function is limited to the mask area, and during inference, only the mask area is used for model prediction.
- Decropping: The model is trained to extend the image in four directions or only one direction. In both cases, the masked area is set to 50% of the image. During training, a side is chosen uniformly at random to be masked, or all 4 sides are chosen to be masked. When masking along one side, further uniform random selection is performed.
- JPEG Recovery: Since decompression of lower quality is a very challenging task, an exponential distribution is used to sample the quality factor (5, 30) during training; Specifically, the sampling probability Q of the quality range is set to $\infty_e - Q/10$.

Reference [18] introduces the Double Diffusion Implicit Bridge (DDIB), an image translation method based on diffusion models that eliminates the need for joint training across source and target domains. DDIB performs image translation using two diffusion models, each trained independently within its own domain. The process consists of two steps: first, the source-domain diffusion model is applied to obtain the latent encoding of the source image; second, the target-domain diffusion model decodes this representation to generate the translated image.

Specifically, DDIB is developed based on the DDIM method. DDIM invented a specific parameterization of the diffusion process, creating a smooth, deterministic, and reversible mapping between the image and its latent representation. This mapping is obtained using a solution to a Probability Flow (PF) Ordinary Differential Equation (ODE), which is the cornerstone of DDIB. Translating a source-target pair using DDIB requires two different PF-ODEs: The source PF-ODE transforms the input image into the latent space; The target ODE then synthesizes the image of the target domain.

Importantly, the trained diffusion models are specific to each domain and do not depend on domain pair information. In fact, DDIB can save a trained model for a domain for future use as a source or target in new domain pairs. DDIB's pairwise translation only requires a linear number of diffusion models (as in Ref. [12], this can be further reduced with conditional models), and does not require scanning two datasets simultaneously during training. Unlike StarGAN, which utilizes shared domains, DDIB's latent space stores all information about the data.

Essentially, between the data and latent distribution, DDIB's Probability Flow ODE includes a solution to a particular Schrödinger Bridge Problem (SBP) with linear or degenerate drift. Moreover, except for the discretization error introduced by the ODE solver, DDIB guarantees exact cycle consistency.

A numerical ODE solver is used to construct the image x at different time steps. Empirical evidence shows that when reconstructing x at $t = 0$, the Score-based Generative Model (SGM) exhibits relatively low discretization error. For simplicity, let $v_\theta = \frac{dx}{dt}$ denote the velocity field parameterized by θ, and let ODESolve represent the mapping from $x(t_0)$ to $x(t_1)$:

$$\text{ODESolve}(x(t_0); v_\theta, t_0, t_1) = x(t_0) + \int_{t_0}^{t_1} v_\theta(t, x(t))\, dt \tag{16.44}$$

This abstracts away the specifics of the underlying model (whether score-based or diffusion) as well as the choice of numerical integrator.

First, given a source model represented as a vector field $v_\theta^{(s)}$, DDIB first applies ODESolve in the source domain to obtain the encoding of the image $x(s)$ at the end time $t = 1$, called the latent code (related to the diffusion model of the domain). Then, the source latent code is fed as the initial condition (the target latent code at $t = 1$) into the ODESolve with the target model $v_\theta^{(t)}$ to obtain the target image $x(t)$. As mentioned earlier, DDIM can implement ODESolve, which has a relatively small discretization error.

16.7 Text-to-Image/Video Synthesis

Due to its enormous potential applications, visual language modeling has recently attracted a lot of attention. Text-to-image generation is the task of generating corresponding images from descriptive text. A tool that generates realistic images from text can enable humans to create rich and diverse visual content in an easy way. The ability to edit images using natural language further allows iterative refinement and fine-grained control of the generated content.

Unconditional image models can synthesize photorealistic images, sometimes with enough realism that they cannot be distinguished from real images. Within this research field, diffusion models have become a promising generative model, achieving state-of-the-art sample quality.

To achieve photorealism in class-conditioned settings, classifier-guided methods allow diffusion models to condition on the labels of classifiers. The classifier is first trained on noisy images, and during the diffusion sampling process, the gradients of the classifier are used to guide the samples toward the label. Non-classifier-guided methods adopt another form of guidance, interpolating between predictions of diffusion models with and without labels.

Reference [16] reported a training diffusion model with 3.5 billion parameters, adapted to natural language descriptions with a text encoder. The samples generated by this diffusion model without classifier guidance were more favored than those of DALL-E, even though the latter used expensive CLIP reranking technology. In addition, this diffusion model can perform image restoration through fine-tuning, achieving powerful text-driven image editing. This system is named GLIDE (guided language to image diffusion for generation and editing), and the filtered small model is referred to as GLIDE (filtered).

To achieve non-classifier guidance with generic text prompts, sometimes during training, the text title is replaced with an empty sequence (also known as Ø). Then, the modified prediction $\hat{\epsilon}$ is guided toward the title c:

$$\hat{\epsilon}_\theta(x_t \mid c) = \epsilon_\theta(x_t \mid \emptyset) + s \cdot (\epsilon_\theta(x_t \mid c) - \epsilon_\theta(x_t \mid \emptyset)) \tag{16.45}$$

Non-classifier guidance has two attractive features: ① It allows a single model to utilize its own knowledge during the guidance process, rather than relying on the knowledge of a separate (sometimes smaller) classification model; ② It simplifies guidance when adjusting information that is difficult for classifiers (such as text) to predict.

CLIP (contrastive language-image pre-training) is a pre-training model that predicts caption-image pairs, learning state-of-the-art image representations from scratch on a dataset of 400 million (image-text) pairs collected from the internet. After pre-training, it uses natural language references to learn visual concepts (or described visual concepts), enabling zero-shot transfer of the model to downstream tasks.

The CLIP model consists of two independent parts: an image encoder $f(x)$ and a caption encoder $g(c)$. During training, batches of (x, c) pairs are sampled from a large dataset, and the model optimizes the contrastive cross-entropy loss. If the image x is paired with the given caption c, this loss encourages a high dot product $f(x) \cdot g(c)$; if the image and caption correspond to different pairs in the training data, it encourages a low dot product.

As CLIP provides scores close to the image and title, some methods use this to guide generative models (like GANs) toward user-defined text-captions. The same idea is also applied to diffusion models, replacing the classifier with the CLIP model in classifier-guided methods. Specifically, the gradient of the dot product of image-caption encoding with respect to the image is used to perturb the mean of the reverse process:

$$\hat{\mu}_\theta(x_t \mid c) = \mu_\theta(x_t \mid c) + s \cdot \sum_\theta (x_t \mid c)) \nabla_{x_t} (f(x_t) \cdot g(c)) \tag{16.46}$$

Similar to classifier guidance, CLIP is trained on noisy images x_t to obtain the correct gradient during the reverse process.

Reference [22] proposed a text-to-image diffusion model Imagen, which has better language understanding ability and unprecedented photorealism. Imagen builds on the ability of large Transformer language models to understand text and relies on the strength of diffusion models to generate high-fidelity images. Increasing the

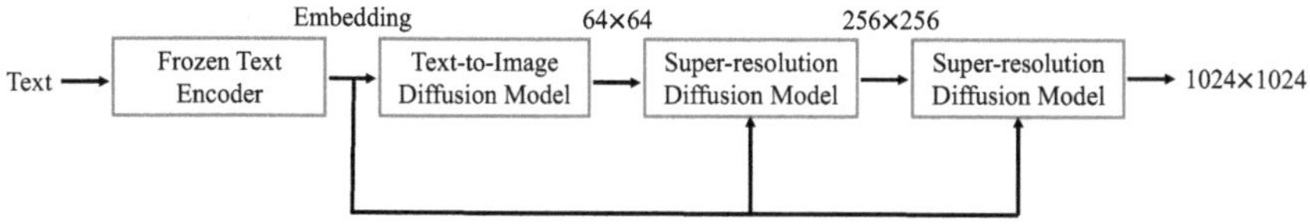

Fig. 16.5 Imagen visualization

scale of the language model in Imagen improves sample fidelity and image-text alignment more than increasing the image diffusion model. To further evaluate text-image models, DrawBench, a challenging benchmark for text-image models, was established.

Figure 16.5 shows a visualization of Imagen, which consists of two parts: a text encoder that maps text to an embedding sequence; a series of conditional diffusion models that map these embeddings to progressively higher resolution images.

Text-image models require powerful semantic text encoders to capture the complexity and combinability of arbitrary natural language text inputs. Text encoders trained on paired image-text data are the standard in current text-image models and can be trained from scratch or pre-trained on image-text data.

Large language models can be another way to encode text for generating text-image models. The latest advancements in large language models have led to leaps in text understanding and generation capabilities. Language models are trained on a corpus of pure text that is much larger than paired image-text data. Imagen examines these pre-trained text encoders, such as CLIP. For simplicity, the weights of these text encoders are frozen.

In the diffusion model, Imagen mainly relies on the classifier-free guidance method for effective text modulation. Throughout the sampling process, due to the iterative application of the diffusion model to its own output, the sampling process can produce unnatural images, sometimes even diverging. To solve this problem, Imagen uses static and dynamic thresholds for per-unit clipping to prevent pixel saturation. The dynamic threshold can significantly improve realism and image-text alignment, especially when the guidance weight is very large.

Inspired by the cascading diffusion model, Imagen uses a pipeline of a 64×64 base model and two text-conditioned super-resolution diffusion models to upsample the generated 64×64 image to a 256×256 size image, and then upsample it to a 1024×1024 image. The cascading diffusion model with noise modulation enhancement is very effective in gradually generating high-fidelity images. In addition, by adjusting the noise level, the super-resolution model perceives the superimposed noise, significantly improves the sample quality, and helps to improve the robustness of the super-resolution model, dealing with the artifacts produced by the low-resolution model. Imagen uses noise modulation enhancement for both super-resolution models, which is crucial for generating high-fidelity images.

Given a modulated low-resolution image and enhancement level (Gaussian noise or blur intensity), use enhancement to destroy the low-resolution image and adjust the diffusion model accordingly. During the training process, the enhancement level is randomly selected; during the inference process, different values are scanned to find the best sample quality.

Imagen uses the UNet architecture for the basic 64×64 text-image diffusion model. The network adjusts the text embedding by mixing the embedding vectors and adds it to the diffusion time step embedding, similar to the class embedding adjustment method. Cross-attention mechanism is added to the multi-resolution text embedding to further adjust the entire sequence of text embedding.

The super-resolution 64×64–256×256 model uses the UNet model. To improve memory efficiency, inference time, and convergence speed, some modifications were made, called the Efficient UNet variant. The super-resolution 256×256–1024×1024 model is trained with 64×64–256×256 crops of 1024×1024 images. For this, the self-attention layer is removed, but the text cross-attention layer is retained, which is a key detail. During the inference process, the model receives the complete 256×256 low-resolution image as input and returns the upsampled 1024×1024 image as output.

The Imagen system is extended to Imagen Video, and a text-conditioned video generation system based on a series of video diffusion models is proposed in Ref. [27]. Given a text prompt, it uses a basic video generation model and a series of spatial and temporal cross video super-resolution models to generate high-definition videos. In addition, the progressive distillation method is applied to the video model, and classifier-free guidance is used for fast, high-quality sampling.

For the entire cascading pipeline of Imagen Video: starting from the input of text prompts, generating a 5.3-second long, 1280×768 pixel size video at a speed of 24FPS; in fact, text embedding is injected into all models, not just the base model. The entire cascading pipeline has a parameter-frozen text encoder, a basic video diffusion model, 3 SSRs (spatial super-resolution) and 3 TSRs (time super-resolution) models, a total of 7 video diffusion models, and a total of 11.6B diffusion model parameters. The data used to train these models is processed to the appropriate spatial and temporal resolution. At generation time, the SSR model increases the spatial resolution of all input frames, while the TSR model increases the temporal resolution by filling in intermediate frames between input frames.

A real video synthesis model Phenaki, which is given a series of text prompts, is proposed in Ref. [28]. It introduces a model that learns video representation, compressing the video into a small representation of discrete tokens. This tokenizer uses temporal causal attention, which can handle variable-length videos. To generate video tokens from text, a bidirectional masked Transformer that depends on precomputed text tokens is used. The generated video tokens are then detokenized to create the actual video.

To solve the data problem, a large amount of image-text pair corpus and a small amount of video-text examples are used for joint training, achieving generalization outside the video dataset. Compared with previous video generation methods, Phenaki generates arbitrarily long videos in the open domain based on a series of prompts (i.e., time-varying text or a story).

Inspired by previous autoregressive (AR) text-image and text-video work, Phenaki designs two main components: an encoder–decoder model that compresses video into discrete embeddings (i.e., tokens), and a Transformer model that converts text embeddings into video tokens. To obtain text embeddings, Phenaki uses the pre-trained language model T5X.

16.8 Improvements to the Diffusion Model

Reconstructing data distribution requires sampling. In each sampling step, the samples generated by random noise are refined to be closer to the original distribution. However, the sampling of diffusion models requires too many steps, making it very time-consuming.

The following are some measures for improvement.

- **Acceleration.** This includes training scheme design, training-free sampling, hybrid modeling, and the unification of score-based and diffusion approaches.
- **Diversification of distributions.** This includes modeling in continuous, discrete, and constrained spaces.
- **Likelihood optimization.** This includes improved ELBO formulations and variable interval optimization strategies.
- **Dimensionality reduction.**

As mentioned before, classical diffusion models have three main drawbacks: low sampling efficiency, suboptimal likelihood estimation, and poor data generalization ability.

The following are some methods for enhancement.

- Acceleration of sampling. This includes SDEE/ODE solvers, non-Markov processes, and partial sampling (in the reverse process).
- Likelihood maximization. This includes noise scheme optimization, learnable reverse variance, and target design, etc.
- Data generalization. This includes unified feature space and data-dependent transformation kernels, etc.

Due to the good characteristics of diffusion models, combining diffusion models with traditional generative models has made new improvements to the original generative models, such as Generative Adversarial Networks (GAN), Autoregressive (AR) models, Normalizing Flows (NF), Variational Autoencoders (VAE), and Energy-Based Modeling (EBM), etc.

Diffusion models have more in common with VAEs. A diffusion model can be seen as a deep hierarchical Markov VAE, which means that the noise addition and recovery processes correspond to the encoding and decoding processes of the VAE. The differences are: ① The latent representation of the VAE contains compressed information of the original image, while the diffusion model completely destroys the data after the last step of the forward process; ② The latent representation of the diffusion model has the same dimension as the original data, while the VAE works better when the dimension is reduced; ③ The mapping to the VAE latent space is trainable, while the forward process of the diffusion model is not.

In addition, DDPM shares decoders across multiple layers, and all latent variables have the same dimensionality as the sample data. In a continuous-time setting, the score matching objective can be further approximated by the ELBO of a deep hierarchical VAE. In this way, optimizing the diffusion model can be viewed as training an infinitely deep hierarchical VAE, which demonstrates that diffusion models can be interpreted as a continuous spectrum of hierarchical VAEs. The ELBO is realized through a special score matching target obtained by diffusing the latent space.

AR models represent images as pixel sequences, and their generation process generates images pixel by pixel conditioned on previously generated pixels. This method implies a one-way bias, which obviously represents the limitations of this type of generative model.

The AR Diffusion Model (ARDM) proposes a different AR model that learns how to generate data of any order.

ARDM extends the order-agnostic AR model and discrete diffusion model. Unlike AR models, ARDM does not require causal masking of representations, so it can be trained with effective targets, similar to diffusion probability models. In the testing process, ARDM can generate in parallel, so it can be applied to generative tasks of any scale.

Normalizing Flows (NF) are similar to diffusion models in that they map data distribution to Gaussian noise. However, Normalizing Flows learn reversible and differentiable functions to map in a deterministic way. In contrast to diffusion models, these characteristics imply additional constraints on network architecture and a learnable forward process.

DiffFlow combines the advantages of flow-based models and diffusion models to propose a new generative modeling algorithm. Compared with diffusion probability models, it can not only obtain sharper boundaries than NF but also learn more general distributions with fewer discrete steps.

In Normalizing Flows, both the forward and reverse processes are deterministic, they are inverse processes of each other, so they are combined into one process; diffusion models have a fixed forward process and a trainable reverse process, both of which are random. In DiffFlow, both the forward and reverse processes are trainable and random.

Energy-Based Models (EBMs) focus on providing estimates of unnormalized density functions, referred to as energy functions. A popular strategy for training these models is score matching. For sampling, in addition to other strategies, Markov Chain Monte Carlo (MCMC) methods can be employed based on the score function. In this view, score-based diffusion models can be regarded as a special case within the energy-based framework, where only the score function is required for both training and sampling.

The Ref. [32] proposes a diffusion recovery likelihood method, which learns samples controllably from a series of EBMs in the reverse process of the diffusion model. Given a noise board at a higher noise level, each EBM is trained through recovery likelihood, with the aim of maximizing the conditional probability of data under a specific noise level. EBM maximizes recovery likelihood because it is easier to handle than marginal likelihood, just as sampling from conditional distribution is easier than sampling from marginal distribution. This model can generate high-quality samples, and long-term MCMC samples from the conditional distribution still resemble real images.

In terms of the quality of generated samples, before the rise of diffusion models, GAN was the most advanced generative model. GAN is also difficult to train due to its adversarial objective and often suffers from mode collapse. In contrast, diffusion models have a stable training process and offer more diversity because they are based on likelihood. Despite these advantages, compared to GAN, diffusion models are still inefficient, requiring multiple network evaluations during the inference process.

The instability of GAN training is mainly caused by the non-overlap between the input data distribution and the generated data distribution. A practical solution is to inject noise into the discriminator input. Slow sampling is caused by the Gaussian assumption of the denoising step and is only applicable to small steps. Therefore, using a conditional GAN to model each denoising step allows for larger steps. Conversely, GAN can help improve the sampling speed of diffusion models.

For Diffusion-GAN, the top row images represent the forward diffusion process of real images, while the bottom row images represent the forward diffusion process of generated fake images. The discriminator learns to distinguish a diffused real image and a diffused fake image at all diffusion steps.

Unlike the basic GAN, Diffusion-GAN compares the noise boards of real images and generated images, which are sampled from the Gaussian mixture distribution of diffusion steps with the help of time step-related discriminators. Each component under this distribution has a different noise-data ratio. This has two benefits: First, it stabilizes training by alleviating the problem of gradient vanishing, which occurs when the data and generator distributions differ too much; Second, it enhances by creating different noise boards for the same image, which can improve data efficiency and the diversity of the generator.

16.9 Diffusion Model for Policy and Planning Learning

Based on the application of diffusion models in image and video generation, and world model learning as neural simulators, it can be understood that diffusion models play a role in content synthesis in simulation of autonomous driving [36, 39, 44].

In addition, diffusion models can also work in policy and planning learning of autonomous driving. In the field of robotics, diffusion models have been used in imitation learning [35, 41–43], showing strong action modeling capabilities. As a special robotic application, autonomous driving can also see the application of diffusion models in policy and planning learning [38, 40, 46].

At present, diffusion-based planning frameworks can generally be divided into two categories according to whether they generate actions or states: action-based diffusion models [35] perform well in well-defined tasks but often lack versatility when adapting to complex or new tasks with flexible interaction requirements; in contrast, state-based diffusion methods [37] often produce unrealistic "ghost states."

The following examples illustrate the application of diffusion models in policy and planning learning of autonomous driving.

MotionDiffuser [38] (from Google WayMo) is a diffusion-based representation method for representing the joint distribution of future trajectories of multiple agents. Based on the diffusion model, this method learns a highly multimodal distribution that can capture different future outcomes. In addition, the trajectory representation is compressed through PCA, which improves model performance and allows efficient computation of accurate sample log probabilities.

As shown in Fig. 16.6: The input scene contains agent history, traffic lights, and road maps, and is encoded into a set of conditional tokens C through the Transformer encoder. During training, a set of random noise is sampled independently and identically from a normal distribution and added to the ground-truth (GT) trajectory. The denoiser predicts a denoised trajectory corresponding to each agent while paying attention to the conditional tokens. The entire model can be trained end-to-end using a simple L2 loss between the predicted denoised trajectory and the GT trajectory. During inference, a set of trajectories can be first sampled for each agent from pure noise with the highest noise level, and then iteratively denoised by the denoiser to generate a reasonable distribution of future trajectories. Optional constraints in the form of arbitrary differentiable loss functions can be injected into the denoising process to enforce constraints.

Diffusion-ES [40] is a method that combines gradient-free optimization with trajectory denoising to optimize a black-box non-differentiable objective while staying in the data manifold. Diffusion-ES samples trajectories from a diffusion model during an evolutionary search and scores them using a black-box reward function. It mutates high-scoring trajectories using a truncated diffusion process that applies

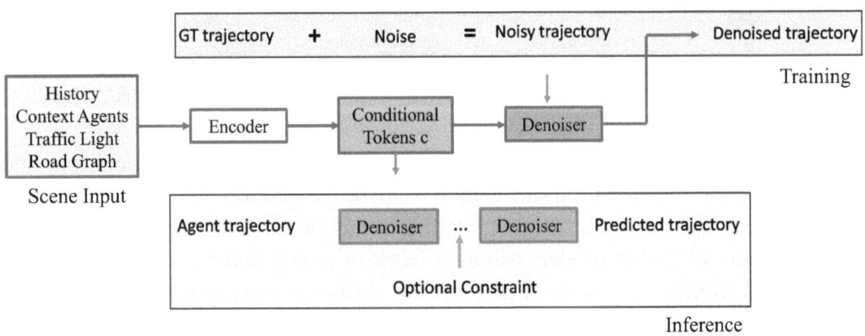

Fig. 16.6 The overview of multi-agent motion trajectory prediction based on the diffusion model

a small number of noise adding and denoising steps, allowing for more efficient exploration of the solution space.

DiffusionDrive [45] proposes a truncated diffusion strategy that combines previous multimodal anchors and truncated diffusion scheduling, enabling the model to learn the denoising process from the anchored Gaussian distribution to the multimodal driving action distribution. In addition, it designs an efficient cascade diffusion decoder to enhance the interaction with the conditional scene context.

The overall architecture of DiffusionDrive is shown in Fig. 16.7. (a) DiffusionDrive can integrate various existing perception modules and different sensor inputs. (b) The designed diffusion decoder takes the noisy trajectory sampled from the anchored Gaussian distribution as input and gradually denoises it in a cascade manner through enhanced interaction with the conditional scene context to generate the final prediction.

Diffusion Planner [46] is a closed-loop diffusion planner based on Diffusion Transformer (DiT) [34] that can effectively model multimodal driving behavior and ensure trajectory quality without any rule-based refinement. Based on the diffusion model, Diffusion Planner jointly models the prediction and planning tasks in the same architecture, learns the gradient of the trajectory scoring function, and adopts a flexible classifier guidance mechanism to achieve collaborative behavior between vehicles.

As shown in Fig. 16.8, the interaction and fusion process of Diffusion Planner is summarized.

1. Connecting the future vehicle trajectory with the current state of each vehicle as constraints to guide the model, providing a clear starting point to simplify the planning task. During the execution of the model, the information integration from different vehicles is achieved through a multi-head self-attention mechanism.

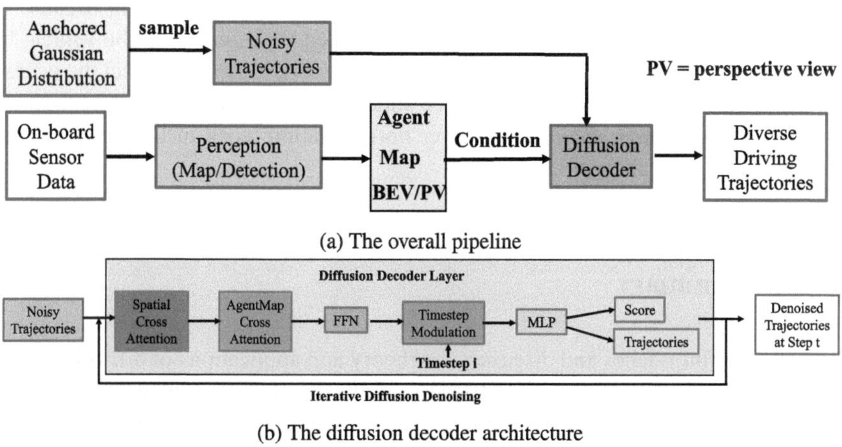

Fig. 16.7 The overall architecture of DiffusionDrive

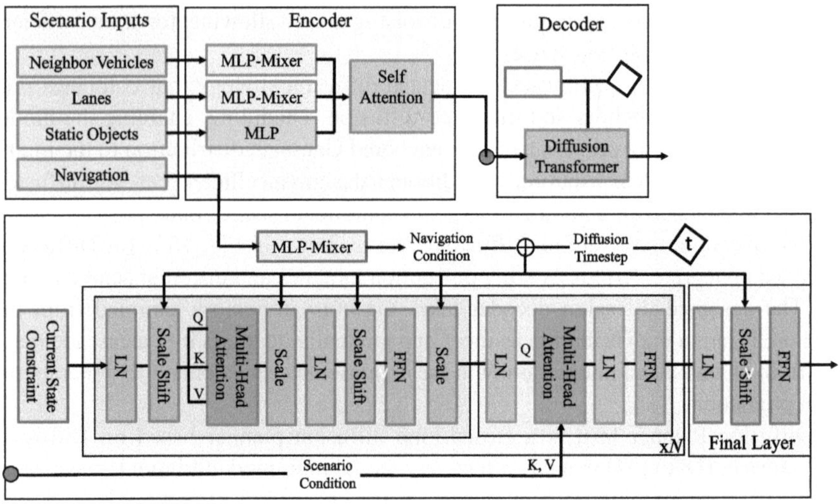

Fig. 16.8 The overview of diffusion planner

2. The historical state of neighboring vehicles and lane information are represented by vectors. Since these vectors are sparse, directly fusing them makes training challenging. To solve this problem, an MLP-Mixer network is used to extract information-dense representations. This more unified and simplified solution iteratively passes the vectors to the MLP mixing layer and then operates on the vector and feature dimensions simultaneously. Static object information is also considered, including coordinates, heading, size, and category. Finally, all representations are concatenated and fed into the original Transformer encoder for further aggregation to obtain the encoder representation.
3. Navigation information is crucial for autonomous driving planning because it provides the necessary guidance for the intended route, enabling the vehicle to make smart decisions. First, an MLP-Mixer network is used to extract the necessary guidance representation. It is then added to the diffusion time step condition and passes through the adaptive layer normalization block in DiT to guide the trajectory generation of all tokens.

16.10 Summary

This chapter introduces and discusses the theory and applications of diffusion models, including Langevin dynamics-based score-based generative models, denoising diffusion probabilistic models, denoising diffusion implicit models, and the unified

SDE framework. Application areas cover image and video synthesis, image-to-image translation, text-to-image and text-to-video generation, as well as policy and planning learning in autonomous driving. Finally, several recent improvement methods for diffusion models are summarized.

References

1. J SOHL-DICKSTEIN, E WEISS, N MAHESWARANATHAN, and S GANGULI. Deep unsupervised learning using nonequilibrium thermodynamics. Int. Conf. on Machine Learning, Lille, France, July 2015: 2256-2265.
2. Y CHOI, M CHOI, M KIM, J-W HA, S KIM, and J CHOO. StarGAN: Unified generative adversarial networks for multi-domain image-to-image translation. IEEE Conference on Computer Vision and Pattern Recognition, Salt Lake City, UT, USA, June 2018: 8789-8797.
3. A BROCK, J DONAHUE, and K SIMONYAN. Large scale GAN training for high fidelity natural image synthesis. arXiv preprint arXiv:1809.11096, 2018.
4. Y SONG,S ERMON, Generative Modeling by Estimating Gradients of the Data Distribution,arXiv preprint arXiv:1907.05600,2019
5. J HO, A JAIN, P ABBEEL, Denoising Diffusion Probabilistic Models,arXiv preprint arXiv:2006.11239, 2020
6. J. SONG, C MENG, S ERMON, Denoising diffusion implicit models. arXiv preprint arXiv:2010.02502, 2020.
7. Y SONG, J S-DICKSTEIN, D P. KINGMA, A KUMAR, S ERMON, B POOLE, Score-Based Generative Modeling through Stochastic Differential Equations, arXiv preprint arXiv:2011.13456, 2020
8. P ESSER, R ROMBACH, and B OMMER. Taming transformers for high-resolution image synthesis. arXiv preprint arXiv:2012.09841, 2020
9. A RAMESH, M PAVLOV, G GOH, S GRAY, C VOSS, A RADFORD, M CHEN and I SUTSKEVER, Zero-shot text-to-image generation. arXiv preprint arXiv:2102.12092, 2021.
10. A RADFORD, J W KIM, C HALLACY, A RAMESH, G GOH, S AGARWAL, G SASTRY, A ASKELL, P MISHKIN, J CLARK, G KRUEGER, and I SUTSKEVER, Learning transferable visual models from natural language supervision. arXiv preprint arXiv:2103.00020, 2021.
11. C. SAHARIA, J. HO, W. CHAN, T. SALIMANS, D. J. FLEET, and M. NOROUZI, Image super-resolution via iterative refinement, arXiv preprint arXiv:2104.07636, 2021.
12. P DHARIWAL, A NICHOL, Diffusion Models Beat GANs on Image Synthesis, arXiv preprint arXiv:2105.05233, 2021
13. J HO, C SAHARIA, W CHAN, D J FLEET, M NOROUZI, and T SALIMANS. Cascaded diffusion models for high fidelity image generation. arXiv preprint arXiv:2106.15282, 2021.
14. C SAHARIA, W CHAN, H CHANG, C A LEE, J HO, T SALIMANS, D J FLEET, M NOROUZI,Palette: Image-to-Image Diffusion Models, arXiv preprint arXiv:2111.05826, 2021
15. A NICHOL, P DHARIWAL, A RAMESH, P SHYAM, P MISHKIN, B MCGREW, I SUTSKEVER, M CHEN, GLIDE: Towards Photorealistic Image Generation and Editing with Text-Guided Diffusion Models, arXiv preprint arXiv:2112.10741, 2021
16. R ROMBACH, A BLATTMANN, D LORENZ, P ESSER, B OMMER, High-Resolution Image Synthesis with Latent Diffusion Models, arXiv preprint arXiv:2112.10752, 2021
17. D KINGMA, T SALIMANS, B POOLE, and J HO. Variational diffusion models. arXiv preprint arXiv:2107.00630, 2021.
18. X SU, J SONG, C MENG, S ERMON, Dual Diffusion Implicit Bridges For Image-to-Image Translation, arXiv preprint arXiv:2203.08382, 2022
19. A ROBERTS, H W CHUNG, A LEVSKAYA, G MISHRA, J BRADBURY, D ANDOR, S NARANG, B LESTER, C GAFFNEY, A MOHIUDDIN. Scaling up models and data with T5X and Seqio. arXiv preprint arXiv:2203.17189, 2022.

20. J HO, T SALIMANS, A GRITSENKO, W CHAN, M NOROUZI, D J. FLEET, Video Diffusion Models, arXiv preprint arXiv:2204.03458, 2022
21. A RAMESH, P DHARIWAL, A NICHOL, C CHU, and M CHEN. Hierarchical Text-Conditional Image Generation with CLIP Latents. arXiv preprint arXiv:2204.06125, 2022.
22. C SAHARIA, W CHAN, S SAXENA, L LI, J WHANG, E DENTON, S K S GHASEMIPOUR, B K AYAN, S. S MAHDAVI, R G LOPES, T SALIMANS, J HO, D FLEET, and M NOROUZI. Photorealistic Text-to-Image Diffusion Models with Deep Language Understanding. arXiv preprint arXiv:2205.11487, 2022.
23. J HO, and T SALIMANS, Classifier-free diffusion guidance. arXiv preprint arXiv:2207.12598, 2022.
24. H CAO, C TAN, Z GAO, G CHEN, P-A HENG and S Z. LI, A Survey on Generative Diffusion Model, arXiv preprint arXiv:2209.02646, 2022
25. L YANG, Z ZHANG, S HONG, R XU, Y ZHAO, Y SHAO, W ZHANG, M-H YANG, B CUI, Diffusion Models: A Comprehensive Survey of Methods and Applications, arXiv preprint arXiv:2209.00796, 2022
26. D WATSON, W CHAN, R MARTIN-BRUALLA, J HO, A TAGLIASACCHI, M NOROUZI, Novel View Synthesis With Diffusion Models, arXiv preprint arXiv:2210.04628, 2022
27. J HO, W CHAN, C SAHARIA, J WHANG, R GAO, A GRITSENKO, D P. KINGMA, B POOLE, M NOROUZI, D J. FLEET, T SALIMANS, Imagen Video: High Definition Video Generation with Diffusion Models, arXiv preprint arXiv:2210.02303, 2022
28. R VILLEGAS, M BABAEIZADEH, P-J KINDERMANS, H MORALDO, H ZHANG, M T SAFFAR, S CASTRO, J KUNZE, D ERHAN, Phenaki: Variable Length Video Generation From Open Domain Textual Description, arXiv preprint arXiv:2210.02399, 2022
29. C-W HUANG, J H LIM, and A C COURVILLE. A variational perspective on diffusion-based generative models and score matching. arXiv preprint arXiv:2106.02808, 2021.
30. E HOOGEBOOM, A GRITSENKO, J BASTINGS, B POOLE, R VAN DEN BERG, and T SALIMANS. Autoregressive Diffusion Models. Int. Conf. Learning Representation, arXiv preprint arXiv:2110.02037 2021.
31. Q ZHANG and Y CHEN. Diffusion Normalizing Flow. Int. Conf. Neural Information Processing Systems, arXiv preprint arXiv:2110.07579, 2021.
32. R GAO, Y SONG, B POOLE, Y N WU, and D P. KINGMA. Learning Energy-Based Models by Diffusion Recovery Likelihood, arXiv preprint arXiv:2012.08125, 2021.
33. Z WANG, H ZHENG, P HE, W CHEN, and M ZHOU. Diffusion-GAN: Training GANs with Diffusion. arXiv preprint arXiv:2206.02262, 2022.
34. W Peebles and S Xie. "Scalable diffusion models with transformers". arXiv 2212.09748, 2022
35. C Chi et al., "Diffusion Policy: Visuomotor Policy Learning via Action Diffusion", arXiv 2303.04137, 2023
36. E Pronovost, K Wang, Nick Roy, "Generating Driving Scenes with Diffusion", arXiv 2305.18452, 2023
37. F Ni, J Hao, Y Mu, and et al. "MetaDiffuser: Diffusion model as conditional planner for offline Meta-RL". arXiv 2305.19923, 2023
38. C Jiang, A Cornman, C Park, and et al. "MotionDiffuser: Controllable multi-agent motion prediction using diffusion". arXiv 2306.03083, IEEE CVPR, 2023
39. X Li et al., "DrivingDiffusion: Layout-Guided multi-view driving scene video generation with latent diffusion model", arXiv 2310.07771, 2023
40. B Yang et al., "Diffusion-ES: Gradient-free Planning with Diffusion for Autonomous Driving and Zero-Shot Instruction Following", arXiv 2402.06559, 2024
41. T-W Ke, N Gkanatsios, and K Fragkiadaki. "3d diffuser actor: Policy diffusion with 3d scene representations". arXiv 2402.10885, 2024
42. Y Ze, G Zhang, K Zhang, and et al. "3d diffusion policy". arXiv 2403.03954, 2024
43. Octo Model Team, "Octo: An Open-Source Generalist Robot Policy", arXiv 2405.12213, 2024

44. E Ma et al., "Unleashing Generalization of End-to-End Autonomous Driving with Controllable Long Video Generation" (Delphi), arXiv 2406.01349, 2024
45. B Liao et al., "DiffusionDrive: Truncated Diffusion Model for End-to-End Autonomous Driving", arXiv 2411.15139, 2024
46. Y Zheng, R Liang, K Zheng, and et al., "Diffusion-Based Planning for Autonomous Driving with Flexible Guidance", arXiv 2501.15564, 2025

Chapter 17
Large Scale Models in Autonomous Driving

Since the Turing test was proposed in the 1950s, humans have begun to explore the mastery of language intelligence by machines. Language is essentially a complex and sophisticated human expression system governed by grammatical rules. Developing artificial intelligence algorithms that can understand and master language is a key challenge. As a major method for language understanding and generation, language modeling has been widely studied in the past two decades, from statistical language models to neural language models. Recently, a pre-trained language model (PLM) was proposed by pre-training the Transformer model on a large-scale corpus, which has shown powerful capabilities in solving various natural language processing (NLP) tasks.

In order to distinguish language models with different parameter sizes, the research community coined the term "Large Language Model" (LLM) [15] for language models with considerable size (e.g., containing tens or hundreds of billions of parameters). Recently, academia and industry have made great progress in the research of LLM. One notable progress is the launch of ChatGPT (a powerful AI chatbot developed based on LLM) [8], which has attracted widespread attention from the society.

The technological developments in LLMs have important implications for the entire AI community, revolutionizing the way AI algorithms are developed and used in areas such as embodied intelligence, agents, and robotics, including autonomous vehicles.

This chapter mainly introduces the application of large language models in autonomous driving. Section 17.1 introduces the pre-training and fine-tuning of large language models; Sect. 17.2 discusses its emergent capabilities, namely prompting, in-context learning (ICL), instruction following, model alignment, and reasoning of chain-of-thought (CoT); Sect. 17.3 generalizes the large language model to the visual-language domain, namely the vision-large language model; Sect. 17.4 further generalizes it to the multimodal large language model, which is convenient for application in complex scenarios in real environments; Sect. 17.5 introduces the world

Y. Huang and Z. Yang, *Autonomous Driving Handbook*,
https://doi.org/10.1007/978-981-95-3725-9_17

model that can model and predict, as well as embodied intelligence technology with environmental interaction capabilities; Sect. 17.6 introduces the optimization and scaling of the inference/test-time of the large scale model; Sect. 17.7 introduces the application of the large scale model in autonomous driving with examples; Sect. 17.8 summarizes the contents of this chapter.

17.1 Pre-training and Fine-Tuning of Large Language Model

17.1.1 Foundation Model and Large Language Model

The foundation model [7], technically speaking, is implemented by transfer learning and extension. The idea of transfer learning is to apply the "knowledge" learned from one task to another task. As shown in Fig. 17.1, the foundation model usually follows this paradigm: the model is pre-trained on an alternative task and then adapted to the downstream task of interest through fine-tuning.

A typical example of a foundation model, the Large Language Model (LLM), is a large-scale, pre-trained statistical language model based on a neural network. The recent success of LLMs is the result of the accumulation of language model research and development, which can be divided into four historical waves: statistical language model (SLM), neural language model (NLM), pre-trained language model (PLM) and LLM.

The key role of pre-training is to encode common sense from a large-scale corpus into huge model parameters. For training LLM, there are two commonly used pre-training tasks, namely language modeling (LM) and denoising autoencoder (DAE).

17.1.2 Pre-training of Large Language Model

Pre-training lays the foundation for LLM capabilities. By pre-training on a large-scale corpus, LLM can acquire basic language understanding and generation skills. In this process, the scale and quality of the pre-training corpus are crucial for LLM

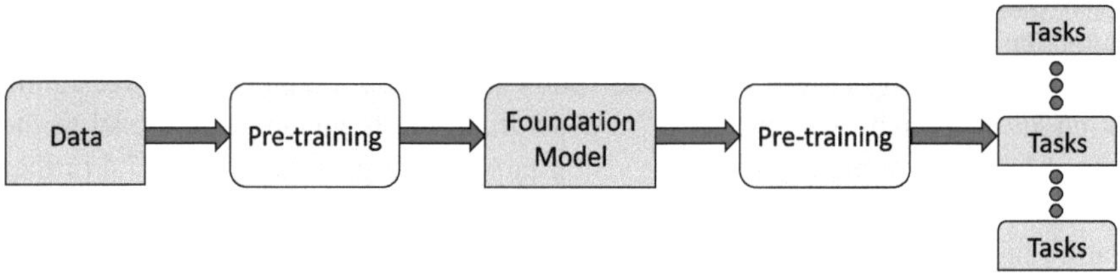

Fig. 17.1 Foundation model

to gain powerful capabilities. In addition, in order to effectively pre-train LLM, the model architecture, acceleration methods, and optimization techniques need to be carefully designed.

Due to its excellent parallelism and capabilities, the Transformer architecture has become the de facto backbone for developing various LLMs, enabling language models to scale to tens or hundreds of billions of parameters. In general, the mainstream architectures of existing LLMs can be roughly divided into three categories, namely encoder-decoder, causal decoder, and prefix decoder.

The original Transformer model is built on an encoder-decoder architecture consisting of two sets of Transformer blocks, acting as encoders and decoders. The encoder encodes the input sequence using stacked multi-head self-attention layers to generate its latent representations, while the decoder cross-attends these representations and autoregressively generates the target sequence.

The causal decoder architecture combines a unidirectional attention mask to ensure that each input token can only pay attention to past tokens and itself. The decoder processes input and output tokens in the same way. As a representative language model of this architecture, the GPT series of models are developed based on the causal decoder architecture. The prefix decoder architecture (also known as the non-causal decoder) modifies the masking mechanism of the causal decoder so that bidirectional attention is performed on the prefix tokens and unidirectional attention is performed only on the generated tokens. In this way, just like the encoder-decoder architecture, the prefix decoder can bidirectionally encode the prefix sequence and autoregressively predict the output token one by one, where the same parameters are shared during encoding and decoding.

In Transformer-based Large Language Models (LLMs), a mixture of experts (MoE) model [31] selects different parameters for each input example. Each MoE layer is usually composed of a set of "expert networks" and a "gating network." The gating network is usually in the form of a linear network with a softmax activation function, and its role is to guide the input to the appropriate expert network. The location of the MoE layer is designed to select the feed-forward network (FFN) within each Transformer block, usually after the self-attention (SA) sublayer.

There are two types of MOE: dense and sparse. For each input, the linear softmax gating will select all experts, i.e., dense MoE, or top-experts, i.e., sparse MoE, to perform conditional calculations. The expert layer returns the output of the selected expert multiplied by the gate value (the softmax of the output of the gating function).

LLMs scale significantly (by several orders of magnitude) in model size, data size, and total computation. Numerous studies have shown that scaling can greatly improve the model capabilities of LLMs. Therefore, it is useful to establish a quantitative method to characterize the scaling effect.

In 2020, Kaplan (from OpenAI team) first proposed a power-law relationship model for neural language models [3], which establishes the relationship between model performance and three main factors: model size (N), dataset size (D), and training computation amount (C).

Another representative study is that Hoffmann (from Google DeepMind team) proposed an alternative law [9] to guide the computational optimization training of

LLM. They conducted rigorous experiments by changing a larger range of model sizes (70M to 16B) and data sizes (5B to 500B tokens), and fitted similar scaling laws but with different coefficients.

As the size of models and data increases, it becomes increasingly challenging to efficiently train LLMs under limited computing resources. In particular, two major technical issues need to be addressed, namely, improving training throughput and loading larger models into GPU memory. Several widely used approaches to address these two challenges include 3D parallelism, ZeRO, and mixed precision training [12].

3D parallelism is actually a combination of three commonly used parallel training technologies, namely data parallelism, pipeline parallelism, and tensor parallelism.

Data parallelism is one of the most basic methods to improve training throughput. It copies the model parameters and optimizer state to multiple GPUs, and then distributes the entire training corpus to these GPUs. In this way, each GPU only needs to process the data assigned to it and perform forward and backward propagation to obtain gradients. The gradients calculated on different GPUs will be further aggregated to obtain the gradients of the entire batch to update the model in all GPUs. In this way, since the calculation of gradients is performed independently on different GPUs, the data parallelism is highly scalable, so that the training throughput can be improved by increasing the number of GPUs. Pipeline parallelism aims to distribute different layers of LLM to multiple GPUs. Especially in the case of Transformer models, pipeline parallelism loads consecutive layers onto the same GPU to reduce the cost of hidden state or gradient transferring calculations between GPUs. However, simple pipeline parallel implementations may result in low GPU utilization because each GPU must wait for the previous GPU to complete the calculation, resulting in unnecessary bubble overhead. To reduce these bubbles in pipeline parallelism, GPipe and PipeDream proposed techniques of filling multiple batches of data and asynchronous gradient updates to improve pipeline efficiency.

Tensor parallelism, which aims to decompose LLM for multi-GPU loading. Unlike pipeline parallelism, tensor parallelism focuses on decomposing the tensors (parameter matrices) of LLM.

ZeRO technology [2], proposed by the DeepSpeed library, mainly solves the problem of memory redundancy in data parallelism. ZeRO technology aims to keep only a small part of the data on each GPU, while the rest of the data can be retrieved from other GPUs when needed. Specifically, ZeRO provides three solutions, depending on how the three parts of the data are stored, namely optimizer state partitioning, gradient partitioning, and parameter partitioning. Empirical results show that the first two solutions do not increase the communication overhead, and the third solution increases the communication overhead by about 50%, but the memory saved is proportional to the number of GPUs. PyTorch implements a similar technology to ZeRO, called Fully Sharded Data Parallelism (FSDP).

In order to pre-train extremely large language models, some studies have begun to use 16-bit floating point numbers (FP16) instead of the original 32-bit floating point numbers (FP32) to reduce memory usage and communication overhead. In addition, since the number of FP16 computing units of popular NVIDIA GPUs (such as A100) is twice that of FP32, the computing efficiency of FP16 can be further improved. However, existing studies have found that FP16 may lead to a loss of computational accuracy, thereby affecting the final model performance. To alleviate this problem, an alternative called Brain Floating Point (BF16) has been used for training, which allocates more exponent bits and fewer significant bits than FP16. For pre-training, BF16 generally performs better than FP16 in terms of representation accuracy.

17.1.3 Fine-Tuning of Large Language Model

After pre-training, the capabilities of LLM can be further tuned according to specific goals. There are two main methods for tuning LLM, namely instruction tuning and alignment tuning. The former method mainly aims to enhance (or unlock) the capabilities of LLM, while the latter method aims to align the behavior of LLM with human values or preferences. In addition, there needs effective fine-tuning and adaptive quantization of model under limited resources.

Essentially, instruction fine-tuning is a method of fine-tuning a pre-trained LLM on a collection of formatted examples in natural language form, which is highly related to supervised fine-tuning and multi-task prompt training. To perform instruction fine-tuning, instruction-formatted examples need to be collected or constructed first. Then, the LLM is fine-tuned with these formatted examples in a supervised learning manner (e.g., trained using a sequence-to-sequence loss). After instruction fine-tuning, the LLM can show excellent ability to generalize to unseen tasks.

Typically, an instruction-based instance consists of a task description (called an instruction), an optional input, the corresponding output, and a small amount of demonstration (optional). As an important public resource, existing research community has released a large amount of labeled data formatted in natural language. There are three different methods for formatting instruction-based instances and building instruction-based instances: task datasets, daily conversation data, and synthetic data.

LLM models sometimes exhibit unexpected behaviors, such as fabricating false information, pursuing inaccurate goals, and producing harmful, misleading, and biased expressions, i.e., hallucinations. For LLM, the language modeling objective pre-trains model parameters through word prediction, lacking consideration of human values or preferences. To avoid these unexpected behaviors, artificial alignment is proposed to make the behavior of LLM match human expectations. However, unlike the original pre-training and adaptation tuning (such as instruction tuning), this alignment needs to consider unusual criteria (such as usefulness, honesty, and harmlessness).

In existing works, the main method of generating human feedback data is human annotation. This highlights the critical role of selecting appropriate human labelers. In order to provide high-quality feedback, human labelers should have qualified education and excellent English proficiency. When human labelers annotate the output of LLM, it is helpful to specify detailed instructions and provide immediate guidance to human labelers, which can further standardize the labelers' annotations.

Reinforcement Learning with Human Feedback (RLHF) fine-tunes the LLM using the collected human feedback data, which helps improve the alignment criteria (e.g., helpfulness, honesty, and harmlessness). RLHF employs reinforcement learning (RL) algorithms, such as proximal policy optimization (PPO), to adapt the LLM to human feedback by learning a reward model.

Parameter Efficient Fine-tuning (PEFT) [14] aims to reduce the number of trainable parameters while maintaining good performance as much as possible. There are four parameter efficient fine-tuning methods for Transformer-based language models, including adapter tuning [1], prefix tuning [4], prompt tuning [18], and low-rank adaptation (LoRA).

In addition, due to the huge number of model parameters, the inference of LLMs is memory-intensive, making them very expensive to deploy in real applications. A popular model compression approach, model quantization, can reduce the memory footprint of LLMs so that large LLMs can be used in resource-constrained environments, which may also reduce inference latency.

17.2 Prompting, ICL and CoT of Large Language Model

17.2.1 Prompting

After pre-training or adaptive tuning, the application of LLMs is mainly to design appropriate prompting strategies to solve various tasks [10]. By choosing appropriate prompts, the LLM model behavior can be manipulated so that the pre-trained model itself can be used to predict the desired output, sometimes even without any additional task-specific training. The so-called prompt engineering works on the principle of finding the most appropriate prompts that allow the LM to solve the task at hand.

A prompt can consist of text, images, sounds, or other media. Prompts are usually constructed using a prompt template, which is a function that contains one or more variables that are replaced with some media (usually text) to create a prompt. This prompt can then be viewed as an instance of the template.

The most commonly used components of prompts include:

- Directive. This is the core intention of the prompt, sometimes simply called the "intention." Instructions can also be implicit.
- Example. Also called an example or sample, it serves as a demonstration to guide the LLM in completing the task.
- Output format. Usually LLM outputs information in some specific format.
- Style instruction. It is a type of output formatting that modifies the output in style rather than structure.
- Persona. Also called personality, it can improve writing and text style.
- Additional information. It is sometimes called "context."

17.2.2 In-Context Learning

A typical prompting approach is In-context Learning (ICL) [11], which formulates the task description and/or demonstration in the form of natural language text. In addition, Chain of Thought (CoT) prompts can enhance in-context learning by incorporating a series of intermediate reasoning steps into the prompt.

ICL takes structured natural language prompts containing a task description and possibly some task instances as demonstrations. With these task instances, LLM can grasp and perform new tasks without explicit gradient updates. Instruction tuning and following aims to teach the model to follow natural language (including prompts, positive or negative examples, and constraints, etc.) to achieve better multi-task learning on training tasks and generalize on unseen tasks.

The performance of ICL depends heavily on the demonstration, including the demo format, the order of demo examples, etc. Demo design strategies can be divided into two groups: demo organization and demo format.

17.2.3 Chain-of-Thoughts and Extension

Chain of Thoughts (CoT) is an improved prompting strategy [22] that aims to improve the performance of LLM on complex reasoning tasks such as arithmetic reasoning, commonsense reasoning, and symbolic reasoning. Unlike ICL, which simply constructs prompts with input-output pairs, CoT incorporates intermediate reasoning steps that lead to the final output into the prompt.

The essence of CoT is to decompose complex problems into a series of intermediate subtasks. By processing these subtasks step by step, LLMs are able to focus on important details and assumptions, which greatly improves their performance in a wide range of reasoning tasks. In addition, the intermediate steps of CoT provide a more transparent reasoning process, which facilitates the interpretation and evaluation of LLMs.

Few-shot CoT is a special case of ICL, which enhances each demonstration ⟨Input, Output⟩ to ⟨Input, CoT, Output⟩ by incorporating a CoT reasoning step. To apply this strategy, there are two key issues, namely how to design a suitable CoT prompt and how to utilize the generated CoT to derive the final answer.

Zero-shot CoT, unlike few-shot CoT, does not include human-annotated task demonstrations in the prompt. Instead, it directly generates reasoning steps and then uses the generated CoT to derive the answer. In zero-shot CoT, LLM first generates reasoning steps with the prompt "Think step by step" and then derives the final answer with the prompt "Therefore, the answer is." This strategy can significantly improve performance when the model scale exceeds a certain size, but it is ineffective for small-scale models, showing a clear emergent ability pattern.

Inspired by the CoT sequential thinking structure, many Chain-of-X (CoX) [29] or X-of-Thoughts (XoT) [22] methods have been developed to address various challenges involving different domains and tasks of LLM. CoX is defined as a generalized form of the CoT method for diverse tasks beyond LLM reasoning, such as chains of feedback (CoF), chains of models (CoM), and chains of augmentation (CoA). The X in CoX is called the "node" of the chain structure. In addition to the thinking in the CoT prompt, the X in CoX can take various forms tailored for specific tasks, including intermediate, augmentation, feedback, and even models.

The idea of CoX is to construct a sequence of problem-related components that either contribute in combination to a solution or iteratively refine the output of a complex task. Likewise, CoX defines a structured format that needs to go beyond prompting strategies like CoT and can be adapted to a variety of algorithmic frameworks or structures for diverse tasks involving LLMs.

The most primitive CoT, is a chain structure that describes the intermediate reasoning steps in natural language, and the structural variations of the original chain structure include chain structure variations, tree structure variations, and graph structure variations. As the model transitions from linear chains to hierarchical trees and complex graph structures, such as tree of thoughts (ToT), graph of thoughts (GoT), and algorithm of thoughts (AoT), the interaction of thoughts becomes more and more complex, gradually enhancing the ability to solve complex problems. However, with the increase in the complexity of the topological structure, the related methods impose more constraints on task selection, resulting in a significant decrease in their generalizability and making their application difficult. Generalizing methods based on complex topological structures to general fields, is a major challenge for future research.

17.3 Vision-Large Language Model

LLMs face two challenges in their applications: (1) the limited supply of high-quality text data; and (2) the inherent limitations of unimodal architectures in capturing and processing real-world information, which requires understanding the complex relationships between different modalities. These limitations have motivated efforts to explore and develop visual-language models (VLMs) [38], which combine visual (e.g., images, videos) and textual inputs to provide a more comprehensive understanding of visual-spatial relationships, objects, scenes, and abstract concepts. VLMs extend the representation boundaries previously limited to unimodal approaches and support richer, more contextualized views of the world, such as visual question answering (VQA). The basic architecture of the vision-language model, including the vision/text encoder and decoder.

The vision encoder plays a crucial role in projecting the visual component into an embedding feature that is aligned with the Large Language Model (LLM) embedding and can be used for tasks such as text or image generation. It is trained to extract rich visual features from image or video data, enabling integration with language representations. Many vision encoders used in VLMs are pre-trained on large-scale multimodal or image data: these encoders are jointly trained on image-text pairs, allowing them to effectively capture visual and language relationships. Examples include CLIP [5], which aligns image and text embeddings via contrastive learning, and BLIP [13], which leverages guided pre-training for robust language-image alignment. Although these encoders were originally unimodal, such as ResNet or Vision Transformers (ViTs), they can be adapted to multimodal tasks.

The text encoder projects the tokenized text sequence into an embedding space, similar to how the visual encoder processes images. Models such as CLIP and BLIP use both image encoders and text encoders. These models use contrastive learning to align image and text embeddings in a shared latent space, effectively capturing cross-modal relationships.

Text decoders leverage the LLM as the primary text generator, using a vision encoder to project image features. These models typically use a minimal visual projection mechanism, allowing a powerful language decoder to generate rich contextual outputs. Training a VLM from scratch typically requires a separate text decoder, while using an LLM as a backbone typically uses the original decoder from the LLM.

The cross-attention mechanism enables dynamic interaction between visual and textual features by allowing tokens from one modality (visual) to affect tokens from another modality (text). A cross-attention layer is typically used to integrate cross-modal information by computing an attention score between each pair of visual and textual tokens. Not all models use the cross-attention mechanism.

Recent work has been done to enhance the fusion of visual and textual features. Treating all modalities as tokens is a newer approach that reads and encodes visual inputs (images and videos) into tokens similar to textual tokens.

The alignment problem in multimodal models refers to the contextual bias between different modalities. The misalignment problem of VLMs can lead to hallucinations. Many efforts have been made to alleviate this problem, either by leveraging the reasoning ability of VLMs for self-reflection or designing projectors across different modalities.

Since current large-scale VLMs are difficult and expensive to train, efficient training and fine-tuning of vision-language models has been a very hot research topic. Recently, more and more studies have focused on the pre-training process of vision-language models, trying to understand the impact of different module settings or supervision on the final performance of VLMs.

At the same time, the specific purpose for which the VLM needs to be applied does not necessarily require the multi-task performance of the VLM's multi-functionalities, but rather requires excellent expertise in one or two tasks. In general, LoRA methods help manipulate the VLM by changing fewer parameters and reducing computing resources. Methods such as reinforcement learning with human or artificial intelligence feedback (RLHF/RLAIF) are also widely used to fine-tune the VLM by integrating the knowledge of humans or other VLMs into the fine-tuning process.

17.4 Multi-modality Large Language Model

Multimodal Large Language Models (MLLMs) [21] have recently become a new research topic, using powerful Large Language Models (LLMs) as brains to perform multimodal tasks. The surprising emergent capabilities of MLLMs, such as image-based storytelling and OCR-free mathematical reasoning, are rarely seen in traditional methods, suggesting that this is a potential path to artificial general intelligence (AGI).

A typical MLLM consists of three core parts: a pre-trained modality encoder, such as image/audio, which is responsible for receiving and pre-processing optical/acoustic signals; a pre-trained LLM, similar to the human brain, which is responsible for understanding, reasoning and processing signals; and a modality interface connecting the two, i.e., the connector, which plays the role of aligning different modalities. Some MLLMs also contain a generator to output other modalities besides text. Connectors can be roughly divided into three categories: projection-based, query-based, and fusion-based. The first two categories use token-level fusion to process features into tokens and send them together with text tokens; the last category implements feature-level fusion within LLM. The MLLM architecture consists of an input layer, an encoder, a decoder, and an output layer. They learn to transform the input text into multiple modal representations, which are then fused in the decoder to generate the final output.

Referring to LLM, a mature MLLM training is divided into three stages: pre-training, instruction tuning, and alignment tuning. Each stage targets different types of data and achieves specific goals.

17.5 World Model and Embodied Intelligence

17.5.1 World Model

The basic purpose of a world model is to understand the dynamics of the world and predict future scenarios [28, 36]. For example, in the field of autonomous driving, a world model is defined as a model that takes multimodal data (such as language, images, and trajectories) as input and continuously outputs the future state of the world in the form of vehicle perception data.

A world model is an implicit representation of the external world, so building a model of environment variations in order to make more informed decisions ultimately aims to predict the evolution of future states. Transforming the external reality into a model that represents these elements as latent variables can facilitate implicit understanding. In addition, with the prominence of large language models (LLMs), previous efforts focused on traditional decision-making tasks have been significantly enhanced by the ability to describe world knowledge in detail.

The world model is a future prediction of the external world. It is the most direct model to simulate the external world, and mainly uses visual video data. These methods emphasize the authenticity of the generated video and more realistically reflect the future state of the physical world. The more important development direction is a truly interactive physical world, and the work becomes a transformation from visual to spatial representation and from video to embodiment.

17.5.2 Embodied Intelligence

Embodied intelligence is an interdisciplinary field of artificial intelligence, robotics, and cognitive science. It mainly studies how to enable robots to have human-like perception, planning, decision-making, and behavioral capabilities [32]. Embodied intelligence emphasizes the perception-action loop, using physical entities to perceive the environment, making plans and decisions based on task objectives, and using action manipulation capabilities to complete tasks.

AI agents are AI entities that perceive the environment, make decisions, and take actions. The recent development of large scale models has enabled embodied entities to have rich knowledge of the physical world, providing them with natural language interaction, environmental perception, and task planning capabilities. At the same time, they can be combined with the traditional framework of embodied intelligence to improve the generalization of agent policy learning.

An AI agent mainly consists of an analysis module, a memory module, a planning module, and an action module [19, 25]. The purpose of the analysis module is to identify the role of the agent. The memory and planning modules place the agent in a dynamic environment, enabling it to recall past behaviors and plan future actions. The action module is responsible for converting the agent's decisions into specific outputs. Among these modules, the analysis module affects the memory and planning modules, and these three modules together affect the action module.

When AI agents are embodied as physical entities, they are able to effectively transfer their capabilities from the virtual space to the physical world, thus becoming embodied agents [32]. To complete tasks, embodied agents usually involve the following processes: (1) decomposing abstract and complex tasks into specific subtasks, i.e., high-level embodied task planning; and (2) effectively utilizing embodied perception and embodied interaction models, or utilizing strategies of foundation models, to gradually implement these subtasks, i.e., low-level embodied action planning.

Embodied perception requires multimodal perception and reasoning capabilities to understand the three-dimensional relationships in the scene, predict and perform complex tasks based on sensory information. Perception is multimodal, including vision, hearing, and tactile.

Unlike traditional vision, embodied visual perception is active, including state estimation, scene perception, and environment exploration. Visual Language Navigation (VLN) requires the embodied agent to understand complex and diverse visual observations while interpreting instructions of different granularities. The input of VLN usually consists of two parts: visual information and natural language instructions. The visual information can be a video of past trajectories or a set of historical observation images. The natural language instructions include the goals that the embodied agent needs to achieve or the tasks that it is expected to complete.

Tactile sensors provide detailed information about the object's properties, such as texture, stiffness, and temperature, enabling embodied agents to fully complete the high-precision task at hand.

Embodied interaction refers to scenarios where embodied agents interact with humans and the environment in physical or simulated space. Typical embodied interaction tasks include embodied question-answering and embodied grasping.

Embodied control learns through interaction with the environment and uses a reward mechanism to optimize behavior to obtain the optimal strategy, thus avoiding the shortcomings of traditional physical modeling methods. The Vision-Language-Action (VLA) model is an end-to-end model of embodied intelligence [30] and a multimodal large language model. VLA was recently proposed by the Google's robot model, RT-2, which was developed to solve the instruction-following task in embodied intelligence. In language-conditioned robotics tasks, the policy must be able to understand language instructions, visually perceive the environment, and generate appropriate actions, which requires the multimodal capabilities of VLA. Compared with early deep reinforcement learning methods, VLA-based policies show superior diversity, flexibility, and generalization in complex environments.

17.6 Inference-Time Optimization for LLM Reasoning

Achieving human-level intelligence requires improving the transition from fast, intuitive System 1 reasoning to slower, more deliberate System 2 reasoning. While System 1 excels at rapid, heuristic decision making, System 2 relies on logical reasoning to make more accurate judgments and reduce bias. Basic large language models (LLMs) excel at fast decision making but lack the depth of complex reasoning because they have not yet fully embraced the step-by-step analysis characteristic of true System 2 thinking [40]. Recently, reasoning LLMs like OpenAI's o1/o3 and DeepSeek's R1 have demonstrated expert-level performance in areas such as mathematics and coding, closely resembling the deliberate reasoning of System 2 and demonstrating human-like cognitive abilities.

Moreover, encouraging the LLM to "think" with more tokens during test-time reasoning can further significantly improve reasoning accuracy. Therefore, the combination of training-time and test-time scaling presents a new frontier on the road to large reasoning models [39].

One of the most important capabilities that emerge in LLMs when model parameters are large enough is to enable human-like reasoning. While zero-shot reasoning may still be unreliable for some tasks, researchers have discovered various prompting techniques to enhance these capabilities. These techniques can be roughly divided into three main approaches: step-by-step reasoning, multi-path exploration, and decomposition-based methods. Step-by-step reasoning methods, exemplified by the Chain of Thought (CoT) prompt, show that explicitly showing intermediate reasoning steps can significantly improve problem-solving ability. Even a simple prompt like "let's think step by step" can effectively guide the reasoning process. Multi-path exploration methods go beyond linear reasoning and consider multiple potential solution paths simultaneously. Decomposition-based methods have been shown to be particularly effective for complex problems.

In addition, one work that reflects the law of inference-time scaling is the intelligent-agent workflow. The intelligent-agent workflow can enhance the reasoning ability of LLM without any additional training, but usually requires more test-time computation.

As the challenges posed by complex reasoning grow, methods such as reinforcement learning (RL) and direct preference optimization (DPO), offer a more effective approach than supervised fine-tuning, using reward models to combine the model's output with human-like reasoning, thereby promoting more coherent, responsible, and contextualized outputs. During inference, reward feedback is usually only available after all reasoning steps are completed and the final solution is obtained, which is called the outcome reward model (ORM).

Another reward approach, based on the process reward model (PRM), represents a significant advancement in LLM reasoning, emphasizing the evaluation of intermediate steps rather than focusing solely on the final state results. PRM's rewards are distributed throughout each reasoning step rather than focusing on the final result. By providing subtle feedback throughout the reasoning trajectory, PRM enables the model to optimize behavior to better match human preferences and complex task requirements. This approach is critical for tasks involving sequential decision making, where intermediate steps or decisions are critical to the final goal. Evolution of PRM improves the role of reasoning by providing step-level rewards during complex tasks.

The scaling law of reasoning LLMs reflects the progress from weak supervision to strong supervision, while traditional CoT fine-tuning faces challenges in scaling effectively. Self-improvement uses the model's exploration ability to perform self-supervision and gradually improve the performance of LLMs in reasoning tasks.

Training-based self-improvement in LLMs, can be classified according to the exploration strategy. The exploration phase focuses on data collection to promote subsequent training improvements, and those methods are significantly different. Instead, test-time self-improvement in LLMs, uses the consistency of the model's internal knowledge to correct illusions during reasoning.

17.7 Large Scale Model Applied for Autonomous Driving

Regarding autonomous driving, the following part introduces some classic large model application works, focusing on the large model theoretical direction mentioned above.

17.7.1 A Language Agent Driver

Reference [23] proposed a new paradigm for autonomous driving that uses a large language model (LLM) as a cognitive agent to integrate human-like intelligence into the autonomous driving system. The approach, called Agent Driver, introduces a general tool library accessible through function calls, a cognitive memory for decision-making that includes common sense and experience knowledge, and an inference engine capable of chain-of-thought (CoT) reasoning, task planning, motion planning, and self-reflection.

The overall architecture of Agent-Driver is shown in Fig. 17.2. The method takes sensory data as input and introduces neural modules to process these sensory data and extract environment information about detection, prediction, occupancy, and maps. On top of the neural modules, a tool library is proposed, in which a set of functions are designed to further abstract the neural output and return text-based messages. For each driving scenario, LLM selectively activates the required neural modules by calling specific functions in the tool library, ensuring that the necessary environment information is collected with less redundancy. After collecting the necessary environment information, LLM uses this data as a query to search for relevant traffic regulations and the most similar past driving experience in cognitive memory. Finally, the retrieved traffic rules and driving experience, as well as the previously

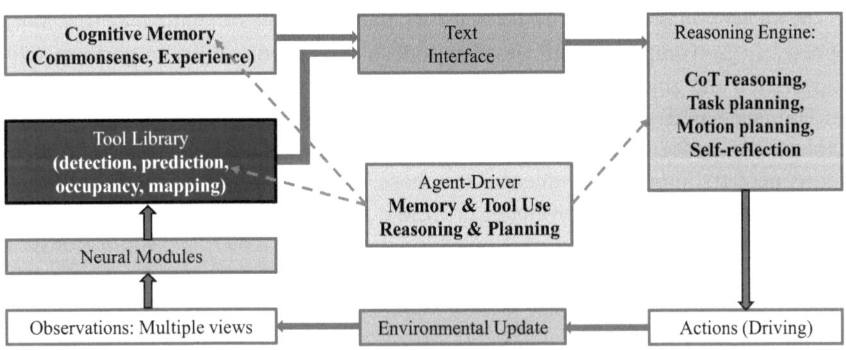

Fig. 17.2 Agent-driver

collected environment information, are used as input to the LLM-based reasoning engine. The reasoning engine performs multiple rounds of reasoning based on the input and ultimately designs a safe and comfortable driving trajectory.

The Agent Driver architecture leverages the dynamic perception and prediction capabilities brought by the tool library, human knowledge from cognitive memory, and the powerful decision-making capabilities of the reasoning engine. This synergistic integration produces a more human-like driving system and enhances decision-making capabilities.

The cornerstone of the tool library is four neural modules, namely detection, prediction, occupancy and map modules, which process observation data and generate detected bounding boxes, future trajectories, occupancy grids and maps respectively. Neural modules cover various tasks of perception and prediction, and extract environment information from observations. However, this information is highly redundant and most of it does not affect the decision-making process. In order to dynamically extract the necessary information from the neural module output, a tool library is proposed, in which a set of functions are designed to summarize the neural output into text-based messages and call dynamic functions to establish the information collection process.

How to utilize traffic rules and driving experience? This method adopts an interactive approach with cognitive memory. Specifically, cognitive memory stores text-based common sense and driving experience. For each driving scenario, function calls are used to collect environment information as queries to search for similar past experiences in cognitive memory to help decision-making. Cognitive memory contains two sub-memories: common sense memory and experience memory.

In order to effectively search for the most similar past driving scenarios in the experience memory, the method adopts a two-stage search algorithm. In the first stage, the query and each record in the memory are input, encoded as an embedding, following the idea of a vector database, and then the top-K similar records are retrieved through a K-NN search in the embedding space. Since driving scenarios are very diverse, embedding-based search is inherently limited by the encoding method, resulting in insufficient generalization capabilities. To overcome this challenge, the second stage is combined with LLM-based fuzzy search. The task of LLM is to rank these records according to their relevance to the query. The ranking is based on the implicit similarity evaluation of LLM, leveraging its generalization and reasoning capabilities.

17.7.2 DriveVLM-Dual: Combination of VLM and Traditional Pipeline

DriveVLM is an autonomous driving system that leverages a vision-language model (VLM) to enhance scene understanding and planning capabilities [27]. In addition, recognizing the limitations of VLM in terms of spatial reasoning and heavy

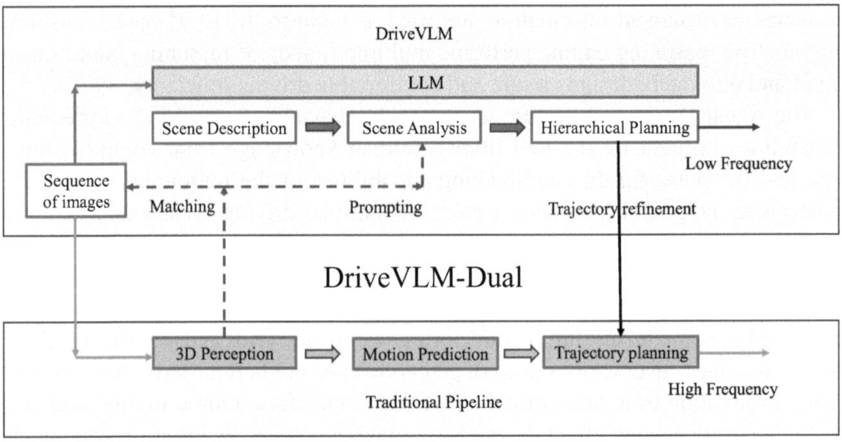

Fig. 17.3 The pipeline of DriveVLM and DriveVLM dual

computational requirements, they further proposed DriveVLM-Dual, a hybrid system that synergizes the advantages of DriveVLM with the traditional autonomous driving pipeline.

Figure 17.3 shows the pipeline of DriveVLM and DriveVLM-Dual. DriveVLM accepts image sequences as input and gives scene description, scene analysis, and hierarchical planning results through the Chain of Thought (CoT) mechanism. DriveVLM-Dual further integrates traditional 3D perception and trajectory planning modules to achieve spatial reasoning capabilities and real-time trajectory planning.

DriveVLM-Dual incorporates two core strategies: key-target analysis with integrated 3D perception, and high-frequency trajectory refinement. As a hybrid framework, DriveVLM-Dual combines the capabilities of DriveVLM with the traditional autonomous driving pipeline. It leverages 3D perception outputs as language prompts to strengthen 3D scene understanding and employs a real-time motion planner to refine trajectory waypoints for improved accuracy and responsiveness.

In order to facilitate the training of DriveVLM, there is the need to build a data mining and annotation pipeline to obtain a scene understanding planning (SUP-AD) dataset. Specifically, it collects samples from a large database by mining long-tail objects and challenging scenes, and then select a key frame from each sample and further perform scene annotation.

This annotator function includes scene description, scene analysis, and planning, with the exception of waypoints that are automatically marked in the vehicle IMU recording. It makes a video annotation tool with the following features: (1) The annotator can slide the progress bar back and forth to replay any part of the video. (2) When annotating key frames, the annotator can draw bounding boxes and language descriptions on the image. (3) The annotator can choose from a list of action and decision candidates when annotating driving planning. Each annotation is carefully

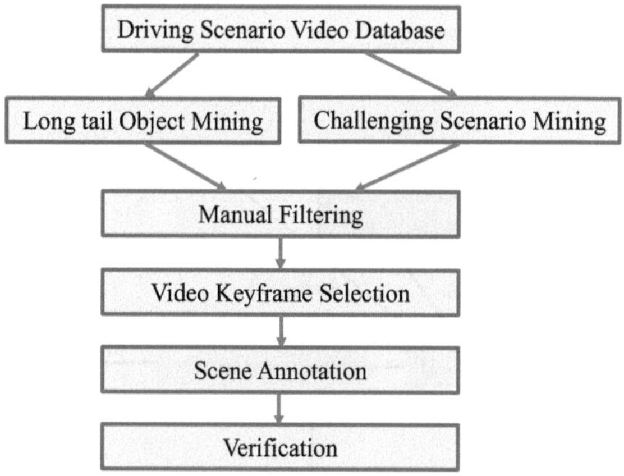

Fig. 17.4 The data mining and annotation pipeline

verified for accuracy and consistency by 3 annotators, ensuring a reliable dataset for model training. The data mining and annotation pipeline is shown in Fig. 17.4.

17.7.3 Senna: Combining VLM and E2E Model

Senna [34] is an autonomous driving system that combines a large vision-language model (Senna-VLM) and an end-to-end model (Senna-E2E). Senna separates high-level planning from low-level trajectory prediction. Senna-VLM generates planning decisions in natural language, while Senna-E2E predicts precise trajectories. Senna-VLM uses a multi-image encoding method and multi-view cues to efficiently understand the scene. In addition, it introduces planning-oriented question answering and a three-step training strategy to improve the planning performance of Senna-VLM while retaining common sense.

The overall architecture of Senna is shown in Fig. 17.5. The input scene information includes multi-view image sequences, user instructions, and navigation commands. User instructions are input to Senna-VLM as prompts, and other instructions are sent to Senna-VLM and Senna-E2E at the same time. Senna-VLM encodes image and text information into image and text tokens, respectively, which are then processed by LLM. LLM generates high-level decisions, which are encoded into high-dimensional features through the Meta-action encoder. Senna-E2E predicts the final planning trajectory based on the scene information and the meta-action features generated by Senna-VLM. A series of planning-oriented QAs are designed to train Senna-VLM. These QAs do not require human annotations and can be generated on a large scale entirely through an automatic labeling process.

Senna-VLM consists of four components. The video encoder takes a multi-view image sequence I as input and extracts image features, which are then further encoded and compressed by the driving vision adapter to obtain image tokens. The text encoder

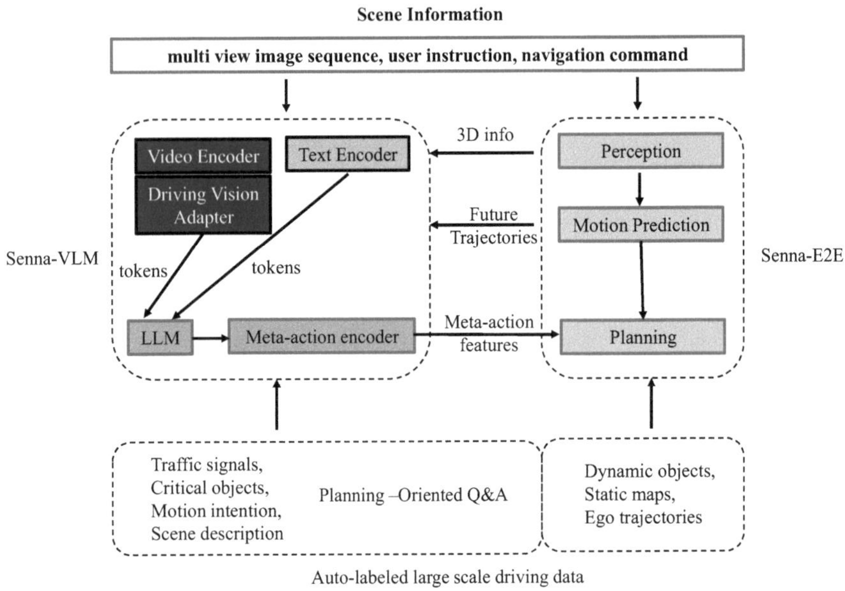

Fig. 17.5 Senna architecture

encodes user instructions and navigation commands into text tokens. Both image and text tokens are input into the LLM, which is used to predict high-level decisions. In practice, Vicuna-v1.5-7 is used as the LLM. Finally, the meta-action encoder encodes the decision and outputs meta-action features. Note: The driving vision adapter module maps image features to the LLM feature space, and also performs additional encoding and compression on the image features to reduce the number of image tokens.

The input of Senna-E2E includes multi-view image sequences, navigation commands, and meta-action features. It consists of three modules: a perception module for detecting dynamic objects and generating local maps; a motion prediction module for predicting the future trajectory of dynamic objects; and a planning module that predicts the planning trajectory V using a set of planning tokens that interact with scene features through an attention mechanism. The meta-action features are integrated as additional interaction tokens of a Senna-E2E. Since the meta-action features are in the form of embedded vectors, Senna-VLM can be easily combined with other end-to-end models.

17.7.4 Google's EMMA

Google Waymo proposed EMMA [35], an end-to-end multi-modal large model for autonomous driving. EMMA is built on a multi-modal large language model that

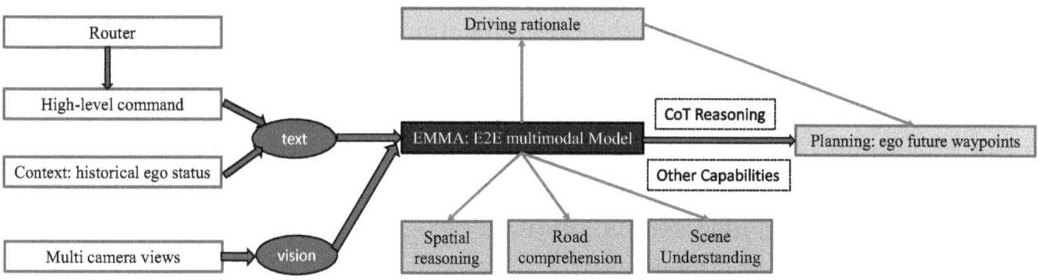

Fig. 17.6 Illustration of EMMA

directly maps raw camera sensor data to a variety of driving-specific outputs, including planner trajectories, perception objects, and road map elements. EMMA maximizes the use of world knowledge from pre-trained large language models by representing all non-sensor inputs (e.g., navigation instructions and ego vehicle states) and outputs (e.g., trajectories and 3D positions) as natural language text. This approach allows EMMA to jointly handle various driving tasks in a unified language space and generate outputs for each task using task-specific prompts.

EMMA is built on top of Google's multimodal language model Gemini, which processes interleaved text and visual inputs to produce text output. Language output is generated by predicting the next token, using the world knowledge it acquires to complete autonomous driving tasks. Shown in Fig. 17.6, the self driving task is mapped into the Gemini-based EMMA model.

The model incorporates three key inputs to match these human driving behaviors:

1. Surrounding view camera video (V): provides comprehensive environment information.
2. High-level intent commands (T_{intent}): originate from the router and include commands such as "go straight," "turn left," "turn right," etc.
3. Historical ego vehicle state set (T_{ego}): represented as a set of waypoint coordinates in the bird's-eye view (BEV) space. All waypoint coordinates are represented as plain text without dedicated tokens. This can also be extended to include higher-order ego vehicle states such as speed and acceleration.

In EMMA, CoT reasoning is incorporated into the end-to-end planner trajectory generation, requiring the model to articulate its decision rationale $O_{\text{rationale}}$ while predicting the final future trajectory waypoints $O_{\text{trajectory}}$.

The model generates future trajectories for motion planning, represented as a set of future trajectory waypoints O of the ego vehicle in the same BEV space, where all output waypoints are also represented as plain text. Driving Principles captions are generated using automated tools without any additional human labeling, ensuring the scalability of the data generation pipeline. Specifically, off-the-shelf perception and prediction expert models are used to identify key agents, and then the Gemini model and designed visual and textual cues are used to generate comprehensive scene and agent behavior descriptions.

Furthermore, while end-to-end motion planning is the ultimate core task, a comprehensive autonomous driving system requires additional capabilities. Specifically, it must perceive the 3D world and recognize surrounding objects, road maps, and traffic conditions. To achieve this goal, EMMA is formulated as a generalist model that can handle multiple driving tasks through hybrid training: spatial reasoning (3D object detection), road map estimation (road recognition), and scene understanding (driving related, such as congestion).

17.7.5 World Model: DrivingWorld

DrivingWorld [37] is a GPT-style world model for autonomous driving with multiple spatio-temporal fusion mechanisms. This design effectively models spatial and temporal dynamics, thereby facilitating high-fidelity, long-term video generation. Specifically, a next-state prediction strategy is proposed to simulate the temporal coherence between consecutive frames, and a next token prediction strategy is applied to capture spatial information within each frame. To further improve the generalization ability, a token prediction mask strategy and a reweighting strategy are proposed to alleviate the long-term drift problem and achieve precise control. The pipeline of DrivingWorld is shown in Fig. 17.7.

In this method, a tokenizer maps multimodal states into a unified discrete space, thereby achieving accurate and controllable multimodal generation. In order to generate time-consistent embeddings for images, a time-aware vector quantization tokenizer is proposed. The proposed vehicle posture tokenizer discretizes the posture trajectory and integrates it into the DrivingWorld model.

Specifically, to capture temporal dependencies, a self-attention layer is inserted before and after VQGAN quantization, where the attention mechanism operates along the temporal dimension. This enables the model to capture long-term temporal relationships between frames, thereby improving the coherence and consistency of the generated sequences.

The world model aims to understand past state inputs, simulate the dynamics of the real world and predict future states. In this context, it predicts upcoming driving scenarios and plans feasible future trajectories. To do this, the world model concatenates historical state tokens into a long sequence where 2D image tokens are unfolded into a 1D form in a zig-zag order. Therefore, the goal is to predict the next

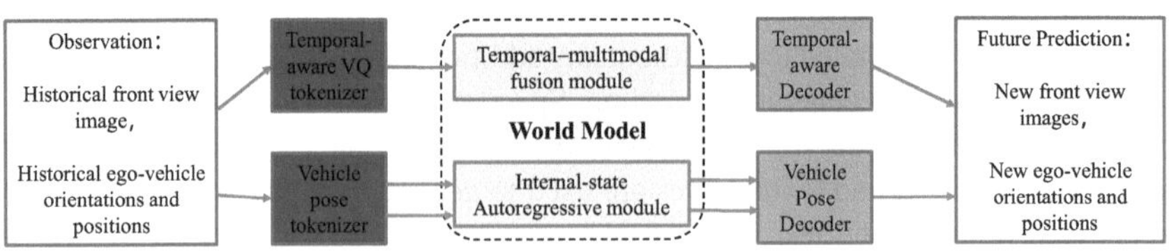

Fig. 17.7 DrivingWorld pipeline

state based on the past observation sequence, capturing temporal and multimodal dependencies. Note: All discrete tokens from different modalities are mapped to a shared latent space by their respective learnable embedding layers before being fed into the world model, and all subsequent processes are carried out within this latent space.

The method adopts a next-state prediction pipeline, which consists of two modules: one module integrates time domain and multimodal information for next-state feature generation, namely the time domain-multimodal fusion module, and the other module is the internal state autoregression module for high-quality internal state token generation.

After the world model predicts the next state token, the decoder can be used to generate the corresponding vehicle relative orientation, relative position, and reconstructed image in that state. This process can map the predicted latent representation back to the physical output, including spatial and visual data.

During training, the world model uses past ground-truth tokens as conditions to predict the next token. However, during inference, the model must condition on previously generated tokens, which may contain defects. Training with only perfect ground-truth images can lead to content drift problems during inference, causing the generated output to quickly degrade and eventually fail. To address this issue, this method proposes a random masking strategy (RMS) where some tokens from the ground-truth are randomly dropped out. Each token has a 50% chance of being replaced by another random token in this frame, and this drop-out is applied to the entire sequence of conditional images with a probability of 30%.

Another point is that, it proposes a balanced attention calculation that prioritizes the vehicle state token in the attention mechanism instead of paying equal attention to all tokens, which can achieve more precise control.

17.8 Summary

This chapter discusses the application of large scale models in autonomous driving, including large language models, vision-large language models, multimodal large language models, intelligent agents, world models, embodied intelligence, and inference/test-time optimization of large scale models reasoning.

References

1. N Houlsby et al., "Parameter-Efficient Transfer Learning for NLP (Adapter Tuning)", arXiv:1902.00751, 2019
2. S Rajbhandari, J Rasley, O Ruwase, Y He, "ZeRO: Memory Optimizations Toward Training Trillion Parameter Models", arXiv:1910.02054, 2019
3. J Kaplan et al., "Scaling Laws for Neural Language Models", arXiv:2001.08361, 2020

4. X Li, P Liang, "Prefix-Tuning: Optimizing Continuous Prompts for Generation", arXiv:2101.00190, 2021
5. A Radford, J W Kim, C Hallacy, et al. "Learning transferable visual models from natural language supervision" (CLIP). ICML 2021.
6. E Hu et al., "LORA: Low-Rank Adaptation of Large Language Models", arXiv:2106.09685, 2021
7. R Bommasani et al., "On the Opportunities and Risks of Foundation Models", arXiv:2108.07258, 2021
8. L Ouyang, J Wu, X Jiang et al., Training language models to follow instructions with human feedback (GPT-3.5/InstructGPT), arXiv:2203.02155, 2022
9. J Hoffmann et al., "Training Compute-Optimal Large Language Model" (scaling law), arXiv:2203.15556, 2022
10. P Liu, W Yuan, J Fu, and et al. "Pre-train, prompt, and predict: A systematic survey of prompting methods in natural language processing". ACM Computing Surveys, 55(9):1–35, 2023.
11. Q Dong et al., "A Survey on In-context Learning", arXiv:2301.00234, 2023
12. K Nagrecha, "Systems for Parallel and Distributed Large-Model Deep Learning Training" (Review), arXiv:2301.0269, 2023
13. J Li et al., "BLIP-2: Bootstrapping Language-Image Pre-training with Frozen Image Encoders and Large Language Models", arXiv:2301.12597, 2023
14. V Lialin, V Deshpande, A Rumshisky, "Scaling Down to Scale Up: A Guide to Parameter-Efficient Fine-Tuning", arXiv:2303.15647, 2023
15. W X Zhao, K Zhou, J Li, et al., "A Survey of Large Language Models", arXiv:2303.18223, 2023
16. A Kirillov, E Mintun, N Ravi, et al., "Segment anything" (SAM). arXiv:2304.02643, 2023
17. M Oquab, T Darcet, T Moutakanni, et al., "DINOv2: Learning Robust Visual Features without Supervision", arXiv:2304.07193, 2023
18. S Zhang et al., "Instruction Tuning for Large Language Models: A Survey", arXiv:2308.10792, 2023
19. L Wang et al., "A Survey on Large Language Model based Autonomous Agents", arXiv:2308.11432, 2023
20. Z Xi, W Chen, X Guo, and et al. "The Rise and Potential of Large Language Model Based Agents: A Survey", arXiv:2309.07864, 2023
21. C Li, Z Gan, Z Yang, and et al. "Multimodal Foundation Models: From Specialists to General-Purpose Assistants" (survey), arXiv:2309.10020, 2023
22. Z Chu et al., "A Survey of Chain of Thought Reasoning: Advances, Frontiers and Future", arXiv:2309.15402, 2023
23. J Mao, J Ye, Y Qian, M Pavone, Y Wang, "A Language Agent for Autonomous Driving" (Agent-Driver), arXiv:2311.10813, 2023
24. Y Huang, Y Chen, Z Li, "Applications of Large Scale Foundation Models for Autonomous Driving" (survey), arXiv:2311.12144, 2023
25. Y Cheng, C Zhang, Z Zhang, and et al. "Exploring Large Language Model Based Intelligent Agents: Definitions, Methods, and Prospects" (survey), arXiv:2401.03428, 2024
26. S Minaee, T Mikolov, N Nikzad, and et al. "Large Language Models: A Survey", arXiv:2402.06196, 2024
27. X Tian, J Gu, B Li, et al., "DriveVLM: The Convergence of Autonomous Driving and Large Vision-Language Models", arXiv:2402.12289, 2024
28. Y Guan et al., "World Models for Autonomous Driving: An Initial Survey", arXiv:2403.02622, 2024
29. Y Xia, R Wang, X Liu, and et al., "Beyond Chain-of-Thought: A Survey of Chain-of-X Paradigms for LLMs", arXiv:2404.15676, 2024
30. Y Ma, Z Song, Y Zhuang, and et al. "A Survey on Vision-Language-Action Models for Embodied AI", arXiv:2405.14093, 2024
31. W Cai, J Jiang, F Wang, and et al., "A Survey on Mixture of Experts", arXiv:2407.06204, 2024

32. Y Liu, W Chen, Y Bai and et al. "Aligning Cyber Space with Physical World: A Comprehensive Survey on Embodied AI", arXiv:2407.06886, 2024
33. Y Zheng, L Yao, Y Su, and et al., "A Survey of Embodied Learning for Object-Centric Robotic Manipulation", arXiv:2408.11537, 2024
34. B Jiang, S Chen, B Liao, et al., "Senna: Bridging Large Vision-Language Models and End-to-End Autonomous Driving", arXiv:2410.22313, 2024
35. J Hwang, R Xu, H Lin, et al., "EMMA: End-to-End Multimodal Model for Autonomous Driving", arXiv:2410.23262, 2024
36. J Ding, Y Zhang, Y Shang, et al., "Understanding World or Predicting Future? A Comprehensive Survey of World Models", arXiv:2411.14499, 2024
37. X Hu, W Yin, M Jia, and et al., "DrivingWorld: Constructing World Model for Autonomous Driving via Video GPT", arXiv:2412.19505, 2024
38. Z Li, X Wu, H Du, et al., "Benchmark Evaluations, Applications, and Challenges of Large Vision Language Models: A Survey", arXiv:2501.02189, 2025
39. F Xu, Q Hao, Z Zong, et al., "Towards Large Reasoning Models: A Survey of Reinforced Reasoning with Large Language Models", arXiv:2501.09686, 2025
40. Z Li, D Zhang, M Zhang, et al., "From System 1 to System 2: A Survey of Reasoning Large Language Models", arXiv:2502.17419, 2025

Chapter 18
Ethics and Legal Aspects

The rapid development of technology is also accompanied by many challenges, especially in terms of ethical decision-making and legal responsibility. Autonomous driving technology has caused widespread controversy and deep reflection. This chapter will explore in depth the ethical and legal issues faced in the development of autonomous driving technology, including ethical dilemmas, legal responsibilities, privacy, and social impact, aiming to provide a comprehensive perspective to think about the relationship between technological progress and human values & legal systems. At the same time, we will also look forward to the future direction of autonomous driving systems in terms of ethics and law by analyzing actual cases and global policies.

18.1 Driving Dilemmas

With the rapid development of artificial intelligence (AI) technology, autonomous driving technology is gradually moving from the laboratory to the real world, becoming a key force in changing the way transportation is carried out. From private passenger cars to public transportation and logistics freight, autonomous driving systems are expected to reduce traffic accidents, improve travel efficiency, and save energy and reduce emissions. However, the rapid development of technology is also accompanied by many challenges, especially in terms of ethical decision-making and legal responsibility. Autonomous driving technology has caused widespread controversy and deep reflection [1, 2].

In traditional driving behaviors, human drivers need to make quick decisions in emergency situations, and these decisions are usually influenced by personal morality, experience, and intuition. However, when driving behavior is handed over to machines, the system must make decisions through preset algorithmic rules. This raises a core question: When faced with moral dilemmas, how can machines make

Y. Huang and Z. Yang, *Autonomous Driving Handbook*,
https://doi.org/10.1007/978-981-95-3725-9_18

ethical choices? For example, in a typical "trolley problem" scenario, should the autonomous driving system prioritize protecting the passengers in the car or avoid causing greater harm to pedestrians? This ethical choice involves not only technical implementation but also the choice of social values and the establishment of ethical consensus [3–5].

At the same time, autonomous driving technology also poses a severe challenge to the existing legal system. At present, the legal frameworks of countries around the world are mostly centered around human drivers, and the gradual popularization of autonomous driving has blurred the boundaries of responsibility. When an autonomous vehicle confronts an accident, who should be blamed? Is it the manufacturer, the software developer, or the user of the vehicle? How does the legal system need to adapt to technological development to encourage innovation while ensuring public safety? There is no unified answer to these questions [6–9].

In addition, the data collection and processing of autonomous driving systems also brings challenges in terms of privacy protection and network security. Autonomous driving relies on sensors, cameras, and cloud data analysis. In this process, the risk of user privacy leakage increases, and the system itself faces the threat of malicious attacks. How to find a balance between technology and privacy protection is another problem that needs to be solved urgently [10–12].

Therefore, this chapter will explore in depth the ethical and legal issues faced in the development of autonomous driving technology, including ethical dilemmas, legal responsibilities, privacy security, and social impact, aiming to provide with a comprehensive perspective to think about the relationship between technological progress and human values & legal systems. At the same time, we will also look forward to the future direction of autonomous driving systems in terms of ethics and law by analyzing actual cases and global policies.

18.2 Ethical Issues

As autonomous driving technology is gradually applied to the real world, the ethical issues involved in machine decision-making have become a focus of widespread attention. Unlike human drivers, autonomous driving systems need to rely on algorithms and data to make decisions, which lets some classic ethical dilemmas more prominent. Especially in emergency situations, the decisions of autonomous driving systems not only affect the safety of life and property but also society's acceptance of technological ethics. This section will explore the ethical issues facing autonomous driving from the perspectives of ethical dilemmas, algorithm transparency, and global challenges [1, 4, 7, 8].

In ethics, the "trolley problem" is a classic moral dilemma, which refers to the problem of how to make a choice in a dilemma situation. Similar dilemmas frequently occur in autonomous driving scenarios. For example, when an autonomous car faces an emergency, the system must decide whether to protect the passengers in the car or give priority to avoiding causing greater harm to pedestrians.

Here let's take a case analysis: Suppose an autonomous car is driving at a normal speed and suddenly a pedestrian crosses the road. At this point, the system faces two choices: (1) Choose to hit the pedestrian to protect the safety of the passengers in the car, but it may cause the pedestrian's death; (2) Choose to turn sharply to avoid the pedestrian, but it may cause the vehicle to lose control and endanger the lives of the passengers in the car. In this situation, how should the autonomous driving system make decisions? Should it protect the interests of the majority or prioritize the protection of individuals? Utilitarianism believes that machines should choose decisions that cause the least overall harm, while deontology advocates that decisions should not be made at the expense of others. This issue is not only a challenge in terms of technical implementation, but also the trade-offs of social ethical concepts.

The MIT "Moral Machine" experiment collected data on ethical choices from participants worldwide and revealed that perspectives on ethical decision-making vary considerably across countries and cultural contexts [14]. For instance, in many Western countries, systems were more inclined to prioritize the protection of pedestrians, whereas in regions with strong collectivist traditions, safeguarding vehicle passengers was often considered more ethical. These cultural differences add a layer of ethical complexity to the design and deployment of autonomous driving systems.

The ethical decision-making of autonomous driving relies on the rule setting of the algorithm, and the "transparency" of such decision-making is one of the core issues of social concern, as shown in Fig. 18.1.

- Algorithm transparency: Currently, the decision-making process of most autonomous driving systems is a "black box" operation, and it is difficult for the outside world to understand its specific decision-making logic. This opacity has caused the public to question whether the machine's decision-making is "fair" or "reasonable." For example, is there bias in the system's decision-making between different groups of people? Are the rights of other groups sacrificed when protecting specific groups?

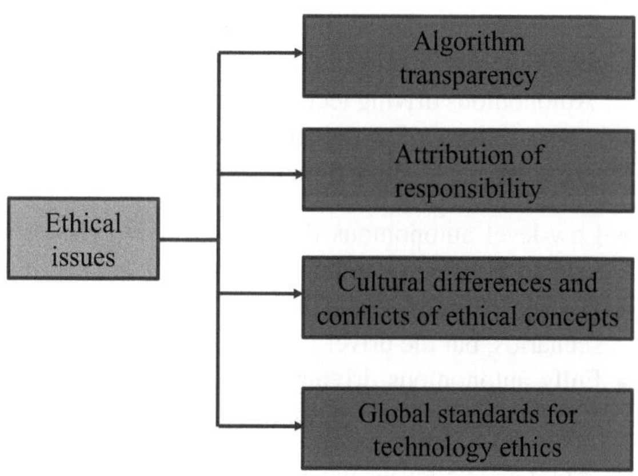

Fig. 18.1 Ethical issues of decision-making in autonomous driving

- Attribution of responsibility: Another issue brought about by algorithmic decision-making is the definition of responsibility. When an accident occurs, who should be responsible for the ethical choices of the system? Is it the algorithm designer, the car manufacturer, or the user of the vehicle? For example, if a decision made by an autonomous vehicle results in the death of a pedestrian, how should the victim's family seek legal redress? This ambiguity in ethical responsibility poses a serious challenge to the legal system. Autonomous driving technology is global, but ethical standards vary due to differences in culture, law, and social values. This difference poses huge challenges to the ethical level of globalized technology.
- Cultural differences and conflicts of ethical concepts: There are significant differences in the priority of the life value and the understanding of accident responsibility among different countries.
- Global standards for technology ethics: In order to promote the application of autonomous driving technology around the world, it is particularly important to develop unified ethical standards. This requires technology companies, legal institutions and ethics circles to work together to reach a widely accepted ethical consensus.

18.3 Legal Issues

The rapid development of autonomous driving technology has brought unprecedented challenges to the existing legal system. In traditional traffic accidents, the responsibility is usually borne by the human driver. However, as autonomous vehicles gradually replace human drivers, the issue of liability attribution in the event of an accident becomes more complicated: when the machine takes over the driving task, who should be held responsible for the accident? Is it the vehicle owner, manufacturer, or software developer? This issue involves multiple legal, technical, and ethical considerations [2, 4, 5]. This section will explore the core issues of legal liability attribution, the development status and challenges of global laws and regulations, and changes in insurance and compensation mechanisms.

Autonomous driving technology is divided into multiple levels (such as SAE L1-L5). The division of responsibilities between human drivers and systems at different levels is different, which directly affects the division of responsibilities in accidents:

- Low-level autonomous driving (L1–L2): The driver still needs to monitor the vehicle at all times and bear the main responsibility.
- High-level autonomous driving (L3): The system can take over driving in some scenarios, but the driver needs to intervene when necessary.
- Fully autonomous driving (L4–L5): The system completely takes over driving without human intervention.

At L4 and L5 levels, when the vehicle is fully autonomous, the role of the driver is weakened, the traditional "human error" principle is difficult to apply, and the issue

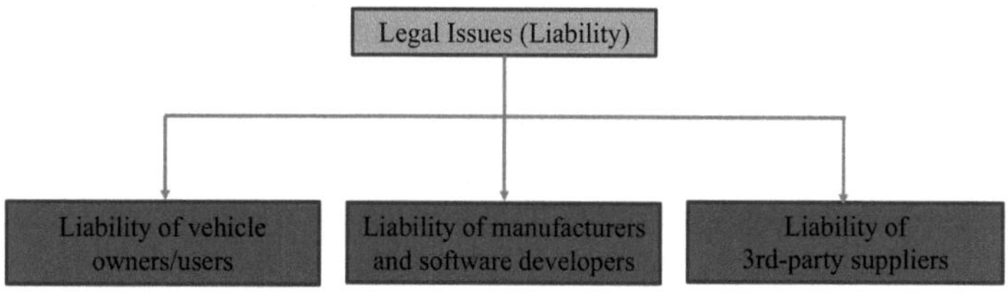

Fig. 18.2 Legal issues: Liability of autonomous driving vehicles

of liability becomes particularly prominent. It mainly involves the following entities, as shown in Fig. 18.2.

1. Liability of vehicle owners/users: If the vehicle owner fails to maintain the vehicle as required, resulting in a failure of the autonomous driving system, does the owner need to bear part of the responsibility? Weakening of the "duty of care": In autonomous driving above L4, the driver hardly needs to intervene. Will this affect his legal liability?
2. Liability of manufacturers and software developers: If the accident is caused by a design defect or software failure in the system, should the manufacturer and developer bear the main responsibility? Take a look at the case: In 2018, an Uber autonomous vehicle hit and killed a pedestrian in Arizona, USA. The investigation results showed that the system's failure to identify the pedestrian was the main cause of the accident.
3. Liability of third-party suppliers: The implementation of the autonomous driving system depends on the cooperation of multiple parties, including sensor suppliers, AI algorithm providers, etc. Once a problem occurs in a certain link, how to define the proportion of responsibility of each party?

Some countries are trying to adapt to the legal challenges brought by autonomous driving technology, but there are differences in legal frameworks and progress in different countries.

1. In the United States, there is a gradually established legal framework. The National Highway Traffic Safety Administration (NHTSA) of the United States has released the "Automated Driving System 2.0: A Vision for Safety," which encourages companies to innovate technology, but emphasizes that manufacturers need to ensure safety. However, laws of states are different: California, for example, requires that human drivers must be present for autonomous driving tests, while Arizona allows completely unmanned testing.
2. In Europe, it attaches great importance to safety and division of responsibilities. The European Union has issued the "General Data Protection Regulation (GDPR)," which imposes strict requirements on data security and privacy protection in autonomous driving. In 2017, Germany became the first country in the world to legislate to allow L3 autonomous driving, clarifying the division of responsibilities: when an accident occurs, the manufacturer must prove that the system is not at fault, otherwise it must bear the responsibility.

3. At China, policies and regulations are gradually improving. The Ministry of Industry and Information Technology has released the "Technical Roadmap for Intelligent Connected Vehicles" to promote the development of autonomous driving technology. China has gradually opened up autonomous driving tests, requiring companies to bear full responsibility for accidents during testing, and future regulations will further clarify the legal responsibilities of different entities.

   There are some challenges and problems as follows:

- Legal lag: The development of technology far exceeds the speed of legal improvement, and the existing legal framework is difficult to adapt to emerging issues.
- Difficulty in determining responsibility: In autonomous driving accidents, system failures, human factors, and third-party influences are intertwined, making the investigation and determination of accident responsibility extremely complicated.
- Cross-border legal conflicts: There are differences in legal standards in different countries. International autonomous driving companies face the problem of legal coordination when conducting business in multiple places.

The popularization of autonomous driving technology will subvert the existing automobile insurance and compensation system, which is mainly reflected in the following aspects:

1. Changes in insurance model: for traditional insurance, compensation is mainly based on the driver's fault. For autonomous driving insurance, in the future, insurance will shift more to vehicle manufacturers or system suppliers, and compensation will be made based on the principle of product liability. Take a look at the case: The UK has proposed a "double insurance system" for autonomous driving vehicles, that is, in the event of an accident, the insurance company will first compensate the victim, and then seek compensation from the manufacturer or software developer.
2. Compensation standards and responsibility sharing: How to measure the degree of fault of the autonomous driving system? Is it based on data analysis at the time of the accident, or does it rely on the determination of a third-party organization? Does compensation for autonomous driving accidents need to introduce new standards, such as system stability, software security, and other indicators?
3. Consumer rights protection: In autonomous driving accidents, the protection of the rights of consumers as technology users is crucial. The law needs to ensure that victims can obtain timely compensation when the responsibility is unclear.

18.4 Privacy and Data Security

The operation of autonomous driving systems relies on the collection, transmission, storage, and processing of a large amount of data, including environmental data captured by sensors, vehicle operation data and user personal information. These data are key to ensuring the safety and efficiency of autonomous driving, but they also bring about issues such as privacy leakage, data abuse and network security risks [10–13]. This section will explore the main challenges faced by autonomous driving technology in terms of privacy and data security, the existing legal framework and protection measures, and future development directions.

The autonomous driving system needs to collect and process data through a variety of sensors and technologies, mainly including the following categories:

1. Environmental data: Information captured by devices such as radar, lidar, cameras and GPS, including road conditions, traffic signs, and pedestrian locations.
2. Vehicle operation data: System operation data such as vehicle speed, direction, acceleration, braking status, and battery power.
3. User data: Private information of vehicle users, including location information, driving habits, travel routes, and entertainment preferences.
4. Cloud and communication data: Autonomous vehicles need to exchange data with the cloud, other vehicles or infrastructure in real time through vehicle-to-everything (V2X), which involves data transmission and sharing issues.

The integration of these data enables autonomous driving to analyze complex scenarios and make decisions in real time, but it also exposes personal privacy and system security to risks.

In the scenario of autonomous driving, personal privacy protection faces the following major challenges:

1. The problem of "informed consent" for data collection: Autonomous vehicles need to continuously monitor the surrounding environment, which means that not only the behavior of users in the car is recorded but also the information of unrelated pedestrians and other vehicles may be captured. How to ensure that the data subjects (such as users and pedestrians) are informed and agree that their data is collected? In particular, how to protect the privacy rights of individuals who do not actively provide data?
2. Data use and abuse risks: The data collected by autonomous driving companies may be used for other commercial purposes, such as advertising push or personalized services, or even illegally used by third parties. Can users control their own data? How to define the boundaries of data use?
3. Location information and behavior analysis: Autonomous vehicles can record users' travel routes, stop locations and time for a long time, so as to conduct in-depth analysis of personal behavior. If the data is leaked, the user's daily activities and privacy habits may be illegally tracked, posing serious security risks.

Autonomous driving technology is highly dependent on networks and data processing, making it vulnerable to cyber-attacks, which are mainly manifested in the following aspects:

1. System intrusion and remote control: By invading the vehicle system, hackers can remotely control the vehicle's acceleration, braking, direction, and other functions, which seriously threatens life safety. In 2015, two security researchers

successfully hacked into a Jeep Cherokee and remotely controlled the vehicle's braking and acceleration systems, revealing the vulnerability of autonomous driving data security.

2. Data tampering and misleading: Attackers may tamper with the data collected by sensors and mislead the vehicle system; For example, by interfering with lidar signals or tampering with road signs, the vehicle misjudges the traffic environment and causes an accident.
3. Data theft and privacy leakage: During the data transmission between the autonomous driving system and the cloud, data may be intercepted or stolen, resulting in personal privacy leakage or illegal acquisition of corporate secrets.
4. Infrastructure security: Autonomous vehicles rely on V2X communication and intelligent transportation infrastructure. Once the infrastructure is attacked, the entire city's transportation system may face the risk of paralysis.

To address the challenges of autonomous driving data privacy and security, major countries and regions around the world are establishing relevant laws and standards:

1. EU's General Data Protection Regulation (GDPR): it clarifies the rights of data subjects, including the right to know, the right to access, and the right to delete, and requires companies to handle personal data transparently and legally. Autonomous driving companies must ensure the anonymization and de-identification of user data to avoid data abuse and privacy leakage.
2. In United States, data privacy protection mainly relies on state laws, such as California's Consumer Privacy Act (CCPA), which gives users data access and control rights. The National Highway Traffic Safety Administration (NHTSA) of the United States issued a safety guide requiring autonomous driving companies to ensure system security and data protection.
3. In China, the Data Security Law and the Personal Information Protection Law provide a legal basis for autonomous driving data protection and clarify the compliance requirements for personal information processing. For intelligent connected vehicles, regulators require companies to ensure that data is stored locally to prevent cross-border data leakage.
4. Technical standards: The International Organization for Standardization (ISO) issued the ISO/SAE 21434 standard, which focuses on automotive network security and provides guidance on security design, threat modeling, and risk assessment. The automotive industry is also promoting data encryption, anonymization technology, and blockchain technology to improve data security.

In the face of the challenges of autonomous driving data privacy and security, the future development directions mainly include:

1. Data anonymization and minimization principle: Promote data de-identification processing to ensure that user privacy is not directly identified. Follow the principle of data minimization, only collect and store necessary data, and reduce the risk of privacy leakage.

2. Application of blockchain technology: Blockchain technology can achieve transparency and traceability of data sharing, while protecting the integrity and security of data.
3. Strengthen network security defense: Strengthen the encrypted transmission of vehicle system and cloud data to prevent data from being stolen and tampered. Introduce artificial intelligence technology to monitor and respond to cyber-attack threats in real time.
4. Global data protection standards: Promote the establishment of unified data security and privacy protection standards, coordinate regulations in different countries and regions, and ensure that autonomous driving technology operates in compliance with regulations worldwide.

18.5 Moral and Ethical Impacts on Society

As an innovation with far-reaching social impact, autonomous driving technology has not only brought about a technological revolution, but also triggered a wide range of ethical and legal issues. These issues are concentrated in many fields such as public safety, employment equity, urban space allocation, environmental protection, and social ethics [4, 5]. This session explores the challenges and processing methods of autonomous driving technology at the social level from the perspective of ethics and law.

The ethical and legal considerations of traffic safety include the following aspects:

1. The ethical boundaries of safety responsibility: The biggest advantage of autonomous driving technology is to reduce human driving errors, but system errors are still inevitable; In emergency situations, how autonomous vehicles make ethical decisions remains an unsolved problem; For example: In the "trolley problem" scenario, should the vehicle prioritize protecting the passengers in the car or avoid harming pedestrians? Who should be responsible for the system's decision? Is it the vehicle manufacturer, software developer, or data provider?
2. Safety standards under the legal framework: Existing traffic safety regulations are mostly based on human driving behavior, and the standards for autonomous driving systems have not yet been fully clarified. For example: In safety compliance, the law needs to define the minimum standard for the "safety" of autonomous driving systems and clarify which safety indicators are acceptable. For responsibility after an accident, when an autonomous driving vehicle has an accident, how does the law determine the responsibility? Is it due to system defects or the user's failure to properly maintain the vehicle? Take a look at the case: In 2018, an Uber autonomous vehicle hit and killed a pedestrian. The accident involved the system's failure to identify the pedestrian and the human safety officer's failure to intervene in time. The attribution of responsibility for the accident triggered extensive legal and ethical discussions.

3. System transparency and fairness: The decision-making of the autonomous driving system is based on complex algorithms. The transparency of the algorithm directly affects society's trust in its fairness. How to ensure that the system decision is unbiased and can equally protect all road users is an important ethical and legal issue.

While promoting social progress, autonomous driving technology has also had a significant impact on the employment structure, especially for those who rely on human driving. The impacts include the following aspects:

1. Social injustice caused by unemployment: The popularization of autonomous driving will lead to large-scale unemployment of freight drivers, taxi drivers, public transportation drivers, etc., most of whom are low-income workers. For ethical considerations, should technological innovation give priority to "the overall interests of society" and ignore individual rights? How to protect the survival rights of these workers? For legal advice, the government needs to introduce supporting regulations, such as levying an "autonomous driving technology tax" for unemployment insurance, retraining and employment subsidies.
2. Fair distribution of new opportunities: New jobs created by autonomous driving technology (such as system maintenance and algorithm development) usually require highly skilled talents, while low-skilled workers find it difficult to adapt quickly. The government and enterprises need to assume moral and legal responsibilities to provide skills training and career transformation support for the affected groups to avoid the technological gap exacerbating social inequality.

Autonomous driving technology will change the urban traffic pattern and bring about ethical and legal issues in space allocation and infrastructure construction, such as:

1. Fair distribution of public resources. The popularization of autonomous driving vehicles may make private travel more convenient, reduce the attractiveness of sharing mobility, and cause more road congestion. For legal considerations, the government needs to reasonably guide the allocation of traffic resources through legislation to ensure that public transportation is given priority and avoid the monopoly of social resources by high-income groups.
2. Reuse of urban space. Autonomous driving may reduce parking demand and release urban space. How to redistribute these spaces, whether public interests or commercial development are given priority, involves ethical and legal decisions. From ethical perspective, spatial allocation should give priority to serving public welfare, such as increasing green space and public facilities, rather than pursuing commercial profits.
3. Fair development of urban and rural areas. Autonomous driving technology is closely related to infrastructure construction; Underdeveloped areas may not be able to enjoy the benefits of technology due to insufficient resources, forming a "technological gap." As legal advice, the government should promote the balanced construction of autonomous driving infrastructure to ensure that urban and rural residents can equally enjoy the convenience brought by technology.

Autonomous driving technology has potential positive and negative impacts on environmental protection. How to balance technological development and environmental responsibility is a key issue:

1. Ethical responsibility for environmental improvement. Autonomous vehicles reduce fuel consumption and carbon emissions by optimizing routes and energy-saving driving, thereby promoting sustainable development. For ethical challenge, enterprises and governments should take environmental protection as the core goal of technological development rather than simply pursuing economic benefits.
2. Environmental pressure caused by abuse of technology. The convenience of autonomous driving may prompt people to increase their travel needs, leading to increased energy consumption and greater environmental pressure. As legal and policy guidance, the government can guide technology development in an environmentally friendly direction through legal measures such as carbon emission taxes and green travel subsidies.
3. Corporate responsibility under the legal framework. Enterprises need to assume social responsibility for environmental protection and ensure that autonomous vehicles are combined with electrification to reduce dependence on fossil fuels.

The popularization of autonomous driving technology involves ethical and legal issues such as technology accessibility, fair resource allocation, and protecting the rights of vulnerable groups:

1. Protection of the rights of vulnerable groups. Autonomous driving technology provides a convenient way of travel for groups such as the elderly and the disabled, which helps to improve social fairness. As legal advice, the government needs to formulate relevant regulations to ensure that the design of autonomous driving vehicles takes into account barrier-free needs and promote the inclusive development of technology.
2. Data and privacy fairness. Autonomous driving systems rely on a large amount of data collection, which involves issues of personal privacy and fairness in data use. For legal protection, the government should legislate to clarify the boundaries of data collection and use, protect individual rights and interests, and prevent data abuse.
3. The moral goal of technology accessibility. The high cost of technology may limit autonomous driving services to high-income groups and widen social inequality. As policy guidance, the government should encourage shared travel modes through subsidy policies to ensure that technology accessibility is accessible to the general public.

18.6 Summary

As a revolutionary application of artificial intelligence, autonomous driving technology is profoundly changing all aspects of human society. From morality, law to social ethics, the popularization of this technology has triggered a series of complex and urgent issues that require joint response and resolution by multiple parties.

In summary, the development of autonomous driving technology is not only a frontier exploration of scientific and technological innovation but also a major test for human society at the moral, legal and ethical levels. In the face of this technological wave, the government, enterprises, academia, and the public need to work together to promote the healthy development of autonomous driving technology through a sound legal framework, clear ethical standards, and social governance mechanisms, achieve a benign interaction between technology and society, and ensure that technology truly serves human welfare and social progress.

References

1. A Hevelke and J Nida-Rümelin. "Responsibility for crashes of autonomous vehicles: An ethical analysis". Science and Engineering Ethics, 21(3), 619-630, 2015.
2. V IlkovÁ, A Ilka. "Legal aspects of autonomous vehicles – an overview". Int. Conf. on Process Control, 2017.
3. J Borenstein, J R Herkert, and K W Miller. "The Ethics of Autonomous Cars". The Atlantic, 2017.
4. I Feroz, N Ahmad, M W Iqbal, N A Main, S K Shahzad. "People perception of autonomous vehicles: Legal and ethical issues". International J. of Advanced and Applied Sciences, 2019.
5. M Ryan. "The Future of Transportation: Ethical, Legal, Social and Economic Impacts of Self-driving Vehicles in the Year 2025". Science and Engineering Ethics, 2020.
6. A.Martinho, N. Herber, M. Kroesen, & C Chorus. "Ethical issues in focus by the autonomous vehicles industry". Transport Reviews, 2021.
7. N Reed, T Leiman, P Palade, M Martens, and L. Kester. "Ethics of automated vehicles: Breaking traffic rules for road safety". Ethics Inf. Technol., 2021.
8. J Rhim, J-H Lee, M Chen and A Lim. "A Deeper Look at Autonomous Vehicle Ethics: An Integrative Ethical Decision-Making Framework to Explain Moral Pluralism". Frontiers in Robotics and AI, 2021.
9. S. O. Hansson, M. Belin, B. Lundgren. "Self-Driving Vehicles–an Ethical Overview". Philos. Technol., 2021
10. C. Xie, Z. Cao, Y. Long, and et al. "Privacy of Autonomous Vehicles: Risks, Protection Methods, and Future Directions". arXiv:2209.04022, 2022
11. A Mishra, J Cha, and S Kim. "Privacy-preserved in-cabin monitoring system for autonomous vehicles". Computational Intelligence and Neuroscience, 2022
12. L Trinh, P Pham, H Trinh, and et al. "PP4AV: A benchmarking Dataset for Privacy-preserving Autonomous Driving". IEEE/CVF Winter Conference on Applications of Computer Vision, 2023.
13. J Chong. "A Literature Review of Data Ethics in Autonomous Vehicles". National High School J. Science, 2024.
14. MIT Media Lab. *Moral Machine*. 2016. Available: https://www.moralmachine.net/.

MIX
Papier aus verantwortungsvollen Quellen
Paper from responsible sources
FSC® C105338

If you have any concerns about our products,
you can contact us on
ProductSafety@springernature.com

In case Publisher is established outside the EU,
the EU authorized representative is:
Springer Nature Customer Service Center GmbH
Europaplatz 3, 69115 Heidelberg, Germany

Printed by Libri Plureos GmbH
in Hamburg, Germany